Lecture Notes in Networks and Systems

Volume 404

The series "Lecture Notes in Networks and Systems" publishes the latest developments in Networks and Systems—quickly, informally and with high quality. Original research reported in proceedings and post-proceedings represents the core of LNNS.

Volumes published in LNNS embrace all aspects and subfields of, as well as new challenges in, Networks and Systems.

The series contains proceedings and edited volumes in systems and networks, spanning the areas of Cyber-Physical Systems, Autonomous Systems, Sensor Networks, Control Systems, Energy Systems, Automotive Systems, Biological Systems, Vehicular Networking and Connected Vehicles, Aerospace Systems, Automation, Manufacturing, Smart Grids, Nonlinear Systems, Power Systems, Robotics, Social Systems, Economic Systems and other. Of particular value to both the contributors and the readership are the short publication timeframe and the world-wide distribution and exposure which enable both a wide and rapid dissemination of research output.

The series covers the theory, applications, and perspectives on the state of the art and future developments relevant to systems and networks, decision making, control, complex processes and related areas, as embedded in the fields of interdisciplinary and applied sciences, engineering, computer science, physics, economics, social, and life sciences, as well as the paradigms and methodologies behind them.

Indexed by SCOPUS, INSPEC, WTI Frankfurt eG, zbMATH, SCImago.

All books published in the series are submitted for consideration in Web of Science.

For proposals from Asia please contact Aninda Bose (aninda.bose@springer.com).

More information about this series at https://link.springer.com/bookseries/15179

Subhadip Basu · Dipak Kumar Kole ·
Arnab Kumar Maji · Dariusz Plewczynski ·
Debotosh Bhattacharjee
Editors

Proceedings of International Conference on Frontiers in Computing and Systems

COMSYS 2021

Springer

Editors
Subhadip Basu
Department of Computer Science
and Engineering
Jadavpur University
Kolkata, India

Dipak Kumar Kole
Department of Computer Science
and Engineering
Jalpaiguri Government Engineering College
Jalpaiguri, West Bengal, India

Arnab Kumar Maji
Department of Information Technology
North-Eastern Hill University
Shillong, Meghalaya, India

Dariusz Plewczynski
Faculty of Mathematics and Information
Science
Warsaw University of Technology
Warsaw, Poland

Debotosh Bhattacharjee
Department of Computer Science
and Engineering
Jadavpur University
Kolkata, India

ISSN 2367-3370 ISSN 2367-3389 (electronic)
Lecture Notes in Networks and Systems
ISBN 978-981-19-0104-1 ISBN 978-981-19-0105-8 (eBook)
https://doi.org/10.1007/978-981-19-0105-8

This Springer imprint is published by the registered company Springer Nature Singapore Pte Ltd.
The registered company address is: 152 Beach Road, #21-01/04 Gateway East, Singapore 189721, Singapore

Organizing Committee

The Second International Conference on Frontiers in Computing and Systems—COMSYS-2021 was held during September 29 to October 1, 2021, at the Department of Electronics and Communication, North-Eastern Hill University (NEHU), Shillong 793022, Meghalaya, India.

Chief Patron

Prof. Prabha Shankar Shukla, Honorable Vice-Chancellor, NEHU, Shillong, India

Patron

Prof. L. Joyprakash Singh, The Dean, School of Technology, NEHU, Shillong, India
Prof. Mita Nasipuri, Department of Computer Science and Engineering, Jadavpur University, Kolkata, India

General Chair

Prof. Debotosh Bhattacharjee, Department of Computer Science and Engineering, Jadavpur University, Kolkata, India
Prof. Dariusz Plewczynski, Center of New Technologies (CeNT), University of Warsaw, Warsaw, Poland

Organizing Chairs

Prof. Dipak Kumar Kole, Department of Computer Science and Engineering, Jalpaiguri Government Engineering College, Jalpaiguri, India
Dr. Arnab Kumar Maji, Department of Information Technology, NEHU, Shillong, India

Co-organizing Chair

Dr. Juwesh Binong, Department of Electronics and Communication Engineering, NEHU, Shillong, India

Program Chairs

Prof. Subhadip Basu, Department of Computer Science and Engineering, Jadavpur University, Kolkata, India
Dr. Rupaban Subadar, Department of Electronics and Communication Engineering, NEHU, Shillong, India

Track Chairs

Prof. Ram Sarkar, Department of Computer Science and Engineering, Jadavpur University, Kolkata, India
Dr. Sayan Chatterjee, Department of Electronics and Communication Engineering, Jadavpur University, Kolkata, India
Dr. Ayan Seal, Department of Computer Science and Engineering, PDPM Indian Institute of Information Technology, Design and Manufacturing, Jabalpur, India
Prof. Nibaran Das, Department of Computer Science and Engineering, Jadavpur University, Kolkata, India
Dr. Somanath Tripathy, Department of Computer Science and Engineering, Indian Institute of Technology, Patna, India
Dr. Sujata Pal, Department of Computer Science and Engineering, Indian Institute of Technology, Ropar, India

Publication Chairs

Dr. Nilanjan Dey, Department of Information Technology, JIS University, Kolkata, India
Prof. Debashis De, Department of Computer Science and Engineering, Maulana Abul Kalam Azad University of Technology, Kolkata, India

Publicity Chairs

Dr. Sujoy Saha, Department of Computer Science and Technology, National Institute of Technology, Durgapur, India
Dr. Pankaj Sarkar, Department of Electronics and Communication Engineering, NEHU, Shillong, India
Dr. Ayatullah Faruk Mollah, Department of Computer Science and Technology, Aliah University, India

Web Chairs

Dr. Apurba Sarkar, Department of Computer Science and Technology, Indian Institute of Engineering Science and Technology (IIEST), Shibpur, India
Dr. Amitabha Nath, Department of Information Technology, NEHU, Shillong, India

Tutorial Chairs

Dr. Jacek Sroka, Faculty of Mathematics, Informatics and Mechanics, University of Warsaw, Poland
Dr. Malay Kule, Department of Computer Science and Technology, Indian Institute of Engineering Science and Technology (IIEST), Shibpur, India

Industrial Chair

Dr. Nipanka Bora, Department of Electronics and Communication Engineering, NEHU, Shillong, India

Sponsorship Chair

Mr. Asif Ahmed, Department of Electronics and Communication Engineering, NEHU, Shillong, India

Finance Chairs

Dr. Swagata Mandal, Department of Electronics and Communication Engineering, Jalpaiguri Government Engineering College, Jalpaiguri, India
Dr. Rajkishur Mudoi, Department of Electronics and Communication Engineering, NEHU, Shillong, India

Registration Chair

Dr. Subhash Chandra Arya, Department of Electronics and Communication Engineering, NEHU, Shillong, India

Special Session Chair

Dr. Sushanta Kabir Dutta, Department of Electronics and Communication Engineering, NEHU, Shillong, India

Accommodation Chair

Dr. Jaydeep Swarnakar, Department of Electronics and Communication Engineering, NEHU, Shillong, India

International Advisory Committee

Prof. Punam K. Saha, University of Iowa, United States
Prof. Bhargab B. Bhattacharya, Indian Institute of Technology, India
Prof. Isabelle Debled-Renesson, University of Lorraine, France
Prof. Jemal Abawajy, Deakin University, Australia
Prof. Ondrej Krejcar, Univerzita Hradec Králové, Czech Republic

Prof. Massimo Tistarelli, University of Sassari, Italy
Prof. Ali Bin Selamat, Malaysia-Japan International Institute of Technology, Malaysia
Prof. Gordon Chan, University of Alberta, Canada
Dr. Naohiro Furukawa, Hitachi R&D, Australia
Prof. Anastasios Koulaouzidis, Pomeranian Medical University, Poland
Prof. Lipo Wang, Nanyang Technological University, Singapore
Prof. Andrey S. Krylov, Moscow Lomonosov State University, Russia
Prof. Seyedali Mirjalili, Torrens University, Australia
Prof. Zong Woo Geem, Gachon University, South Korea
Prof. Friedhelm Schwenker, University of Ulm, Germany
Prof. Erik Valdemar Cuevas Jimenez, Universidad de Guadalajara, Mexico

Technical Program Committee

Dr. Ajoy Kumar Khan, Mizoram University, India
Dr. Alessandro Bevilacqua, University of Bologna, Italy
Dr. Anandarup Mukherjee, University of Cambridge, UK
Dr. Anindita Ganguly, IISC Bangalore, India
Dr. Anindya Halder, North-Eastern Hill University, India
Dr. Aparajita Ojha, PDPM IIITDM Jabalpur, India
Dr. Apurba Sarkar, IIEST, Shibpur, India
Dr. Arindam Kar, Indian Statistical Institute, India
Dr. Asish Bera, Edge Hill University, UK
Dr. Atanu Kundu, Heritage Institute of Technology, India
Dr. Bibhash Sen, National Institute of Technology, Durgapur, India
Dr. Biswapati Jana, Vidyasagar University, India
Dr. Bubu Bhuyan, North-Eastern Hill University, India
Dr. Chandan Giri, IIEST, Shibpur, India
Dr. Christian Kollmann, Medical University of Vienna, Austria
Dr. Claudio Zito, Technology Innovation Institute, UAE
Dr. Consuelo Gonzalo Martín, Universidad Politécnica de Madrid, Spain
Dr. Dakshina Ranjan Kisku, National Institute of Technology, India
Dr. Diego Alberto Oliva Navarro, Universidad de Guadalajara, Mexico
Dr. Ernestina Menasalvas, Universidad Politécnica de Madrid, Spain
Dr. Esssam H. Houssein, Minia University, Egypt
Dr. Euisik Yoon, University of Michigan, USA
Dr. Gordon Chan, University of Alberta Edmonton, Canada
Dr. Hiroaki Hanafusa, Hiroshima University, Japan
Dr. Huong T. Vu, The University of Warwick, UK
Dr. Ilora Maity, Aalto University, Helsinki, Finland
Dr. Imon Mukherjee, Indian Institute of Information Technology, Kalyani, India
Dr. Indrajit Bhattacharjee, Kalyani Government Engineering College, India

Dr. Indrajit Ghosh, Ananda Chandra College, India
Dr. Ioannis Pratikakis, Democritus University of Thrace, Greece
Dr. Jin Hee Yoon, Sejong University, South Korea
Dr. Joanna Jaworek-Korjakowska, AGH University of Science and Technology, Poland
Dr. João Luís Garcia Rosa, University of Sao Paulo (USP), Brazil
Dr. João Manuel R. S. Tavares, Universidade do Porto (FEUP), Portugal
Dr. Juan D. Velasquez, University of Chile, Chile
Dr. Jugal Kalita, University of Colorado, USA
Dr. Khwairakpam Amitab, North-Eastern Hill University, Shillong, India
Dr. Malay Kule, IIEST, Shibpur, India
Dr. Michał Jasiński, Wroclaw University of Science and Technology, Poland
Dr. Mrinal Kanti Bhowmik, Tripura University, Tripura, India
Dr. Nanda Dulal Jana, National Institute of Technology, Durgapur, India
Dr. Naoto Hori, University of Texas, Austin, USA
Dr. Narendra D. Londhe, National Institute of Technology, Raipur, India
Dr. Neggaz nabil, Oran University of Science and Technology, Algeria
Dr. Ngo Phuc, Université de Lorraine, France
Dr. Nishatul Majid, Fort Lewis College, USA
Dr. Oishila Bandyopadhyay, IIIT Kalyani, India
Dr. Ozan Keysan, Middle East Technical University, Turkey
Dr. Paramartha Dutta, Visva-Bharati University, India
Dr. Partha Sarathi Paul, IIIT Delhi, India
Dr. Partha Sarathi Roy, University of Wollongong, Australia
Dr. Paulo Quaresma, The University of Évora, Portugal
Dr. Pawan Kumar Singh, Jadavpur University, India
Dr. Pradyut Sarkar, Maulana Abul Kalam Azad University of Technology, India
Dr. Rafael Kleiman, McMaster University, Canada
Dr. Rajat Kumar Pal, University of Calcutta, Kolkata
Dr. Ranjit Ghoshal, St. Thomas College of Engineering and Technology, India
Dr. Ratna Mandal, IEM Kolkata, India
Dr. Robert A. Taylor, University of New South Wales, Australia
Dr. Sakurai Kouichi, Kyushu University, Japan
Dr. Samarjeet Borah, SMIT, Sikkim Manipal University, India
Dr. Samir Malakar, Asutosh College, Kolkata, India
Dr. Samuelson W. Hong, Oriental Institute of Technology, Taiwan
Dr. Sandhya Arora, MKSSS's Cummins College of Engineering for Women, India
Dr. Sandip Rakshit, American University of Nigeria, Nigeria
Dr. Santanu Das, Jalpaiguri Government Engineering College, India
Dr. Santanu Pal, University of Saarland, Germany
Dr. Santanu Phadikar, Maulana Abul Kalam Azad University of Technology, India
Dr. Santanu Sarkar, IIT Madras, India
Dr. Sema Candemir, The Ohio State University, USA
Dr. Serestina Viriri, University of KwaZulu-Natal, South Africa
Dr. Shibaprasad Sen, Future Institute of Engineering and Management, India

Dr. Sitti Rachmawati Yahya, National University Jakarta
Dr. Somnath Mukhopadhyay, Assam University, Silchar
Dr. Soumen Bag, Indian Institute of Technology, Dhanbad, India
Dr. Soumen Kumar Pati, Maulana Abul Kalam Azad University of Technology, India
Dr. Soumya Pandit, University of Calcutta, India
Dr. Soumyabrata Dey, Clarkson University, New York, USA
Dr. Sourav De, Cooch Behar Government Engineering College, India
Dr. Subhas Barman, Jalpaiguri Government Engineering College, India
Dr. Subhas Chandra Sahana, North-Eastern Hill University, India
Dr. Sufal Das, North-Eastern Hill University, India
Dr. Sujatha Krishamoorthy, Wenzhou-Kean University, China
Dr. Sung-Yun Park, University of Michigan, USA
Dr. Sungmin Eum, Booz Allen Hamilton/US Army Research Laboratory, USA
Dr. Swarup Roy, Central University of Sikkim, India
Dr. Tamghana Ojha, National Research Council, Italy
Dr. Teresa Goncalves, University of Evora, Portugal
Dr. Tetsushi Koide, Hiroshima University, Japan
Dr. Tomas Klingström, Swedish University of Agricultural Sciences, Sweden
Dr. Vijay Mago, Lakehead University, Canada
Dr. Vijayalakshmi Saravanan, University at Buffalo, USA

Preface

COMSYS-2021, the Second International Conference on Frontiers in Computing and Systems, was organized from September 30 to October 1, 2021, at the North-Eastern Hill University, Shillong, Meghalaya, India, affectionately abbreviated as NEHU. Like its previous edition, COMSYS-2021 offered a unique platform for scientists and researchers in computing and systems to interact and exchange scientific ideas and present their novel contributions in front of a distinguished audience, fostering business and research collaborations.

COMSYS-2021 was hosted in the hybrid mode, physically organized by NEHU, a public-funded Central University in India, located at a hilltop in the picturesque city of Shillong, Meghalaya, and hosted online through the professionally managed Floor conference management software. Many thanks to the local organizing committee and numerous student volunteers for managing every minute detail related to organizing the conference.

Core objective of this conference was to offer an intellectually stimulating ambience for the scientists and researchers active in computing and systems. COMSYS-2021 provided a unique platform for the delegates to exchange new ideas and establish academic and research relations in the scientific fields related to image, video and signal processing, AI, machine learning and data science, VLSI, devices and systems, computer networks, communication and security, biomedical and bioinformatics, IoT, and cloud and mobile computing. COMSYS-2021 conference proceedings constitutes significant contributions to the knowledge in these scientific fields.

We received 153 submissions from different institutions of India and abroad. After thorough review and plagiarism checking, 58 papers were accepted for oral presentations and spread over 10 technical sessions, of which 9 were organized in online mode, using the Floor virtual platform, and one in physical mode at NEHU, Shillong. In addition, the COMSYS-2021 technical program included four keynote lectures by eminent scientists and academicians from the USA, Germany, Poland, and Slovakia, two engaging tutorial sessions, one from academia and one from industry, and a panel discussion session with three industry experts from Australia, USA, and India.

COMSYS-2021 received considerable global and national attention, with technical program committee members and reviewers from 30+ different countries voluntarily participating in the technical process. In addition, delegates from 12 countries outside India and 10 different states in India attended the conference. Around 30% of all participants are female.

Each submission was reviewed by at least two reviewers, and after rigorous evaluation, 58 papers were selected for presentation and subsequent publication in the conference proceedings. We checked plagiarism using iThenticate software, once at the time of submission and once after acceptance, at the time of final preparation of the camera-ready copy. We convey our sincere gratitude to Springer for providing the opportunity to publish the proceedings of COMSYS-2021 in the prestigious series of Lecture Notes in Networks and Systems. We sincerely hope that the articles will be helpful to the researchers pursuing the field of computing and systems.

COMSYS-2021 was inaugurated by Chief Guest Prof. P. S. Shukla, Honorable Vice-Chancellor, NEHU, on September 29, 2021, in the presence of distinguished dignitaries from renowned institutions of India and abroad. The Computer Chapter—IEEE Kolkata Section has also extended technical cooperation to COMSYS-2021.

In a word, it is always a team effort that defines a successful conference. We look forward to seeing all of you at the next edition of COMSYS.

Kolkata, India
Jalpaiguri, India
Shillong, India
Warsaw, Poland
Kolkata, India
October 2021

Subhadip Basu
Dipak Kumar Kole
Arnab Kumar Maji
Dariusz Plewczynski
Debotosh Bhattacharjee

Acknowledgments

Any successful conference is always a team effort, requiring earnest support and cooperation from all quarters. COMSYS-2021 being no exception, and more specifically during this pandemic period, it would not have been possible to be organized in hybrid mode without the active cooperation from the faculty and student volunteers from NEHU and other institutes in India and abroad. COMSYS-2021 was inaugurated by Prof. P. S. Shukla, Honorable Vice-Chancellor, NEHU, on September 29, 2021, in the presence of distinguished dignitaries from renowned institutions of India and abroad.

We humbly acknowledge the voluntary contributions of all the committee members of COMSYS-2021, especially the technical program committee members and reviewers for extensive review and shortlisting the final set of accepted papers. We also express our gratitude to the distinguished session chairs, the eminent keynote and tutorial speakers, and the honorary panelists at the industry-connect session. Last but not least, we are indebted to all the participants from 10 different states in India and 12 countries outside India, for attending COMSYS-2021.

We convey our sincere gratitude to Springer for publishing the proceedings of COMSYS-2021 in the prestigious series of Lecture Notes in Networks and Systems, and also to the Computer Chapter—IEEE Kolkata Section for extending technical cooperation to COMSYS-2021.

We look forward to seeing all of you at the next edition of COMSYS.

Editorial Board
COMSYS-2021

Contents

Editors and Contributors

About the Editors

Subhadip Basu received his Ph.D. degree from Jadavpur University in 2006 and did his Postdoctoral researches from University of Lowa, USA, in 2010-11 and from University of Warsaw, Poland, during 2012–14. He is a faculty member of the Computer Science and Engineering Department of Jadavpur University since 2006, where he is currently working as a full professor. He is also Honorary Research Scientist in the Department of Electrical and Computer Engineering, University of Lowa, USA, since 2016. He has published around 200 research articles in various international journals, conference proceedings and book chapters in the areas of pattern recognition and related applications, bioinformatics, biomedical image analysis, etc. He has co-edited seven books, co-invented two US patents and guest edited special issues in Briefings in Functional Genomics (OUP) and in Pattern Recognition Letters (Elsevier)). He has served as a technical program committee member and also as an organizing member of several international conferences in India and abroad. He has worked in many international institutes of repute, including Hitachi Central Research Laboratory, Japan, Bournemouth University, UK, University of Lorraine, France, Nencki Institute of Experimental Biology, Poland, etc. He is the recipient of the DAAD Fellowship from Germany, Research Award from UGC, Government of India, BOYSCAST, and FASTTRACK Young Scientist Fellowships from DST, Government of India, HIVIP Fellowship from Hitachi, Japan, and EMMA Fellowships from the European Union. He is a senior member of IEEE and life member of IUPRAI (Indian Unit for IAPR).

Dipak Kumar Kole received the Ph.D. degree in Engineering from Bengal Engineering and Science University, which is currently known as IIEST, Shibpur, India, in 2012. He also received an M.Tech. and B.Tech. in Computer Science and Engineering and B.Sc. in Mathematics Honours from Calcutta University. He has approximately 19 years of professional experience. He is a faculty member of the Computer Science and Engineering Department of Jalpaiguri Government Engineering College since

2014, where he is currently working as a full professor. His research interest includes synthesis and testing of reversible circuits, social network analysis, digital watermarking and agriculture engineering. He published more than 61 research articles in various international journals, conference proceedings and book chapters in the areas of VLSI, reversible circuits, social network analysis, agriculture engineering, image and video processing and cryptography.

Arnab Kumar Maji received B.E. degree in Information Science and Engineering from Visvesvaraya Technological University (VTU) in 2003 and M.Tech. in Information Technology from Bengal Engineering and Science University, Shibpur (Currently IIEST, Shibpur) in 2005. He received his Ph.D. from Assam University, Silchar (A Central University of India) in the year 2016. He has approximately 17 years of professional experience. He is currently working as Associate Professor in the Department of Information Technology, North Eastern Hill University, Shillong (A Central University of India). He has published more than 32 nos. of articles in different reputed International Journals and Conferences, more than 22 nos. of articles as book chapter and authored 02 nos. of books with several international Publishers like Elsevier, Springer, IEEE, MDPI, IGI Global, and McMilan International etc. 04 nos. of Ph.D. scholars are currently pursuing Doctoral Degrees under his active supervision. He has also guided successfully 15 nos. of M.Tech. thesis and 03 nos. of Ph.D. scholars. He is also reviewer of several reputed international journals and guest editor of one Springer journal. His Research interests include Machine Learning, Image Processing, and Natural Language Processing.

Dariusz Plewczynski interests are focused on functional and structural genomics. Functional genomic attempts to make use of the vast wealth of data produced by high-throughput genomics projects, such as the structural genomics consortia, human genome project, 1000 genomes project, ENCODE and many others. The major tools that are used in this interdisciplinary research endeavor include statistical data analysis (GWAS studies, clustering, machine learning), genomic variation analysis using diverse data sources (karyotyping, confocal microscopy, aCGH microarrays, next-generation sequencing: both whole genome and whole exome), bioinformatics (protein sequence analysis, protein structure prediction), and finally biophysics (polymer theory and simulations) and genomics (epigenetics, genome domains, three-dimensional structure analysis of chromatin). He is presently involved in several big data informatics projects at Faculty of Mathematics and Information Sciences at Warsaw University of Technology, biological experiments at the Centre of New Technologies at University of Warsaw (his second affiliation), collaborating closely with The Jackson Laboratory for Genomic Medicine (an international partner of the TEAM project), and The Centre for Innovative Research at Medical University of Bialystok (UMB). He was actively participating in two large consortia projects, namely 1000 Genomes Project (NIH) by bioinformatics analysis of genomic data from aCGH arrays and NGS (next-generation sequencing, deep coverage) experiments for structural variants (SV) identification; and biophysical modeling of chromatin three-dimensional conformation inside human cells using HiC and ChIA-PET techniques within the 4D Nucleome project funded by the NIH in the USA. His goal is to combine the SV data with three-dimensional cell nucleus structure for better understanding of normal genomic variation among human populations, the natural selection process during human evolution, mammalian cell differentiation, and finally the origin, pathways, progression and development of cancer and autoimmune diseases.

Debotosh Bhattacharjee is working as a full professor in the Department of Computer Science and Engineering, Jadavpur University, with fifteen years of post-Ph.D. experience. His research interests pertain to the applications of machine learning techniques for face recognition, gait analysis, hand geometry recognition, and diagnostic image analysis. He has authored or co-authored more than 280 journals, conference publications, including several book chapters in the areas of biometrics and medical image processing. Two US patents have been granted on his works. He has been granted sponsored projects by the Government of India funding agencies like Department of Biotechnology (DBT), Department of Electronics and Information Technology (DeitY), University Grants Commission (UGC) with a total amount of around INR 2 Crore. For postdoctoral research, he has visited different universities abroad like the University of Twente, The Netherlands; Instituto Superior Técnico, Lisbon, Portugal; University of Bologna, Italy; ITMO National Research University, St. Petersburg, Russia; University of Ljubljana, Slovenia; Northumbria University, Newcastle Upon Tyne, UK, and Heidelberg University, Germany. He is Life Member of Indian Society for Technical Education (ISTE, New Delhi), Indian Unit for Pattern Recognition and Artificial Intelligence (IUPRAI), Senior Member of IEEE (USA) and Fellow of West Bengal Academy of Science and Technology.

Contributors

Bara' W. Al-Mistarehi Civil Engineering Department, Faculty of Engineering, Jordan University of Science and Technology, Irbid, Jordan

Ahmad H. Alomari Civil Engineering Department, Faculty of Engineering, Yarmouk University, Irbid, Jordan

Md. Amir Sohail Associate Data Scientist, Cognizant Technology Solutions India Pvt. Ltd., Pune, India

Khwairakpam Amitab Department of Information Technology, NEHU, Shillong, Meghalaya, India

V. S. Anoop Rajagiri College of Social Sciences (Autonomous), Kochi, Kerala, India

Momen Ayasrah Master Civil Engineer & Research Assistant, Department of Civil Engineering, Jordan University of Science and Technology, Irbid, Jordan

Navjeet Bagga Pandit Dwarka Prasad Mishra Indian Institute of Information Technology Design and Manufacturing, Jabalpur, Madhya Pradesh, India

Bhabajyoti Baishya Department of Computer Science and Engineering, Girijananda Chowdhury Institute of Management and Technology, Guwahati, Assam, India

Bunil Kumar Balabantary Department of Computer Science and Engineering, National Institute of Technology Meghalaya, Shillong, India

Anjan Bandyopadhyay Department of ECE, NIT Durgapur, Durgapur, India

Sakhi Bandyopadhyay Department of Computer Science, Vidyasagar University, Midnapore, India

Gouravmoy Banerjee Department of Computer Science, Ananda Chandra College, Jalpaiguri, West Bengal, India

Kingshuk Banerjee R&D Center, Hitachi India Pvt Ltd, Bangalore, Karnataka, India

Aleksandra Bartnik Faculty of Physics, University of Warsaw, Warsaw, Poland

Saikat Basu Department of Computer Science and Engineering, Maulana Abul Kalam Azad University of Technology, Kolkata, West Bengal, India

Subhadip Basu Department of Computer Science and Engineering, Jadavpur University, Kolkata, India

Catherina Basumatary Tezpur University, Tezpur, Assam, India

Peter Bellmann Institute of Neural Information Processing, Ulm University, James-Franck-Ring, Ulm, Germany

Debotosh Bhattacharjee Department of Computer Science and Engineering, Jadavpur University, Kolkata, India;
Center for Basic and Applied Science, Faculty of Informatics and Management, University of Hradec Kralove, Hradec Kralove, Czech Republic

Alexy Bhowmick Department of Computer Science and Engineering, Assam Don Bosco University, Azara, Guwahati, Assam, India

Bubu Bhuyan Department of Information Technology, North Eastern Hill University, Shillong, Meghalaya, India

Juwesh Binong Department of Electronics and Communication Engineering, North Eastern Hill University, Shillong, Meghalaya, India

Kaushik Bose Govt. General Degree College, Narayangarh, India

André Brechmann Leibniz-Institute for Neurobiology, Combinatorial NeuroImaging, Magdeburg, Germany

Satyabrat Malla Bujar Baruah Department of Electronics and Communication Engineering, Tezpur University, Napam, Tezpur, Assam, India

Bibekananda Buragohain Department of Computer Science and Engineering, Girijananda Chowdhury Institute of Management and Technology, Guwahati, Assam, India

Debangshu Chakraborty Department of Computer Science, Ananda Chandra College, Jalpaiguri, India

Debasish Chakraborty RRSC-East, ISRO, Kolkata, India;
Indian Institute of Technology, Kharagpur, West Bengal, India

Debashish Chakravarty Indian Institute of Technology, Kharagpur, West Bengal, India

Tanmoy Chaku Jalpaiguri Government Engineering College, Jalpaiguri, West Bengal, India

Akash Chatterjee Department of Computer Science, Asutosh College, Kolkata, India

Kingshuk Chatterjee Government College of Engineering and Ceramic Technology, Kolkata, India

Sayan Chatterjee Electronics and Telecommunication Engineering, Jadavpur University, Kolkata, West Bengal, India

Swarnesha Chatterjee Government College of Engineering & Ceramic Technology, Kolkata, India

Sheli Shina Chaudhuri Electronics and Telecommunication Engineering, Jadavpur University, Kolkata, West Bengal, India

Mateusz Chiliński Laboratory of Bioinformatics and Computational Genomics, Faculty of Mathematics and Information Science, Warsaw University of Technology, Warsaw, Poland;
Laboratory of Functional and Structural Genomics, Centre of New Technologies, University of Warsaw, Warsaw, Poland

Partha Chowdhuri Department of Computer Science, Vidyasagar University, Midnapore, West Bengal, India

Priyanjana Chowdhury Tezpur University, Tezpur, Assam, India

Swarup Dandapat Electronics and Telecommunication Engineering, Jadavpur University, Kolkata, West Bengal, India

Swarnav Das Barman Tezpur University, Tezpur, Assam, India

Dhritiman Das Department of Computer Science and Engineering, Girijananda Chowdhury Institute of Management and Technology, Guwahati, Assam, India

Ranjita Das National Institute of Technology Mizoram, Aizawl, India

Sangeeta Das Electronics and Communication Engineering Department, School of Technology, North-Eastern Hill University, Shillong, India

Soumyarshi Das Netaji Subhash Engineering College, Kolkata, West Bengal, India

Susovan Das Department of Computer Science, Asutosh College, Kolkata, India

Tanuja Das Department of Information Technology, Gauhati University, Guwahati, Assam, India

Kakali Datta Department of Computer and System Sciences, Visva-Bharati University, Santiniketan, West Bengal, India

Pranayee Datta Department of Electronics and Communication Technology, Gauhati University, Guwahati, Assam, India

Soumya Debnath Indian Institute of Engineering Science and Technology, Shibpur, India

Nandita Deka Department of ECE, NEHU, Shillong, India

Rup Kumar Deka Department of Computer Science and Engineering, Assam Don Bosco University, GuwahatiAssam, Kamrup, India

Vaskar Deka Department of Information Technology, Gauhati University, Guwahati, Assam, India

Paramita Dey Government College of Engineering & Ceramic Technology, Kolkata, India

Ratnadeep Dey Department of Computer Science and Engineering, Jadavpur University, Kolkata, India

Riya Rani Dey Netaji Subhash Engineering College, Kolkata, West Bengal, India

Samiran Dey Department of Computer Science, Asutosh College, Kolkata, India

Neha Dutta Department of Computer Science and Engineering, University of Calcutta, Kolkata, West Bengal, India

Soumi Dutta Institute of Engineering and management, Kolkata, India

Himadri Duwarah Department of Electronics and Communication Technology, Gauhati University, Guwahati, Assam, India

Ashrafia Esha Daffodil International University, Dhaka, Bangladesh

Sachin Gadakh Laboratory of Functional and Structural Genomics, Centre of New Technologies, University of Warsaw, Warsaw, Poland

Anna Gambin Faculty of Mathematics Informatics and Mechanics, University of Warsaw, Warsaw, Poland

Shrouti Gangopadhyay Department of Computer Science and Engineering, Jadavpur University, Kolkata, India

Debayan Ganguly Government College of Engineering and Leather Technology, Kolkata, India

Maitreyee Ganguly Government College of Engineering & Ceramic Technology, Kolkata, India

Jan Gawor DNA Sequencing and Oligonucleotide Synthesis Laboratory, Institute of Biochemistry and Biophysics, Polish Academy of Sciences, Warsaw, Poland

Rowdra Ghatak Department of ECE, NIT Durgapur, Durgapur, India

Chinmoy Ghosh Department of Computer Science and Engineering, Jalpaiguri Government Engineering College, Jalpaiguri, India

Indrajit Ghosh Department of Computer Science, Ananda Chandra College, Jalpaiguri, West Bengal, India

Marta Giziewska Faculty of Physics, University of Warsaw, Warsaw, Poland

Minakshi Gogoi Department of Computer Science and Engineering, Girijananda Chowdhury Institute of Management and Technology, Guwahati, Assam, India

Paweł Gora Faculty of Mathematics, Informatics and Mechanics, University of Warsaw, Warsaw, Poland

Harsh Indian Institute of Engineering Science and Technology, Shibpur, India

Md. Maruf Hassan Daffodil International University, Dhaka, Bangladesh

Shyamanta M. Hazarika Biomimetic Robotics and Artificial Intelligence Lab. Department of Mechanical Engineering, Indian Institute of Technology Guwahati, Guwahati, Assam, India

Uddipan Hazarika Department of Electronics and Communication Engineering, Tezpur University, Napam, Tezpur, Assam, India

Nsikak Imoh American University of Nigeria, Yola, Nigeria

Anand Singh Jalal Department of Computer Engineering and Applications, GLA University, Mathura, India

Biswapati Jana Department of Computer Science, Vidyasagar University, Midnapore, West Bengal, India

Sunanda Jana Department of Computer Science and Engineering, Haldia Institute of Technology, Haldia Purba Medinipur, West Bengal, India

M. K. Jindal Department of Computer Science and Applications, Panjab University Regional Centre, Muktsar, India

Karolina Jodkowska Laboratory of Functional and Structural Genomics, Centre of New Technologies, University of Warsaw, Warsaw, Poland;
Centre for Advanced Materials and Technologies (CEZAMAT), Warsaw University of Technology, Warsaw, Poland

Anwesh Kabiraj Government College of Engineering and Leather Technology, Kolkata, India

Viktor Kessler Institute of Neural Information Processing, Ulm University, James-Franck-Ring, Ulm, Germany

Taiser S. Khedaywi Civil Engineering Department, Faculty of Engineering, Jordan University of Science and Technology, Irbid, Jordan

Dipak Kumar Kole Jalpaiguri Government Engineering College, Jalpaiguri, West Bengal, India

Christian Kollmann Center for Medical Physics and Biomedical Engineering, Medical University Vienna, Vienna, Austria

Ondrej Krejcar Center for Basic and Applied Science, Faculty of Informatics and Management, University of Hradec Kralove, Hradec Kralove, Czech Republic

Malay Kule Indian Institute of Engineering Science and Technology, Howrah, West Bengal, India

Arnab Kumar Maji Department of Information Technology, North-Eastern Hill University, Shillong, Meghalaya, India

Bhupender Kumar Department of Information Technology, North Eastern Hill University, Shillong, Meghalaya, India

Munish Kumar Department of Computational Sciences, Maharaja Ranjit Singh Punjab Technical University, Bathinda, Punjab, India

Ritesh Kumar Department of Information Technology, NEHU, Shillong, Meghalaya, India

Sharath Kumar R&D Center, Hitachi India Pvt Ltd, Bangalore, Karnataka, India

Aiden Langba Department of Computer Science and Engineering, Assam Don Bosco University, Azara, Guwahati, Assam, India

S. Lekshmi Rajagiri College of Social Sciences (Autonomous), Kochi, Kerala, India

Sk Mahmudul Hassan North Eastern Hill University, Shillong, Megahalya, India

Ritajit Majumdar Advanced Computing & Microelectronics Unit, Indian Statistical Institute, Kolkata, India

Koushik Majumder Maulana Abul Kalam Azad University of Technology, Kolkata, West Bengal, India

Sounak Majumder Indian Institute of Engineering Science and Technology, Shibpur, India

Samir Malakar Department of Computer Science, Asutosh College, Kolkata, India

Ankan Mallick Indian Institute of Engineering Science and Technology, Shibpur, India

Ankita Mandal Institute of Engineering and management, Kolkata, India

Ratna Mandal Institute of Engineering and management, Kolkata, India

Satyendranath Mandal Department of Information Technology, Kalyani Government Engineering College, Nadia, India

Shyamal Mandal Department of Biomedical Engineering, North Eastern Hill University, Shillong, Meghalaya, India

Nestor Mariyasagayam R&D Center, Hitachi India Pvt Ltd, Bangalore, Karnataka, India

Md. Maruf Mallick Senior Associate, Cognizant Technology Solutions India Pvt. Ltd., Kolkata, India

Hamebansan Mawlong Department of Computer Science and Engineering, Assam Don Bosco University, Azara, Guwahati, Assam, India

Maad M. Mijwil Computer Techniques Engineering Department, Baghdad College of Economic Sciences University, Baghdad, Iraq

Sandeep Deva Misra Tezpur University, Tezpur, Assam, India

Soma Mitra Brainware University, Barasat, Kolkata, India

Ayatullah Faruk Mollah Department of Computer Science and Engineering, Aliah University, Kolkata, India

Swarnali Mondal Indian Institute of Engineering Science and Technology, Shibpur, India

Agnieszka Motyka Faculty of Physics, University of Warsaw, Warsaw, Poland

Rajkishur Mudoi Department of Electronics and Communication Engineering, North-Eastern Hill University, Shillong, Meghalaya, India

Subhadip Mukherjee Department of Computer Science, Kharagpur College, Kharagpur, India

Somnath Mukhopadhyay Department of Computer Science and Engineering, Assam University, Silchar, India

Prasenjit Mura Department of Computer Science, Vidyasagar University, Midnapore, West Bengal, India

Subrata Nandi National Institute of Technology, Durgapur, India

Mita Nasipuri Department of Computer Science and Engineering, Jadavpur University, Kolkata, India

Sanghamitra Nath Tezpur University, Tezpur, Assam, India

Yuichi Nonaka R&D Group, Hitachi Ltd, Tokyo, Japan

Doreen Nongrum Department of Electronics and Communication Engineering, North-Eastern Hill University, Shillong, Meghalaya, India

Partha Pakray National Institute of Technology Silchar, Silchar, Assam, India

Amartya Pal Indian Institute of Engineering Science and Technology, Shibpur, India

Debojyoti Pal Government College of Engineering and Leather Technology, Kolkata, India

Pabitra Pal BSTTM, IIT Delhi, New Delhi, India

Rajat Kumar Pal Department of Computer Science and Engineering, University of Calcutta, Kolkata, West Bengal, India

Ramen Pal Department of Computer Science and Engineering, Assam University, Silchar, India

Amarnath Pathak National Institute of Technology Mizoram, Aizawl, India

Soumi Paul Department of Computer Science and Engineering, Jadavpur University, Kolkata, India

Michal Pietal Faculty of Computer and Electrical Engineering, Rzeszow University of Technology, Rzeszów, Poland

Dariusz Plewczynski Laboratory of Bioinformatics and Computational Genomics, Faculty of Mathematics and Information Science, Warsaw University of Technology, Warsaw, Poland;
Laboratory of Functional and Structural Genomics, Centre of New Technologies, University of Warsaw, Warsaw, Poland

Fidalizia Pyrtuh Department of Electronics and Communication Engineering, North-Eastern Hill University, Shillong, Meghalaya, India

Sandip Rakshit American University of Nigeria, Yola, Nigeria

P. Rangababu Department of Electronics and Communication Engineering, National Institute of Technology Meghalaya, Shillong, India

Voore Subba Rao Department of Physics and Computer Science, Dayalbagh Educational Institute, Agra, India

Manikandan Ravikiran R&D Center, Hitachi India Pvt Ltd, Bangalore, Karnataka, India

Abhirup Ray RCC Institute of Information Technology, Kolkata, West Bengal, India

Rafika Risha Daffodil International University, Dhaka, Bangladesh

Kaushiki Roy Department of Computer Science and Engineering, Jadavpur University, Kolkata, India

Sarbani Roy Jadavpur University, Kolkata, India

Soumik Roy Department of Electronics and Communication Engineering, Tezpur University, Napam, Tezpur, Assam, India

Fairriky Rynjah North Eastern Hill University, Shillong, India

Punam K. Saha Departments of Electrical and Computer Engineering and Radiology, University of Iowa, Iowa City, IA, USA

Ramesh Saha Department of Information Technology, Gauhati University, Guwahati, Assam, India

Shilpa Saha Department of Computer Science, Asutosh College, Kolkata, India

Sudipta Saha Maulana Abul Kalam Azad University of Technology, Kolkata, West Bengal, India

Sujoy Saha National Institute of Technology, Durgapur, India

Hubha Saikia Department of Electronics and Communication Engineering, North-Eastern Hill University, Shillong, Meghalaya, India

Kandarpa Kumar Saikia Department of BioEngineering and Technology, Gauhati University, Guwahati, Assam, India

Toushik Santra Netaji Subhash Engineering College, Kolkata, West Bengal, India

Kamal Sarkar Computer Science and Engineering Department, Jadavpur University, Kolkata, India

Pankaj Sarkar Electronics and Communication Engineering Department, School of Technology, North-Eastern Hill University, Shillong, India

Sweta Sarkar Netaji Subhash Engineering College, Kolkata, West Bengal, India

Uditendu Sarkar National Informatics Centre, Ministry of Electronics and Information Technology, Government of India, Jalpaiguri, West Bengal, India

Saheli Sarkhel Netaji Subhash Engineering College, Kolkata, West Bengal, India

Friedhelm Schwenker Institute of Neural Information Processing, Ulm University, James-Franck-Ring, Ulm, Germany

Ayan Seal PDPM Indian Institute of Information Technology, Design and Manufacturing, Jabalpur, Madhya Pradesh, India

Shibashish Sen R&D Center, Hitachi India Pvt Ltd, Bangalore, Karnataka, India

Kaustav Sengupta Laboratory of Functional and Structural Genomics, Centre of New Technologies, University of Warsaw, Warsaw, Poland;
Faculty of Mathematics Informatics and Mechanics, University of Warsaw, Warsaw, Poland

Monika Sethi Department of Information Technology, Goswami Ganesh Dutta Sanatan Dharma College, Chandigarh, India

Mosammath Shahnaz Department of Computer Science and Engineering, Aliah University, IIA/27 New Town, Kolkata, India

Himanshu Sharma Department of Computer Engineering and Applications, GLA University, Mathura, India

Neelotpal Sharma Department of BioEngineering and Technology, Gauhati University, Guwahati, Assam, India

Pallabi Sharma Department of Computer Science and Engineering, National Institute of Technology Meghalaya, Shillong, India

Utpal Sharma Tezpur University, Tezpur, Assam, India

L. J. Singh North Eastern Hill University, Shillong, India

Prabhash Kumar Singh Department of Computer Science, Vidyasagar University, Midnapore, West Bengal, India;
Department of Computer and System Sciences, Visva-Bharati University, Santiniketan, West Bengal, India

K. Srinivas Department of Electrical Engineering, Dayalbagh Educational Institute, Agra, India

Jacek Sroka Faculty of Mathematics, Informatics and Mechanics, University of Warsaw, Warsaw, Poland

Rupaban Subadar Department of ECE, NEHU, Shillong, India

Aleazer Viannie Sunn Department of Computer Science and Engineering, Assam Don Bosco University, Azara, Guwahati, Assam, India

Bronson Syiem North Eastern Hill University, Shillong, India

Suparba Tapna Department of Electronics and Communication Engineering, Durgapur Institute of Advanced Technology and Management, Rajbandh, Durgapur, India

Narasimha Rao Vajjhala University of New York Tirana, Tirana, Albania

Aditi Verma Indian Institute of Engineering Science and Technology, Shibpur, India

Siddharth Vohra R&D Center, Hitachi India Pvt Ltd, Bangalore, Karnataka, India

Shrasti Vyas PDPM Indian Institute of Information Technology, Design and Manufacturing, Jabalpur, Madhya Pradesh, India

Munshi Yusuf Alam Budge Budge Institute of Technology, Kolkata, India

Natalia Zawrotna Laboratory of Functional and Structural Genomics, Centre of New Technologies, University of Warsaw, Warsaw, Poland

Aleksandra Żurko Faculty of Physics, University of Warsaw, Warsaw, Poland

Artificial Intelligence

Experimental Face Recognition Using Applied Deep Learning Approaches to Find Missing Persons

Nsikak Imoh, Narasimha Rao Vajjhala, and Sandip Rakshit

Abstract The spike in challenges to security as well as information and resource management across the globe has equally borne the rising demand for a better system and technology to curb it. A news release from the International Committee of the Red Cross (ICRC) in 2020 revealed over 40,000 people were declared missing in Africa. A staggering percentage of that number, a little over 23,000, is documented in Nigeria alone. Despite the numerous factors surrounding missing persons globally, at more than 50% of the original figure, it is unsurprising that most of the cases in Nigeria are attributed to the insurgency and security mishap that has plagued the country for almost a decade. Some of the cases remain unsolved for years, causing the victims to remain untraceable, thereby taking up a different identity and existence, especially if they went missing. Current solutions to find missing persons in Nigeria revolve around word of mouth, media and print announcements, and more recently, social media. These solutions are inefficacious, slow, and do not adequately help find and identify missing persons, especially in situations where time is a determining factor. The use of a facial recognition system with deep learning functionality can help Nigerian law enforcement agencies, and other human rights organizations and friends and families of the missing person speed up the search and find process. Our experimental system combines facial recognition with deep learning using a convoluted neural network. In this study, the authors have used high-standard facial calibration and modeling for feature extraction. These extracted features form the face encodings that are after that compared to a given image.

Keywords Deep learning · Face recognition · Artificial intelligence · Neural network · Machine learning · Convolutional neural networks

N. Imoh · S. Rakshit
American University of Nigeria, Yola, Nigeria
e-mail: nsikak.imoh@aun.edu.ng

S. Rakshit
e-mail: sandip.rakshit@aun.edu.ng

N. R. Vajjhala (✉)
University of New York Tirana, Tirana, Albania
e-mail: narasimharao@unyt.edu.al

S. Basu et al. (eds.), *Proceedings of International Conference on Frontiers in Computing and Systems*, Lecture Notes in Networks and Systems 404,
https://doi.org/10.1007/978-981-19-0105-8_1

1 Introduction

The cases of missing persons of all genders and ages are a common phenomenon worldwide, and most victims are women and children. The factors that surround these cases may vary from place to place. However, the common reasons revolve around kidnapping, natural disasters, security mishaps, illness, and voluntary or involuntary abscondment. Most technologically advanced countries have implemented systems and databases to accelerate finding and identifying missing persons. Notably, more countries are adding facial recognition technology with deep learning to their tech arsenal enabling law enforcement agencies to locate missing persons to become more productive [1].

A case study where facial recognition has been successfully used to find missing persons is in India. A news publication in India Today revealed that on average, "174 children go missing every day, half of them remain untraceable" [2]. Also, "more than one lakh children" approximately 111,569 "were reported to have gone missing till 2016, and 55,625 of them remained untraceable till the end of the year" [2]. However, there was a breakthrough when Indian police launched a facial recognition system to trace and identify thousands of missing children within four days [3]. A similar breakthrough was reported in China, where a couple could locate their child that was missing 30 years ago using facial recognition with deep learning capability [4].

2 Background and Review of Literature

2.1 *Problem Statement*

Nigeria has been credited to have the highest caseloads of missing persons globally by the International Committee of the Red Cross (ICRC) based on data reported [5]. However, many more are unaccounted for and go unreported for years, and in some cases, never attended to [5]. This is primarily due to the lack of an efficient database of the citizens and the poor technological infrastructure required to identify these victims. Given the recent call for citizen accountability and database by the Nigerian Communications Commission (NCC), Nigeria could benefit from using a facial recognition system to tackle the issues of missing persons. Our experimental facial recognition system can help law enforcement agencies find missing persons in any location across Nigeria by running an automatic detection system at major roadblocks, public facilities, and giving consent to private volunteers. This is done by taking a given image of a missing person to train the system and store it in the data store. When the recognition system runs, the camera scans the various faces around, carries out automatic detection and recognition, and compares the result to the trained images in the data store. When a face fits that of a missing person, an automated notification containing essential identification details is sent. Optionally, giving the right consent, the person's location is tracked using Google Earth API and sent along in the notification, making it easy to locate the missing person.

2.2 *Existing Systems*

The implementation of facial recognition in tech devices has seen tremendous growth in the last decade. This increased adoption can be seen in different sectors and industries such as banking, medicine, marketing, finance, hospitality, transportation [6–10]. As a result, there have been different approaches to its use. These approaches are usually tailored to a particular use case, just like the system we are proposing. The closest existing system to implement a similar use case to ours is an enterprise SAAS called FaceFirst. With the mission, "FaceFirst is creating a safer and more personalized planet with facial recognition technology. We empower organizations to detect and deter real-time threats, transform team performance, and strengthen customer relationships". FaceFirst tailors its product to different industries to leverage a custom face recognition technology. However, we are unable to tell the underlying software implementation.

3 Proposed System

The proposed system uses deep learning to improve the facial recognition process. Deep learning is a branch of artificial intelligence that has increased modern recognition technology use due to its accuracy rate [11]. Our system uses deep learning to analyze the subject's face from a photograph and feature extraction to generate one-of-a-kind facial mathematical representations using a trained dataset of images. These extracted features are then stored in a database. Rather than use of available facial training models like DeepFacial and FacialNet with their respective recognition algorithms, we chose to move with the neural network machine learning algorithm due to the complex functionality of the use case of our recognition system [11]. One good pro of using neural network is training models that pick out complex, unique features that other algorithms will find challenging. And, because the networks are not linearly placed, it is adopted as the best algorithm for complex facial recognition. However, using a pure neural network algorithm proved to be too complicated and demanding [8]. This is due to the need for high computing power and a large amount of data [12].

A workable and better alternative would be using a subclass of neural network known as a convolutional neural network (CNN) [13]. With the possession of at least one convolution layer, CNN excels in picking up native information such as "surrounding words in a text" [12] or, in this case, "neighbor pixels in an image" [12]. They also excel at minimizing the complexity of the regular neural network in model training [14]. As a result, the training time for the datasets is faster, and the required data samples are a lot less [12]. As stated by Syafeeza et al. [12], "CNN is unique due to the architecture itself. It performs segmentation, feature extraction, and classification in one processing module with minimal preprocessing tasks on the input image". With this in mind, CNN has been proven to be one of the most reliable algorithms in deep learning for face detection [15] and face recognition [16]. It significantly reduces the physical manpower required in building its capabilities. It does not depend so much on preprocessing, making it easy to grasp the flow of events and quickly implement the needed training. And most significantly, its degree of accuracy is the best among all algorithms for image detection and recognition.

In this experimental facial recognition project, we used dual "4-layer CNN" architectures. The issues of front-facing images such as occlusion, illumination, and varying facial expressions are the primary focus of the first CNN architecture. The second CNN architecture handles the other secondary occlusion issues, illumination, varying facial expressions, and, more generally, face pose.

The training dataset was generated from over 10,000 images gotten from different free stock image Web sites and appropriately labeled based on facial pose (straight face, left side, right side, face down, face up), illumination (silhouette, bright, dark), occlusion (face masks, eye wears, facial cosmetic, clear), and visible and noticeable expressions (laugh, smile, sad, normal, anger). Although there could be more factors for labeling datasets during training, we observed these basic ones from the images we sourced. However, some of the images were repeated intentionally due to the presence of different factors. For instance: A person smiling with their eyes closed can be interpreted differently from a person who smiles with their eyes open and a slight grim. Also, some of the images had groups of faces comprising of 3 to about 50 people. We cropped each face and trained them separately as a single face rather than a grouped face image to get all of these faces for training.

4 Methodology

When the law enforcement agency receives a report regarding a missing person, the relevant image with fitting training templates will be requested and entered into the system. Afterward, a camera, preferably a high-resolution camera, is mounted in a police road stop, public place, or just about any place that serves as the center of focus. Having a consistent illumination with minimal motion will help to facilitate this process. With this, the police will scan the area of focus. The system detects the faces around, converts them to mathematical data representation, and compares them to our database's trained data. If a match is found, the system sends a notification to the police with a snapshot of the person and the GPS coordinates and location data. Our proposed facial recognition system will handle the finding and identification process in the following steps outlined in the flowchart in Fig. 1.

The first step of the recognition process involves detecting the face using a convolutional neural network (CNN). There have been numerous approaches using this method. The proposed system uses an approach laid out by Zhang et al. [17] called

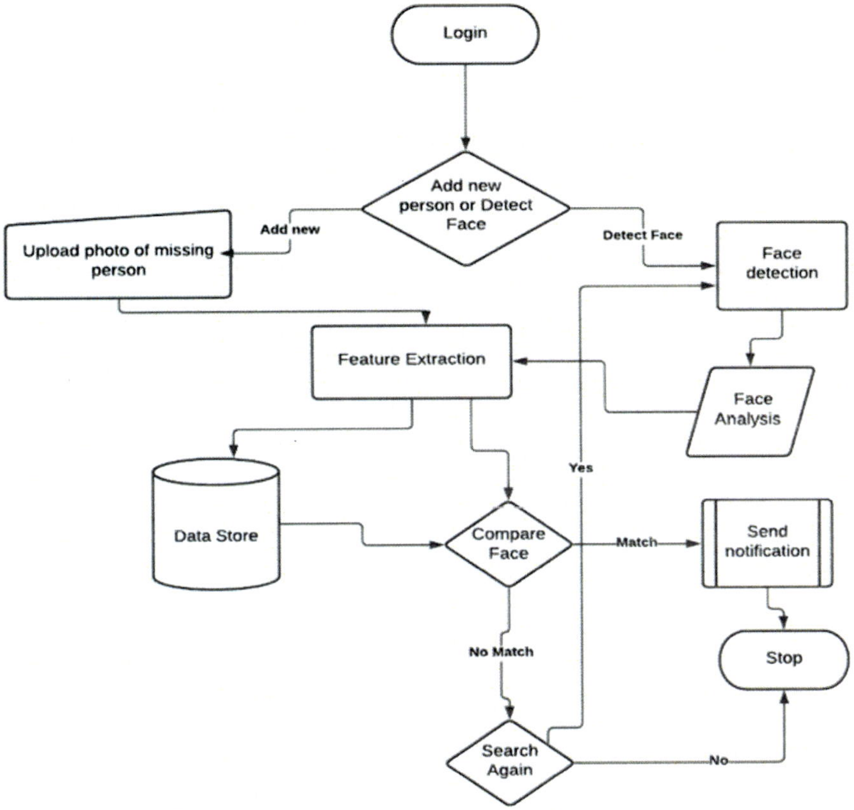

Fig. 1 Steps of the proposed facial recognition

the "multi-task cascaded convolutional neural network" (MTCNN). This approach is suitable as it can recognize landmark features of the face such as the eyes and mouth. It works in three stages: "In the first stage, it produces candidate windows quickly through a shallow CNN. Then, it refines the windows to reject many non-faces windows through a more complex CNN. Finally, it uses a more powerful CNN to refine the result and output facial landmarks positions" [17]. Figure 2 shows the three steps.

In the image above, we can see that a pyramid shape consisting of different image sizes is generated by rescaling the input image. Next, the contending facial regions are highlighted after the first model, "proposal network or P-Net", processes it. The box highlighting the facial regions is underscored by the second model, "refine network or R-Net". And at the final step, the facial landmarks are put forward by the third model, "output network or O-Net". Together, these three models are trained to handle three essential functions "face classification, bounding box regression, and facial landmark localization" [14]. The output of each model is given as an input to the next. As a result, they are not directly working hand in hand, which

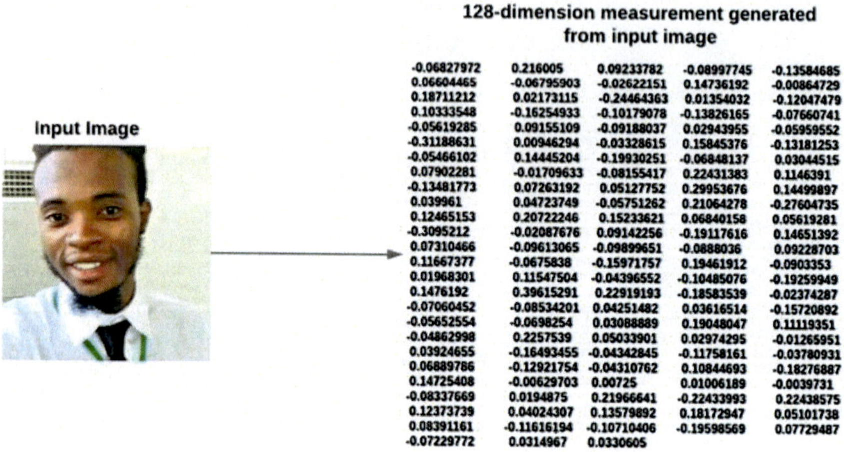

Fig. 2 128-dimension measurement from a sample image

provides room for different processes to be carried when needed [1]. Working under MIT License for open source, we used the MTCNN Python library by Iván de Paz Centeno with the GitHub username "ipazc". This library uses the machine learning library, Tensorflow, and the OpenCV computer vision library to implement the multi-task cascaded convolutional neural network (MTCNN) architecture. Even though it comes with its default highly effective pre-trained model, we could utilize the custom dataset we trained.

After the faces have been detected, the system estimates and selects 68 unique landmark points on every face. Using this landmark, we can successfully create the other image templates by scaling and rotating the input image using the extracted features using the OpenCV library to maintain the position of the eyes and the mouth. Afterward, we get a 128-dimension measurement from the image, which is compared with a different image of the same person. Doing this, we get the same 128-dimension measurements for several images of the same person. This forms a unique faceprint of each individual even though they may have a look-alike as shown in Fig. 2.

We used Django Rest Framework to create a RestFul API that enables the users to interact with the models and controllers at the backend from the UI. When a face is detected and the feature extraction has taken place, a classifier powered by support vector machine (SVM) to analyze the data from the input, compare the result in the database, and produce the most likely match carries out the face recognition. With the aid of the Google Earth API, we can get the location of the individual through the longitude and latitude coordinates collected during the recognition process. This data is then sent along with a screenshot to the users of the system.

5 Results

The result we got from the prototype built was better than expected. The proposed system used an HD camera with high resolution. The system recognized a face in less than 100 ms. Not only that, testing on a subject with no pose or occlusion limitation produces an accuracy of 99.6%. We also noted a 92% recognition accuracy with mild to medium pose limitation and an 85% recognition accuracy. Test results with occlusion limitation stood at a range of 66–82% recognition accuracy depending on the extent of facial occlusion. Considering the limited dataset used for training the model, this tremendously outperforms recognition systems built with other recognition algorithms.

One of the primary challenges that plague facial recognition systems of all kinds is face pose. There are different ways other systems have taken to solve this problem. One solution involves taking many images and making a template labeling of different poses during the training. Another solution uses several face poses of a particular subject for the training dataset. We used a hybrid of both solutions by taking a few images of a subject and making a template of different poses by rotating the images. Another challenge of facial recognition is the differences in illumination. "Illumination variations in images are an essential factor in facial recognition. The intensity of the color in a pixel can vary depending on lighting conditions" [13]. This means that there is a high chance that different images of a single subject may produce different calibrations during the training. To solve this, we created a template of different illuminating conditions for each image. By leveraging this technique, we can successfully separate variations in illumination from the shape and texture of facial images.

Another challenge of a facial recognition system is occlusion. This occurs when a part of the face is obstructed or covered by natural facial defect or facial hair and the use of accessories such as eyewear, nose masks, hijabs, and cosmetics. This influences the calibrations during the training. Although not tested yet as all trained images were not obstructed, our solution is to set a template for each generic type of facial occlusion and get images that fit that template for the training. Varying facial expressions is also one of the main challenges of facial recognition systems. A system might not correctly identify similar images due to the facial expressions in one or the other. Like the occlusion, our hypothetical solution is to set a template for each generic type of facial expression and get images that fit that template for the training. Besides those above, face recognition faces different issues, such as camera quality and device memory capacity. Still, they do not affect the flow or process of recognition.

6 Conclusion

The cases of missing persons of all genders and ages are a common phenomenon worldwide, and most victims are women and children. Most technologically advanced countries have implemented systems and databases to accelerate finding and identifying missing persons. Notably, more countries are adding facial recognition technology with deep learning to their tech arsenal enabling law enforcement agencies to locate missing persons to become more productive. Nigeria has been credited to have the highest caseloads of missing persons globally by the International Committee of the Red Cross (ICRC) based on data reported. Nigeria could benefit from the use of a facial recognition system to tackle the issues of missing persons. There are several limitations of facial recognition systems. The primary ones, such as face pose, illumination, occlusion, and varying facial expressions, affect the recognition process directly. Others, such as memory of the device and camera quality, do not directly affect the recognition process. The world of technology is ever-evolving. More businesses and industries adopting face recognition systems will give rise to upgrades and more viable solutions to the current limitations. This will help law enforcement agencies accelerate the task of finding missing and identifying persons.

References

1. Shrestha R, Panday SP (2020) Face recognition based on shallow convolutional neural network classifier. In: Proceedings of the 2020 2nd international conference on image, video and signal processing. Association for Computing Machinery, Singapore, pp 25–32
2. Nigam C (2019) Delhi's shame! 19 children go missing daily, in India Today. New Delhi, pp 1–13
3. Cuthbertson A (2018) Indian police trace 3,000 missing children in just four days using facial recognition technology. In: The independent. London, pp 1–7
4. Wray M (2020) Parents reunite with son kidnapped 30 years ago, thanks to facial recognition technology. In: Global news. New York
5. Ewang A (2020) Nigeria's rising number of missing persons, Human Rights Watch, HRW
6. Bae H, Kim S (2005) Real-time face detection and recognition using hybrid-information extracted from face space and facial features. Image Vis Comput 23:1181–1191
7. Chowdhury S et al (2010) A hybrid approach to face recognition using generalized two-dimensional fisher's linear discriminant method. In: 2010 3rd international conference on emerging trends in engineering and technology
8. Dagher I, Al-Bazzaz H (2019) Improving the component-based face recognition using enhanced Viola–Jones and weighted voting technique. Modelling Simul Eng 8234124
9. Dospinescu O, Popa I (2016) Face detection and face recognition in android mobile applications. Informatica Economica 20(1):20–28
10. Hanmandlu M, Singhal S (2017) Face recognition under pose and illumination variations using the combination of Information set and PLPP features. Appl Soft Comput 53:396–406
11. Al-Waisy AS et al (2018) A multimodal deep learning framework using local feature representations for face recognition. Mach Vis Appl 29(1):35–54
12. Syafeeza AR et al (2014) Convolutional neural network for face recognition with pose and illumination variation. Int J Eng Technol 6(1):44–57

13. Dunstone T, Yager N (2009) An introduction to biometric data analysis. In: Dunstone T, Yager N (eds) Biometric system and data analysis: design, evaluation, and data mining. Springer US, Boston, MA, pp 3–26
14. LeCun Y, Bengio Y, Hinton G (2015) Deep learning. Nature 521(7553):436–444
15. Lawrence S et al (1997) Face recognition: a convolutional neural-network approach. IEEE Trans Neural Netw 8(1):98–113
16. Khalajzadeh H, Mansouri M, Teshnehlab M (2013) Hierarchical structure based convolutional neural network for face recognition. Int J Comput Intell Appl 12(3):1350018
17. Zhang K et al (2016) Joint face detection and alignment using multitask cascaded convolutional networks. IEEE Signal Process Lett 23(10):1499–1503

MultiNet: A Diffusion-Based Approach to Assign Directionality in Protein Interactions Using a Consensus of Eight Protein Interaction Datasets

Kaustav Sengupta, Anna Gambin, Subhadip Basu, and Dariusz Plewczynski

Abstract Protein–protein interaction network (PPIN) plays a major role in information processing and decision making in cells. The PPIN works as a skeleton for cell signaling and functionality. Understanding the flow of information in a cell can enhance our understanding of functional outcomes and flow of signals. To utilize the whole potential of PPIN, we need to direct the edges of the networks. In recent years, a deluge of PPIN became available for analysis but to understand the full picture, these PPINs need to be oriented, or directed, as the direction of signal or information flow in human PPIN is still mostly unknown. In this paper, we propose MultiNet, a method based on the well-known diffusion-based approach which assigns directionality to PPIN created from eight different networks to cover the most of the human genome MultiNet achieves the highest AUC score of 0.94 over protein DNA interaction test set and performs better than the current *state-of-the-art* algorithms using networks from single sources.

Keywords Protein–protein interaction network (PPIN) · Directed networks · Diffusion · Drug targets · Pathways · Network biology · Network orienting

---

K. Sengupta · D. Plewczynski (✉)
Center of New Technologies, University of Warsaw, Warsaw, Poland
e-mail: d.plewczynski@cent.uw.edu.pl

K. Sengupta
e-mail: k.sengupta@cent.uw.edu.pl

K. Sengupta · A. Gambin
Faculty of Mathematics Informatics and Mechanics, University of Warsaw, Warsaw, Poland
e-mail: aniag@mimuw.edu.pl

S. Basu
Department of Computer Science and Engineering, Jadavpur University, Kolkata 700032, India
e-mail: subhadip.basu@jadavpuruniversity.in

D. Plewczynski
Faculty of Mathematics and Information Science, Warsaw Technical University, Warsaw, Poland

S. Basu et al. (eds.), *Proceedings of International Conference on Frontiers in Computing and Systems*, Lecture Notes in Networks and Systems 404,
https://doi.org/10.1007/978-981-19-0105-8_2

1 Introduction

In recent years with the advancement of high throughput methods such as yeast two hybrids [1, 2], mass spectrometry [3], tandem affinity purification [4], a plethora of protein interaction data became available. These datasets provide information about interacting proteins and their functionality [5, 5]. The protein–protein interactions (PPIs) are also crucial for cellular information flow and decision making [7]. These networks are seldom oriented or directed based on various factors such as drug effects, cancer genomics, or signaling pathways. Most of the currently available PPI networks lack the directions on information flow and logic of connections. Until recent times, the interaction networks such as PPI or protein DNA networks were mostly studied based on their topological properties and how such properties effects the signature or essential proteins [8–10] or, alternatively, using graph theoretic approaches to identify a specific subnetwork that gives us information about the phenotype [8, 11].

The first step beyond such topological models is to assert the directionality of the network by predicting the direction of the flow of signals [12]. Research by Cao et al. [13] shows that such directed networks can be beneficial for better function prediction. The oriented networks can be used in various tasks such as drug development [14], effect of gene expression [15], study of diseases related to genomic rearrangement [10, 16–18], effects of chemical inhibitors associated with diseases and predicting signaling pathways [19]. Still till date only a small percentage of human interactions are directed—the expected value of directed PPI in the human proteome is around 40% [20]. The existing orientation methods are mainly applied to yeast [21, 22]. The notions of the cause and effect in such studies are mainly based on the expression levels of genes and perturbation experiments. The methods of Vinayagam et al. [7], Silverbush, and Sharan [23] are the only two methods available to orient the human PPI. In the paper by Vinayagam et al. [7], they hypothesized all membrane receptors that are connected to transcription factors and use the shortest path approach. In Silverbush and Sahran [23], the authors use diffusion-based algorithms to omit the use of the shortest path but they use protein interactions from only single sources. In the present work to cover most of the human proteome as we merge PPI networks from multiple sources and create a consensus network called MultiNet by merging eight different datasets. Then, using diffusion-based algorithms proposed in Silverbush and Sharan 2019, we achieve higher orientation accuracy over MultiNet than current state-of-the-art algorithms working over networks curated from single sources. Moreover, such consensus networks are easily associated with other biological networks, like gene–gene association and chromatin interaction networks, exhibiting non-random properties as shown in Halder et al. [10].

2 Material and Methods

MultiNet: We used eight different datasets to create the meta networks. We collected a set of interacting proteins from DIP [24], STRING [25], MINT [2], IntAct [26], HINT [27], BioGRID [28], HPRD [29], HuRI [30]. To make the PPI networks highly reliable we require each interaction to fulfill two conditions: 1) All the interactions must be experimentally verified and 2) the interactions must be reported by at least two publications. In this way, we converted around 91% of the entire human proteome. The statistics of the eight different networks is shown in Table 1.

2.1 *Diffusion-Based Orientation*

For determining the directionality, we follow the algorithm proposed in Silverbush and Sharan 2019 [23]. However, we modified the algorithm to run over consensus networks by merging networks from various data sources, as it gives us a full picture of the whole human proteome. The algorithm is based on the diffusion to identify the directionality of the PPI network. The directionality is computed from the cause and effect networks, inferred from the drug and pathway networks. While most of the other current algorithms use only the shortest path between source and target nodes, this algorithm considers all possible paths between source and target. For the purpose of the algorithm, the edges can either be weighted, or unweighted. If the input network is unweighted, the weight of each edge is assigned to 1, which is the case for our MultiNet. On this network, we apply the diffusion algorithm; its main steps are:

1. We infer the diffusion from both sides, i.e., ones from the source protein and ones from the target protein (as explained in Sect. 2.1).

Table 1 Statistics of nodes and edges from various networks used to construct the MultiNet

Database	*No. of nodes*	*No of edges*
DIP [24]	1877	2125
STRING [25]	17,517	120,147
MINT [2]	6510	18,991
HINT [27]	3133	5262
BioGRID [28]	5756	15,743
IntAct [26]	14,062	120,147
HPRD [29]	8635	28,594
HuRI [30]	8275	52,569
Total	**65,765**	**363,578**

Tolat no of Proteins and Interactions present in the above datasets

2. Two scores, one from source node and one from target node, are merged together to obtain a single direction.
3. We use the combined score for different drugs or pathways and merge them to get the final direction.

2.2 *Diffusion Process and Edge Scoring*

The diffusion process is calculated over each protein as follows; the score is defined as $F(v)$ described in Silberbush et al., and the formula is given in Eq. (1).

$$F(v) = \alpha\left[\sum\nolimits_{u\epsilon N(v)} F(u)w(u,v)\right] + (1-\alpha)P(v) \tag{1}$$

Here to balance the prior terms and the network value, we use the parameter $\boldsymbol{\alpha}$. w is the average outgoing edge weight, according to the formula in Eq. (2), $P(v)$ is the prior value initially set to 0 and $N(u)$ stands for the set of nodes neighboring to u.

$$w(u,v) = \frac{\mathrm{weight}(u,v)}{\sum_{k\epsilon N(u)} \mathrm{weight}(u,k)} \tag{2}$$

At each iteration, the score is pushed to the neighborhood of the protein node proportionally to the average outgoing edge weight, until we achieve a convergence as described in Cowen et al. The convergence is achieved if the changes of the score values from the previous iteration is very small, i.e., if the square root of change is less than, we set it at 10^{-5} as suggested in Silverbush and Sharan 2019. See Fig. 1 for the flowchart of the method.

2.2.1 Inference of Directionality

We collect the drug response data from three databases, namely KEGG DRUG [31], DrugBank [32], DCDB databases [33]. Following the paper of Iskar et al. 2010 [34], we performed the normalization and filtering and thus identified 21,424 differentially expressed genes associated with 551 drugs at a rate of 38.88 genes per drug on average. Then for each drug (experiment), we run the diffusion algorithm as described in the earlier section. So, this gives us a vector in which each row is an edge and the column is a specific experiment (a drug) and its score is stored in the matrix. This score gives us the contribution of the edge to the path connecting source and target for a given experiment. Then we use this score for each edge accumulated over all experiments to generate a final score of the direction by using a linear regression model trained using the scores from the matrix as features.

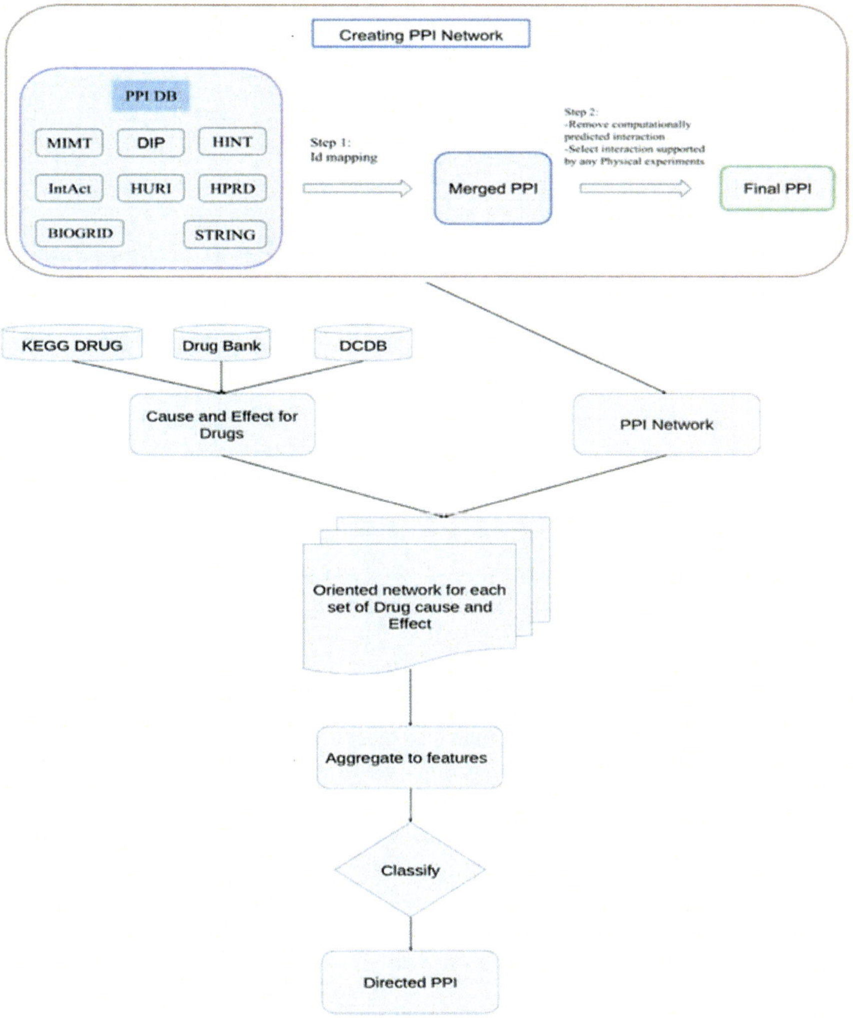

Fig. 1 Control flow through the proposed method

3 Result and Discussion

The performance evaluation of the algorithm and our MultiNet network we used the same test set as described in Silverbush and Sharan [23]. From Table 2, it can be also concluded that using MultiNet for directionality prediction outperforms the other methodologies using the same algorithms in the terms of direction prediction accuracy values. The validation was done using true directionality data from five different databases providing directionality (i) signal transduction in mammalian cells (STKE) from database of cell signaling [35] (ii) EGFR signaling pathway (EGFR) [36] (iii)

Table 2 Direction prediction accuracy scores between various methods

Network Type	Topology only [40]	Vinayagam et al. [7]	BioGRID [23]	MultiNet
PDI	0.71	0.72	0.90	0.94
KPIs	0.62	0.64	0.92	0.92
STKE	0.47	0.71	0.74	0.75
EGFR	0.51	0.58	0.85	0.86
E3	0.49	0.67	0.85	0.82

protein DNA interactions (PDI) from ChEA database [37] (iv) kinase/phosphate to substrate interactions (KPIs) [38] and (v) ubiquitination interactions (E4FR) downloaded from hUbiquitome database [39]. The performance was enhanced due to the several reasons which are as follows: (1) Our networks are a combination of networks from various sources, thus covering a wide percentage of the human genome. (2) We only take into account the interactions which are supported by at least two wet laboratory experiments, thus increasing the reliability of the network.

4 Conclusion

It can be concluded that the integration of several datasets underlying the MultiNet increases the prediction accuracy of the proposed method. Incorporating weights of interactions inferred from protein class, topology, co-expression similarity, sequence or homology information, pathway information might help to boost its accuracy. The proposed method is not confined to the PPI network of humans but may also be applicable to chromatin interaction networks and PPI networks of other organisms. The directed networks thus obtained can be weighted and used for community detection which in turn will help in functional module identification and function prediction, thus showing a future direction in this regard.

Acknowledgements This work has been supported by the Polish National Science Centre (2019/35/O/ST6/02484), Foundation for Polish Science co-financed by the European Union under the European Regional Development Fund (TEAM to DP). Research was funded by the Warsaw University of Technology within the Excellence Initiative: Research University (IDUB) programme. Computations were performed thanks to the Laboratory of Bioinformatics and Computational Genomics, Faculty of Mathematics and Information Science, Warsaw University of Technology using Artificial Intelligence HPC platform financed by the Polish Ministry of Science and Higher Education (decision no. 7054/IA/SP/2020 of 2020-08-28). SB was partially supported by the Department of Biotechnology (DBT), Government of India project, BT/PR16356/BID/7/596/2016.

References

1. Fields S, Song O-K (1989) A novel genetic system to detect protein–protein interactions. Nature 340:245–246
2. Stelzl U, Worm U, Lalowski M, Haenig C, Brembeck FH, Goehler H et al (2005) A human protein-protein interaction network: a resource for annotating the proteome. Cell 122:957–968
3. Gstaiger M, Aebersold R (2009) Applying mass spectrometry-based proteomics to genetics, genomics and network biology. Nat Rev Genet 10:617–627
4. Ho Y, Gruhler A, Heilbut A, Bader GD, Moore L, Adams S-L et al (2002) Systematic identification of protein complexes in *Saccharomyces cerevisiae* by mass spectrometry. Nature 415:180–183
5. Halder AK, Bandyopadhyay SS, Chatterjee P, Nasipuri M, Plewczynski D, Basu S (2020) JUPPI: a multi-level feature based method for PPI prediction and a refined strategy for performance assessment. IEEE/ACM Trans Comput Biol Bioinform. https://doi.org/10.1109/TCBB.2020.3004970
6. Halder AK, Dutta P, Kundu M, Basu S, Nasipuri M (2017) Review of computational methods for virus–host protein interaction prediction: a case study on novel Ebola–human interactions. Brief Funct Genomics 17:381–391
7. Vinayagam A, Gibson TE, Lee H-J, Yilmazel B, Roesel C, Hu Y et al (2016) Controllability analysis of the directed human protein interaction network identifies disease genes and drug targets. Proc Natl Acad Sci U S A 113:4976–4981
8. Sengupta K, Saha S, Chatterjee P, Kundu M, Nasipuri M, Basu S (2018) Ranked gene ontology based protein function prediction by analysis of protein–protein interactions. information and decision sciences. Springer Singapore, pp 419–427
9. Sengupta K, Saha S, Chatterjee P, Kundu M, Nasipuri M, Basu S (2019) Identification of essential proteins by detecting topological and functional clusters in protein interaction network of *Saccharomyces cerevisiae*. Int J Nat Comput Res 8:31–51
10. Halder AK, Denkiewicz M, Sengupta K, Basu S, Plewczynski D (2020) Aggregated network centrality shows non-random structure of genomic and proteomic networks. Methods 181–182:5–14
11. Saha S, Sengupta K, Chatterjee P, Basu S, Nasipuri M (2017) Analysis of protein targets in pathogen–host interaction in infectious diseases: a case study on Plasmodium falciparum and Homo sapiens interaction network. Brief Funct Genomics 17:441–450
12. Sharan R (2013) Toward a role model. EMBO Rep 14:948
13. Cao M, Pietras CM, Feng X, Doroschak KJ, Schaffner T, Park J et al (2014) New directions for diffusion-based network prediction of protein function: incorporating pathways with confidence. Bioinformatics 30:i219–i227
14. Kansal AR (2004) Modeling approaches to type 2 diabetes. Diabetes Technol Ther 6:39–47
15. Rao VS, Srinivasa Rao V, Srinivas K, Sujini GN, Sunand Kumar GN (2014) Protein-protein interaction detection: methods and analysis. Int J Proteomics, pp 1–12. https://doi.org/10.1155/2014/147648
16. Sadowski M, Kraft A, Szalaj P, Wlasnowolski M, Tang Z, Ruan Y et al (2019) Spatial chromatin architecture alteration by structural variations in human genomes at the population scale. Genome Biol 20:148
17. Szałaj P, Tang Z, Michalski P, Pietal MJ, Luo OJ, Sadowski M et al (2016) An integrated 3-dimensional genome modeling engine for data-driven simulation of spatial genome organization. Genome Res 26:1697–1709
18. Christopher R, Dhiman A, Fox J, Gendelman R, Haberitcher T, Kagle D et al (2004) Data-driven computer simulation of human cancer cell. Ann NY Acad Sci 1020:132–153
19. Bhalla Upinder SR, Iyengar R (1999) Emergent properties of networks of biological signaling pathways. Sci pp 381–387. https://doi.org/10.1126/science.283.5400.381
20. Silberberg Y, Kupiec M, Sharan R (2014) A method for predicting protein-protein interaction types. PLoS One. 9:e90904

21. Gitter A, Klein-Seetharaman J, Gupta A, Bar-Joseph Z (2011) Discovering pathways by orienting edges in protein interaction networks. Nucleic Acids Res 39:e22
22. Silverbush D, Sharan R (2014) Network orientation via shortest paths. Bioinformatics 30:1449–1455
23. Silverbush D, Sharan R (2019) A systematic approach to orient the human protein–protein interaction network. Nat Commun 10:1–9
24. Xenarios I, Salwínski L, Duan XJ, Higney P, Kim S-M, Eisenberg D (2002) DIP, the Database of Interacting Proteins: a research tool for studying cellular networks of protein interactions. Nucleic Acids Res 30:303–305
25. Szklarczyk D, Franceschini A, Wyder S, Forslund K, Heller D, Huerta-Cepas J et al (2014) STRING v10: protein–protein interaction networks, integrated over the tree of life. Nucleic Acids Res 43:D447–D452
26. Kerrien S, Aranda B, Breuza L, Bridge A, Broackes-Carter F, Chen C et al (2012) The IntAct molecular interaction database in 2012. Nucleic Acids Res 40:D841–D846
27. Das J, Yu H (2012) HINT: high-quality protein interactomes and their applications in understanding human disease. BMC Syst Biol 6:92
28. Chatr-Aryamontri A, Breitkreutz B-J, Oughtred R, Boucher L, Heinicke S, Chen D et al (2015) The BioGRID interaction database: 2015 update. Nucleic Acids Res 43:D470–D478
29. Hermjakob H, Montecchi-Palazzi L, Bader G, Wojcik J, Salwinski L, Ceol A et al (2004) The HUPO PSI's Molecular Interaction format—a community standard for the representation of protein interaction data. Nat Biotechnol 22:177–183
30. Luck K, Kim D-K, Lambourne L, Spirohn K, Begg BE, Bian W et al (2020) A reference map of the human binary protein interactome. Nature 580:402–408
31. Du J, Yuan Z, Ma Z, Song J, Xie X, Chen Y (2014) KEGG-PATH: Kyoto encyclopedia of genes and genomes-based pathway analysis using a path analysis model. Mol Biosyst 10:2441–2447
32. Wishart DS, Feunang YD, Guo AC, Lo EJ, Marcu A, Grant JR et al (2018) DrugBank 5.0: a major update to the DrugBank database for 2018. Nucleic Acids Res 46:D1074–D1082
33. Liu Y, Wei Q, Yu G, Gai W, Li Y, Chen X (2014) DCDB 2.0: a major update of the drug combination database. Database 2014:bau124
34. Iskar M, Campillos M, Kuhn M, Jensen LJ, van Noort V, Bork P (2010) Drug-induced regulation of target expression. PLoS Comput Biol 6.https://doi.org/10.1371/journal.pcbi.1000925
35. Igarashi T, Kaminuma T (1996). Development of a cell signalling networks database. In: Proceedings of the Pacific symposium of Biocomputing, pp 187–197
36. Samaga R, Saez-Rodriguez J, Alexopoulos LG, Sorger PK, Klamt S (2009) The logic of EGFR/ErbB signaling: theoretical properties and analysis of high-throughput data. PLoS Comput Biol 5:e1000438
37. Lachmann A, Xu H, Krishnan J, Berger SI, Mazloom AR, Ma'ayan A (2010) ChEA: transcription factor regulation inferred from integrating genome-wide ChIP-X experiments. Bioinformatics 26:2438–2444
38. Hornbeck PV, Chabra I, Kornhauser JM, Skrzypek E, Zhang B (2004) PhosphoSite: a bioinformatics resource dedicated to physiological protein phosphorylation. Proteomics 4:1551–1561
39. Du Y, Xu N, Lu M, Li T (2011) hUbiquitome: a database of experimentally verified ubiquitination cascades in humans. Database 2011:bar055
40. Erten S, Bebek G, Ewing RM, Koyutürk M (2011) DADA: degree-aware algorithms for network-based disease gene prioritization. BioData Min 4:19

An Irrigation Support System Using Regressor Assembly

Gouravmoy Banerjee, Uditendu Sarkar, and Indrajit Ghosh

Abstract In India, the agriculture sector uses about 84% of the total available water for irrigation. Most of the irrigation is surface-based, and nearly 40% of the irrigation water is wasted due to the unscientific practices adopted by the rural farmers to estimate the actual water demand of their crops. This practice leads to groundwater scarcity and high production costs of the crops. For precise irrigation, the estimation of water needed by a crop is the most significant factor. For proper estimation of irrigation water, the Food and Agriculture Organization recommends the Penman–Monteith method as the most scientific one. However, this method is very complex, involving several region-specific and climatic parameters. It is practically impossible for rural farmers to estimate the daily need for irrigation water using the Penman–Monteith method. As a practical alternative, an automated system is most desirable for rural farmers. This paper presents an irrigation support system to estimate the irrigation water required for a crop correctly. The system has been designed using an assembly of multiple regressors to get the best possible outcome. The outperformed regressor(s) is selected based on the performance evaluated using five standard metrics. This system will benefit rural farmers to estimate the exact quantity of irrigation water needed for a crop. The proposed system is an effective solution to encounter groundwater scarcity and lower the production cost of the crops.

Keywords Irrigation support system · Agriculture · Regressors · Regressor assembly

G. Banerjee · I. Ghosh (✉)
Department of Computer Science, Ananda Chandra College, Jalpaiguri, West Bengal 735101, India

U. Sarkar
National Informatics Centre, Ministry of Electronics and Information Technology, Government of India, Jalpaiguri, West Bengal 735101, India

S. Basu et al. (eds.), *Proceedings of International Conference on Frontiers in Computing and Systems*, Lecture Notes in Networks and Systems 404,
https://doi.org/10.1007/978-981-19-0105-8_3

1 Introduction

Agriculture is the biggest consumer of water in India, utilizing about 84% of the total available water [1]. About 60% of the irrigation in India is based on groundwater [2], and a huge amount of irrigation water is wasted due to imprecise irrigation [3]. The irrigation water applied by the farmers is based on a rough estimate of the actual water demand of the crop. In most cases, unscientific practices of irrigation are adopted by the rural cultivators, and excessive water is used to irrigate the crop without proper estimation of the actual water requirement of the crop. These improper practices increase the production cost of the crops and invite groundwater scarcity in the near future [4].

Estimation of the water need of a crop is region-specific and depends on several climatic factors such as temperature, relative humidity, wind speed, solar radiation, and total precipitation [5]. The other two essential parameters are reference evapotranspiration (ET_0) and the crop factor (K_c) [6]. To calculate the value of reference evapotranspiration (ET_0), the Food and Agriculture Organization (FAO) recommends the Penman–Monteith (P-M) method as the best one, but it is a very complex method involving several input parameters [3]. It is practically impossible for the rural layman farmers to correctly estimate the daily irrigation water requirement for a crop using the Penman–Monteith method. As a practical alternative, an automated system is most desirable for rural farmers.

Since 1983, several artificial intelligent systems have been suggested to solve different problems in the dynamic agriculture domain [7, 8]. These intelligent systems provide a feasible alternative to the costly and scarce expert advisory. Several fuzzy logic-based intelligent irrigation systems have been reported worldwide, which include Bahat et al. [9], Kia et al. [10], Liai Gao et al. [11], Giusti and Marsili-Libelli [12], Alfin and Sarno [13], Jaiswal and Ballal [14]. Some neuro-fuzzy hybrid techniques were suggested to design the irrigation support systems, such as Pulido-Calvo et al. [15], Mohapatra and Lenka [16], and Navaro-Hellin et al. [17]. However, all these systems designed for irrigation were either based on fuzzy logic or neuro-fuzzy hybrid systems that are dependent on expert knowledge. This expert knowledge is very scarce and hard to implement using unambiguous and consistent rule bases.

On the other hand, the different climatic and crop-related input parameters for scientific estimation of the quantity of water needed for precise irrigation are inherently numerical. Some complete and real-world datasets of these deciding parameters are also available from several benchmarked sources. Thus, as an improved alternative, regressor-based systems can be designed with better efficiency for precise irrigation. To date, only a single work has been reported by Goldstein et al. where linear, polynomial, and gradient boosted tree regression techniques were used to estimate the irrigation water [18]. However, the significant limitations of this system were that it was designed using only three regressors, and the performance was not properly evaluated.

The present work proposes an irrigation support system for the precise estimation of the irrigation water required for a crop. The system has been designed with sixteen single and four ensemble regressors following the recommendations made by the FAO. The ensemble regressors have been incorporated because, in agriculture, ensemble machine learning techniques yield better accuracy than any single machine learning technique [8]. The estimated quantity of irrigation water has been obtained as output for the same set of input parameters. The performances of all the regressors have been evaluated in terms of five standard metrics: root means square error (RMSE), mean square error (MSE), mean absolute error (MAE), the coefficient of determination R^2, and explained variance score (EVS) to choose the best one for a crop in a particular region [19, 20]. This model is designed to assist rural farmers in estimating the exact quantity of irrigation water needed by their crops. This system will be beneficial for both encountering water scarcity and lowering the production cost of the crops.

2 Materials and Methods

2.1 *Mathematical Model for Irrigation Water Estimation*

For proper estimation of the quantity of water needed for a crop, the Food and Agriculture Organization (FAO) recommends the Penman–Monteith method as the best method [5]. The Penman–Monteith method suggests the reference evapotranspiration (ET_0), which is a general estimate of the water required, measured in mm/day as presented in Eq. 1.

$$ET_0 = \frac{0.408 \times \Delta \times (R_a - G) + \gamma \times \frac{900}{T+273} \times u_2 \times (e_s - e_a)}{\Delta + \gamma \times (1 + 0.34 \times u_2)} \tag{1}$$

where R_a is the net radiation (MJ m^{-2} day^{-1}), G is heat flux density of soil (MJ m^{-2} day^{-1}), T is the mean daily temperature (°C), u_2 is the wind speed at 2 m above the ground surface (ms^{-1}), e_s is saturation vapor pressure (kPa), e_a is actual vapor pressure (kPa), Δ is the slope of the vapor pressure curve (kPa $°C^{-1}$), and γ is the psychometric constant (kPa $°C^{-1}$) [5]. All these parameters used in the P-M equation are derived from several atmospheric data such as temperature, humidity, effective rainfall, and wind speed using other 24 equations [5].

The reference evapotranspiration (ET_o) is then multiplied with the crop factor (K_c) to get the crop evapotranspiration (ET_c). The value of crop factor (K_c) is different for different crops provided by the FAO [6]. The crop evapotranspiration (ET_c) is the estimated quantity of water needed by a healthy crop with proper fertilization.

$$ET_c = ET_o \times K_c \tag{2}$$

Thus, the required irrigation water (I_w) is the difference between the crop evapotranspiration (ET_c) and the total effective precipitation (P_e) in the given area [6].

$$I_w = ET_c - P_e \tag{3}$$

2.2 Dataset

Several case studies have been conducted for different crops, out of which three are presented for radish (*Raphanus sativus*) cultivated in three farms in three distinct locations: Gairkata, Kranti, and Jalpaiguri in West Bengal, India. These three locations have been chosen for four significant reasons; most of the irrigation systems are groundwater-based, and a significant drop in groundwater level has been reported [21], a downward trend in the annual rainfall has been observed [22], river canal irrigation project achieved only 10% of its intended target [23], and the total crop water demand is expected to increase by 32% in the year 2050 as compared to 2020 [24].

The input climatic data have been collected for the past twelve years, starting from 2009 to 2020 during a cropping season of 45 days for these three specific locations. The climatic data, such as temperature, relative humidity, effective rainfall, and wind speed, were collected from NASA Langley Research Center (LaRC) POWER Project data access portal [25]. The percentage of cloud cover was obtained from the World Weather Historical database [26]. For each instance, the irrigation water demand (I_w) has been calculated using ET_c, ET_o, K_c, and P_e for radish crop adhering to FAO recommended methods. This irrigation water demand was used as the target output. The dataset contains 1620 instances that have been partitioned into two parts: 70% for training and validation and the rest 30% for testing.

2.3 Data Preprocessing

The values of the input parameters are collected in terms of different scales. Thus, normalization is required to scale down the gap between the maximum and the minimum values for each parameter. For the present system, the normalization has been done using one of the most popular method, min–max scaling [27], as given in Eq. 4.

$$NV = \left(\frac{OV - MIN}{MAX - MIN} \right) \tag{4}$$

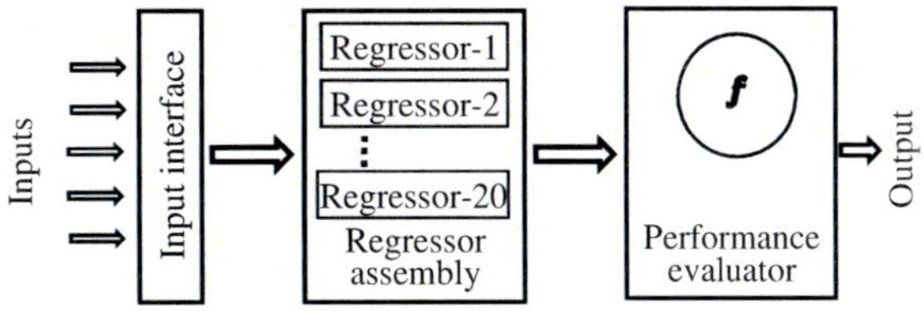

Fig. 1 Block diagram of the system

where NV is the normalized value, OV is the original value of the parameter, and MIN and MAX are the corresponding minimum and maximum values, respectively, for a particular parameter.

2.4 *System Design*

The system consists of three modules: input interface, regressor assembly, and the performance evaluator. The block diagram is presented in Fig. 1.

The input interface accepts all the input parameters. The inputs are then fed to each regressor in the regressor assembly. The regressor assembly consists of twenty regressors from ten families. The families, the regressors used, and their respective tunable hyperparameters are presented in Table 1. The hyperparameters of each of the regressors were tuned using an exhaustive grid search strategy by fivefold cross-validation.

The performance evaluation module evaluates the performances of each regressor in terms of five error metrics and gives the output of the best-performing regressor as the estimated quantity of water needed for the precise irrigation for a crop. All the system modules were coded using Python (*Ver. 3.7*) standard library functions.

2.5 *Performance Evaluation*

To measure the performances of the models, the five most popular metrics: root mean squared error (RMSE), mean squared error (MSE), mean absolute error (MAE), the coefficient of determination R^2, and explained variance score (EVS) have been used [19, 20]. The RMSE, MSE, and MAE are extensively used as popular metrics to assess the performance of various machine learning models [19]. The RMSE, MAE, and R^2 scores have some successful applications in similar types of comparative analysis [28, 29]. The experimental results obtained for the performance evaluation and the noteworthy outcomes marked in bold letters are provided in Table 2.

Table 1 Regressors with families and tunable hyperparameters

Families	Regressors	Tunable hyperparameters
Linear regressor	1. Ordinary least squares regressor (OLS)	–
	2. Ridge regressor (RID)	Regularization parameter (alpha), solver, and error tolerance
	3. Elasticnet regressor (ELS)	Coefficient selection method and error tolerance
	4. Stochastic gradient descent regressor (SGD)	Learning rate and penalty parameter
	5. Orthogonal pursuit matching regressor (OPM)	Error tolerance
	6. Passive–Aggressive regressor (PAG)	Maximum step size and error tolerance
	7. Huber regressor (HUB)	Epsilon (maximum number of outliers), regularization parameter (alpha), and tolerance
Gaussian regressor	8. Gaussian process regressor (GPR)	Kernel function
Bayesian regressor	9. Bayesian ridge regressor (BYR)	Error tolerance
Neural network regressor	10. Multilayer perceptron regressor (MLP)	Hidden layer size, training algorithm, and activation function
Support vector machine	11. Support vector regressor (SVR)	Kernel function, error tolerance, and regularization parameter
Cross decomposition regressor	12. PLS regressor (PLS)	Number of components and error tolerance
Nearest neighbor regressor	13. K-nearest neighbors regressor (KNN)	Number of neighbors
Bagging regressors	14. Decision tree-based bagging regressor (BDT)	Number of estimators
	15. Support vector-based bagging regressor (BSV)	Number of estimators
Decision tree regressor	16. Decision tree regressor (DST)	Cost of pruning and maximum tree depth
Ensemble regressors	17. Gradient boost regressor (GBR)	Number of estimators
	18. Random forest regressor (RFR)	Number of estimators
	19. AdaBoost regressor (ADA)	Number of estimators
	20. Extreme gradient boost regressor (XGB)	Number of estimators

Table 2 Comparative performances of the regressors

	Gairkata			Kranti			Jalpaiguri			Avg. Err.	Gairkata		Kranti		Jalpaiguri		Avg. Coeff.
Model	**RMSE**	**MSE**	**MAE**	**RMSE**	**MSE**	**MAE**	**RMSE**	**MSE**	**MAE**		R^2	**EVS**	R^2	**EVS**	R^2	**EVS**	
ELS	0.2788	0.0778	0.2609	0.2666	0.0711	0.2480	0.2613	0.0683	0.2431	0.1973	0.0000	0.0000	0.0000	0.0000	0.0000	0.0000	0.0000
SGD	0.0904	0.0082	0.0698	0.0857	0.0073	0.0663	0.0841	0.0071	0.0651	0.0538	0.8949	0.8949	0.8967	0.8967	0.8964	0.8964	0.8960
PAG	0.0786	0.0062	0.0666	0.0745	0.0056	0.0643	0.0756	0.0057	0.0650	0.0491	0.9205	0.9206	0.9219	0.9229	0.9162	0.9223	0.9207
RIG	0.0783	0.0061	0.0659	0.0737	0.0054	0.0619	0.0724	0.0052	0.0608	0.0477	0.9211	0.9211	0.9237	0.9237	0.9233	0.9233	0.9227
HUB	0.0781	0.0061	0.0659	0.0734	0.0054	0.0615	0.0721	0.0052	0.0605	0.0476	0.9216	0.9216	0.9243	0.9243	0.9238	0.9239	0.9232
PLS	0.0780	0.0061	0.0665	0.0732	0.0054	0.0620	0.0719	0.0052	0.0610	0.0477	0.9218	0.9218	0.9246	0.9246	0.9242	0.9242	0.9235
BYR	0.0780	0.0061	0.0663	0.0732	0.0054	0.0619	0.0720	0.0052	0.0609	0.0477	0.9218	0.9218	0.9245	0.9245	0.9241	0.9241	0.9235
GPR	0.0779	0.0061	0.0664	0.0732	0.0054	0.0620	0.0719	0.0052	0.0610	0.0477	0.9219	0.9219	0.9246	0.9246	0.9242	0.9242	0.9236
OLS	0.0779	0.0061	0.0664	0.0732	0.0054	0.0620	0.0719	0.0052	0.0610	0.0477	0.9219	0.9219	0.9246	0.9246	0.9242	0.9242	0.9236
OMP	0.0779	0.0061	0.0664	0.0732	0.0054	0.0620	0.0719	0.0052	0.0610	0.0477	0.9219	0.9219	0.9246	0.9246	0.9242	0.9242	0.9236
SVR	0.0713	0.0051	0.0630	0.0707	0.0050	0.0639	0.0707	0.0050	0.0642	0.0465	0.9345	0.9358	0.9296	0.9314	0.9267	0.9279	0.9310
BSV	0.0617	0.0038	0.0522	0.0620	0.0039	0.0533	0.0620	0.0039	0.0534	0.0396	0.9511	0.9535	0.9459	0.9498	0.9437	0.9485	0.9487
KNN	0.0408	0.0017	0.0237	0.0270	0.0007	0.0143	0.0382	0.0015	0.0224	0.0189	0.9785	0.9786	0.9898	0.9898	0.9786	0.9786	0.9823
ADA	0.0327	0.0011	0.0277	0.0307	0.0009	0.0257	0.0292	0.0009	0.0247	0.0193	0.9863	0.9882	0.9867	0.9883	0.9875	0.9886	0.9876
MLP	0.0108	0.0001	0.0074	0.0171	0.0003	0.0104	0.0125	0.0002	0.0082	0.0074	0.9985	0.9985	0.9959	0.9959	0.9977	0.9977	0.9974
BDT	0.0105	0.0001	0.0069	0.0089	0.0001	0.0060	0.0093	0.0001	0.0061	0.0053	0.9986	0.9986	0.9989	0.9989	0.9987	0.9987	0.9987
DST	0.0102	0.0001	0.0060	0.0092	0.0001	0.0057	0.0088	0.0001	0.0054	0.0051	0.9987	0.9987	0.9988	0.9988	0.9989	0.9989	0.9988
RAF	0.0099	0.0001	0.0064	0.0097	0.0001	0.0063	0.0091	0.0001	0.0061	0.0053	0.9987	0.9987	0.9987	0.9987	0.9988	0.9988	0.9987
GBR	0.0054	0.0000	0.0041	0.0053	0.0000	0.0040	0.0051	0.0000	0.0039	0.0031	0.9996	0.9996	0.9996	0.9996	0.9996	0.9996	0.9996
XGB	**0.0010**	**0.0000**	**0.0007**	**0.0010**	**0.0000**	**0.0007**	**0.0010**	**0.0000**	**0.0007**	**0.0006**	**0.9999**	**0.9999**	**0.9999**	**0.9999**	**0.9999**	**0.9999**	**0.9999**

2.6 Pseudocode

Step 1. **Begin**
Step 2. Read the climatic & crop parameters as supplied by the user
Step 3. Calculate P_e, K_c, ET_0 ET_c and I_w
Step 4. Divide the complete dataset into training and testing sets in 70:30 ratio
Step 5. **For** *each regressor* R_i *in the regressor assembly* **repeat** *steps 6–9:*
Step 6. **Fit** *the regressor* R_i *using the training dataset*
Step 7. Find the optimum value for the hyperparameters using grid search
Step 8. **Test** R_i *using the testing dataset*
Step 9. Calculate the values of the error metrics and similarity coefficients
Step 10. Let R_s *be the regressor with the smallest (highest) error (coefficient) value*
Step 11. Display the value of the required irrigation water I_w *as estimated by the best regressor* R_s *Step 12.* **End**

3 Results and Discussion

The results of the performance evaluation for twenty regressors applied in the three case studies are presented in Table 2. The performance of each regressor was measured in terms of five metrics: RMSE, MSE, MAE, R^2, and EVS. Out of these five metrics, RMSE, MSE, and MAE are error-based and the rest two, R^2 and EVS, are based on coefficients. For better understanding, the average values of error metrics and the coefficient-based metrics are presented in columns 11 and 18, respectively. It is evident from columns 11 and 18 that both the error-based and the coefficient-based metrics lead to the same observation.

It is revealed that for extreme gradient boost regressor (XGB), the average value of the error-based metrics is the lowest, and that of the coefficient-based metrics is the highest for each of these three case studies. Thus, the XGB outperformed all the others. The performance of the gradient boost regressor (GBR) is the second-best. The Elasticnet regressor (ELS) is the worst-performing one and therefore, not applicable for such problems. Average performance is observed for other regressors. It reveals that all the ensemble regressors performed better than the single regressors except the decision tree (DST). The family of linear regression models exhibited poor performance for the current problem.

4 Conclusion

The proposed system is a precise irrigation support system. The proposed system estimates the optimum quantity of irrigation water needed for a crop depending on the region-specific climatic parameters. Instead of a single regressor, an assembly of regressors has been used. After evaluating the performance of each regressor, the outcome of the best-performing one is accepted. The best recommendation is ensured by selecting the best regressor in terms of the performance evaluated using five well-established metrics.

The Food and Agriculture Organization (FAO) recommends the P-M method as the best protocol for the precise estimation of irrigation water required for a crop. The process uses region-specific climatic parameters that are practically impossible to be handled by layman farmers. This proposed system gives the farmers the opportunity to estimate the actual requirement of irrigation water for their crops. This will lower the production cost of the crop and minimize the misuse of groundwater resources.

The model proposed in this paper is a prototype developed for satisfying the irrigation requirements of radish. However, the proposed model may be modified in the future for irrigating any other crops as per user requirements. Furthermore, the authors are currently working on developing an automated irrigation system using IoT and artificial intelligence.

References

1. Dhawan V (2017) Water and agriculture in India. Background paper for the South Asia expert panel during the Global Forum for Food and Agriculture (GFFA), Germany
2. Gandhi VP, Bhamoriya V (2011) Groundwater irrigation in India: growth, challenges, and risks, Indian infrastructure report water: policy and performance for sustainable development. Oxford University Press, New Delhi, pp 90–117
3. Brouwer C, Prins K, Heibloem M (1989) Irrigation water management, training manual No. 4. Food and Agriculture Organisation (FAO), Rome
4. Gupta SK, Deshpande RD (2004) Water for India in 2050: first-order assessment of available options. Curr Sci 86(9):1216–1224
5. Allen RG, Pereira LS, Raes D, Smith M (1998) Crop evapotranspiration—guidelines for computing crop water requirements—FAO Irrigation and drainage paper 56. Food and Agriculture Organisation (FAO), Rome
6. Brouwer C, Heibloem M (1986) Irrigation water management, training manual No. 3. Food and Agriculture Organisation (FAO), Rome
7. Baker DN, Lambert JR, McKinion, JM (1983) GOSSYM: a simulator of cotton crop growth and yield. Agricultural Experiment Station, South Carolina, Technical bulletin, USA
8. Banerjee G, Sarkar U, Das S, Ghosh I (2018) Artificial intelligence in agriculture: a literature survey. Int J Sci Res Comput Sci Appl Manag Stud 7(3):1–6
9. Bahat M, Inbar G, Yaniv O, Schneider M (2000) A fuzzy irrigation controller system. Eng Appl Artif Intell 13(2):137–145
10. Kia PJ, Far AT, Omid M, Alimardani R, Naderloo L (2009) Intelligent control based fuzzy logic for automation of greenhouse irrigation system and evaluation in relation to conventional systems. World Appl Sci J 6(1):16–23

11. Gao L, Zhang M, Chen G (2013) An intelligent irrigation system based on wireless sensor network and fuzzy control. J Networks 8(5):1080–1087
12. Giusti E, Marsili-Libelli S (2015) A Fuzzy Decision support system for irrigation and water conservation in agriculture. Environ Model Softw 63:73–86
13. Alfin AA, Sarno R (2017) Soil irrigation fuzzy estimation approach based on decision making in sugarcane industry. In: International conference on information & communication technology and system (ICTS-2017). Surabaya, Indonesia, pp 137–142
14. Jaiswal S, Ballal MS (2020) Fuzzy inference based irrigation controller for agricultural demand side management. Comput Electron Agric 175:105537
15. Pulido-Calvo I, Gutierrez-Estrada JC (2009) Improved irrigation water demand forecasting using a soft-computing hybrid model. Biosys Eng 102(2):202–218
16. Mohapatra AG, Lenka SK (2015) Neural network pattern classification and weather dependent fuzzy logic model for irrigation control in WSN based precision agriculture. In International conference on information security & privacy (ICISP 2015). Nagpur, India, pp 499–506
17. Navarro-Hellin H, Martinez-del-Rincon J, Domingo-Miguel R, Soto-Valles F, Torres-Sanchez R (2016) A decision support system for managing irrigation in agriculture. Comput Electron Agric 124:121–131
18. Goldstein A, Fink L, Meitin A, Bohadana S, Lutenberg O, Ravid G (2018) Applying machine learning on sensor data for irrigation recommendations: revealing the agronomist's tacit knowledge. Precision Agric 19(3):421–444
19. Botchkarev A (2019) A new typology design of performance metrics to measure errors in machine learning regression algorithms. Interdiscip J Inf Knowl Manag 14:45–79
20. Keprate A, Chandima Ratnayake RM (2020) Artificial intelligence based approach for predicting fatigue strength using composition and process parameters. In: 39th international conference on ocean, Offshore and Arctic Engineering (OMAE-2020), American Society of Mechanical Engineers, Virtual Conference, pp 1–10
21. Ground Water Year Book of West Bengal and Andaman & Nicobar Islands (2016–2017). Technical Report, Series D. Ministry of Water Resources. Government of India, Kolkata, India (2017)
22. Das L, Prasad H, Meher JK (2018) 20th century district-level spatio-temporal annual rainfall changes over West Bengal. J Clim Change 4(2):31–39
23. Mukherjee B, Saha DU (2016) Teesta Barrage Project–a brief review of unattained goals and associated changes. Int J Sci Res 5(5):2027–2032
24. Banerjee S, Biswas B (2020) assessing climate change impact on future reference evapotranspiration pattern of West Bengal, India. Agric Sci 11(9):793–802
25. POWER data access viewer. NASA Langley Research Center (LaRC). https://power.larc.nasa.gov/data-access-viewer/. Accessed 3 Apr 2021
26. Historical Forecast Weather (HFW). Zoomash Ltd. https://www.worldweatheronline.com/hwd/hfw.aspx. Accessed 2 Apr 2021
27. Kotsiantis SB, Kanellopoulos D, Pintelas PE (2006) Data preprocessing for supervised leaning. Int J Comput Sci 1(2):111–117
28. Wu T, Zhang W, Jiao X, Guo W, Hamoud YA (2020) Comparison of five boosting-based models for estimating daily reference evapotranspiration with limited meteorological variables. Plos One 15(6):e0235324
29. Fan J, Yue W, Wu L, Zhang F, Cai H, Wang X, Lu X, Xiang Y (2018) Evaluation of SVM, ELM and four tree-based ensemble models for predicting daily reference evapotranspiration using limited meteorological data in different climates of China. Agric For Meteorol 263:225–241

Does Random Hopping Ensure More Pandal Visits Than Planned Pandal Hopping?

Debojyoti Pal, Anwesh Kabiraj, Ritajit Majumdar, Kingshuk Chatterjee, and Debayan Ganguly

Abstract India is a country with rich cultural diversities where festivals keep occurring. In this paper, we introduce the pandal hopping problem where travellers intend to acquire maximum rewards, which may be considered as the satisfaction attained by visiting a pandal. Visiting one pandal from another incurs some cost, but reward is awarded only when a new pandal is visited. The goal is to acquire maximum reward within the associated cost. In this paper, we compare different variations of random walk algorithms such as self-avoiding, exponential, log-normal, quadratic distribution and greedy algorithm. We finally establish that a self-avoiding random walk is best suited to solve such problems.

Keywords Random walk · Greedy algorithm · Node traversal

1 Introduction

In this paper, we introduce the pandal hopping problem. Durga puja is one of the most widely celebrated religious and cultural festival in India, particularly in Kolkata. Several pandals with the deity of Goddess Durga are created throughout the city. People of this city, as well as those from abroad, gather here to visit these pandals. In general, there are some pandals widely recognized for their decor and brilliance. Most people want to visit these. However, many unknown and small pandals often show splendid architecture and decor. Every person wants to visit good pandals within the time limit they have, and the pandals well known for their decor are often far apart. Therefore, the police, and other clubs, design pandal maps of the city. In this paper, we study how a person should visit pandals so that they can visit many good

D. Pal · A. Kabiraj · D. Ganguly (✉)
Government College of Engineering and Leather Technology, Kolkata, India
e-mail: debayan3737@gmail.com

R. Majumdar
Advanced Computing & Microelectronics Unit, Indian Statistical Institute, Kolkata, India

K. Chatterjee
Government College of Engineering and Ceramic Technology, Kolkata, India

S. Basu et al. (eds.), *Proceedings of International Conference on Frontiers in Computing and Systems*, Lecture Notes in Networks and Systems 404,
https://doi.org/10.1007/978-981-19-0105-8_4

pandals within their specified cost limit. We show that instead of rigorous planning, randomly visiting pandals can provide a better reward (i.e. satisfaction).

We model this problem as follows: Let $G = (V, E)$ be a complete undirected graph, which is the map of the city, where $|V| = N$ is the number of pandals. With each vertex, we define a reward function $R : V \to \mathbb{R}$. Similarly associated with each edge is a cost function $C : E \to \mathbb{R}$ which mimics the time (or the actual cost) required to visit some pandal B from another pandal A via edge (A, B). A person starts from a random location and has a budget of T. The objective is to maximize the reward for that person within this budget. In other words

$$
\begin{aligned}
\max \quad & \sum_i r_i \cdot v_i \\
\text{subto} \quad & \sum_i 1 \le N \\
& \sum_j c_j \le T \\
& r_i \in \mathbb{R}, v_i \in \{0, 1\}
\end{aligned}
$$

The problem, therefore, is that a person can visit the same pandal multiple times, each time the cost of the edge gets deducted from its budget (hence $\sum_j 1$ is not upper bounded by N). However, for each pandal, the reward will be awarded only once (as enforced by $v_i \in \{0, 1\}$).

In this paper, we study various random walk algorithms and greedy approach. We have applied random walks with different variations such as (i) completely random and (ii) self-avoiding walk (will not return to its last position), exponential distribution (edges leading to higher rewarding nodes are assigned exponentially high probability). We find that the self-avoiding random walk proves to be the most efficient in collecting higher reward than the other models in the same time duration.

The rest of the paper is organized as follows: In Sect. 2, we describe the random walk and used distributions. Sections 3 and 4, respectively, describe the methodology and observations. We conclude in Sect. 5

2 One-Dimensional Random Walk

The one-dimensional random walk is basically a random walk on a graph [1], where at each step, the traveller jumps from one pandal to another neighbouring pandal. The probabilities of jumping to one of its immediate neighbours are the same, i.e. $\frac{1}{n}$ [2], where n is the number of neighbours. To define this walk formally, we take independent random variables $z_1, z_2, \ldots$ where z varies from 1 to n. We set $S_n = \sum z_j$. The series $\{S_n\}$ is called the simple random walk [1] on $\mathbb{Z}$. Now $E(S_n) = (\sum r_{ij}) \times (k/n)$ where r_{ij} is the reward for travelling from node i to node j, and k is the number of iterations of random walk used upon the complete graph. For random walk, a random walk can visit n vertices in O(nlogn) steps [3].

3 Random Walk and Used Distributions

We have used four different types of distribution to test the efficiency of each one. We begin with variations of random walk like self-avoiding, exponential, log-normal, and then we end up with U-quadratic distribution which is not a completely random walk [4]. We also show the performance of a greedy algorithm, in which the walker always moves to that neighbour which has the maximum reward.

3.1 Log-Normal Distribution

A log-normal distribution is a distribution that becomes a normal distribution if one converts the values of the variable to the natural logarithms. This is often used in the stock market model where stock prices are taken to be following log-normal random walk distribution. It introduces more variation thereby increasing chances of more reward collection [5].

3.2 Exponential Distribution

This is an almost greedy algorithm where the node with highest reward is given exponentially higher probability of getting selected.

3.3 Quadratic Distribution

The U-quadratic distribution is a continuous probability distribution defined by a unique convex quadratic function with lower limit a and upper limit b. This distribution is a useful model for symmetric bi-modal processes [4].

3.4 Greedy Algorithm

The traveller searches for the node with the highest reward and directly travels to that particular nodes. This can visit all the n vertices in descending order of reward in $O(n^2 \log n)$ steps [6].

4 Methodology

4.1 Data

For simulation purpose, we have considered a complete graph with 50 nodes. The cost for travelling from one node u to another node v $1 \leq C(u, v) \leq 9$. The reward associated with each vertex v is $40 \leq R(v) \leq 80$. The underlying graph does not contain self-loop. An example adjacency matrix of 5×5 complete graph is shown below:

$$\begin{pmatrix} 0 & 1 & 1 & 1 & 1 \\ 1 & 0 & 1 & 1 & 1 \\ 1 & 1 & 0 & 1 & 1 \\ 1 & 1 & 1 & 0 & 1 \\ 1 & 1 & 1 & 1 & 0 \end{pmatrix}$$

4.2 Procedure

In Algorithm 1, we show the procedure of random walk used in this paper. In Figs. 2, 3 and 4, we show the average-, best- and worst-case scenarios for hundred people using Algorithm 1 (Fig. 1).

Algorithm 1 Proposed Pandal Hopping Algorithm

Input: Graph(G), Distribution(D), Budget(B)
Output: Collected Reward (r)

$u \leftarrow$ randomly selected start node.
$r \leftarrow 0$.
curr $\leftarrow u$.
while $\exists V \in$ neighbour(curr) | C(curr,V) $\leq$ B **do**
  Select $V \in$ neighbour(curr) based on distribution D such that C(curr,V) $\leq$ B.
  if V not previously visited **then**
    $r := r + R(V)$.
  end if
  $B := B-$ C(curr,V).
  $curr := V$.
end while

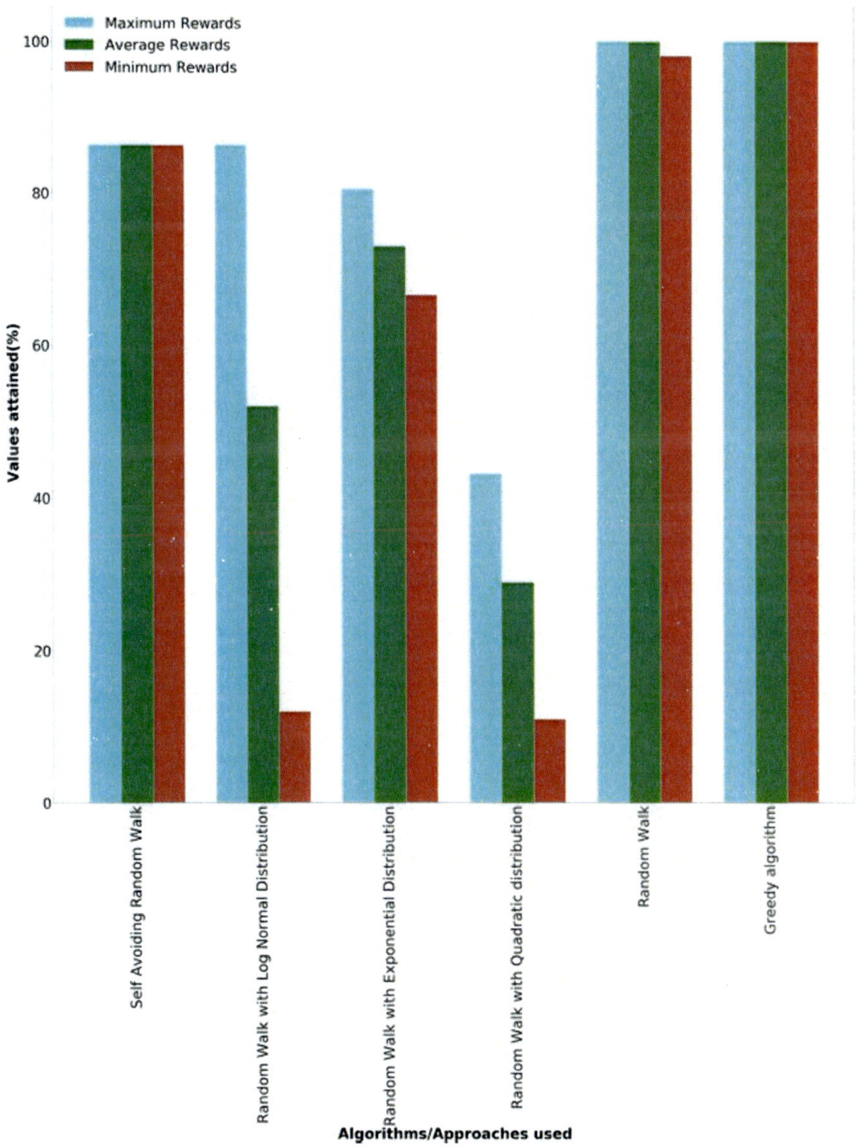

Fig. 1 Bar plot showing how much per cent of total reward is collected by different random walks

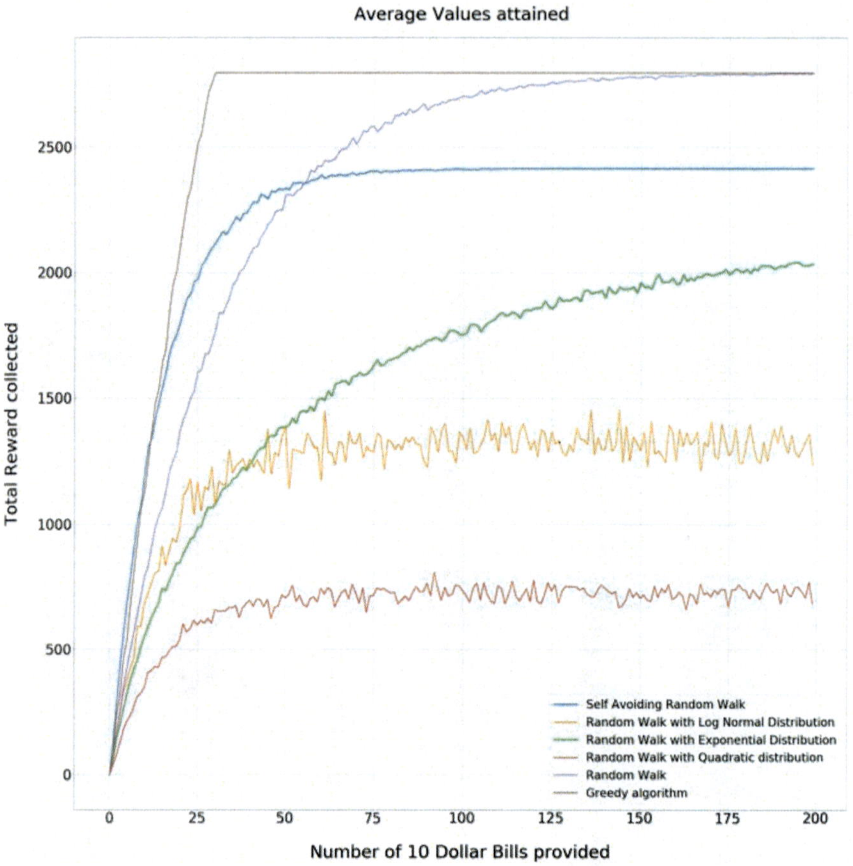

Fig. 2 Line graph showing average amount (for 100 people on a 50 node complete graph) of rewards collected by different random walks corresponding to the money provided

5 Observations

5.1 Average Values Attained During Random Walk

In Fig. 2, we observe that the normal random walk achieves the highest reward of 2759 at around 1750 dollars in an average random walk. The greedy algorithm rises above self-avoiding random walk at around 17% of the cost, signifying that random walk, on average, obtains the same reward but often for a higher price. Initially, the self-avoiding random walk has a steeper slope than the other random walks, but it starts to saturate after incurring 34% cost. We can infer that the self-avoiding random walk is an excellent alternative to the greedy algorithm in the initial stages, i.e. when incurring less than 7% of money but eventually dies out in the later stages compared

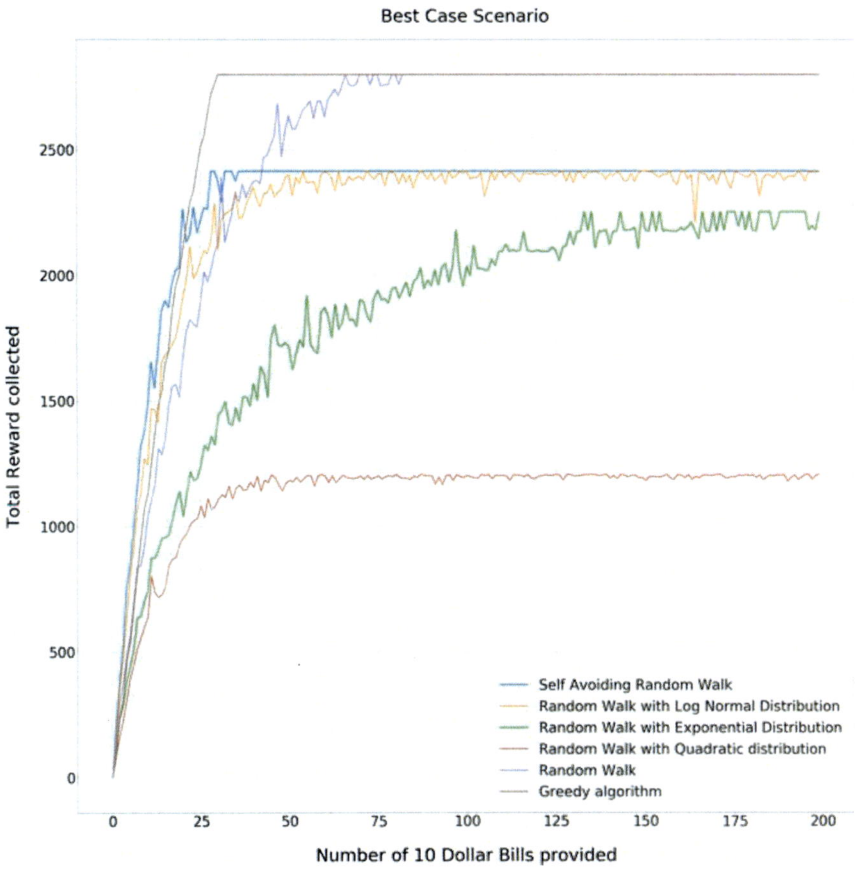

Fig. 3 Line graph showing maximum amount (for 100 people on a 50 node complete graph) of rewards collected by different random walks corresponding to the money provided

to the average rewards collected. The random walk on a quadratic distribution being the least efficient in terms of reward collection followed by the random walk on a log-normal distribution.

5.2 *Best-Case Scenario of Random Walk*

In Fig. 3, we can see that among various walks, a regular random walk achieves the highest reward among all others. The self-avoiding random walk shows the steepest rise before saturating in the 350 dollar region. The log-normal distribution shows a very relative performance to self-avoiding. The random walk with an exponential distribution has a greedy approach which shows a gradual rise through the entire graph however, falling behind the random walk performed with log-normal distribution. The quadratic distribution shows the worst result among all the variations. The random walk and the greedy algorithm take top place, followed by the self-avoiding random walk.

5.3 Worst-Case Scenario of Random Walk

From Fig. 4, it is observed that the self-avoiding random walk fares well in the initial phases with the greedy algorithm. However, it falls apart after 7% of cost is incurred. The random walk under-performs in the initial stage, but it shows an overwhelming result in the latter phases, especially after 35% cost is incurred. We can categorize

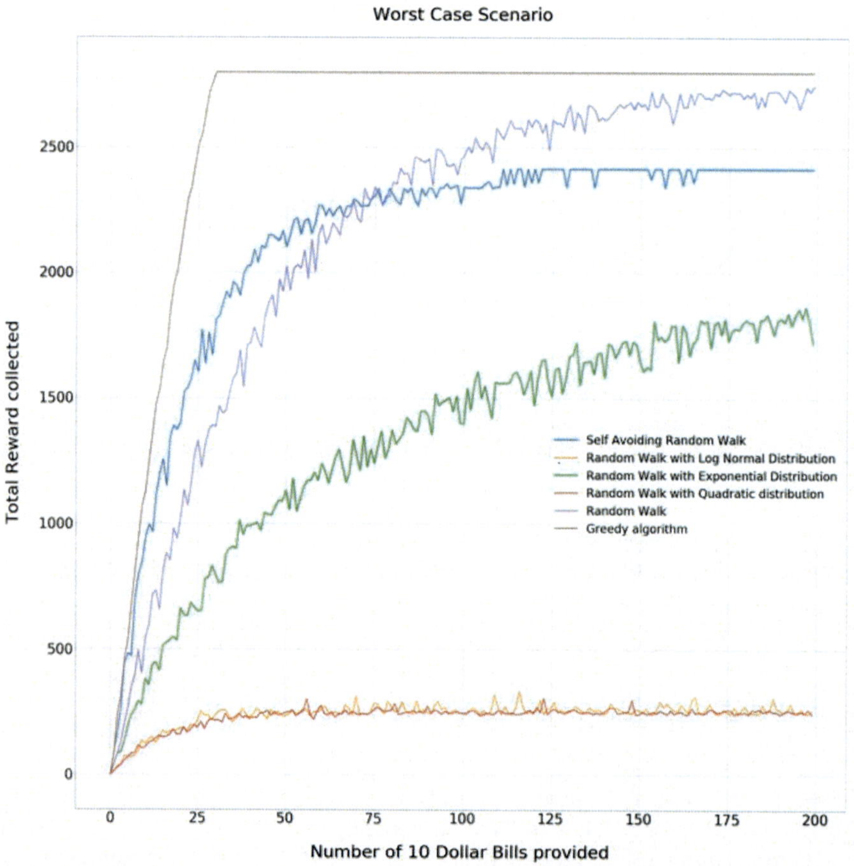

Fig. 4 Line graph showing minimum amount (for 100 people on a 50 node complete graph) of rewards collected by different random walks corresponding to the money provided

the greedy algorithm, the random walk and the self-avoiding random walk as the best version in the worst-case scenario. The worst being the random walk performed on exponential and quadratic distributions.

5.4 *Change in Rewards Collected with the Increase in Size of Graph*

The size of the complete graph is varied with the number of nodes ranging from 20 to 100. The maximum reward, collected by self-avoiding random walk, from each graph is plotted as shown in Fig. 5. We expect that the self-avoiding random walk will

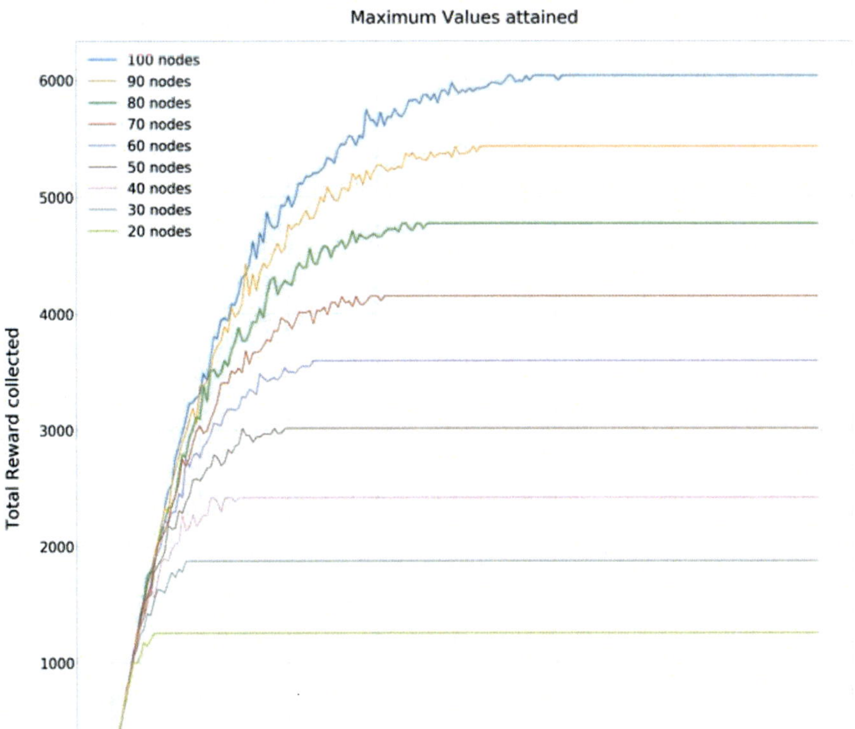

Fig. 5 Line plot showing how saturation point for self-avoiding random walk shifts right with the increase in number of nodes

be able to collect more reward with increasing number of nodes. And, in accordance to that, it is noticed that the saturation point for self-avoiding walk shift more towards right with an increase in the number of nodes.

6 Conclusion

Thus, we can conclude the following from our observations

1. Random walk and self-avoiding random walk show a very steep curve in the reward collecting graph, almost competing with the greedy algorithm in the initial phase. However, they fall behind after roughly 17% cost is incurred. This signifies that these two options are more suitable solutions since it involves less time complexity.

2. We also notice that the saturation point shifts more towards the right on increasing the number of nodes. This signifies that the more nodes we deal with random walk will compete with the traditional greedy algorithm for a longer time, i.e. it will collect reward very near to greedy algorithm. On the other hand, the complexity of greedy algorithm increases much more than that of a random walk with increase in nodes. 2. We also notice that the saturation point shifts more towards the right on increasing the number of nodes. This signifies that the more nodes we deal with random walk will compete with the traditional greedy algorithm for a longer time, i.e. it will collect reward very near to greedy algorithm. On the other hand, the complexity of greedy algorithm increases much more than that of a random walk with increase in nodes.

References

1. Lovász L et al (1993) Random walks on graphs: a survey. Comb Paul Erdos Eighty 2(1):1–46
2. Ross SM (2014) Introduction to probability models. Academic Press
3. Liu W, Lü L (2010) Link prediction based on local random walk. Europhys Lett (EPL) 89(5):58007
4. Imhof J-P (1961) Computing the distribution of quadratic forms in normal variables. Biometrika 48(3/4):419–426
5. Heyde CC (1963) On a property of the lognormal distribution. J Roy Stat Soc Seri B (Methodol) 25(2):392–393
6. Heidari M, Asadpour M, Faili H (2015) SMG: fast scalable greedy algorithm for influence maximization in social networks. Phys A Stat Mech Its Appl 420:124–133

Consensus-Based Identification and Comparative Analysis of Structural Variants and Their Influence on 3D Genome Structure Using Long- and Short-Read Sequencing Technologies in Polish Families

Mateusz Chiliński, Sachin Gadakh, Kaustav Sengupta, Karolina Jodkowska, Natalia Zawrotna, Jan Gawor, Michal Pietal, and Dariusz Plewczynski

Abstract Structural variants (SVs) such as deletions, duplications, insertions, or inversions are alterations in the human genome that may be linked to the development of human diseases. A wide range of technologies are currently available to detect and analyze SVs, but the restrictions of each of the methods are resulting in lower

M. Chiliński · D. Plewczynski (✉)
Laboratory of Bioinformatics and Computational Genomics, Faculty of Mathematics and Information Science, Warsaw University of Technology, Koszykowa 75, 00-662 Warsaw, Poland
e-mail: dariuszplewczynski@cent.uw.edu.pl

M. Chiliński
e-mail: m.chilinski@cent.uw.edu.pl

M. Chiliński · S. Gadakh · K. Sengupta · K. Jodkowska · N. Zawrotna · D. Plewczynski
Laboratory of Functional and Structural Genomics, Centre of New Technologies, University of Warsaw, Banacha 2c, 02-097 Warsaw, Poland
e-mail: s.gadak@cent.uw.edu.pl

K. Sengupta
e-mail: k.sengupta@cent.uw.edu.pl

K. Jodkowska
e-mail: k.jodkowska@cent.uw.edu.pl

K. Jodkowska
Centre for Advanced Materials and Technologies (CEZAMAT), Warsaw University of Technology, Poleczki 19, 02-822 Warsaw, Poland

J. Gawor
DNA Sequencing and Oligonucleotide Synthesis Laboratory, Institute of Biochemistry and Biophysics, Polish Academy of Sciences, Pawinskiego 5a, 02-106 Warsaw, Poland
e-mail: gaworj@wp.pl

M. Pietal
Faculty of Computer and Electrical Engineering, Rzeszow University of Technology, Powstańców Warszawy 12, 35-959 Rzeszów, Poland
e-mail: m.pietal@prz.edu.pl

S. Basu et al. (eds.), *Proceedings of International Conference on Frontiers in Computing and Systems*, Lecture Notes in Networks and Systems 404,
https://doi.org/10.1007/978-981-19-0105-8_5

total accuracy. Over the past few years, the need to develop a reliable computational pipeline has arisen to merge and compare the SVs from various tools to get accurate SVs for downstream analysis. In this study, we performed a detailed analysis of long-read sequencing of the human genome and compared it with short-read sequencing using Illumina technology in terms of the distribution of structural variants (SVs). The SVs were identified in two families with three members each (mother, father, son) using fifteen independent SV callers. Then we utilized ConsensuSV algorithm to merge the results of these SVs callers to identify the reliable list of SVs for each member of the families. Furthermore, we studied the influence of SVs on chromatin interaction-based paired-end tags (PETs). Finally, while we compared the length and number-wise distribution between long-read-based and short-read-based SVs and their respective mapping on PETs. We conclude that SVs detected by our algorithm over sequencing data using ONT are superior compared to Illumina across all SV sizes and lengths, as well as the number of mapped to PETs.

Keywords Structural variants (SVs) · Paired-end tags (PETs) · Long-read · Short-read · Sequencing technologies · Oxford nanopore technology (ONT) · Genomics

1 Introduction

The order of bases (A, T, G, C) of DNA contains the information of human development and its inheritance. Therefore, the requirement to detect such sequences is crucial for biological research. The structure of DNA was discovered by Watson and Crick in 1953 [1]; however, the first sequencing technologies were developed nearly 25 years later, and the ability to understand the genetic code is the key to understanding terrestrial life and possible causes as well as treatments for many genetic diseases [2]. Sanger [3] and Maxam [4] were instrumental to the development of first sequencing technologies. Afterward, the new sequencing technologies were termed as next-generation sequencing (NGS), currently called as short-read sequencing. However, this technology had its drawbacks due to being unable to solve repetitive areas of the complex genomes. Those problems led to the creation of new sequencing technology—called long-read sequencing. That technology was created with the ability of producing reads with length of up to several kilobases, currently even exceeding a mega base, solving genome assembly problems and repetitive regions of complex genomes. Oxford nanopore technology (ONT) [5] being the forefront in long-read sequencing has demonstrated its utility in many studies. That technology is also suitable for high-quality de novo assembly and detection of structural variation for human-sized genomes [6]. The study of the variation of the DNA in the human population was introduced with the HapMap Project [7]. The human genome is diverse due to variation caused by sequence and structural changes impacting the regulatory process [8]. The 1000 Genome Project reports that a classic human genome varies at 4 to 5 million sites compared to reference genome [9]. Most of those variants

Table 1 Sequencing data available for the families used in the study

Family/technology	Illumina	Oxford nanopore technologies
Polish family I	X	
Polish family II	X	X

are single nucleotide polymorphisms (SNPs), short indels (insertion/deletions up to size of 50 base pairs) and structural variants (SVs—large rearrangements of DNA region spanning beyond 50 to kilobase pair). Therefore, studying the implications of SNPs, indels, and SVs is important to understand its impact on structural and functional genomics. The goal of our study is not only to show the superiority of long-read sequencing in terms of inferring large size structural variants compared to short-read one, but also to study how these larger rearrangements affect the organization of the human genome, via mapping them to anchors of chromatin loops identified by ChIA-PET experiments (Chromatin Interaction Analysis by Paired-End Tag Sequencing) [10].

2 Materials and Methods

2.1 *Overview of Samples*

We used the sequencing data created in our wet lab for 2 polish families as study cases. Each of these families includes biological parents and a male child. The details of the families are shown in Table 1.

2.2 *Structural Variation Calling*

For the short-reads data, we used twelve different golden standard SV callers, including BreakDancer [11], BreakSeq2 [12], CNVnator [13], delly [14], lumpy [15], GenomeSTRiP [16], Manta [17], Pindel [18], SVelter [19], tardis [20], novoBreak [21], and wham [22]. For the long-read analysis, we used three state-of-the-art independent SV callers: Sniffles [23], CuteSV [24], and SVIM [25]. The structural variants identified were then merged using ConsensusSV algorithm.

Table 2 Summary of the samples and families used in the study as well as the numbers of SVs detected in each of the samples

Family	Polish family I		Polish family II			
sequencing technology	Illumina		Oxford nanopore technology		Illumina	
Sample name	Coverage	Number of SVs	Coverage	Number of SVs	Coverage	Number of SVs
Father	7.44 X	1618	13.52 X	6564	29.03 X	3609
Mother	5.34 X	1316	11.55 X	5638	28.53 X	3580
Son (healthy)	3 X	792	19.16 X	9747	29.33 X	3603

2.3 *Mapping to 3D Genomics*

Structural variants were mapped to ChIA-PET interactions of the GM12878 cell line (reads mapped using hg38) from 4D Nucleome project, created by Yijun Ruan's laboratory [26].

3 Results and Discussion

3.1 *Families and Phenotypes*

In our study, we have discovered the numbers corresponding to the structural variants present in the two analyzed families. Details discovered in each individual are summarized in Table 2. The overall design of the study is shown in Fig. 1.

3.2 *Consensus SVs and Their Comparison Between Short-Read and Long-Read Sequencing Technologies*

Using sequencing data shown in Table 1 and following the procedure presented in Fig. 1, we have created a set of SVs unique to a sample that is larger than 50 bp. For each individual, a consensus set of SVs was created with ConsensuSV tool. However, as the sequencing was done using two different technologies (ONT and Illumina), therefore we compared the obtained sets and their distribution of size and length. We have performed two-sample Kolmogorov–Smirnov test to check if the distributions of Illumina and ONT, SVs can be considered as taken from the same contiguous distribution (H0). We found that there is no evidence to reject H0, as p-value from the test was 0.83 for the father and mother samples and 0.81 for sample of the child. The obtained results show that across different SVs sizes and quantities, despite low

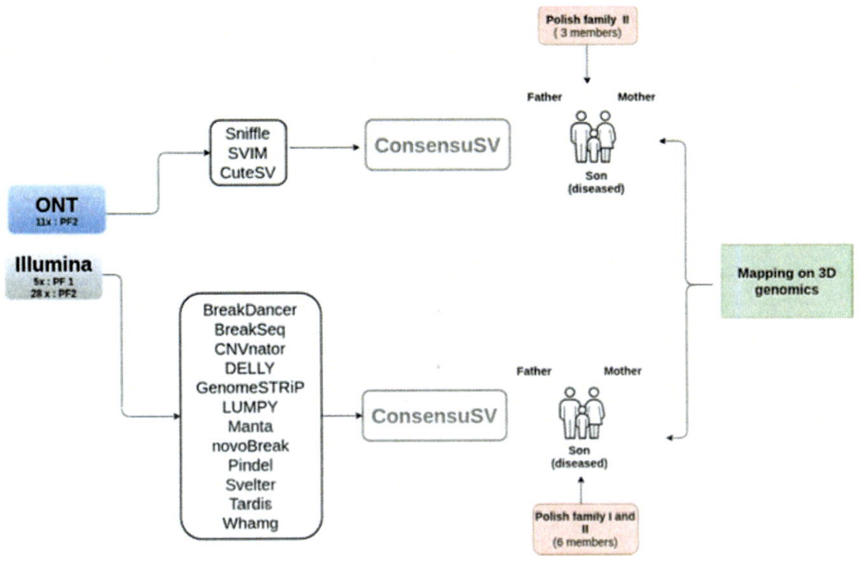

Fig. 1 Overall design of the study

coverage, long-read-based technology detects more SVs than the short-read Illumina (Fig. 2).

3.3 Mapping to 3D Genomics

We obtained the interactions from the ChIA-PET experiments conducted on the GM12878 cell line [26]. We merged data from all available replicates. Next, the structural variants were mapped to the anchors using each sample of the second analyzed family (sequenced using both technologies). The anchors are unique DNA sequences that are specific to certain protein binding sites, in our case it is CTCF, and pair of anchors form the paired-end tags (PET). Thus even an anchor getting affected by SV ultimately affects the PET. We found a significant number of anchors overlapping with structural variants in both Illumina and ONT cases.

Figure 3 presents an example of an anchor affected by SV detected by both sequencing technologies. The high-quality variants are essential for the further investigation of the 3D genomics.

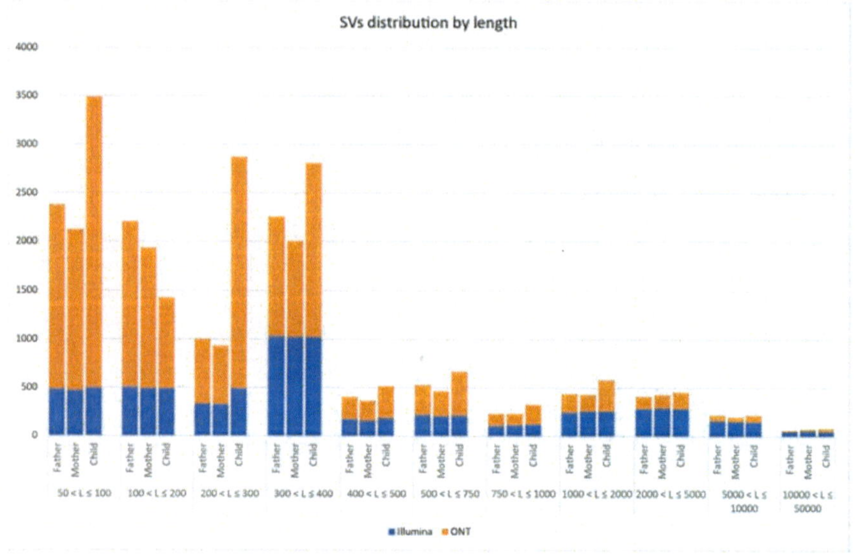

Fig. 2 Structure variation size distribution in second polish family, using Illumina and ONT. The numbers are independent from each other

Table 3 Count of technology-wise inferred SVs mapped on PET and shared PET

ConsensuSV (polish family II)					
Sequencing technology	Oxford Nanopore Technology		Illumina		Shared mapped PETs
Sample name	Number of SVs	No of mapped PET	Number of SVs	No of mapped PET	
Father	6564	1653	3609	1173	418
Mother	5638	1354	3580	1181	337
Son	9747	2422	3603	1214	531

4 Conclusion

This study seeks attention toward the possible impact of genetic variants on spatial genome organization. The performed analysis shows clear advantages of long-read sequencing (ONT) in terms of detecting more high-quality SVs, which generally cannot be resolved using short-read sequencing. It shows long-reads enable us to detect not only more SVs, but also those larger in size in comparison with the short-reads.

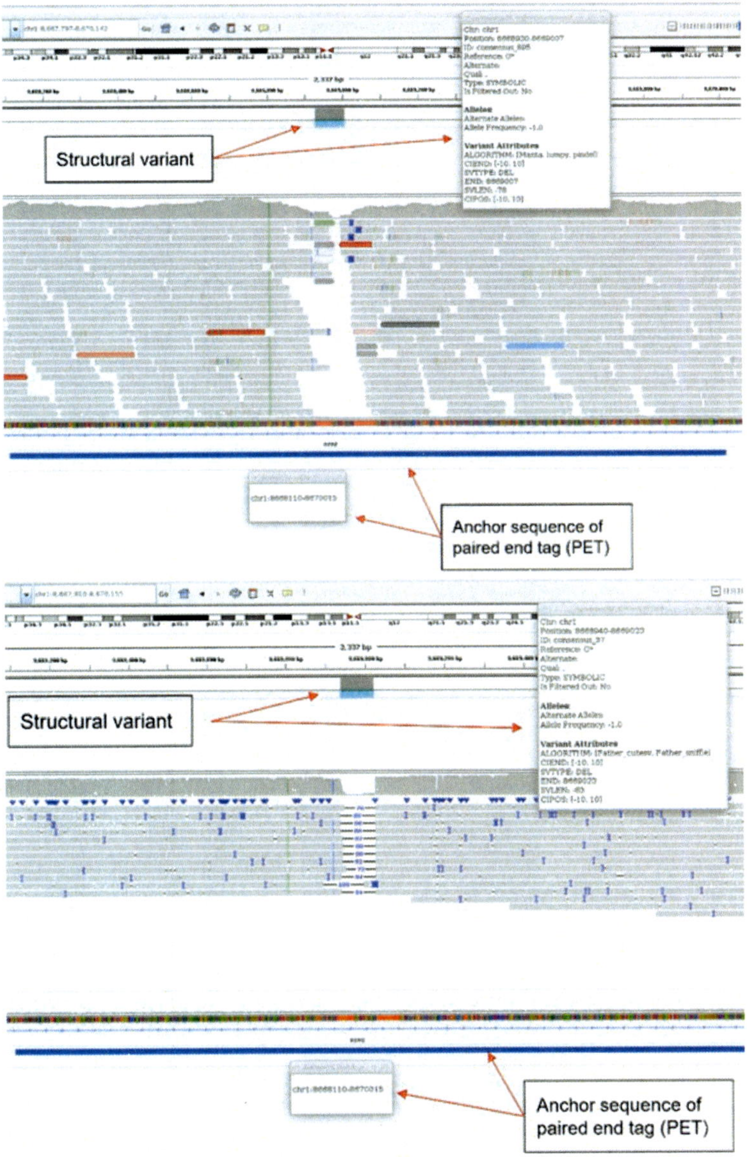

Fig. 3 SVs from father sample (ONT and Illumina) mapped on the same chromatin loop anchor. Top part of the figure represents ONT, bottom part Illumina

Acknowledgements This work has been supported by National Science Centre, Poland (2019/35/O/ST6/02484 and 2020/37/B/NZ2/03757); Foundation for Polish Science co-financed by the European Union under the European Regional Development Fund (TEAM to DP). The work has been co-supported by European Commission Horizon 2020 Marie Skłodowska-Curie ITN Enhpathy grant "Molecular Basis of Human enhanceropathies" and National Institute of Health USA 4DNucleome grant 1U54DK107967-01 "Nucleome Positioning System for Spatiotemporal Genome Organization and Regulation." Research was co-funded by Warsaw University of Technology within the Excellence Initiative: Research University (IDUB) programme. Computations were performed thanks to the Laboratory of Bioinformatics and Computational Genomics, Faculty of Mathematics and Information Science, Warsaw University of Technology using the Artificial Intelligence HPC platform financed by the Polish Ministry of Science and Higher Education (decision no. 7054/IA/SP/2020 of 2020-08-28).

References

1. Crick F, Watson J (1953) A structure for deoxyribose nucleic acid. Nature 171:3
2. Kchouk M, Gibrat J-F, Elloumi M (2017) Generations of sequencing technologies: from first to next generation. Biol Med 9
3. Sanger F, Nicklen S, Coulson AR (1977) DNA sequencing with chain-terminating inhibitors. Proc Natl Acad Sci 74:5463–5467
4. Maxam AM, Gilbert W (1977) A new method for sequencing DNA. Proc Natl Acad Sci 74:560–564
5. Jain M, Olsen HE, Paten B, Akeson M (2016) The Oxford nanopore MinION: delivery of nanopore sequencing to the genomics community. Genome Biol 17:239. https://doi.org/10.1186/s13059-016-1103-0
6. Amarasinghe SL, Su S, Dong X et al (2020) Opportunities and challenges in long-read sequencing data analysis. Genome Biol 21:1–16
7. Consortium IH (2005) A haplotype map of the human genome. Nature 437:1299
8. Haraksingh RR, Snyder MP (2013) Impacts of variation in the human genome on gene regulation. J Mol Biol 425:3970–3977
9. Auton A, Abecasis GR, Altshuler DM et al (2015) A global reference for human genetic variation. Nature 526:68–74. https://doi.org/10.1038/nature15393
10. Fullwood MJ, Ruan Y (2009) ChIP-based methods for the identification of long-range chromatin interactions. J Cell Biochem 107:30–39
11. Chen K, Wallis JW, McLellan MD et al (2009) BreakDancer: an algorithm for high-resolution mapping of genomic structural variation. Nat Methods 6:677–681. https://doi.org/10.1038/nmeth.1363
12. Abyzov A, Li S, Kim DR et al (2015) Analysis of deletion breakpoints from 1,092 humans reveals details of mutation mechanisms. Nat Commun 6:7256. https://doi.org/10.1038/ncomms8256
13. Abyzov A, Urban AE, Snyder M, Gerstein M (2011) CNVnator: an approach to discover, genotype, and characterize typical and atypical CNVs from family and population genome sequencing. Genome Res 21:974–984. https://doi.org/10.1101/gr.114876.110
14. Rausch T, Zichner T, Schlattl A et al (2012) DELLY: structural variant discovery by integrated paired-end and split-read analysis. Bioinformatics 28:i333–i339. https://doi.org/10.1093/bioinformatics/bts378
15. Layer RM, Chiang C, Quinlan AR, Hall IM (2014) LUMPY: a probabilistic framework for structural variant discovery. Genome Biol 15:R84. https://doi.org/10.1186/gb-2014-15-6-r84
16. Handsaker RE, Van Doren V, Berman JR et al (2015) Large multiallelic copy number variations in humans. Nat Genet 47:296–303. https://doi.org/10.1038/ng.3200

17. Chen X, Schulz-Trieglaff O, Shaw R et al (2016) Manta: rapid detection of structural variants and indels for germline and cancer sequencing applications. Bioinformatics 32:1220–1222. https://doi.org/10.1093/bioinformatics/btv710
18. Ye K, Schulz MH, Long Q et al (2009) Pindel: a pattern growth approach to detect break points of large deletions and medium sized insertions from paired-end short reads. Bioinformatics 25:2865–2871. https://doi.org/10.1093/bioinformatics/btp394
19. Zhao X, Emery SB, Myers B et al (2016) Resolving complex structural genomic rearrangements using a randomized approach. Genome Biol 17:126. https://doi.org/10.1186/s13059-016-0993-1
20. Soylev A, Kockan C, Hormozdiari F, Alkan C (2017) Toolkit for automated and rapid discovery of structural variants. Methods 129:3–7. https://doi.org/10.1016/j.ymeth.2017.05.030
21. Chong Z, Ruan J, Gao M et al (2017) novoBreak: local assembly for breakpoint detection in cancer genomes. Nat Methods 14:65–67. https://doi.org/10.1038/nmeth.4084
22. Kronenberg ZN, Osborne EJ, Cone KR et al (2015) Wham: Identifying Structural Variants of Biological Consequence. PLoS Comput Biol 11:e1004572–e1004572. https://doi.org/10.1371/journal.pcbi.1004572
23. Sedlazeck FJ, Rescheneder P, Smolka M et al (2018) Accurate detection of complex structural variations using single-molecule sequencing. Nat Methods 15:461–468. https://doi.org/10.1038/s41592-018-0001-7
24. Jiang T, Liu Y, Jiang Y et al (2020) Long-read-based human genomic structural variation detection with cuteSV. Genome Biol 21:189. https://doi.org/10.1186/s13059-020-02107-y
25. Heller D, Vingron M (2019) SVIM: structural variant identification using mapped long reads. Bioinformatics 35:2907–2915. https://doi.org/10.1093/bioinformatics/btz041
26. Dekker J, Belmont AS, Guttman M et al (2017) The 4D nucleome project. Nature 549:219–226. https://doi.org/10.1038/nature23884

On the Performance of Convolutional Neural Networks with Resizing and Padding

Mosammath Shahnaz and Ayatullah Faruk Mollah

Abstract As performance of convolutional neural networks in image classification experiments largely depends on appropriate feature extraction, resizing and/or padding scenarios need to be studied as it affects feature extraction directly. While direct resizing may alter the texture of an image, appropriate padding is not straightforward. In this paper, possible resizing and/or padding scenarios have been empirically studied to have more insights into preprocessing in image classification experiments. It is found that appropriate padding while maintaining the aspect ratio appears to be reasonably prospective than direct resizing. Specifically, padding with estimated background appears to be well-performing in experiments where textures are significant. In all the experiments carried out, resizing with some kind of padding is found to yield the highest classification accuracy. Fairly significant improvement in classification accuracy, i.e., 2–3% in the range of over 90% is obtained with suitable padding.

Keywords CNN · Image resizing · Padding · Background estimation · Image preprocessing

1 Introduction

In recent times, machine learning techniques have geared up many disciplines of study and areas of applications. Deep learning, on the other hand, has emerged as one of the state of the art in pattern recognition tasks. Deep architectures are found to yield far better accuracy than traditional machine learning techniques. Convolutional neural network (CNN) is one of the most impressive and baseline forms of deep learning. CNN is widely applied in image classification tasks, wherein a CNN model learns from a set of training images through automatic feature extraction and predicts classes of new samples. Every CNN architecture works with input images of the same shape [1]. So, it is obvious that images need to undergo a resizing method before

M. Shahnaz (✉) · A. F. Mollah
Department of Computer Science and Engineering, Aliah University, IIA/27 New Town, Kolkata 700160, India

S. Basu et al. (eds.), *Proceedings of International Conference on Frontiers in Computing and Systems*, Lecture Notes in Networks and Systems 404,
https://doi.org/10.1007/978-981-19-0105-8_6

feeding into a CNN as all images in a dataset may not be of the same size. In direct resizing of an image, aspect ratio may change which may presumably alter the textural characteristics of the image. Ideally, resizing should be done without losing object characteristics. A possible approach may be to resize while maintaining the aspect ratio and applying appropriate padding. Consequently, padding technique and padding position come into consideration. Empirical study on the way of resizing and associated considerations is therefore a necessity.

Many works employing CNN are available in literature. However, studies on different techniques for improving the performance of CNN are not available in plenty. Role of network architecture, different layers, activation functions, dropout rate, etc. have been studied to some extent. However, preprocessing of images which may also affect the performance of CNN is not adequately studied. Chan et al. [2] have studied the effect of image enhancement operators on the performance of convolutional neural network. Pal et al. [3] have shown the application of Zero Component Analysis (ZCA) in improving CNN performance. On the other hand, use of CNN for image denoising is not uncommon [4]. Though on a different type of image samples, i.e., ECS signal images, effect of image quality, compression, formatting as well as resizing is studied [5]. Geometric normalization is applied for performance improvement by Koo et al. [6]. From the related literature, it is evident that image features are dependent on various properties of images like color, quality, shape, texture, etc. While it is known that image quality enhancement may lead to better accuracy of CNN, studies on how changes in aspect ratio or padding affect the accuracy are still left.

A few works have been reported on the issue of resizing and padding in connection to the performance of CNN. Resizing of images using interpolation and zero-padding has been discussed. These two methods of resizing had no effect on accuracy but on training time [8]. Nam et al. also experimented by resizing the images with zero-padding and making them as the size of the largest image of the dataset [9]. They have experimented on recognizing Japanese handwritten scripts. Ghosh et al. [10] reported a study on reshaping inputs for CNN. They have experimented using 6 different datasets and 25 different types of resizing methods including interpolation, cropping, padding, tiling and mirroring. In this paper, it is shown how accuracy varies with different resizing methods. In another paper [11], different padding methods have been discussed with image datasets where a winner is found about resizing method.

Although, many different experiments have been carried out on the resizing methods, we cannot find a padding method that can give a very prominent winner. Depending on different datasets and classification problems different resizing methods have performed well. More studies are needed to come out with more insights into such considerations in regard to resizing and padding methods for CNN. In this paper, applicable resizing and/or padding scenarios have been studied and empirically assessed in text/non-text and scripts classification problems with CNN. In this study, the direct resizing method where images are resized directly to the target size without maintaining the aspect ratio, as well as various padding possibilities while retaining the aspect ratio are considered. Additionally, the effect of padding with estimated background color is also included. Such studies may undoubtedly give more insights to researchers in the field of deep learning.

2 Methodology

Uniform shape or dimensions of image samples is an important prerequisite to many deep learning networks such as CNN. However, in practical situations, different image samples have different dimensions in terms of number of rows and columns. Hence, appropriate resizing is a dire necessity. Direct resizing in a brute force way may alter texture of the image. On the other hand, padding of different considerations may render marginal loss of information content present in the image. In this respect, four different resizing and/or padding possibilities have been studied in this paper as described below. The same have been pictorially demonstrated in Figs. 1 and 2 for two different target sizes.

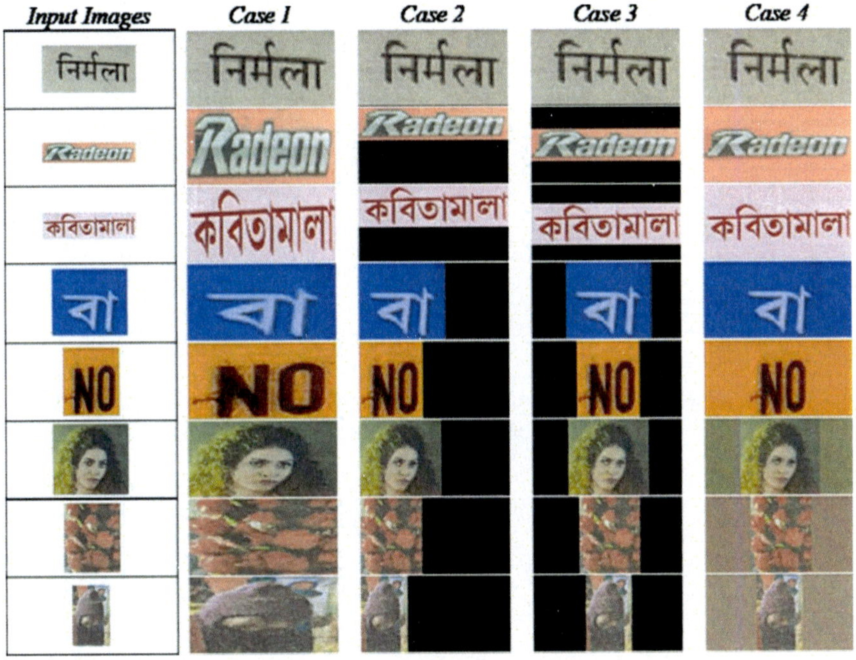

Fig. 1 Sample original and resized images when the target size is 50 × 100 pixels

Fig. 2 Original and resized image samples for target size of 100 × 100 pixels

2.1 Case 1: Direct Resizing

Images are resized directly to the target size without taking aspect ratio into account. In this case, aspect ratio of the original image gets changed. It is a lossy resizing method where image textures are not preserved.

2.2 Case 2: Resizing with Padding at Right/Bottom

Images are resized while keeping the aspect ratio same and padding added to right/bottom part of the resized image as necessary. To perform this resizing, two steps are followed: (i) Resizing: (a) Aspect ratio of the original image is calculated, (b) Image is resized keeping the aspect ratio same such that current height <= target height and current width <= target width, (ii) Padding: (a) Black (0,0,0) is selected as padding color, (b) If height/width of the resized image is less than target height/width, then padding is added to right/bottom side of the modified image to reach the target size. This is a loss-less conversion and keeps the features of the original image.

2.3 *Case 3: Resizing with Padding at Both Sides*

In this case, images are resized while keeping the aspect ratio same and padding added in a way such that the resized image is kept at the center. This is also a loss-less conversion and it keeps the features of the original image. To perform this resizing, two steps are followed: (i) Resizing: (a) Aspect ratio of the original image is calculated, (b) Image is resized keeping the aspect ratio same such that current height $<=$ target height and current width $<=$ target width; (ii) Padding: (a) Black (0,0,0) is selected as padding color, (b) If height/width of the resized image is less than target height/width, then padding is added to both side of the modified image to reach the target size. If padding is needed to reach the target height, then padding is added to the top and bottom of the modified image equally, and if padding is needed to reach the target width, then padding is added to the right and left side of the modified image equally.

2.4 *Case 4: Resizing with Padding at Both Sides with Selected Padding Color*

Unlike Case 3, padding color is estimated from the background of the original image. In the previous methods, black (0,0,0) was chosen as the padding color. As all images may not have black background, there is a possibility that CNN may extract some inappropriate features because of the padding. So, this method estimates the background color from the original image and then adds padding with that color. To select the background color of an image we consider all border pixels of the image. R,G,B values of each of those pixels are collected and then the averages of all red, green and blue values are found. The color constructed with these average values are selected as padding color.

3 Network Configuration

A deep CNN model is employed to train the machine to classify images into different number of classes. This model consists of 16 layers along with two CNN layers and four fully connected layers. The model accepts images of size (50,100,3). Details of the network model are discussed below and shown in Fig. 3.

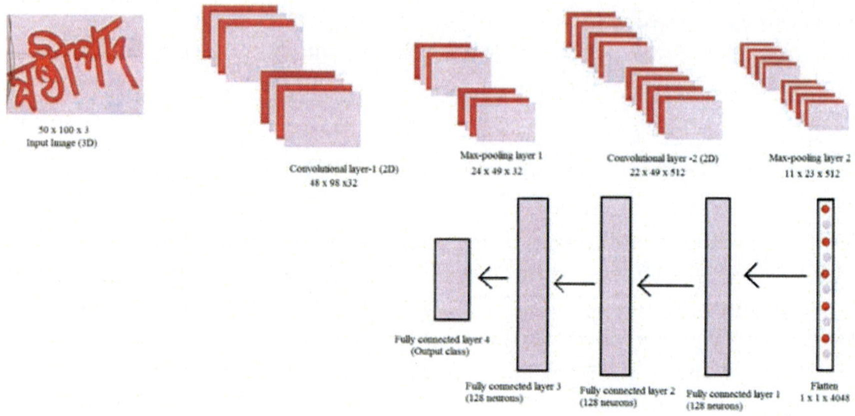

Fig. 3 A visual outline of the employed CNN model

- The very first layer is a convolution layer with a kernel 3 × 3 and 32 filters that creates 32 feature maps. Then, 'relu' activation function was added to remove non-zero values in the next layer. After that a max-pooling layer with pool size 2 × 2 was added to the model to down-sample the image for feature assumption.
- The next three layers are just repetition of the previous three layers with very small changes. The next three layers are—a convolution layer with a kernel 3 × 3 and 16 filters, an activation layer with activation function 'relu' and a max-pooling layer of pool size 2 × 2.
- A dropout layer is added with dropout rate of 0.8 to avoid over-fitting.
- The flatten layer converts the input into 1-D matrix.
- A dense layer with 128 units was added and then an activation layer with activation function 'relu' was added to remove non-zero values. These two layers are added twice.
- Then, a dropout layer with rate 0.7 was added again.
- Finally, a fully connected dense layer with units same as the number of classes was added as output layer.

4 Experimental Results

In this paper, three sets of experiments have been performed to assess the resizing and/or padding methods under consideration. A component level multi-utility image dataset, named AUTNT [7] is used for these experiments. It contains component level image samples of non-texts as well as texts belonging to three different scripts, i.e., Bengali, Devanagari and Latin. Separate test samples are also present. The four methods discussed in Sect. 2 are implemented in (i) a two class experiment where text and non-text object classification is done, (ii) a three class experiment where script is identified from direct component images and (iii) a four class combined experiment where texts of different scripts and non-text are classified. It is noteworthy to mention that in all these experiments, the CNN model discussed in Sect. 3 is used for 30 epochs.

4.1 Text Non-Text Object Classification

In this set of experiments, all the images of the AUTNT dataset are used. The images have been classified as text or non-text. A total of 8619 images are used in training. Among them, 6314 images are texts of different scripts and 2305 images are non-text. 2152 images are used to predict their classes. Among them, 1576 are text images and 576 are non-text images. Summary of results are shown in Table 1. It may be noted that while method 3 yields the highest accuracy, accuracies of the three other methods are closed by.

4.2 Script Identification

All the text images belonging to three classes, i.e., Bengali, Devanagari and Latin are used. 6314 images are used to train the machine, among which 2013 images are of Bengali script, 3017 images are of Latin script and 1284 images are of Devanagari script. 1576 images are used for prediction, among which 502 images are Bengali script, 753 images are Latin script and 321 images are Devanagari script. In Table 2, a summary of results are presented. It may be noted from Table 2 that padding with estimated background color while retaining the aspect ratio yields the highest classification accuracy of 93.08%, whereas the accuracy of the direct method is only 90.73%. It implies that significant improvement is achieved with case 4 in comparison to the direct method.

4.3 End-To-End Approach

In this set of experiments, images are classified in 4 labels—Bengali text, Devanagari text, Latin text and non-text. Among 8619 training samples, 2013 are of Bengali script, 3017 are of Latin script, 1284 are of Devanagari script and 2305 component images do not contain any text. Among 2152 number of images used for prediction, 502 images are Bengali scripted, 753 images are Latin scripted, 321 images are of Devanagari script and 576 images are non-text. Summary of the obtained results is shown in Table 3. In these experiments, case 2, i.e., padding at right/bottom while maintaining aspect ratio gives the highest classification accuracy.

Table 1 Summary of results of text non-text object classification experiments for different resizing and/or padding methods

Method	Training		Validation		Precision	Recall	F-Score	Accuracy	RMSE	Cohen's Kappa
	Acc	Loss	Acc	Loss						
Case 1	0.9903	0.0292	0.9095	0.2516	0.9734	0.9735	0.9734	0.9735	0.1628	0.9319
Case 2	0.9817	0.0455	0.8915	0.2866	0.9725	0.9726	0.9724	0.9726	0.1656	0.9292
Case 3	0.9858	0.0399	0.9234	0.2087	0.9764	0.9763	0.9763	**0.9763**	0.154	0.9398
Case 4	0.9830	0.0512	0.8973	0.2877	0.9739	0.974	0.9739	0.9740	0.1613	0.9331

Table 2 Performance of script classification under the four resizing and/or padding methods

Method	Training		Validation		Precision	Recall	F-Score	Accuracy	RMSE	Cohen's Kappa
	Acc	Loss	Acc	Loss						
Case 1	0.9289	0.1897	0.9208	0.2626	0.9193	0.9073	0.9083	0.9073	0.4714	0.8530
Case 2	0.9100	0.2451	0.9390	0.2152	0.9029	0.8958	0.8972	0.8959	0.4379	0.8365
Case 3	0.9119	0.2320	0.9406	0.2142	0.9223	0.9187	0.9195	0.9187	0.4178	0.8721
Case 4	0.9330	0.1801	0.9493	0.1892	0.9335	0.9308	0.9314	**0.9308**	0.4125	0.8907

Table 3 Summary of results of the end-to-end four class classification problem (Bengali, Latin, Devanagari and non-text)

Method	Training		Validation		Precision	Recall	F-Score	Accuracy	RMSE	Cohen's Kappa
	Acc	Loss	Acc	Loss						
Case 1	0.9253	0.2186	0.7918	0.7427	0.9167	0.9103	0.9098	0.9103	0.5527	0.8766
Case 2	0.9189	0.2370	0.9379	0.237	0.9338	0.9316	0.9320	**0.9317**	0.4236	0.9064
Case 3	0.9153	0.2408	0.9600	0.1427	0.9305	0.9298	0.9298	0.9298	0.4203	0.9039
Case 4	0.9225	0.2145	0.9194	0.2847	0.9251	0.9228	0.9229	0.9228	0.4855	0.8941

4.4 Discussion

From the above experiments, it is evident that direct resizing may not be well-suited for image classification experiments, though it is easy and straightforward. In all the three sets of experiments, some kinds of padding have yielded the best classification accuracy. In the text, non-text object classification problem, padding at center/middle (case 3) has yielded the best accuracy of 97.63%, whereas in script identification, padding at center/middle with estimated background color (case 4) has produced the highest accuracy. It may be realized from the fact that script identification heavily relies upon the textures of samples which is largely retained in padding approach. It may also be noted that due to max-pooling in CNNs, feature extraction largely depends upon the object area rather than the padded pixels. However, if the target size and the object size are far different, amount of padding will be high which may lead to loss of image information details to some extent. Therefore, target size is to be selected so as to represent average image size of the dataset under consideration. It is also evident that padding at center/middle with estimated background color (case 4) appears to be prospective. However, background estimation is not a straightforward task and background of samples may not be uniform at all. Yet, as observed from the current experiments, a reasonable approach of background estimation such as the one employed in this paper, i.e., the average of the object boundary pixels works fairly well.

5 Conclusion

In this paper, resizing and/or padding scenarios in image classification experiments with CNN has been studied. As performance of CNNs depends on appropriate feature extraction which is linked with resizing and/or padding, the empirical findings of this study give more insights into preprocessing in image classification experiments. It is found that direct resizing without retaining the aspect ratio may not be well-suited. Alternatively, appropriate padding while maintaining the aspect ratio appears to be reasonably prospective. Particularly, in experiments where textures are significant such as the case of script classification, padding with estimated background color/intensity appears to be well-performing. In all the experiments considering different resizing and/or padding scenarios, resizing with some kind of padding is found to yield the highest performance. In future, other background estimation techniques may be studied in relation to case 4 of the present study to check if it improves further.

References

1. Guo T, Dong J, Li H, Gao Y (2017) Simple convolutional neural network on image classification. In: IEEE 2nd international conference on big data analysis, pp 721–724
2. Chan X (2019) Image enhancement effect on the performance of convolutional neural network. In Master Thesis at Blekinge Institute of Technology, Sweden
3. Pal KK, Sudeep KS (2016) Preprocessing for image classification by convolutional neural networks. In IEEE international conference on recent trends in electronics, information and communication technology, pp 1778–1781
4. Jifara W, Jiang F, Rho S, Cheng M, Liu S (2019) Medical image denoising using convolutional neural network: a residual learning approach. J Supercomput 75(2):704–718
5. Boltaevich MB, Ogli NRH, Qizi GNS, Ogli MSS (2019) Estimation affects of formats and resizing process to the accuracy of convolutional neural network. In International conference on information science and communications technologies, pp 1–5
6. Koo KM, Cha EY (2017) Image recognition performance enhancements using image normalization. Human-centric Comput Inf Sci 7(33):1–11
7. Khan T, Mollah AF (2019) AUTNT—a component level dataset for text non-text classification and benchmarking with novel script invariant feature descriptors and D-CNN. Multimedia Tools Appl 78(22):32159–32186
8. Hashemi M (2019) Enlarging smaller images before inputting into convolutional neural network: zero-padding versus interpolation. J Big Data 6(98):1–13
9. Nam NT, Hung PD (2019) Padding methods in convolutional sequence model: an application in japanese handwriting recognition. in proceedings of the 3rd international conference on machine learning and soft computing, pp 138–142
10. Ghosh S, Das N, Nasipuri M (2019) Reshaping inputs for convolutional neural network: some common and uncommon methods. Pattern Recogn 93:79–94
11. Tang H, Ortis A, Battiato S (2019) The impact of padding on image classification by using pre-trained convolutional neural networks. In International conference on image analysis and processing, pp 337–344

Analysis of Vegetation Health of the Sundarbans Region Using Remote Sensing Methods

Soma Mitra and Saikat Basu

Abstract The world's largest mangrove region, Sundarban, plays a vital role in sustaining the ecological balance in the Ganga–Brahmaputra deltic region. It spans two neighboring countries, namely India and Bangladesh. In this article, we strive to assess the vegetation health of the Sundarbans region. We rely on the satellite images from Landsat 8 from January 2014 to January 2020. The vegetation health assessment is based on two indices, normalized difference vegetation index (NDVI) and forest canopy density (FCD) analysis of the interest area. NDVI consists of two bands: red and infrared band, showing a negative trend of 0.0085 per year. The image difference technique is employed further to study the vegetation health of the area of interest. NDVI difference marks the coastal regions with a higher depletion rate of vegetation than the regions away from the sea coasts. The histogram of the NDVI difference shows a negative shift from the central zero axis, which is the zone of no change in the vegetation cover. Being an early model to study vegetation health, NDVI has some disadvantages. It tends to quickly saturates at high biomass content and blurs the difference between moderately high plant cover from very high plant cover. So, we have taken another model, FCD, to compare the results of NDVI with it. It highlights the sharp depletion of highly forested cover. Nearly, 80% of the highly forested region has been depleted and joins the medium forested area. The overall mangrove cover is likely to dip more in the ensuing decades if governments of the two countries do not take necessary measures.

Keywords GIS · Sundarban · Forest canopy density · Landsat 8 · Remote sensing · Land coverage · Wet land

---

S. Mitra
Brainware University, Barasat, Kolkata, India
e-mail: som.cs@brainwareuniversity.ac.in

S. Basu (✉)
Department of Computer Science and Engineering, Maulana Abul Kalam Azad University of Technology, Kolkata, West Bengal, India
e-mail: saikat.basu@smst.iitkgp.ernet.in

S. Basu et al. (eds.), *Proceedings of International Conference on Frontiers in Computing and Systems*, Lecture Notes in Networks and Systems 404,
https://doi.org/10.1007/978-981-19-0105-8_7

1 Introduction

Sundarbans is formed at the estuaries of the Ganges and Brahmaputra rivers and is the largest mangrove forest in the world. It plays a substantial role in sustaining an ecological balance, boosting biodiversity, and diminishing the adverse effect on the regional climate. Vegetation cover is of utmost need along with the river banks and in the inland areas to protect the soil cover from natural calamities like hurricanes, cyclones, and tsunamis [1]. Numerous distributaries of Ganga, Brahmaputra, and Meghna, along with the main rivers, contribute to the origin of the Sundarban region in an intertidal zone, encompassing nearly 10,000 km^{2}· [5]. Its 60% of the area falls under Bangladesh, whereas India claims only 40% of its area. In 1989, UNESCO declared it a biosphere reserve and spurred the activity to preserve its biodiversity and to foil the illegal human activity for deforestation.

A sharp decline of the mangrove forest areas had been registered throughout the world. Though a slight increase of forest area was recorded in 1970 to 1990 in later decades, it faced depletion of forest area nearly 2.5% during 1990 to 2000 [6]. Nearly, 0.42% area had been permanently deforested during 1975 to 2006 due to increasing of human habitation and uncontrolled felling down of trees to suffice human needs [3].

A time series analysis of estuarine dynamics had been carried out in 2014, where the rise of mean sea level was evident. Here, satellite images were studied for the period 1999 till 2013. The southwestern islands are facing a gradual erosion along the coastline, whereas some islands are experiencing a solid deposition along the coastal region [9].

Research work had been carried out to establish a quantitative model using multi-temporal data of Saptamukhi and Muriganga inter-fluvial area to measure the shoreline change during the recent decades [4]. The transgression of the sea in this region for several years is evident.

Comparatively, in recent years, a forest canopy density model is utilized to assess the vegetation cover in the Sundarban region. Spatio-temporal analysis of forest cover health shows that areas with greater than 60% forest density registered a gradual decline in forest density [11].

In 2002, a mathematical model was proposed to predict future coastline positions of the Sagar Island. Here, the main concern is to calculate the rate of recession of the coastline; depletion of vegetation cover was not the main concern there [7].

In 2017, a time series analysis of 35 years showed an 11,500 ha or (1.2%) gross erosion in the Sundarban region. A decline in mangrove forest area is registered due to yearly hit of super cyclones, increase in water salinity and deforestation [8].

In 2019, the Continuous Change Detection and Classification (CCDC) method which had been applied Landsat images over 30-years showed that nearly 25% of the area undergoes a fall in forest density due to decline of Heritiera fomes, a species of mangrove tree [2].

Over the previous decade, the research works that have been carried out on the Sudarban area mainly for shoreline change calculation. But these studies lack the assessment of the interior vegetation health using different established vegetation indices. So, in this study, we have taken recent Landsat 8 images from 2014 to 2020 and implement two well-known vegetation indices, e.g., NDVI and FCD, to classify the forest vegetation and to establish a trend of intra-migration of forested land among the different vegetation classes. This paper is an overall assessment of the health of the interior vegetation cover of the Sundarban region and to shade light on the depleting vegetation cover of this region, which could be a great concern for the ecological balance of the lower Ganga basin in ensuing decades.

2 Study Area

The assessment portion is occupying the southmost part of the district of South 24 Parganas of West Bengal in India Fig. 1. It belongs to the Sundarbans Reserved Forest [10]. The whole area is chiefly distributed between mangrove forest 69% and remaining 31% is enveloped with saline water of the sea and the wide estuaries of numerous distributaries of river Hooghly. The bounding latitudes and longitudes of the said area are given in Table 1.

Fig. 1 Sundarbans—the study area

Table 1 Boundary latitude and longitudes of the area

Area	Latitude	Longitude
Upper left	22°24′44.64″N	88°25′28.20″E
Lower left	21°32′6.18″N	88°24′56.79″E
Upper right	22°23′50.00″N	89°38′0.01″E
Lower right	21°31′13.89″N	89°37′2.00″E
Center	21°58′2.70″N	89°1′21.93″E

3 Research Data

This research had used Landsat 8 imagery of 30 mt spatial resolution obtained from Jan. 2014 to Jan. 2020 from the USGS Earth Explorer site using https://earthexplorer.usgs.gov link. All the images were acquired in the December–January session to incorporate a cloud-free scene over 12 months. According to Worldwide Reference System 2 (WRS2), we took the images from path 138, and row 45, and clipped out the reference area.

3.1 *Landsat 8 Band Description*

Two types of sensors are incorporated in Landsat 8 satellite, one is the Operational Land Imager (OLI), and the other one is the Thermal Infrared Sensor (TIRS). It gathers data in 11 bands from two different sensors and incorporates two separate thermal bands, namely bands 10 and 11. As per our requirement, we have used six bands (B2 to B6, B10). Description of the bands is given in Table 2. Assessment of vegetation health mainly depends on colored bands, NIR bands, and thermal bands, so the remaining bands are excluded for ease of calculations.

Table 2 List of landsat 8 bands

Band name	Description	Spatial resolution (m)
Band 10	TIRS	100
Band 6	SWIR	30
Band 5	NIR	30
Band 4	Red	30
Band 3	Green	30
Band 2	Blue	30

4 Methodology

4.1 Image Preprocessing

Seven Landsat 8 images from 2014 to 2020 were preprocessed to extract and assess the salient information from satellite images using a semi-automatic classification plugin in QGIS 3.16. Image preprocessing includes atmospheric correction using dark object subtraction method and conversion of digital numbers to total atmospheric reflectance (TOA reflectance) values. Finally, TOA reflectance values contribute to the final calculations.

4.2 Calculation of NDVI

NDVI is generally used in finding the quality of the green coverage on the part of the land, and it can classify the forested region according to a defined scale Fig. 2. It uses NIR and red bands to qualify the health of green vegetation. The equation of the NDVI calculation is as follows.

$$\text{NDVI} = (\gamma - \mu)/(\gamma + \mu) \quad (1)$$

where γ = NIR Band μ = RED Band
NDVI values range from -1 to +1 per pixel. Highly negative value indicates the presence of wet-bodies whereas +0.5 and onward it signifies a presence of healthy vegetation cover. Values nearer to +1, signify a dense forested area according to Fig. 2.

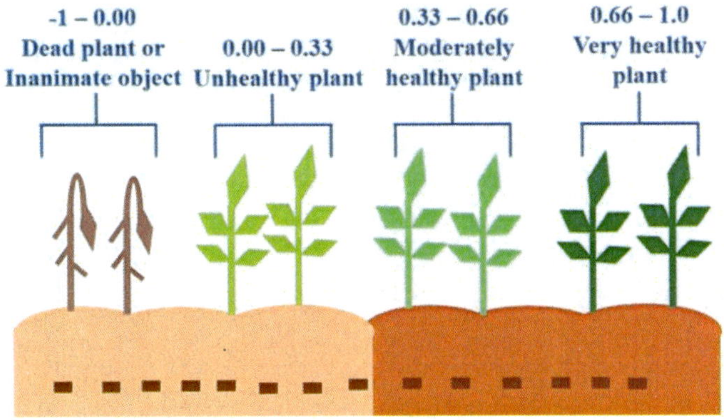

Fig. 2 Range of NDVI

4.3 Calculation of FCD

Other than NDVI, another method forest canopy density (FCD) model is assessed on the same interest area. This index employs density of forest canopy as an essential criterion for the estimation of the woodland conditions. To effectively build the FCD model, four other indices are derived, which are as follows.

- Bare soil index (BI)
- Thermal index (TI)
- Advanced vegetation index (AVI)
- Scaled shadow index (SSI)

The final FCD is scaled between 1 and 100% for each pixel.

5 Result Analysis

5.1 Result Analysis for NDVI

NDVI values are calculated for 2014 to 2020. The following Table 3 depicts the gradual decline of vegetative health from the year 2014. A trend line is drawn over the Mean NDVI line in Fig. 3 which shows a negative trend over the years.
NDVI difference technique has been employed to deduce the vegetation loss and gain areas. In the NDVI difference technique, 2014 NDVI is subtracted from 2020 NDVI pixel by pixel. Image subtraction shows three different regions in Table 4. The histogram in Fig. 4, is showing the pixel frequency in above mentioned three categories in Table 4. A pixel in the image covers an area of $900\,m^2$. In y-axis of the histogram, the number of pixels is counted for a given pixel value. The entire histogram is left-aligned, and pixel frequencies are greater in the negative zone than the zero or the positive zone. So, it can be easily surmised that a great area of this region is facing a depletion of forest cover. The total area could be calculated by

Table 3 Mean NDVI calculation from 2014 to 2020

Year	Mean NDVI
2014	0.7248
2015	0.7391
2016	0.7589
2017	0.6653
2018	0.6927
2019	0.6934
2020	0.6981

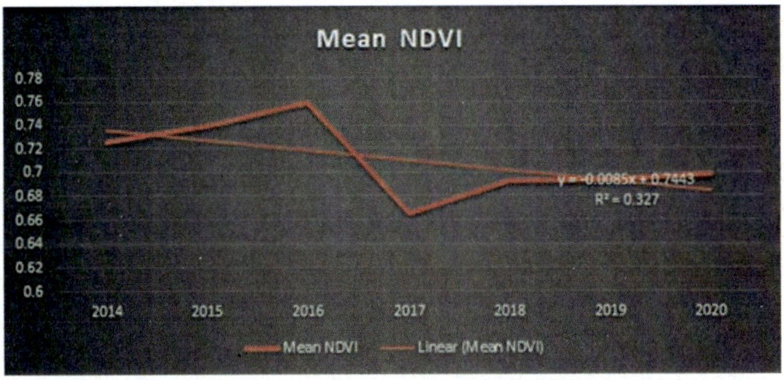

Fig. 3 Mean NDVI and trendLine

Table 4 NDVI difference

NDVI difference	Meaning
0	No change
< 0	Vegetation loss
> 0	Vegetation gain

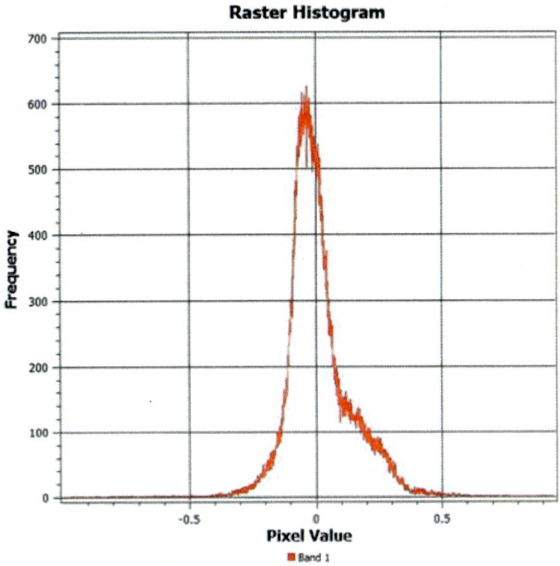

Fig. 4 Histogram of NDVI difference

multiplying the number of pixels of negative values, by 900 m^2. In the course of seven years, 2014–2020, it is evident that the region has a downward tendency of losing its vegetation cover. NDVI index is an age-old technique that has several disadvantages. It eventually saturates at high biomass, reducing the probability of proper calculation of moderately high plant cover from very high plant cover. It is much more prone to atmospheric noise in the NIR and red bands than other sophisticated indices like FCD and GARI.

5.2 *Result Analysis for FCD*

The same Landsat 8 OLI images from Jan. 2014 and Jan. 2020 are used. Based on the results of the FCD model, we can conclude that it is roughly coinciding with the results of the NDVI analysis. The FCD maps of 2014 and 2020 are classified into five classes: Water bodies, scanty vegetation region, low vegetation region, medium region, and high vegetation region. With reference to Table 5, it can be surmised that the high vegetation region is depleted and reduced. Mainly, it is migrated to the medium vegetation region, which is inflated in 2020. There is an insignificant growth in the area of scanty and low vegetation region, but the significant negative change in the vegetative cover is evident. Anthropogenic activity and yearly hitting of super cyclones are the main cause of the forest canopy depletion of this region. If it continues to deplete unchecked, high forest vegetation area may likely to dip further, from where it would be impossible to restore it back.

Table 5 FCD comparison of 2020 and 2014

FCD analysis in 2020	
Class	Area%
Water	1%
Scanty vegetation	8%
Low vegetation	9%
Medium vegetation	73%
High vegetation	9%
FCD analysis in 2014	
Class	Area %
Water	0.50%
Scanty vegetation	3.38%
Low vegetation	4.74%
Medium vegetation	41.66%
High vegetation	49.35%

6 Conclusion

The Sundarbans within the coastal part of the Bay of Bengal is one among the natural wonders of our planet. In this study, we have strived to assess the overall vegetation health of Sundarban region. Both the NDVI and FCD analysis registered a significant decline in vegetation cover over the assessment period. NDVI (normalized difference vegetation index) consisting of two bands, red and infrared band, shows a negative trend of 0.0085 per year. NDVI difference indicates that the coastal areas of isolated islands are experiencing a high rate of vegetation cover erosion. FCD model also highlights the sharp depletion of highly forested cover. Nearly, 80% of the highly forested region has been depleted and joins the medium forested region. Over the decades, the Sundarban region is continually exposed to yearly cyclone hit, gradual rise of sea level, and anthropogenic activities. These factors would be considered in our future research to assess the fate of the Sundarbans region in the forthcoming decades.

References

1. Alongi DM (2008) Mangrove forests: resilience, protection from tsunamis, and responses to global climate change. Estuar Coast Shelf Sci 76(1):1–13
2. Awty-Carroll K, Bunting P, Hardy A, Bell G (2019) Using continuous change detection and classification of landsat data to investigate long-term Mangrove dynamics in the Sundarbans region. Remote Sen 11(23):2833
3. Datta D, Deb S (2012) Analysis of coastal land use/land cover changes in the Indian Sunderbans using remotely sensed data. Geo-Spat Inf Sci 15(4):241–250
4. Ghosh A, Mukhopadhyay S (2016) Quantitative study on shoreline changes and Erosion Hazard assessment: case study in Muriganga-Saptamukhi interfluve, Sundarban, India. Model Earth Syst Environ 2(2):75
5. Ghosh A, Schmidt S, Fickert T, Nüsser M (2015) The Indian Sundarban mangrove forests: history, utilization, conservation strategies and local perception. Diversity 7(2):149–169
6. Giri C, Pengra B, Zhu Z, Singh A, Tieszen LL (2007) Monitoring mangrove forest dynamics of the Sundarbans in Bangladesh and India using multi-temporal satellite data from 1973 to 2000. Estuar Coast Shelf Sci 73(1–2):91–100
7. Hazra S, Ghosh T, DasGupta R, Sen G (2002) Sea level and associated changes in the Sundarbans. Sci Cult 68(9/12):309–321
8. Quader MA, Agrawal S, Kervyn M (2017) Multi-decadal land cover evolution in the Sundarban, the largest mangrove forest in the world. Ocean Coast Manag 139:113–124
9. Raha AK et al (2014) Sea level rise and submergence of Sundarban Islands: a time series study of estuarine dynamics. J Ecol Environ Sci ISSN 0976–9900
10. Rahman MM, Lagomasino D, Lee S, Fatoyinbo T, Ahmed I, Kanzaki M (2019) Improved assessment of mangrove forests in Sundarbans East Wildlife Sanctuary using WorldView 2 and Tan DEM-X high resolution imagery. Remote Sens Ecolo Conserv 5(2):136–149
11. Sahana M, Sajjad H, Ahmed R (2015) Assessing spatio-temporal health of forest cover using forest canopy density model and forest fragmentation approach in Sundarban reserve forest, India. Model Earth Syst Environ 1(4):1–10

You Reap What You Sow—Revisiting Intra-class Variations and Seed Selection in Temporal Ensembling for Image Classification

Manikandan Ravikiran, Siddharth Vohra, Yuichi Nonaka, Sharath Kumar, Shibashish Sen, Nestor Mariyasagayam, and Kingshuk Banerjee

Abstract In this work, we present our study on the influence of intra-class variations and seed selection on image classification using temporal ensembling. Through our experiments, we observe that for Fashion-MNIST and Kuzushiji-MNIST datasets with medium and high intra-class variations, (a) classification accuracy declines by 20% and 30%, (b) raising seed images by 5x improves accuracy by 15% and 8% respectively. Additionally, preliminary investigation on the Fashion-MNIST dataset reveals that with diversified seed selection class, level accuracy improves by 6%. In due process, our study exhibits convergence issues for unsupervised loss when training on datasets with medium and high intra-class variations.

Keywords Semi-supervised learning · Image classification · Temporal ensembling

Manikandan Ravikiran and Siddharth Vohra—Both authors contributed equally. Names are ordered alphabetically.
Siddharth Vohra—Work done as intern at R&D Center, Hitachi India Pvt Ltd.

M. Ravikiran (✉) · S. Vohra · S. Kumar · S. Sen · N. Mariyasagayam · K. Banerjee
R&D Center, Hitachi India Pvt Ltd, Bangalore, Karnataka, India
e-mail: manikandan@hitachi.co.in

S. Kumar
e-mail: sharath@hitachi.co.in

S. Sen
e-mail: shibashish.sen@hitachi.co.in

N. Mariyasagayam
e-mail: nestor@hitachi.co.in

K. Banerjee
e-mail: kingshuk.banerjee@hitachi.co.in

Y. Nonaka
R&D Group, Hitachi Ltd, Tokyo, Japan
e-mail: yuichi.nonaka.zy@hitachi.com

S. Basu et al. (eds.), *Proceedings of International Conference on Frontiers in Computing and Systems*, Lecture Notes in Networks and Systems 404,
https://doi.org/10.1007/978-981-19-0105-8_8

1 Introduction

Deep learning has made giant strides in the task of image classification. However, much of its success is conditional on acquiring labeled images which is laborious. Further, this is complicated by the inherent complexity of the labeling process and low noise requirements. Recently, to address this concern, a variety of approaches have been devised, including semi-supervised learning algorithms [1–11] which typically show higher results by training with few labeled images known as seed images or seeds. The simplest among these algorithms is temporal ensembling [12], which uses an ensemble of outputs of the network in training on different epochs as an unsupervised target label to achieve high accuracies.

However, it fails to address two main issues that are common in practical computer vision applications. Firstly, the study implicitly does not account for intra-class variations that are typical in practical image datasets. This raises a question on *how does temporal ensembling fair for image datasets with diverse intra-class variations?* Secondly, while much of the work on temporal ensembling reports results with varying seed image sizes, none of them explicitly account *how intra-class variations affect the number of seed images used?* Finally, to the best of our knowledge, there are no studies in current literature that *explicitly examines the effect of intra-class variations within seed* images *on the overall results?* Accordingly, through this work, we make the following empirical contributions.

- First, we study the impact of intra-class variations on training behavior and the accuracy of temporal ensembling. To this end, we experiment on image datasets with different intra-class variations to find a sheer drop in the accuracy of temporal ensembling (see Sect. 3.1).
- Second, we examine the impact of the number of seed images on temporal ensembling across datasets with different intra-class variations. Accordingly, we find that with raising seed image size, there is an enhancement in resulting accuracy. However, we do see the net gain to be limited on datasets with high intra-class variations (see Sect. 3.2).
- Finally, we explore the influence of the type of images (seeds) on temporal ensembling to establish that the model trained with the diverse seeds delivers higher results (see Sect. 3.3).

2 Background and Preliminaries

2.1 Temporal Ensembling

Given an input training data x_i with B mini-batches, L indices of the labeled inputs with correct labels $y_i \in \{1,\ldots,C\}$, for C different classes. Temporal ensembling combines multiple outputs from network in training as a noisy target (Q) as shown in Eq. 1. More specifically, temporal ensembling generates outputs with different image additions (q_i) and regularization ($\tilde{q}_i$), allowing the network to converge with noisy labels.

$$\overbrace{l_B(q, \tilde{q}, y)}^{\text{Total Loss}} = \overbrace{-\frac{1}{|B \cap L|} \sum_{i \in (B \cap L)} \log q_i[y_i]}^{\text{Supervised Loss}} + w(t) * \overbrace{\frac{1}{C|B|} \sum_{i \in B} || q_i - \tilde{q}_i ||^2}^{\text{Unsupervised Loss}} \quad (1)$$

This allows neural network to explore feature space without changes in network predictions for varying inputs especially upon coupling with exponential moving average of repetitive noisy predictions. Details on ensembling of labels, weighing of unsupervised components, etc. are presented in [12].

2.2 *Dataset and Hyperparameter Settings*

Since we focus on image classification problem, in this work, we use multiple datasets (Table 1) with common data characteristics but with different intra-class variations (Fig. 1) of *low, medium and high*. These intra-class variations are decided based on Eidolon experiment in line with [13].

Moreover, our experimental setting is the same as [12] where we train a two-layer channel-wise weight normalized CNN with feature maps size 16 & 32, batch size of 100, dropout of 0.5, and learning rate of 0.002 for five episodes by randomly selecting 100 seeds. Thus, the average results of accuracy score are shown here.

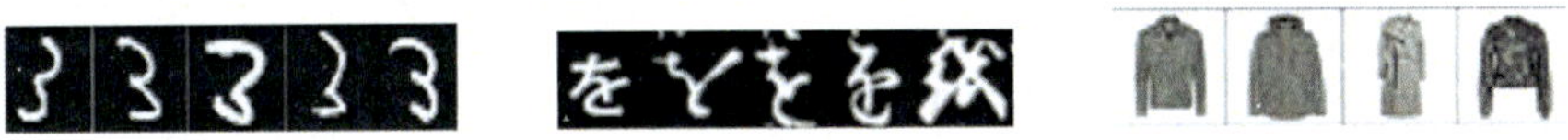

(a) MNIST (class=3) (b) Kuzushiji-MNIST (class=*wo*) (c) Fashion-MNIST (class=coat)

Fig. 1 Intra-class variations across MNIST, Kuzushiji-MNIST, and fashion-MNIST datasets

Table 1 Summary of datasets used in this work with their size and intra-class variation characteristics

Dataset	Images	Content	Intra-class variations
MNIST [14]	60,000 training,	Handwritten digits	Low
Fashion-MNIST [15]	10,000 testing,	Clothing articles	Medium
Kuzushiji-MNIST [16]	28 × 28 gray scale	Kuzushiji characters	High

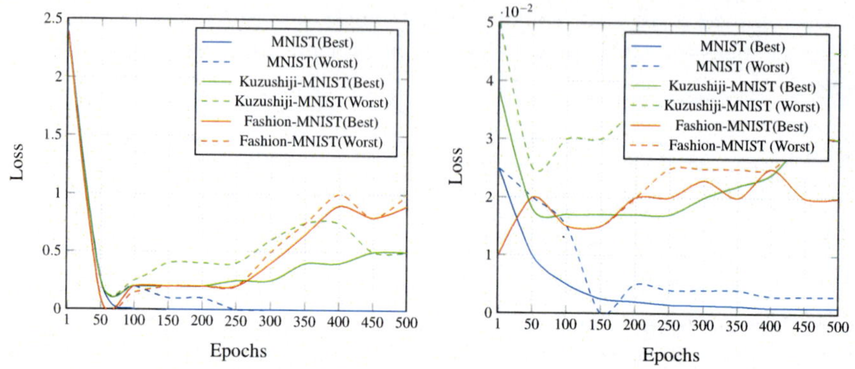

(a) Loss Curves of Supervised Component

(b) Loss Curves of Unsupervised Component

Fig. 2 Temporal ensembling loss curves with MNIST dataset. Results of best seed are shown with thick line and worst seed are shown with dashed line

3 Experiments

In this section, various experiments are presented in detail. Specifically, in Sects. 3.1 and 3.2, we focus on the effect of datasets and the number of seeds on training and testing behaviors. Meanwhile, in Sect. 3.3, we study the impact of seeds themselves through a manual seed selection strategy.

3.1 Effect of Intra-class Variations on Temporal Ensembling

To start with, we study *how does temporal ensembling fair for datasets with diverse intra-class variations?* To do this, we concentrate on MNIST, Kuzushiji-MNIST, and Fashion-MNIST inline with settings from Sect. 2.2. Besides we examine models trained with epochs in range of 100–500 with respective results in Fig. 3. Multiple interesting characteristics are visible upon investigation.

Training Behavior: Firstly, we can see that from MNIST training behavior, the convergence is faster for both supervised and unsupervised loss (Fig. 2). Meanwhile, Kuzushiji-MNIST and Fashion-MNIST show a significant delay in convergence time and increased loss values even with higher epochs for the unsupervised loss component. Additionally, the convergence with Kuzushiji-MNIST and Fashion-MNIST across seeds differs by a large margin.

Testing Accuracy: Meanwhile, comparing accuracy across different datasets in Fig. 3, reveals MNIST to be the peak performer, without variations across epochs. However, for Kuzushiji-MNIST, we can see that longer epochs are warranted for enhancing accuracy, where with 300 epochs, there is a minor drop of 0.15%, which

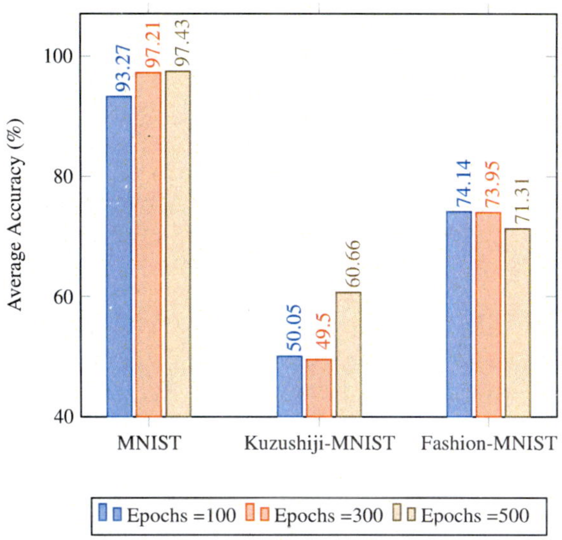

Fig. 3 Average accuracy score various datasets using temporal ensembling

Table 2 Average accuracy score various datasets using temporal ensembling with random reset

Epochs	100	300	500
MNIST	93.67	97.10	97.67
Kuzushiji-MNIST	52.35	52.70	61.76
Fashion-MNIST	71.55	71.55	72.34

increases by 10% when trained for 500 epochs. Meanwhile, for the Fashion-MNIST dataset, we can see that there is a drop in results with increasing epochs.

We suppose the above findings are possibly due to reasons involving (a) low intra-class variations between train-test and in sample classes for MNIST (b) local minima issue with unsupervised labels for Kuzushiji-MNIST and Fashion-MNIST. While the former can be established by eyeballing MNIST, to verify the latter, we perform a preliminary study by modifying temporal ensembling with *random reset (RR)* of unsupervised labels and test the model with Fashion-MNIST dataset based on intuitions from random restart hill-climbing [17]. In random reset, we randomly reset unsupervised predictions to alternate classes during the EMA update. If the labels were indeed stuck in local minima, we believe such random reset will alleviate this. The corresponding results with random reset are as shown in Table 2.

From the results, we can see that an increase of ≈1% was possible with random reset for Fashion-MNIST and Kuzushiji-MNIST datasets, respectively. We believe such an improvement is because; during training, seed images often lead the network to only learns the most discriminative features of the seeds, thereby restricting weight space exploration and random resets help in mitigating this problem. However, we

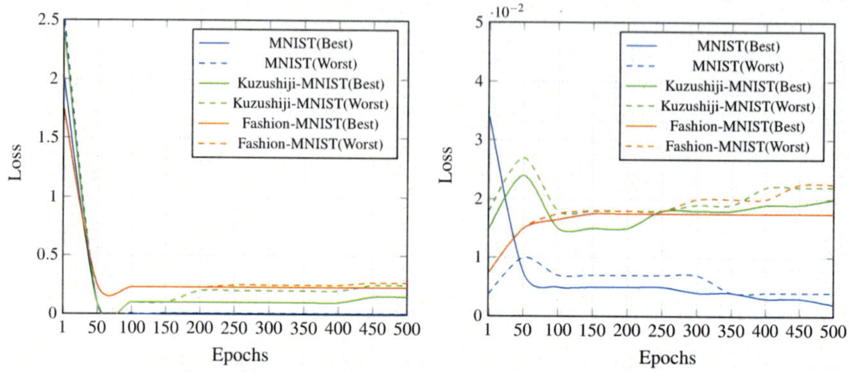

(a) Loss curves of supervised Loss

(b) Loss curves of unsupervised Loss

Fig. 4 Temporal ensembling loss curves with 300 seeds. Results of best seed are shown with thick line and worst seed are shown with dashed line

think further experimentation is needed to verify local minima issues, which we plan to revisit in the future. Overall, all these results together suggest that temporal ensembling is not immediately usable across datasets that offer higher intra-class variations and require adjustments to accommodate intra-class variations and drop of accuracy so encountered.

3.2 *Effect of Number of Seeds on Temporal Ensembling*

Previously, we examined intra-class variations' effect on temporal ensembling. However, as seen, different seeds (*best and worst*) yield distinct behaviors in training and accuracy. Hence, here *we study the effect of the number of seeds used* for training. Also, here, we change seed size in the range of 100–500 and train the models for 500 epochs.

Training Behavior: Fig. 4 shows the training behavior of Kuzushiji-MNIST datasets with 300 seeds. Contrasting this with convergence behavior from Fig. 2, we can see the supervised loss quickly converge to a loss of 0.29 within 150 epochs and saturate, unlike what was seen before. Comparable behavior could also be observed with Fashion-MNIST. However, the unsupervised loss still exhibits convergence issues but comparatively lower than Sect. 3.1.

Testing Accuracy: From Fig. 5, MNIST dataset to show little to no impact, with much of results around 97% with a maximum of 97.93% on a model trained for 500 epochs with 400 seed images. The simplicity of MNIST acts as a core contributor for high results. Moreover, train-test classes showing minimal variability in their respective samples, leading to no impact with higher seed size. Meanwhile, Kuzushiji-MNIST and Fashion-MNIST show drastic improvement on accuracy score with more seeds with a maximum gain of 14.7% and 8%, respectively.

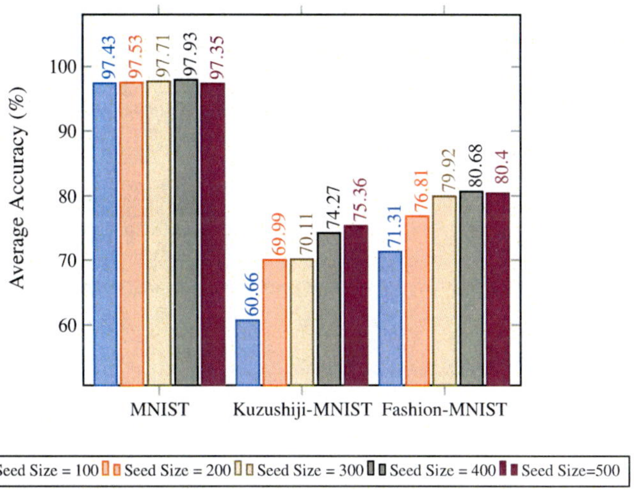

Fig. 5 Accuracy of temporal ensembling with varying seed sizes across datasets used in this work

Fig. 6 Benchmarking temporal ensembling against respective baselines models

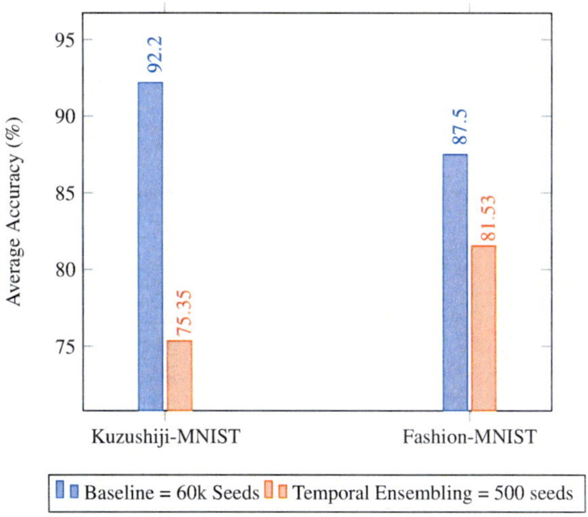

Additionally, we can see that for Fashion-MNIST, the results are close to that state-of-the-art deep learning baseline of 87% (see Fig. 6). Thus, despite 150% fewer images, the net drop in the result is only 7%. Similar behavior could be seen for Kuzushiji-MNIST (Fig. 6). Overall, these findings prove that expanding the number of seeds improves the overall results and training behavior. At the same time, these results also hint at a potential relationship between the individual seeds themselves and results on the test set. As such, in the next section, we study how results are impacted by seed type.

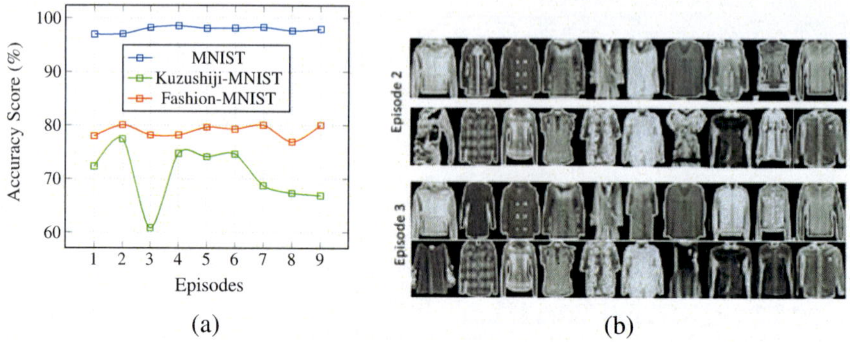

Fig. 7 Summary of training results across episodes with **a** accuracy scores and **b** seeds from episodes 2 and 3

Table 3 Class wise F1-score across episodes 2 and 3 for fashion-MNIST dataset

	Classes									
	0	1	2	3	4	5	6	7	8	9
Episodes	T-shirt/top	Trouser	Pullover	Dress	Coat	Sandals	Shirt	Sneaker	Bag	Ankle Boots
2	0.7550	0.8800	0.7520	0.8020	0.7400	0.8800	0.6300	0.8395	0.8650	0.8620
3	0.6550	0.7770	0.7630	0.7850	0.6810	0.8890	0.6600	0.8650	0.8760	0.8620

3.3 *Effect of Seed Type on Temporal Ensembling*

In this part of the study, we hypothesize that *using seeds with distribution in line with test set will show better results*. To prove this and understand different seeds' effects, we train with 500 seeds for 500 epochs to utilize specific seed examples. Figure 4a shows the performance obtained across various episodes (Fig. 7).

We analyze the impact of seed types on results by manually visualizing the seeds for episodes 2 and 3 of the training process and relating them to class-wise results. For the current explanation, we focus on the Fashion-MNIST dataset owing to its ease of interpretability. We believe the same to hold for Kuzushiji-MNIST and MNIST, which we plan to visit in the future. The class-wise F1-score are as shown in Table 3 for both of these episodes. In addition, since class 4 (coat) and class 6 (shirt) vary broadly across these episodes, we visualize corresponding seed images (Fig. 4b).

Firstly from Fig. 4b, for class 4, we can see that episode 2 has diverse seeds compared to episode 3. Additionally, we can see that for class 6, only five of the seed images are different between the episodes, yet there is a significant increase in results. We believe this is because, with temporal ensembling, additional features are to be learnt to accommodate the fine grained classes, which is possible due to varying seed images.

Overall based on the above, one can argue selecting the number of seeds and seed types has a significant impact on results, where different seeds of sufficient size are

Fig. 8 Example manual seed for class *coat*, with last four representing the randomly sampled seeds

Table 4 Class wise F1-score for fashion-MNIST dataset with manual seed (MS) selection and Random Reset (RR)

	Classes									
	0	1	2	3	4	5	6	7	8	9
Strategy	T-shirt/top	Trouser	Pullover	Dress	Coat	Sandals	Shirt	Sneaker	Bag	Ankle boots
MS	0.6640	0.7670	0.7630	0.8120	0.6610	0.8680	0.6610	0.8850	0.8860	0.8420
MS + RR	0.6730	0.7580	0.7725	0.8220	0.6850	0.8680	0.6780	0.8895	0.8950	0.8650

vital for temporal ensembling. To verify this, we revisit experiments by training for a single episode with 500 epochs and manual seed selection.

In the manual seed selection process, we select 40% (4 seeds per class randomly), and the remaining 60% (6 seeds) of seeds per class were manually chosen similar to Eidolon experiment [13], where we select visually distinctive-looking seeds (Fig. 8). Also, images from the train set that is similar to the testing set were removed apriori to ensure generalization. The corresponding Fashion-MNIST results are as shown in Table 4. Contrasting this with results from Table 4, we can see the results are on par with the 300 seed experiment results, confirming the earlier mentioned findings. The table also presents results with random reset, and we can see that random reset still helps with additional improvement of 1%.

4 Conclusion

In this paper, we analyzed intra-class variations and seed selection for image classification using temporal ensembling. Our study finds that for Fashion-MNIST and Kuzushiji-MNIST datasets with medium and high intra-class variations, there is a drop in accuracy by 20% and 30%. Further, raising seed images by 5x leads to accuracy improvement by 15% and 8%, respectively. Also, examining the effect of seed through manual seed selection shows the approach to be fruitful by 6% for the Fashion-MNIST dataset, encouraging further studies on seed selection approaches.

Nevertheless, (a) the increased unsupervised loss, (b) the effect of larger seed size, (c) the connection between seeds, and (d) data distribution, manual seed selection in the context of multiple datasets. Additionally, much of this work is limited to image classification, while this concept of seeds and semi-supervised learning is equally valid across other vision applications and areas such as Natural Language and Speech processing, warranting expanded experiments for broader applicability.

References

1. Ravikiran M, Muljibhai AE, Miyoshi T, Ozaki H, Koreeda Y, Masayuki S (2020) Hitachi at semeval-2020 task 12: offensive language identification with noisy labels using statistical sampling and post-processing. ArXiv: abs/2005.00295
2. Tarvainen A, Valpola H (2017) Mean teachers are better role models: weight-averaged consistency targets improve semisupervised deep learning results. In: NIPS, pp 1195–1204
3. Ke Z, Wang D, Yan Q, Ren JSJ, Lau RWH (2019) Dual student: breaking the limits of the teacher in semi-supervised learning. In: ICCV. IEEE, pp 6727–6735
4. Park S, Park J, Shin S, Moon I (2018) Adversarial dropout for supervised and semi-supervised learning. In: AAAI. AAAI Press, pp 3917–3924
5. Xie Q, Dai Z, Hovy EH, Luong T, Le Q (2020) Unsupervised data augmentation for consistency training. In NeurIPS
6. Li Y, Guo L, Zhou Z (2021) Towards safe weakly supervised learning. IEEE Trans Pattern Anal Mach Intell 43(1):334–346
7. Wang X, Kihara D, Luo J, Qi G (2021) EnAET: a self-trained framework for semi-supervised and supervised learning with ensemble transformations. IEEE Trans Image Process 30:1639–1647
8. Ghosh A, Thiery AH (2021) On data-augmentation and consistency-based semi-supervised learning. In: ICLR. OpenReview.net
9. Deng J, Yu J (2021) A simple graph-based semi-supervised learning approach for imbalanced classification. Pattern Recogn 118
10. Zhong Z, Zheng L, Kang G, Li S, Yang Y (2020) Random erasing data augmentation. In: AAAI. AAAI Press, pp 13,001–13,008
11. Yang X, Song Z, King I, Xu Z (2021) A survey on deep semi-supervised learning
12. Laine S, Aila T (2017) Temporal ensembling for semi-supervised learning. ArXiv: abs/1610.02242
13. Geirhos R, Janssen DH, Rauber J, Bethge M, Wichmann F (2017) Comparing deep neural networks against humans: object recognition when the signal gets weaker. ArXiv: abs/1706.06969
14. LeCun Y, Cortes C (2010) MNIST handwritten digit database
15. Xiao H, Rasul K, Vollgraf R (2017) Fashion-mnist: a novel image dataset for benchmarking machine learning algorithms. ArXiv: abs/1708.07747
16. Clanuwat T, Bober-Irizar M, Kitamoto A, Lamb A, Yamamoto K, Ha D (2018) Deep learning for classical Japanese literature. ArXiv: abs/1812.01718
17. Russel S, Norvig P (2003) Artificial intelligence: a modern approach. Prentice Hall series in artificial intelligence

Light U-Net: Network Architecture for Outdoor Scene Semantic Segmentation

Aleazer Viannie Sunn, Aiden Langba, Hamebansan Mawlong, Alexy Bhowmick, and Shyamanta M. Hazarika

Abstract We present a novel and efficient semantic segmentation architecture termed *Light U-Net* in this paper. Light U-Net is less complex with regard to the parameter and model size than the original U-Net. We also propose a network and training strategy based on geometric data augmentation for efficient use of the annotated images from small and limited datasets. Following improvements, we show the use of Light U-Net for semantic segmentation of outdoor scenes. Results reveal that our improvements and data augmentation remarkably enhance the proposed network's performance. Results from experiments demonstrate that the proposed architecture outperforms some recent segmentation models achieving mIoU scores of 70.06% and 79.84% on the *Camvid* and *Stanford Background* datasets respectively.

Keywords U-Net · Semantic segmentation · Data augmentation · Encoder–decoder · Light U-Net

1 Introduction

Semantic segmentation, also referred to as scene parsing, is a core problem in computer vision. Semantic segmentation densely labels every pixel in the scene with a meaningful class label, (e.g., *bridge*, *building*, *sky*). The term 'semantic segmentation' is used interchangeably with *scene labeling*, *image parsing*, *full-scene labeling* (FSL), and *multi-class pixel labeling*. Semantic segmentation is vital for numerous

A. V. Sunn · A. Langba · H. Mawlong · A. Bhowmick (✉)
Department of Computer Science and Engineering, Assam Don Bosco University, Airport Road, Azara, Guwahati, Assam 781017, India
e-mail: alexy.bhowmick@ieee.org

S. M. Hazarika
Biomimetic Robotics and Artificial Intelligence Lab, Department of Mechanical Engineering, Indian Institute of Technology Guwahati, Guwahati, Assam 781039, India
e-mail: s.m.hazarika@iitg.ac.in

S. Basu et al. (eds.), *Proceedings of International Conference on Frontiers in Computing and Systems*, Lecture Notes in Networks and Systems 404,
https://doi.org/10.1007/978-981-19-0105-8_9

applications such as autonomous driving, medical image analysis, and remote-sensed image analysis [1]. Semantic segmentation is challenging because it involves solving multiple unalike but correlated problems, namely—object detection, segmentation, recognition, and contextual reasoning in a unified framework.

U-Net [11] is considered the de-facto standard deep learning model for segmenting biomedical images. U-Net's structure is quite similar to *Fully Convolutional Networks* [9] and *SegNet* [1], comprising an encoder–decoder network. U-Net architecture is symmetric, comprising an *Encoder* and *Decoder* sub-network linked by skip connections. The encoder sub-network extracts spatial and contextual features from the input, while the decoder localizes objects and produces the final segmentation map. The U-Net architecture was proposed primarily for biomedical images. U-Net, though popular, leaves room for improvement and can be used for different datasets or domains. We studied the U-Net architecture and developed a novel architecture termed *Light U-Net*, an efficient version of U-Net that we demonstrate in outdoor scene datasets for semantic segmentation. We describe the architecture in Sect. 3.1.

Our main contributions include:

- A novel segmentation architecture *Light U-Net* which is an efficient version of U-Net [11] vis-à-vis the number of parameters and training time.
- We demonstrate data augmentation for augmenting small and limited semantic segmentation benchmarks.
- We experiment with two semantic segmentation datasets comprising of outdoor scenes and report *Light U-Net's* performance on outdoor scene segmentation.

2 Related Work

2.1 *The U-Net Segmentation Architecture*

U-Net [11] is a prominent encoder–decoder architecture that was built upon the *fully convolutional network* [9] for segmenting biomedical images. U-Net incorporates a *contracting* path (left side, *encoder*) that identifies contextual cues, and a symmetric *expansive* path (right side, *decoder*) that identifies precise localization. The contracting path (down-sampling sub-network) is comprised of repeated 3×3 convolutions succeeded by rectified linear activation unit (ReLU) layers and doubling of feature channels. The expansive path (up-sampling sub-network) up-samples the feature channels succeeded by a 2×2 transpose convolution. The cropped feature map from the corresponding contracting path is concatenated, and two 3×3 convolutions are succeeded by a ReLU. Lastly, a 1×1 convolution layer maps each 64-component feature vector to the appropriate number of classes.

U-Net became popular due to its multi-scale skip connections that are a distinctive aspect of the network and enable the architecture to recover the spatial cues lost during pooling (down-sampling). In all the four levels of the encoder–decoder architecture, the convolutional layer output, before the max-pool 2×2 operation in the encoder sub-network, is fed to the decoder sub-network. The output of the up-sampling operation in the corresponding decoder sub-network is concatenated with these feature maps which are then propagated to the subsequent layers. Notwithstanding excellent performance in multimodal medical image segmentation, U-Net architecture lacked in certain aspects which paved the way for several improvements to deal with semantic segmentation problems.

U-Net++ [16] is a new segmentation architecture that employs a sequence of dense, nested skip connections to join the encoder to the decoder sub-network. U-Net++ proved to be effective but computationally more costly than the original U-Net due to a large number of intermediate convolutions. Attention U-Net [12] proposed to utilize salient regions present in medical images by using attention gates integrated in the network. Several implementations of U-Net have since been proposed but not discussed due to space limitations. We attempt to design a lightweight variant of U-Net for meaningful labeling of outdoor scenes.

3 Proposed Method

In this section, we describe the motivation behind the proposed architecture and describe the proposed *Light U-Net* network architecture in Sect. 3.1. We describe the training of the model on the semantic segmentation datasets and data augmentation approach in Sect. 3.2.

3.1 Light U-Net: Proposed Network Architecture

The proposed *Light U-Net* is an *Encoder-Decoder* network with a final pixel-wise classification layer. As illustrated in Fig. 1, the *Encoder* network of Light U-Net comprises of five convolutional layers. There is a corresponding *Decoder* network which has corresponding five convolutional layers. The *Decoder* network output is propagated to a softmax function to generate independent pixel-wise class probabilities.

The *Encoder* is comprised of a series of two 3×3 convolution operations (unpadded) with a bank of filters, each succeeded by a ReLU nonlinearity to produce a feature map. This is followed by a 2×2 max-pooling (stride = 2, non-overlapping window) for down-sampling. This sequence is repeated five times in Light U-Net and involves doubling of feature channels after each down-sampling, as in original U-Net [11]. The additional contraction block here helps to capture higher semantic information from the image. The doubling of the filter size results in a large receptive field for every pixel in the feature map enabling the network to learn complex structures effectively. Lastly, the *Encoder* and the *Decoder* are connected by a sequence of two 3×3 convolution operations.

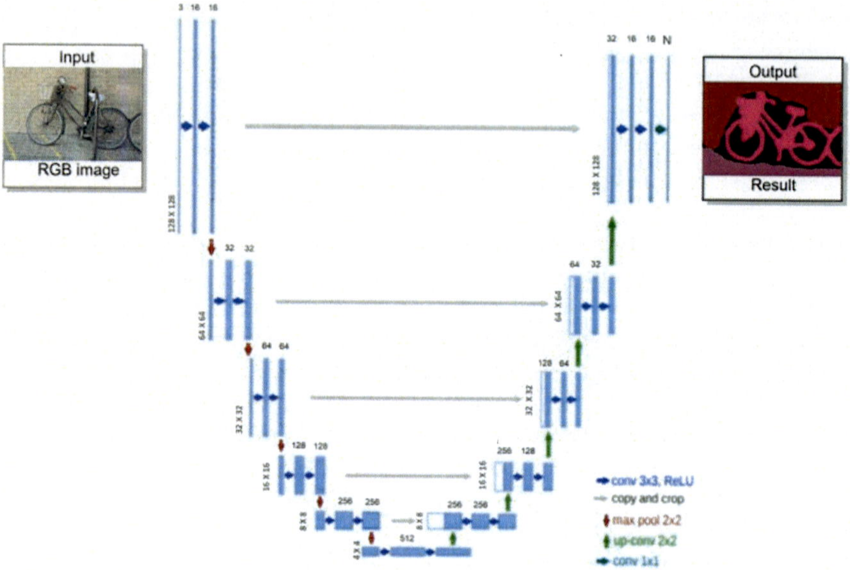

Fig. 1 Proposed *Light U-Net* architecture. We introduce *two* convolutional layers and a max-pooling layer at the contraction and expansion paths. The number of channels is denoted by the value at the top of the box. Each blue box denotes a multi-channel feature map, while a white box denotes a copied feature map. The arrows in the legend denote the different operations

On the contrary, as a first step, the corresponding *Decoder* up-samples its input feature map using a 2×2 transposed convolution ('up-convolution'), which divides the feature channels by two at each stage. This is followed by a series of two 3×3 convolution operations. Akin to the *contracting path*, the sequence of 2×2 up-convolution and two 3×3 convolution operations is replicated five times, which essentially halves the filter size in each stage. Lastly, the 1×1 convolution layer generates the final semantic segmentation of the input. The final convolutional layer uses a softmax activation function for independent pixel-wise classification, whose output is an N channel image of probabilities where N is the number of classes. As compared to the U-Net [11] model, Light U-Net architecture differs as follows:

1. Light U-Net trains on full outdoor scenes resized to 128×128 pixels and has an additional *contraction block*-2 convolution layers and a max-pooling layer each at the *Contraction* and the *Expansion* path to extract more complex features and impact the generalization accuracy.
2. To reduce the network complexity, we decreased the size of filters from 64 to 16 at the initial level convolution layers. The reduction of the number of parameters (to ≈ 7.5 M) resulted in a smaller-sized model compared to the original *U-Net*. The training time of the model is also considerably reduced.
3. We demonstrate the use of *Light U-Net* on outdoor scene data sets with geometric data augmentation, making it suitable for domains with small limited datasets.

3.2 Training

The proposed Light U-Net architecture is developed in Keras and TensorFlow 2.0.[1] As a result of data augmentation, our model performs well even with very few annotated dataset images. The training time is very reasonable: $\approx$ 1 h for *Camvid* [2], and $\approx$ 25 min for *Stanford Background* [6], on a PC with 2.5 GHz CPU, 8 GB RAM, and 4 GB GPU. The network is trained on images resized to 128×128 pixels and batch size of 41 for both training and testing. Because semantic segmentation involves classification of multiple classes, for our loss function, we use *categorical cross entropy*. We use the *Adam optimizer* (learning rate = 10^{-4}) in all cases. We evaluate our results in terms of standard evaluation metrics *Pixel accuracy* (PA) and *mean Intersection-over-Union* (mIoU), which are described in Sect. 4, and compared to state-of-the-art segmentation architectures in Tables 1 and 2.

3.2.1 Datasets

We assess *Light U-Net* on *two* small but challenging datasets: (i) CamVid, or the Cambridge-driving Labeled Video Database,[2] is a small road/driving scene dataset consisting of 701 labeled frames (720×960) captured in the day and dusk modes. It consists of five video sequences taken from a moving vehicle (four sequences shot in daylight, one shot in dusk). The image pixels are labeled with labels from the 32 semantic classes (*void, building, wall, tree, vegetation, fence*, etc.). We adopt the standard training/validation/testing split protocol (368/100/233 images) [1], and (ii) Stanford Background dataset (SBD)[3] has 715 images of urban, rural, and harbor scenes. The images are of 320×240 pixels depicting outdoor scenes and have no less than one foreground object. The dataset is made up of 8 classes. Each pixel is assigned two labels—a semantic class label (sky, tree, road, grass, water, etc.) and a geometric class label (sky, horizontal, and vertical). The dataset is generally partitioned into training 572 images (80%) and 143 testing images (20%) with fivefold cross-validation.

3.2.2 Data Augmentation

Data augmentation is a set of techniques that introduces desired invariances and robustness properties in the network when training data is limited. It is useful in fields where large datasets are still not available, e.g., optical flow, biomedical image processing, etc. By artificially inflating the data, data augmentation modifies the datasets to produce big data characteristics in a limited data domain. To accelerate the training process, we perform data warping augmentations including popular

[1] https://www.tensorflow.org/.

[2] http://mi.eng.cam.ac.uk/research/projects/VideoRec/CamVid/.

[3] http://dags.stanford.edu/projects/scenedataset.html.

Table 1 The table reports comparison with state-of-the-art methods in terms of mean intersection over union (mIoU) and per-pixel accuracy on the *Camvid* dataset

Segmentation models on Camvid			
Method [Reference]	mIoU (%)	Per-pixel (%)	Training time, # parameters (M)
ReSeg [14] (2016)	58.8	88.7	NA
SegNet [1] (2017)	60.10	90.4	NA
FC-DenseNet103 [7] (2017)	66.9	91.5	9.4 M
DeepLabV3Plus [3] (2018)	**82.9**	NA	NA
EDANet [8] (2019)	66.4	90.8	0.68 M
Proposed *Light U-Net*	70.06	**93.65**	≈ 62 mins, 7.5 M

The results that are in bold indicate best performance

Table 2 The table reports comparison with state-of-the-art methods in terms of mean intersection over union (mIoU) and per-pixel accuracy on the *Stanford Background* dataset

Segmentation models on stanford background			
Method [Reference]	mIoU (%)	Per-pixel (%)	Training time, # parameters (M)
Multiscale Net+CRF [5]	NA	81.4	2–5 days, ≈ 0.5 M
Recurrent CNNs [10]	NA	80.2	NA
U-Net [11]	47.51	76.00	≈ 26 mins, 31 M
PSP-Net [15]	74.09	NA	NA
ML-CRNN [4]	65.70	87.2	NA
Proposed *Light U-Net*	**79.84**	**94.31**	≈ 25 mins, 7.5 M

The results that are in bold indicate best performance

geometric transformations that preserve the labels and add them to the dataset. We adopted horizontal and vertical shift augmentation, horizontal and vertical flip augmentation, random rotation augmentation for training data to alleviate the overfitting problem. The *ImageDataGenerator* class in Keras[4] was used to implement the geometric augmentations. There is no agreement when it comes to ascertaining what proportion of original to final dataset size results in the best performance of the model [13]. We restrict our augmentations to the same size as the datasets. In our case, data augmentation helped the model generalize very well and removed bias evident in results.

[4] https://keras.io/.

4 Results and Analysis

To compare the quantitative performance of *Light U-Net* with various encoder–decoder variants, we employ two standard performance measures: *i) Pixel Accuracy* (PA), and *ii) mean Intersection-over-Union* (mIoU). For this section, we denote n_{ij} as the number of pixels belonging to class i (according to the ground truth), and were labeled as class j in prediction; n_{ij} is an element in the confusion matrix $\left(n_{ij}\right)$. Let t_i be the total number of pixels of class i (according to the ground truth), and $K \in \mathbb{N}$ be the number of classes. *Pixel Accuracy* (PA) is the proportion of pixels correctly classified in the dataset. Formally,

$$PA = \frac{\sum_{i=1}^{K} n_{ii}}{\sum_{i=1}^{K} t_i}. \tag{1}$$

Mean intersection over union (mIoU), or *Jaccard Index* (JI), is a sophisticated measure that computes the intersection over union between the predictions and ground truth, averaged over all the classes. Formally,

$$mIoU = \frac{1}{K} \cdot \sum_{i=1}^{K} \frac{n_{ii}}{t_i - n_{ii} + \sum_{j=1}^{K} n_{ji}}. \tag{2}$$

We show *Light U-Net's* application on the semantic segmentation task. The results are recorded after 120 and 70 training epochs for *Camvid* and *Stanford Background*, respectively. Beyond this, the model did not show significant further improvement. We select the iteration in which the pixel accuracy is highest on the validation set. The datasets were configured for fivefold cross-validation tests. We present the numerical results of our analysis in Tables 1 and 2.

The proposed *Light U-Net* achieves mIoU of 70.06% on Camvid and clearly outperforms recent segmentation models in terms of PA and mIoU, though it lags behind DeepLabV3Plus [3] in mIoU (see Table 1). On the Stanford dataset, Light U-Net achieves mIoU of 79.84% which is significantly better than PSP-Net [15], ML-CRNN [4], and original UNet (see Table 2). The results indeed show an increase in mIoU on both datasets as a result of data augmentation and improvements in Light U-Net.

The variation of training and validation accuracy and loss versus number of epochs for Camvid (Fig. 2a, b) and Stanford Background (Fig. 2c, d) are shown. Results on the validation set are relatively better than those on the training set which implies our model does not suffer from overfitting. The loss curves indicate a good fit and show a good learning rate. Figure 3 shows sharp segmentation results obtained. Light U-Net has room for improvement, as it is seen that misclassification of pixels happen when objects are close to each other or overlapping, e.g., *poles* in Fig. 2b results.

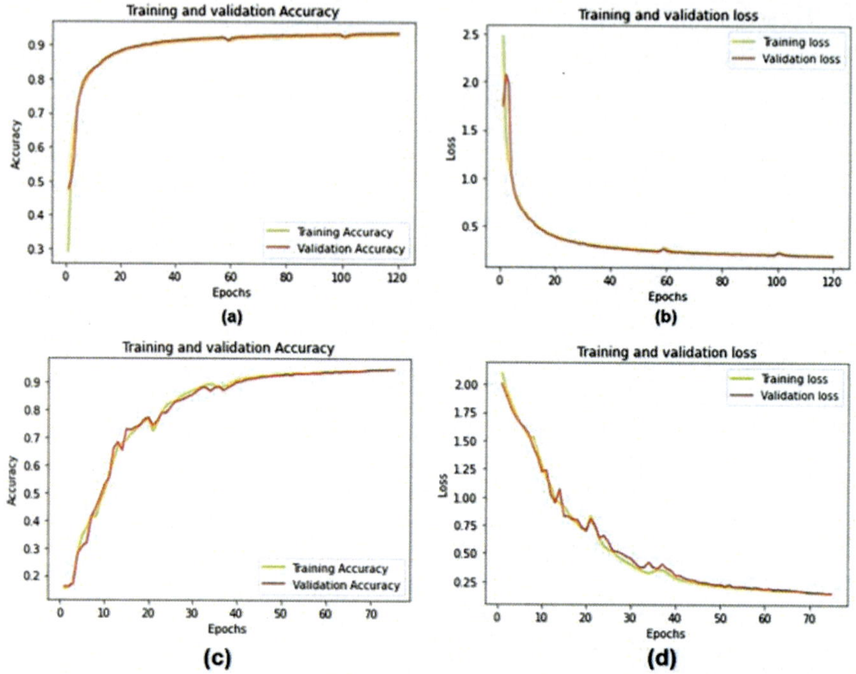

Fig. 2 The variation of accuracy and loss on training and validation sets versus number of epochs on Camvid dataset **a** and **b** and Stanford dataset (**c**, **d**)

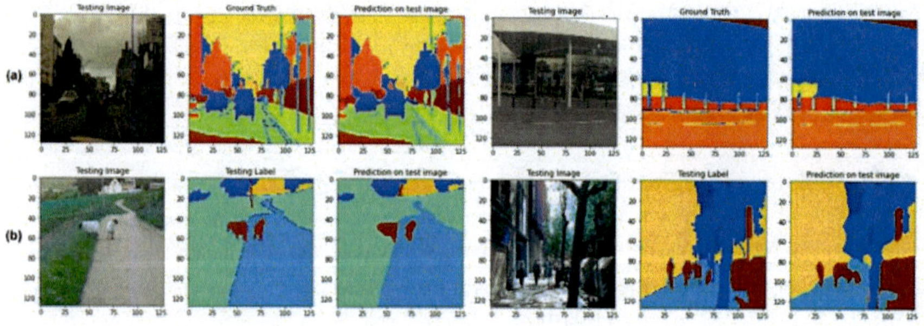

Fig. 3 Sample semantic segmentation results of proposed Light U-Net: **a** *first row*, two examples from Camvid and **b** *second row*, two examples from Stanford background

5 Conclusion

We presented *Light U-Net* an efficient version of the prominent U-Net architecture that performs very well on two standard semantic segmentation benchmarks. *Light U-Net* was motivated by the goal of designing a network with a limited parameter size for semantic segmentation of outdoor scenes. The proposed Light U-Net is significantly smaller in size and more efficient in terms of memory and computation time when compared to U-Net. In our training, the encoder–decoder network maintains a good generalization accuracy, and data augmentation enables the network to train with limited data. Since Light U-Net produced results on par with other state-of-the-art segmentation models, we believe it can be applied easily to other computer vision tasks and larger datasets.

References

1. Badrinarayanan V, Kendall A, Cipolla R (2017) SegNet: a deep convolutional encoder-decoder architecture for image segmentation. IEEE Trans Pattern Anal Mach Intell 39(12):2481–2495
2. Brostow GJ, Fauqueur J, Cipolla R (2009) Semantic object classes in video: a high-definition ground truth database. Pattern Recogn Lett 30(2):88–97
3. Chen LC, Zhu Y, Papandreou G, Schroff F, Adam H (2018) Encoder-decoder with Atrous separable convolution for semantic image segmentation. In: Ferrari V, Hebert M, Sminchisescu C, Weiss Y (eds) Computer vision—ECCV 2018. Lecture notes in computer science. Springer International Publishing, Cham, pp 833–851
4. Fan H, Mei X, Prokhorov D, Ling H (2018) Multi-level contextual RNNs with attention model for scene labeling. IEEE Trans Intell Transp Syst 19(11):3475–3485
5. Farabet C, Couprie C, Najman L, LeCun Y (2013) Learning hierarchical features for scene labeling. IEEE Trans Pattern Anal Mach Intelli 35(8):1915–1929. (Conference Name: IEEE Transactions on Pattern Analysis and Machine Intelligence)
6. Gould S, Fulton R, Koller D (2009) Decomposing a scene into geometric and semantically consistent regions. In: 2009 IEEE 12th international conference on computer vision. IEEE, pp 1–8
7. Jégou S, Drozdzal M, Vazquez D, Romero A, Bengio Y (2017) The one hundred layers Tiramisu: fully convolutional DenseNets for semantic segmentation. In: 2017 IEEE conference on computer vision and pattern recognition workshops (CVPRW), pp 1175–1183
8. Lo SY, Hang HM, Chan SW, Lin JJ (2019) Efficient dense modules of asymmetric convolution for real-time semantic segmentation. In: Proceedings of the ACM multimedia Asia, MMAsia '19. Association for Computing Machinery, New York, NY, USA, pp 1–6
9. Long J, Shelhamer E, Darrell T (2015) Fully convolutional networks for semantic segmentation. In: 2015 IEEE conference on computer vision and pattern recognition (CVPR), pp 3431–3440
10. Pinheiro PO, Collobert R (2014) Recurrent convolutional neural networks for scene labeling. In: Proceedings of the 31st international conference on international conference on machine learning, vol 32, ICML'14. JMLR.org, Beijing, China, pp I–82–I–90
11. Ronneberger O, Fischer P, Brox T (2015) U-Net: convolutional networks for biomedical image segmentation. In: Navab N, Hornegger J, Wells WM, Frangi AF (eds) Medical image computing and computer-assisted intervention—MICCAI 2015. Lecture notes in computer science. Springer International Publishing, Cham, pp 234–241
12. Schlemper J, Oktay O, Schaap M, Heinrich M, Kainz B, Glocker B, Rueckert D (2019) Attention gated networks: learning to leverage salient regions in medical images. Med Image Anal 53:197–207

13. Shorten C, Khoshgoftaar TM (2019) A survey on image data augmentation for deep learning. J Big Data 6(1):60
14. Visin F, Romero A, Cho K, Matteucci M, Ciccone M, Kastner K, Bengio Y, Courville A (2016) ReSeg: a recurrent neural network-based model for semantic segmentation. In: 2016 IEEE conference on computer vision and pattern recognition workshops (CVPRW), pp 426–433
15. Zhao H, Shi J, Qi X, Wang X, Jia J (2017) Pyramid scene parsing network. In: 2017 IEEE conference on computer vision and pattern recognition (CVPR). IEEE, pp 6230–6239. ISSN 1063-6919
16. Zhou Z, Siddiquee MMR, Tajbakhsh N, Liang J (2020) UNet++: redesigning skip connections to exploit multiscale features in image segmentation. IEEE Trans Med Imaging 39(6):1856–1867

Framework of Intelligent Transportation System: A Survey

Ratna Mandal, Ankita Mandal, Soumi Dutta, Munshi Yusuf Alam, Sujoy Saha, and Subrata Nandi

Abstract Intelligent transportation is indeed a prominent aspect of smart city development. Intelligent transportation targets to handle several issues by taking into consideration traffic or human mobility, sensory data, and geographical data generated in cities. It brings together the different concepts of urban sensing, data management and analytics along with various service providing mechanisms into a recurrent process for an unobtrusive and continuous improvement of an individual's transit experience and operations of the city transport authority. In this paper, we first introduce the concept of an intelligent transportation system, along with discussion about the overall framework and key challenges with respect to the perspective of each step of the framework. Second, we classify the data acquisition methods. Third, summarize the typical methodologies needed in intelligent transportation systems for knowledge fusion across heterogeneous sensory data. Finally, we give an outlook on the future of intelligent transportation systems, with suggestion about a few prospective topics of research which have remained unexplored in the research community so far.

Keywords Smart transportation · Intelligent sensing · Smartphone sensor · Survey

1 Introduction

Intelligent transportation which is an inevitable part of smart city is critical to implement in low-economic countries due to several socio-economic factors. Many of

R. Mandal (✉) · A. Mandal · S. Dutta
Institute of Engineering and management, Kolkata, India
e-mail: ratna.mandal.iem@gmail.com

S. Dutta
e-mail: soumi.dutta@iemcal.com

M. Yusuf Alam
Budge Budge Institute of Technology, Kolkata, India
e-mail: munshiyusufalam@bbit.edu.in

S. Saha · S. Nandi
National Institute of Technology, Durgapur, India

S. Basu et al. (eds.), *Proceedings of International Conference on Frontiers in Computing and Systems*, Lecture Notes in Networks and Systems 404,
https://doi.org/10.1007/978-981-19-0105-8_10

the big cities of low or middle economic countries are undergoing the rapid process of urbanization, which has modernized many people's lives and raises eloquent challenges toward environment, energy consumption, and transportation service. A detailed investigation on public transportation service reveals the precarious condition of transportation service in poor economic countries where public buses procure a significant portion of public transportation in urban cities and serve as a cheap and convenient mode of transport for all social classes. An intelligent transportation system should be adaptive enough with the rapid changes occurring due to urbanization process. This should be upgraded according to the need of citizens for comfortable mobility. Only government-owned transport is not plentiful to meet the demand of huge population of developing countries, and it promotes the privatization of public transportation sector, leading to an unhealthy transportation service. The government becomes helpless to control the situation and cannot penalize such activities only by implementing strict law and order. The government lacks sufficient funding assistance to provide a well-regulated transportation service. Pucher and Korattyswaroopam [39] say that passenger revenues are insufficient to cover overall operating costs and capital investment of public transport. Financial support for public transport is very tenuous, depending on annual budgetary appropriations for such developing countries. Tackling these challenges seemed to be impossible decades ago, given the prevalent anomalies in the complex structure of the cities. Nowadays, proliferation of advanced sensing and communication technology drives the establishment of smart transportation good management strategies to ensure comfort, availability, and security of travelers, tackling all the underlying challenges in the context of developing countries. The transport research community looks into the landmark identification from GPS traces. Research on intelligent transportation systems covers a wide range of activities like trajectory data collection [5–7], noise filtration, preprocessing, stay point detection, segmenting on spatiotemporal and semantic features, map matching, travel time estimation, traffic management, speed characterization, navigation, etc.

Sensing technologies and large-scale computation produce a variety of sensory data, which implies rich knowledge not only about the traffic behavior but also city roads, vehicle conditions, and city demography, having an extensive impact on the overall transportation service quality. These sensory data can help to capture intricate as well as effectual settings of cities, thereby formulating a better city transport planning for the time ahead. Opportunities of building more intelligent transportation motivate us to bring forward a broad urban sensing vision, aiming to unravel the knowledge gathered from sensory data collected from various urban ecosystems which are in large volumes and heterogenous in nature and finally aiming to implement this information for finding solution of major issues of transportation service.

There has been plenty of research in intelligent transportation systems based on mobility analysis of vehicles like traffic estimation, travel time estimation, crowd flow prediction, and public transport characterization. Many of these research works also focus on road context analysis like road surface monitoring, congestion detection, road accident detection, street light monitoring, and traffic light monitoring. The broad objective is to develop several smart transportation-related services like driving assistance, driver rating, off-line navigation, context-aware route planning,

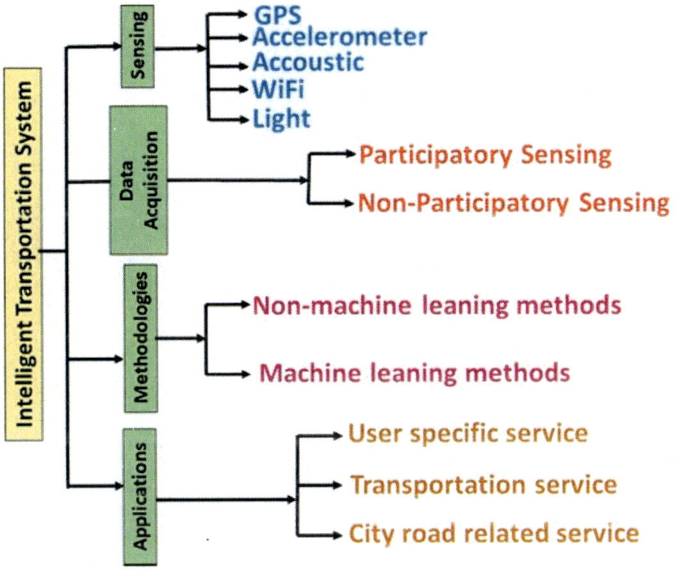

Fig. 1 Framework of existing literature on intelligent transportation service

comfortable public ridding. In this section, we have organized the general framework of the intelligent transportation system as shown in Fig. 1. Intelligent transportation is an advanced application aiming to provide innovative services related to various transport as well as traffic management modes, where smartphone of daily commuters and vehicles is accounted as an essential source of sensing, acquisition of mobility, and heterogeneous sensory data. The given framework addresses the crucial issues that cities come across (e.g., traffic congestion, irregular transportation service, improper travel time estimation) followed by data analysis and applications. Though, people have considerable contribution to transportation service but efficacy of sensing technology in cities of developing countries is still an obscure concept having various open queries. For example, what are the core research problems of intelligent transportation? What are the various sensing technology challenges in poor economic countries? What are the prominent applications in this domain?

In order to handle these issues, we fittingly coin the use of sensing in the intelligent transportation systems. The present paper gives the overall framework, fundamental research problems, methodologies, and respective applications . It will assist the research community in finer understanding and explore this research domain, enhancing quality research results and real systems which can gradually induce smarter transportation systems.

The rest of the paper has been organized as given. In Sect. 2, we introduce different sensors useful for smart transportation, presenting a general framework. Section 3 presents the data acquisition methodologies with key challenges in each method. Section 4 summarizes the key observations of overall survey, and Sect. 5, finally, concludes the paper and points out the few future research directions of the discussed theme.

2 Sensing

Smartphone sensing methods are poised to address their usability, controllability and management using inbuilt sensors in the phone. GPS sensor in the device is used to capture the mobility patterns and localize the objects, whereas IMU sensors, microphones, WiFi, light as shown in Fig. 2 are used to sense the context of specific anomalies in mobility pattern which defers from the original form. The human and vehicle mobility data and urban sensed data are methodically organized by some indexing composition which correspondingly incorporate spatiotemporal informations along with texts to support systematic data analytic. A detailed study on the implementation of sensing techniques for intelligent transportation system gives a clear view of different applications of sensor data captured from the user's smartphone, as shown in Fig. 1.

Based on the characteristics of mobile sensing architectures and exponential use of mobile devices in crowd sourcing, intelligent transportation services are categorized into different types of applications, which are (**A**) traffic management system, which utilizes smartphone sensors mounted over vehicles in order to assist applications related to traffic management, such as detection of congestion and rerouting the traffic, POI identification, and ETA calculation; (**B**) accident detection system, which uses sensor data for vehicle health monitoring; (**C**) driver assistance system to detect and classify driving behavior with the help of the inbuilt gyroscope,accelerometer, magnetometer and GPS of a smartphone; placed in a vehicle; and finally, (**D**) road monitoring system with a significant focus on the detection of road anomalies like the presence of potholes, broken road patch using accelerometer data along with GPS data. Both Mednis et al. [32] and Ghose et al. [19] developed various applications

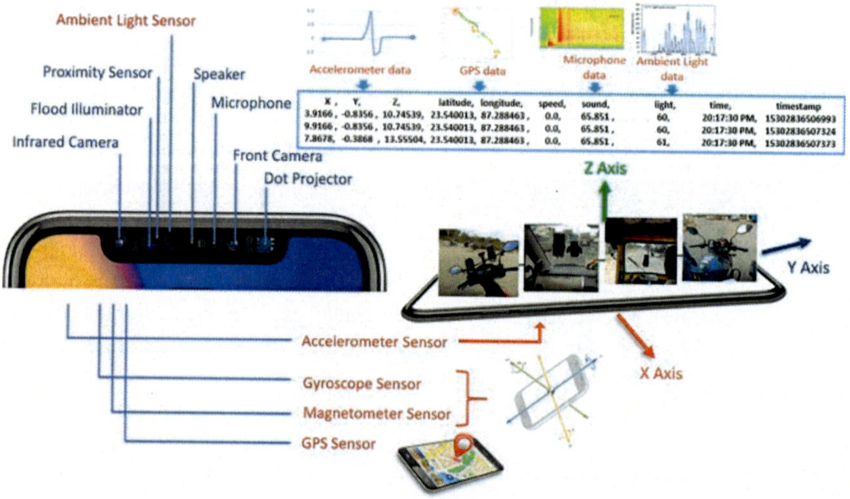

Fig. 2 Sensing objects through smartphone sensors

to monitor condition of roads for a smartphone that transmits the sensor data to a remote server and cautions drivers about the potholes present in the road. Mohan et al. [35] evolved a road and traffic monitoring system, named as Nericell, which also employs smartphone sensors to detect certain conditions. In addition to monitoring of driver behavior monitoring feature, Fazeen et al. [15] also introduced a road condition characterization and mapping feature to the existing system while using the accelerometer and GPS of a smartphone.

The study described above highlights the usage of different sensors in the domain of the transportation system. These sensors are the inbuilt sensors of today's smartphone like (a) *GPS* (b) *Accelerometer* (c) *Gyroscope* (d) *Microphone* (e) *WiFi*. Only, a detailed study on the usage of each sensor separately in the transportation system field proves its worthiness. Each of these sensors and their utilities and limitations is introduced in Table 1,

2.1 Sensing with GPS

With an abundance of GPS sensor-enabled smartphones, the mobility data of humans and vehicles has been a good source of information about peoples' lives and city operations systems in the context of urban spaces. Human mobility analysis is the best way to infer different operational zones like residential, educational, commercial, etc. Similarly, traffic mobility data is also essential for transit planners and traffic management to identify different functional zones and understand their temporal and spatial structures within the city to make better transportation plans. Better transportation service and its planning can be offered only by observing the mobility data collected by GPS sensors of citizens' smartphones and public vehicles, which are the end-user of this service. Their insights are most important to plan and improve transportation service. In recent years, the massive availability of smartphones and seamless network connectivity has steered efforts to collect data related to mobility of humans and working interest patterns within the considered data to develop different services and applications on location basis. According to Toch [48], the application of mobility trace analysis can be classified into different mobility trace analysis contexts where machine learning methods are used. Mobility traces are analyzed for trajectory modeling or user modeling or place modeling.

Trajectory modeling is intended for analyzing the set of spatiotemporal points to infer specific movement patterns like Veloso [52] do statistical analysis of spatiotemporal characteristics of taxi traces between pick up as well as drop-off nodes and vice versa. The author considers various potential features and applies the Bayesian classifier to predict the taxi flow depending on several constraints like the day of the week, weather condition, area, etc. Liu [28] has analyzed mobility patterns of taxi traces to estimate the probability of choosing bus as transport mode by people by regression model on extracted features. The choice of transportation mode depends on travel time, monetary cost, accessibility, and availability in order to improve bus routing. Lee [26] has classified trajectories based on its patterns to facilitate traffic prediction

Table 1 Sensing in intelligent transportation service

Existing works	Objective	Sensor used	Limitations
Toch [48]	Mobility trace analysis	GPS	Mobility pattern from GPS traces vary with individual driving pattern
Veloso [52]			
Liu [28]			
Lee [26]			
Meng [34]			
Kong [25]			
Tan [50]	Energy-aware navigation	Accelerometer	Accuracy of accelerometer sensor data depends on orientation of device. It is difficult to collect accelerometer data by crowdsourcing as there is no scope to train the city commuters about orientation of the probing device
RohitVerma [53]			
Selvathi [42]	Accident prevention and detection		
Fernandes [16]			
Megalingam [33]			
Faiz [14]			
Amin [6]			
Thompson [56]			
Ali [5]			
White [56]			
Vaitkus [51]	Driving behavior analysis		
Akhtar [1]			
Pholprasit [37]			
Saiprasert [41]			
Joubert [22]			
Alam [4]			
Predic [38]			
Ma [29]	Vehicle monitoring and classification		
Zaldivar et al. [58]			
Eriksson et al. [13]	Road surface monitoring		
Mednis et al. [32]			
Ghose et al. [19]			
Mohan et al. [35]			
Fazeen et al. [15]			
Jiagen (Jason) Ding [12]	Vehicle detection	Acoustic sensor	Collection of sound traces with fixed roadside infrastructure is expensive and creates difficulty while road repairing and maintenance. Collection of sound data from probing device or by crowsourcing is not feasible due to the prevalent chaos present in spatiotemporal mobility pattern of the city in low-economic countries
Ishida [21]			
Wang [55]			
Tyagi [49]	Traffic congestion estimation		
Gatto [18]			
Guerrero [20]			
Barbagli [7]			
Sen [44]			
Duan [12]	Driver activity detection	WiFi	Availability of WiFi access points is a major constraint in developing cities
Alam [2]	Street light monitoring	Light	Light sensor is effective only in night hours

and solve the traffic congestion problem. Meng [34] infers the human activity from his/her trajectory by using the dynamic Bayesian network (DBN) model. The model is trained with different activities and corresponds to different types of POIs, thus predicting human activity from its trajectory. Kong [25] identifies the region of long term traffic congestion in the city by analyzing spatiotemporal bus trajectory data to infer the anomalous segment in terms of anomaly index.

2.2 *Sensing with Accelerometer*

There is a high correlation between measurements of the inertia of a vehicle (which can be captured by accelerometer) and direct vehicle control signals like steering wheel angle, brake or acceleration pedal operations, etc. The output equation of the accelerometer is derived to relate linear and angular motion relative to a fixed inertial frame. This phenomenon of accelerometer sensor motivates the researchers to utilize accelerometer sensor in various context of intelligent transportation system.

(i) Energy-Aware Navigation Tan [45] has proposed a navigation model without using gyroscope, whereas Verma et al. [53] have contributed in proposing a navigation system with minimal GPS usage of GPS sensor.

(ii) Accident Prevention and Detection People have worked a lot in the prevention and detection of accidents by continuously monitoring the acceleration of three axes. Selvathi [42], Fernandes [16], Megalingam [33], Faiz [14], Kattukkaran [23], Amin [6], Thompson [47], Ali [5], White [56] have identified the abrupt change in acceleration reading for any of the three axes, to be the event of an accident and have proposed several safety measurements.

(iii) Driving Behavior Analysis Not only monitoring the movement of the vehicle but a good analysis of driving behavior can also avoid chances of accidents because aggressive driving is one of the prominent reasons behind most accident cases. Vaitkus [51] has classified rash driving and regular driving pattern using three-axis accelerometer data as input of the classifier algorithm. Apart from the driver's performance, the efficient driver profiling and driver rating system have different key factors like vehicle health and road anomalies, i.e., presence of bumps, potholes, broken road surface, and absence of streetlights. Being motivated by this fact, researchers worldwide have emphasized using the smartphone inertial sensor (i.e., accelerometer) to infer contextual driving behavior. Many works have focused on continuous monitoring of lateral, longitudinal, and vertical acceleration and identifying any abrupt changes in the 3-axial reading of vehicle-mounted accelerometer sensors or smartphones carried by passengers. Changes in axial reading in accelerometer capture different driving events like a lane change, sudden brake, high speed, crossing speed breakers, or potholes. These events clearly indicate aggressive driving or bad road condition, or poor health of the vehicle. A handful of work Akhtar [1], Pholprasit [37], Saiprasert [41], Joubert [22], Alam [4], Predic [38] considers city road conditions while profiling driving behavior or developing a driver rating system.

(iv) Vehicle Monitoring and Classification Vehicle classification is one major important part of an intelligent system in terms of several useful applications like efficient traffic operation, highway performance monitoring, Ma [29] measures the pavement vibration by the accelerometer which helps to detect the axle of wheels of vehicles and thus classifies the vehicles. Zaldivar et al. [58] use smartphones as an substitution mechanism for built in devices in vehicles in order to retrieve information from the electronic control unit (ECU) of the vehicle wirelessly.
(v) Road Surface Monitoring Among all intelligent transportation-related works, specific focus is also put on the road monitoring system. The condition of the road surface has a major impact on the overall transportation service. Again, researchers have relied on three axial accelerometer sensor data to capture road anomalies like the presence of potholes, bumps, or broken road surface [3] from specific driving signatures. Eriksson et al. [13] developed one of the initial systems to monitor road condition which detect as well as map road defects, such as potholes,with the help of GPS and accelerometer. Mednis et al. [32] and Ghose et al. [19] both developed various applications to monitor condition of roads for a smartphone that transmits the sensor data to a remote server and cautions drivers about the potholes present in the road. Mohan et al. [35] evolved a road and traffic monitoring system, named as Nericell, which also employs smartphone sensors to detect certain conditions. In addition to monitoring of driver behavior monitoring feature, Fazeen et al. [15] also introduced a road condition characterization and mapping feature to the existing system while using the accelerometer and GPS of a smartphone.

2.2.1 Sensing with Acoustic Sensor

Study of smartphone sensors in the field of intelligent transportation systems would be incomplete without mentioning the role of acoustic sensors. Application of acoustic sensors can be broadly classified in terms of two applications like (i) vehicle detection or classification and (ii) traffic congestion estimation.

(i) Vehicle Detection

Ding [11], Ishida [21], Wang [55] have worked on acoustic sensor data for vehicle or vehicle count detection purpose.

(ii) Traffic Congestion Estimation

Tyagi [49], Gatto [18], Guerrero-Ibáñez [20], Barbagli [7], Sen [44], have talked about the usage of acoustic sensor in traffic congestion estimation or traffic monitoring from acoustic sensor data; collected from statically deployed sensor nodes by roadside. Sen [43] presents technical approach for classification of the state of the considered traffic as congested or free flowing by estimating speed of the vehicle, from honks of the considered vehicles, which is an extensive feature with respect to disorderly conditions of roads.

2.3 *Sensing with WiFi*

Huge population burst and poor socio-economic factors are the prime barriers of developing countries building a smart transportation system. City demography is a vital context that sets the maximum limit on the city transportation system's performance. WiFi density is a useful feature to capture the densely populated locations, representing the busy POIs across the city. Intuitively, a well-annotated WiFi map of a city represents the demographic view of the city. Wu [57] assists individuals using smartphone in finding first rated WiFi hotspots from standard information like user experience with the same by motivating individuals to provide such information as input with an incentive scheme. WiFi sensor has a promising role in human activity recognition, and it has been applied to identify driver activity by Duan [12], which is a useful application for intelligent transportation system also.

2.4 *Sensing with Light*

Well functionality of city road transportation system immensely depends on street lights during night hours as traffic speed gets slower in the absence of sufficient street lights with a severe impact on the individual driving pattern of each driver. Investigation of accident cases reveals that most of the accidents occur due to insufficient street lights by the roadside during night hour, specifically in suburban areas of developing countries. Therefore, the perception of the smart city concept also involves the idea of an efficient street light monitoring system considering the optimal energy consumption and other resources. There is some research on monitoring street lights based on lighting parameters (illuminance, luminance, glare, etc.) using either digital video camera; mounted on the vehicle or sensor deployment techniques. A specific context of dark road patch at night hours also needs to be identified while characterizing the overall road context in an intelligent transportation system. [2] have classified street segments based on illumination levels combined with street surface features using smartphone sensors.

3 Data Acquisition

Data acquisition techniques in intelligent transportation systems involve collecting different types of data from different sources likewise: (A) sensible and continuous

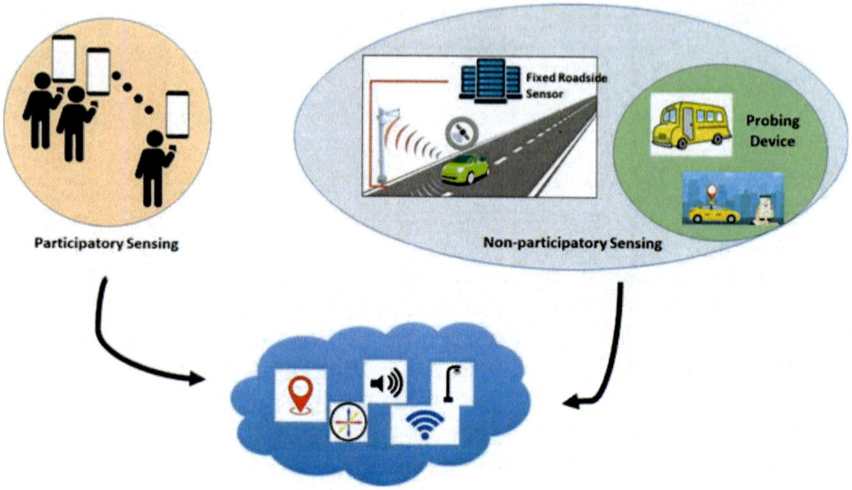

Fig. 3 Sensor data acquisition techniques

collection of sensor data for mobility and context analysis on a citywide scale; (B) reliable ground-truth data to ensure the performance of the proposed system of the intelligent transportation system; (C) geographical data (e.g., number of road intersections) from the digital map like Google Map, Bing Map); (D) survey data collected from citizens of several cities across the globe for better feasibility study based on their valuable responses. This survey also meets the requirement analysis for a particular ITS application by inferring the actual public demand. Monitoring the traffic flow on a road segment is easy, but continuous probing of citywide traffic is challenging as we do not have sensors on every road segment. The deployment process of sensors can be categorized in the way shown in Fig. 3.

3.1 Participatory Sensing

Success of low-cost smartphone sensor data collection highly depends on the voluntary participation of common people who are smartphone users. However, energy-hungry smartphone applications and the possibility of breaching privacy distract interested users from proactive participation. The advantage of this technique of data collection is its low cost. As the participants cannot be controlled during the process, the data is unstructured and is not balanced spatially or temporarily.

3.2 Statically Deployed Sensors

Sensors, to collect different information related to road condition, can be fixedly placed on as well as by the roadside like inductive loop detectors [10] image sensors [27] magnetic sensors [9] or acoustic sensor. The primary cost of installing a vehicle loop detector is huge. The supposition made with respect to orderly traffic on the basis of lanes will make the above said methodology inappropriate for disorderly road conditions specifically for low-economy countries. These techniques will fail to capture the variations in speed profile for different modes of transport and road characters where heavy trucks move with slow speed and two-wheelers move with high speed on the same road simultaneously. Finally, installing and maintaining static loop and magnetic detectors requires digging of roads badly, affecting the traffic condition because substitute routes are generally not available in cities of developing countries like India. Moreover, as part of urbanization, whenever a new construction will take place by roadside, the sensors need to be re-installed. Therefore, on-road fixed sensing infrastructure is an expensive for low-economy countries.

3.3 Sensors Deployed on Moving Object

Alternate way to gather urban sensor data is to utilize the objects that are already available in our urban space. Smartphone-enabled human and on-road vehicles can be a good source of sensor data. Instead of installing costly sensor nodes, an inexpensive way to collect sensor data for different intelligent transportation system applications is to utilize the inbuilt sensors of smartphones, either placed in vehicles or carried out by humans that move on the road. Thiagarajan et al. [46] utilizes GPS-enabled probe vehicles while Mohan [35] collects data by multiple sensor-enabled smartphones in vehicles. According to Sen [44] in India, the usage of GPS receivers modules within vehicles is considerably less. Only a few cab and state transport vehicles have GPS units enabled in them working actively in metro cities. In India, the usage of smartphone is also low compared to the usage of the mobile phones. Most of the users have phones with less specifications and are not very tech-savvy to contribute in participatory sensing activities. It is not easy to attract individuals having smartphones to participate in the sensing process even by including incentive approaches as communication costs are also involved in the process. RoadSoundSense shows trouble free installation as well as maintenance of different sensing units on roadside light poles with an average cost and detects congestion in chaotic traffic without hampering normal city traffic.

4 Key Observation

The above survey summarizes the most popular applications in intelligent transportation system; we first list the basic three categories of transportation scenarios: (1) user (i.e., driver or vehicle) specific applications [4, 8, 11]; (2) transportation service-related application (i.e., bus or taxi service [17, 24, 30, 40], route recommendation/guidance, travel time estimation [53], comfortable ridding [54], exploration of popular point of interests (POIs) [36, 50]; (3) city road-related applications like, context-aware speed profiling of city roads [31], road surface monitoring [2, 35]. People contributing in domain of intelligent transportation system have used mobility data to infer the stop and movement patterns of vehicles, crowd or have analyzed any point of interest from a specific perspectives. But, individual driving pattern, sparsity of expensive mobility data and limited scope of customizing the existing digital map-based services, motivates us to use other sensory data in addition with GPS logs to develop sophisticated transport services even for low-economic countries. Existing survey suggests the use of different sensory data for traffic estimation, congestion detection or road condition detection, but use of all crowdsourced sensory data together, to infer the city context of developing countries are still missing in the existing survey. Formulation of overall sophisticated transportation service requires evaluating a broad range of factors, such as traffic flow, human mobility, points of interest, and road network structures. These complex and anomalous factors turn transportation planning into a very challenging task. Traditionally, transport planners relied on labor-intensive surveys for their decision-making process. But, that information was obtained through the surveys, which might not be sufficient and timely enough. Recently, the widely available human mobility data generated in urban spaces have revealed the underlying problems of a city, specifically for suburban cities of low-economy countries, providing city planners with opportunities to formulate smart transport planning better.

Intelligent transportation system connects the steps of urban sensing, data collection, data analytic, and applications providing a regular process for continuous improvement of people's commuting experience. The success of this process depends on several socio-economic factors of a particular country as some of the existing applications of the intelligent transportation system requires highly expensive infrastructure support which is almost infeasible for low-economic countries due to severe shortfall of budget, huge population burst, lack of space and chaotic traffic condition. Therefore, existing ITS applications cannot be merely replicated in the case of low-economic countries. This particular observation from a literature study brings the motivation toward developing a low-cost, context-sensitive, customized, accurate ITS application for a suburban city of the low-economic country.

5 Conclusion

The intelligent transportation system aims to offer a modernized commuting experience to the citizens of urban space by making use of the crowd-sensed data which is generated in cities (e.g., flow of traffic, mobility of humans and geographical data). Public transportation services in developing countries get disrupted in several ways. Redressal of each of these problems is crucial for the smooth functioning of public transportation. The enormous amount of data that is being generated in urban spaces and advanced sensing, computing and communication technology have provided us with unprecedented opportunities to handle the challenges that city transport services faces. IoT-enabled smart transportation holds a promise to revolutionize the urban transportation and development, where quite a few techniques, such as collection of multi-modal sensory data, knowledge fusion across heterogeneous data sources, exploratory visualization for traffic, road, geographical and other contextual data, integration of algorithms of different domains, are yet to be explored. This paper discusses the concept, framework, and challenges of present state of intelligent transportation area, introduces the representative applications and techniques, and suggests a few research directions which can set develop systems for a better perception of original context and city demands, originated from changing lifestyles by utilizing different spatiotemporal data to improve the overall experience of public transportation service.

References

1. Akhtar N, Pandey K, Gupta S (2014) Mobile application for safe driving. In: 2014 fourth international conference on communication systems and network technologies. IEEE, pp 212–216
2. Alam MY, Imam S, Anurag H, Saha S, Nandi S, Saha M (2018) LiSense: monitoring city street lighting during night using Smartphone sensors. In: 2018 IEEE international conference on data mining workshops (ICDMW)
3. Alam MY, Nandi A, Kumar A, Saha S, Saha M, Nandi S, Chakraborty S (2020) Crowdsourcing from the true crowd: Device, vehicle, road-surface and driving independent road profiling from smartphone sensors. Pervasive Mobile Comput 61:101103
4. Alam M, Saurav S, Mandal R, Saha S, Nandi S, Chakraborty S (2017) A fair and effective driver rating system for developing regions. In: 2017 9th international conference on communication systems and networks (COMSNETS)
5. Ali HM, Alwan ZS (2017) Car accident detection and notification system using smartphone
6. Amin MS, Reaz MBI, Bhuiyan MAS, Nasir SS (2014) Kalman filtered GPS accelerometer-based accident detection and location system: a low-cost approach. Curr Sci 1548–1554
7. Barbagli B, Manes G, Facchini R, Manes A (2012) Acoustic sensor network for vehicle traffic monitoring. In: Proceedings of the 1st international conference on advances in vehicular systems, technologies and applications, pp 24–29
8. Castignani G, Derrmann T, Frank R, Engel T (2015) Driver behavior profiling using smartphones: A low-cost platform for driver monitoring. IEEE Intell Transp Syst Mag 7(1):91–102
9. Cheung SY, Coleri S, Dundar B, Ganesh S, Tan C-W, Varaiya P (2005) Traffic measurement and vehicle classification with single magnetic sensor. Transp Res Record 1917(1):173–181

10. Coifman B, Cassidy M (2002) Vehicle reidentification and travel time measurement on congested freeways. Transp Res Part A Policy Pract 36(10):899–917
11. Ding J, Cheung S-Y, Tan C-W, Varaiya P (2004) Signal processing of sensor node data for vehicle detection. In: Proceedings of the 7th international IEEE conference on intelligent transportation systems (IEEE Cat. No. 04TH8749). IEEE, pp 70–75
12. Duan S, Yu T, He J (2018) Widriver: driver activity recognition system based on WiFi CSI. Int J Wirel Inf Netw 25(2):146–156
13. Eriksson J, Girod L, Hull B, Newton R, Madden S, Balakrishnan H (2008) The pothole patrol: using a mobile sensor network for road surface monitoring. In: Proceedings of the 6th international conference on Mobile systems, applications, and services, pp 29–39
14. Faiz AB, Imteaj A, Chowdhury M (2015) Smart vehicle accident detection and alarming system using a smartphone. In 2015 international conference on computer and information engineering (ICCIE). IEEE, pp 66–69
15. Fazeen M, Gozick B, Dantu R, Bhukhiya M, González MC (2012) Safe driving using mobile phones. IEEE Trans Intell Transp Syst 13(3):1462–1468
16. Fernandes B, Alam M, Gomes V, Ferreira J, Oliveira A (2016) Automatic accident detection with multi-modal alert system implementation for its. Veh Commun 3:1–11
17. Ferris B, Watkins K, Borning A (2010) OneBusAway: results from providing real-time arrival information for public transit. In: Proceedings of the SIGCHI conference on human factors in computing systems, pp 1807–1816
18. Gatto RC, Forster CHQ (2020) Audio-based machine learning model for traffic congestion detection. IEEE Trans Intell Transp Syst
19. Ghose A, Biswas P, Bhaumik C, Sharma M, Pal A, Jha A (2012) Road condition monitoring and alert application: using in-vehicle smartphone as internet-connected sensor. In: 2012 IEEE international conference on pervasive computing and communications workshops. IEEE, pp 489–491
20. Guerrero-Ibáñez J, Zeadally S, Contreras-Castillo J (2018) Sensor technologies for intelligent transportation systems. Sensors 18(4):1212
21. Ishida S, Liu S, Mimura K, Tagashira S, Fukuda A (2016) Design of acoustic vehicle count system using DTW. In: Proceedings of ITS World Congress, pp 1–10
22. Joubert JW, De Beer JW, De Koker N (2016) Combining accelerometer data and contextual variables to evaluate the risk of driver behaviour. Transp Res Part F Traffic Psychol Behav 41:80–96
23. Kattukkaran N, George A, Haridas TM (2017) Intelligent accident detection and alert system for emergency medical assistance. In 2017 international conference on computer communication and informatics (ICCCI). IEEE, pp 1–6
24. Kisilevich S, Mansmann F, Keim D (2010) P-DBSCAN: a density based clustering algorithm for exploration and analysis of attractive areas using collections of geo-tagged photos. In: Proceedings of the 1st international conference and exhibition on computing for geospatial research & application, pp 1–4
25. Kong X, Song X, Xia F, Guo H, Wang J, Tolba A (2018) Lotad: long-term traffic anomaly detection based on crowdsourced bus trajectory data. World Wide Web 21(3):825–847
26. Lee J-G, Han J, Li X, Cheng H (2010) Mining discriminative patterns for classifying trajectories on road networks. IEEE Trans Knowl Data Eng 23(5):713–726
27. Li L, Chen L, Huang X, Huang J (2008) A traffic congestion estimation approach from video using time-spatial imagery. In: 2008 first international conference on intelligent networks and intelligent systems. IEEE, pp 465–469
28. Liu Y, Liu X, Yuan NJ, Duan L, Fu Y, Xiong H, Xu S, Wu J (2017) Intelligent bus routing with heterogeneous human mobility patterns. Knowl Inf Syst 50(2):383–415
29. Ma W, Xing D, McKee A, Bajwa R, Flores C, Fuller B, Varaiya P (2013) A wireless accelerometer-based automatic vehicle classification prototype system. IEEE Trans Intell Transp Syst 15(1):104–111

30. Mandal R, Agarwal N, Nandi S, Das P, Anvit A, Sanyal S, Saha S (2015) Stoppage pattern analysis of public bus GPS traces in developing regions. In: 2015 IEEE international conference on pervasive computing and communication workshops (PerCom workshops). IEEE, pp 276–279
31. Mandal R, Sonowal P, Kumar M, Saha S, Nandi S (2020) Roadspeedsense: context-aware speed profiling from smart-phone sensors. EAI Endorsed Trans Energy Web 7(28). https://doi.org/10.4108/eai.7-1-2020.162802
32. Mednis A, Strazdins G, Zviedris R, Kanonirs G, Selavo L (2011) Real time pothole detection using android smartphones with accelerometers. In: 2011 international conference on distributed computing in sensor systems and workshops (DCOSS). IEEE, pp 1–6
33. Megalingam RK, Nair RN, Prakhya SM (2010) Wireless vehicular accident detection and reporting system. In: 2010 international conference on mechanical and electrical technology. IEEE, pp 636–640
34. Meng C, Cui Y, He Q, Su L, Gao J (2017) Travel purpose inference with GPS trajectories, POIs, and geo-tagged social media data. In: 2017 IEEE international conference on big data (Big Data). IEEE, pp 1319–1324
35. Mohan P, Padmanabhan VN, Ramjee R (2008) Nericell: rich monitoring of road and traffic conditions using mobile smartphones. In: Proceedings of the 6th ACM conference on embedded network sensor systems, pp 323–336
36. Palma AT, Bogorny V, Kuijpers B, Alvares LO (2008) A clustering-based approach for discovering interesting places in trajectories. In: Proceedings of the 2008 ACM symposium on applied computing, pp 863–868
37. Pholprasit T, Choochaiwattana W, Saiprasert C (2015) A comparison of driving behaviour prediction algorithm using multi-sensory data on a smartphone. In: 2015 IEEE/ACIS 16th international conference on software engineering, artificial intelligence, networking and parallel/distributed computing (SNPD). IEEE, pp 1–6
38. Predic B, Stojanovic D (2012) Localized processing and analysis of accelerometer data in detecting traffic events and driver behaviour. J Univers Comput Sci 18(9):1152–1176
39. Pucher J, Korattyswaroopam N, Ittyerah N (2004) The crisis of public transport in india: overwhelming needs but limited resources. J Publ Transp 7(3):5
40. Rizvi A, Sclar E (2014) Implementing bus rapid transit: a tale of two Indian cities. Res Transp Econ 48:194–204
41. Saiprasert C, Thajchayapong S, Pholprasit T, Tanprasert C (2014) Driver behaviour profiling using smartphone sensory data in a V2I environment. In: 2014 international conference on connected vehicles and expo (ICCVE). IEEE, pp 552–557
42. Selvathi D, Pavithra P, Preethi T (2017) Intelligent transportation system for accident prevention and detection. In: 2017 international conference on intelligent computing and control systems (ICICCS). IEEE, pp 442–446
43. Sen R, Raman B, Sharma P (2010) Horn-ok-please. In: Proceedings of the 8th international conference on Mobile systems, applications, and services, pp 137–150
44. Sen R, Siriah P, Raman B (2011) Roadsoundsense: acoustic sensing based road congestion monitoring in developing regions. In: 2011 8th annual IEEE communications society conference on sensor, mesh and ad hoc communications and networks. IEEE, pp 125–133
45. Tan C-W, Park S (2005) Design of accelerometer-based inertial navigation systems. IEEE Trans Instrum Measur 54(6):2520–2530
46. Thiagarajan A, Ravindranath L, LaCurts K, Madden S, Balakrishnan H, Toledo S, Eriksson J (2009) VTrack: accurate, energy-aware road traffic delay estimation using mobile phones. In: Proceedings of the 7th ACM conference on embedded networked sensor systems, pp 85–98
47. Thompson C, White J, Dougherty B, Albright A, Schmidt DC (2010) Using smartphones to detect car accidents and provide situational awareness to emergency responders. In: International conference on mobile wireless middleware, operating systems, and applications. Springer, Berlin, pp 29–42
48. Toch E, Lerner B, Ben-Zion E, Ben-Gal I (2019) Analyzing large-scale human mobility data: a survey of machine learning methods and applications. Knowl Inf Syst 58(3):501–523

49. Tyagi V, Kalyanaraman S, Krishnapuram R (2012) Vehicular traffic density state estimation based on cumulative road acoustics. IEEE Trans Intell Transp Syst 13(3):1156–1166
50. Uddin MR, Ravishankar C, Tsotras VJ (2011) Finding regions of interest from trajectory data. In: 2011 IEEE 12th international conference on mobile data management, vol 1. IEEE, pp 39–48
51. Vaitkus V, Lengvenis P, Žylius G (2014) Driving style classification using long-term accelerometer information. IEEE, pp 641–644
52. Veloso M, Phithakkitnukoon S, Bento C (2011) Urban mobility study using taxi traces. In: Proceedings of the 2011 international workshop on Trajectory data mining and analysis, pp 23–30
53. Verma R, Shrivastava A, Mitra B, Saha S, Ganguly N, Nandi S, Chakraborty S (2016) UrbanEye: an outdoor localization system for public transport. In: IEEE INFOCOM 2016-the 35th annual IEEE international conference on computer communications. IEEE, pp 1–9
54. Verma R, Ghosh S, Saketh M, Ganguly N, Mitra B, Chakraborty S (2018) Comfride: a smartphone based system for comfortable public transport recommendation. In: Proceedings of the 12th ACM conference on recommender systems, pp 181–189
55. Wang K, Wang R, Feng Y, Zhang H, Huang Q, Jin Y, Zhang Y (2014) Vehicle recognition in acoustic sensor networks via sparse representation. In: 2014 IEEE international conference on multimedia and expo workshops (ICMEW). IEEE, pp 1–4
56. White J, Thompson C, Turner H, Dougherty B, Schmidt DC (2011) Wreckwatch: automatic traffic accident detection and notification with smartphones. Mobile Netw Appl 16(3):285
57. Wu F-J, Luo T (2014) WiFiScout: a crowdsensing WiFi advisory system with gamification-based incentive. In: 2014 IEEE 11th international conference on mobile ad hoc and sensor systems. IEEE, pp 533–534
58. Zaldivar J, Calafate CT, Cano JC, Manzoni P (2011) Providing accident detection in vehicular networks through OBD-II devices and Android-based smartphones. In: 2011 IEEE 36th conference on local computer networks. IEEE, pp 813–819

Convolutional Neural Networks-Based VQA Model

Himanshu Sharma and Anand Singh Jalal

Abstract In visual question answering (VQA) task, we generally use a convolutional neural network (CNN) to extract image features, a recurrent neural network (RNN) for question representation. Instead of using a RNN for language representation, we use a CNN for question representation. The advantage of using a CNN for question representation is that CNN is more effective in capturing image and question words interactions which are not expressed by a RNN. Thus, in this paper, we have used three CNNs for VQA task. One CNN extracts visual features, second CNN for extracting question features, and third CNN for combining both extracted feature vectors. Further, we have employed a softmax layer to generate answer for a given question. The proposed VQA model is evaluated on DAQUAR, COCO-QA, and VQA2.0 datasets.

Keywords CNN · RNN · DAQUAR · COCO-QA · VQA2.0

1 Introduction

In visual question answering (VQA), a question is asked related to an image in a natural language, and an answer is generated as an output in that language. There exist many challenges in VQA such as representation of language, object recognition, question-related reasoning, text extracting contained in images and counting. Apart from these problems, there may be many complex problems like answering the question related to the relationship between the objects (What is between the computer and the man?) and problems that may require common reasoning or world knowledge (Why is the boy sad?). An efficient VQA system must handle problems related to computer vision, language representations and have the ability to answer problems related to common sense reasoning.

H. Sharma (✉) · A. S. Jalal
Department of Computer Engineering and Applications, GLA University, Mathura 281406, India
e-mail: himanshu.sharma@gla.ac.in

A. S. Jalal
e-mail: asjalal@gla.ac.in

S. Basu et al. (eds.), *Proceedings of International Conference on Frontiers in Computing and Systems*, Lecture Notes in Networks and Systems 404,
https://doi.org/10.1007/978-981-19-0105-8_11

There are many possible applications of a VQA system in daily life. A VQA system can be very helpful identifying an object, organization of huge amount of unstructured visual data, helping a young child in learning new activities, and utilizing data available on Internet and summarizing data coming from a surveillance feed for an analyst.

Most of the recent state-of-the-art models used convolutional neural network (CNN) for taking out image features and long short-term memory (LSTM) for question representation. In this paper, we apply deep convolutional neural networks (CNNs) to perform the VQA task. The proposed CNN architecture is trained of an image, a question and an answer. The proposed model can answer open-ended questions that have answers of one or more words. The proposed model fuses both the image encoding and a question encoding to produce the answer depended on the question and image.

The remaining part of this paper is organized as follows. In Sect. 2, various VQA models are discussed. Section 3 describes the framework of the proposed model. In Sect. 4, both qualitative and quantitative results are discussed. In Sect. 5, we present promising prospects and future directions in VQA.

2 Related Work

There exists many VQA algorithms that make use of attention mechanism. Anderson et al. [1] used both top-down and bottom-up approach to perform the task of image captioning and VQA. Yang et al. [2] generate multiple attention layers on an image to answer the questions. These attention layers are stacked on one another and represent the multiple levels of reasoning over the image content to answer the question. Lu et al. [3] use attention on both question word and image regions and named it co-attention maps, for filling the space between image and question feature vectors.

Ilievski et al. [4] suggested an adversarial self-learning-based generative attention model. Kim et al. [5] utilized bilinear attention networks (BANs) which compute bilinear attention distributions for using seamless information between language and vision. Osman et al. [6] proposed recursive attention units for producing soft attention for visual question answering task. Ma et al. [7] utilized the concept of memory-augmented neural networks for generating answers that occur rarely in training set more accurately. Nguyen et al. [8] gave a novel model which is symmetric in nature between image and question representations, where image and question word were attended on each other. Su et al. [9] presented the idea of incorporating human knowledge in structured form and visual features in memory networks for VQA task. Yu et al. [10] used a novel approach for the fusion of both image vector and question vector. For improving the convergence rate of MLB, Yu et al. [10] fused feature vector by matrix factorization method and named it as multimodal factorized bilinear (MFB) pooling method by reducing the parameters required. Singh et al. [11] used top-down attention mechanism which is question-embedding guided on the image. Teney et al. [12] suggested dome tips for VQA task such as bottom-up attention for image features, sigmoid function for output, tanh-gated activation, soft training, and many more.

Chen et al. [13] proposed question-guided attention by using CNN configures on attention mechanism. Wu et al. [14] utilized image attributes and outside knowledge to generate captions for an image and visual question answering task. Wu et al. [15] used outside knowledge and commonsense reasoning to answer question on related to an image. Li et al. [16] suggested updating the question encoding by choosing image regions frequently related to the words in a question and thus predict the accurate answer. Kafle et al. [17] used Bayesian structure to guess the type of answer a question can have. Ma et al. [18] used CNNs for visual question answering task. Lu et al. [19] presented a novel way to fuse image region and question vectors by a multimodal multiplicative feature embedding approach. Wu et al. [20] discussed an approach of including high-level semantic concepts for both captioning of images and visual question answering task. Noh et al. [21] used dynamic parameters with CNN, and the weight of these parameters is determined with the help of questions. Ren et al. [22] used visual embeddings and neural networks for visual question answering task with detecting objects or segmenting an image. Song et al. [23] used a model based on cubic visual attention (CVA) by applying a spatial attention and a new channel on object regions to enhance VQA task. Malinowski et al. [24] gave extra view point for VQA task by evaluating the amount of information is included simply in the question component for which they gave a novel human baseline. Sharma et al. [25] use GNN for representing the relationship between the objects and image regions. Sharma et al. [26] use external knowledge for image captioning task which can be further used VQA task in future.

3 Proposed Model

As demonstrated in Fig. 1, the proposed VQA model contains of three independent convolutional neural networks: One of CNN is for extracting image features, one CNN for question representation, and one of the CNN for fusing both the feature vectors obtained above. The fused vector obtained is given as input to the softmax layer to produce the answer.

For extracting image features, we use VGGNet [18]. The image feature map is represented by $\boldsymbol{I_v}$ from the input image I. We extracted features from last pooling layer thus maintaining the spatial input image information. First, the rescaling of image is performed and maintained to 448x448, and then, the image feature vector is taken out from final pooling layer of the VGGNet. The dimension of the pooling layer is $512 \times 14 \times 14$. The numbers of regions in the image are 14x14, and the feature vector is of sizes 512 for every area. Every feature map in $\boldsymbol{I_v}$ contains 32×32 pixel regions corresponding to input image.

Every word in a question is encoded by word embedding. Many convolutional layers and max-poling layers are applied to produce a question representation. Both these layers establish the relationship between the each pair of word in a given

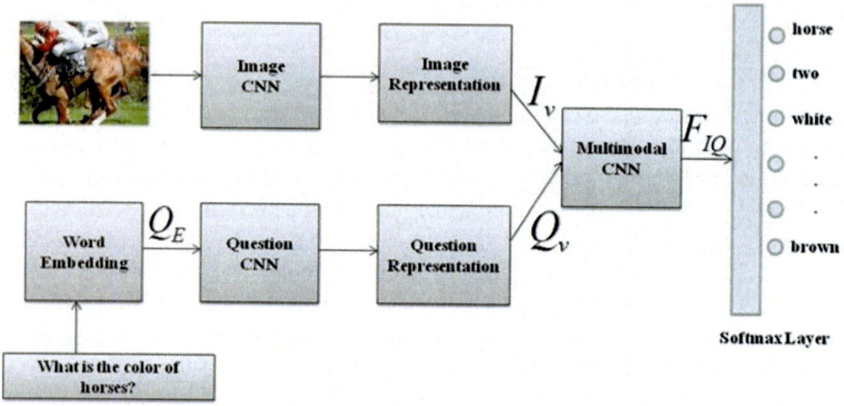

Fig. 1 Framework of the proposed model

question. Using many convolutional and max-pooling layers, we obtain the whole question representation $\boldsymbol{Q_v}$. We use CNN for a question representation as CNN is capable to express the interaction between image and question words more effectively which is not captured well by LSTMS.

We now use the third multimodal CNN to merge both the image vector and question vector to produce their combined representation $\boldsymbol{F_{IQ}}$, to generate the answer. This joint multimodal representation is then passed to a softmax layer for generating answer for a given question.

4 Results

The proposed VQA model is evaluated on three publically available datasets: DAQUAR [27], COCO-QA [22], and VQA2.0 [28]. The results are evaluated both qualitatively and quantitatively. Figures 2, 3, and 4 show results qualitatively. Tables 1, 2, and 3 compare the propose model with state-of-the-art VQA approaches.

5 Conclusion

In this paper, we have used three CNNs for VQA task. One CNN extracts visual features, second CNN for extracting question features, and third CNN for combining both extracted feature vectors. The advantage of using a CNN for question representation is that CNN is more effective in capturing image and question words interactions which are not expressed by a RNN. Further, we have employed a softmax layer to generate answer for a given question. In future, external knowledge base can be

Q: What is the object in front of sofa
A: Table

Q: How many windows are there?
A : 2

Fig. 2 Answers generated by proposed model on DAQUAR [27] dataset

Q: What is color of computer?
A: white

Q: How many women are talking on phone?
A: 2

Fig. 3 Answers generated by proposed model on COCO-QA [25] dataset

used to achieve better results. Use of semantic relationship between objects can also improve the accuracy of existing VQA models.

Q: Is the color of horse white?
A: Yes

Q: What is the color of flower pot?
A: Green

Fig. 4 Answers generated by proposed model on VQA2.0 [28] dataset

Table 1 Comparison of VQA models on VQA2.0 [28] dataset

Model	Val-Acc	Test-std Acc	Test-dev Acc
VQA_Model [1]	63.2	70.34	–
VQA_Model [4]	60.8	64.18	64.01
VQA_Model [5]	–	70.35	70.04
VQA_Model [6]	–	66.85	66.45
VQA_Model [7]	–	62.1	–
VQA_Model [8]	62.94	66.97	66.87
VQA_Model [9]	–	–	66.67
VQA_Model [10]	–	68.02	–
VQA_Model. [11]	64.78	–	–
VQA_Model [12]	63.37	–	–
Proposed Model	64.92	70.01	68.98

Table 2 Comparison of VQA models on COCO-QA [22] dataset

Model	Acc	WUPS 0.9	WUPS 0.0
VQA_Model [13]	58.1	68.44	89.85
VQA_Model [16]	62.5	72.58	91.62
VQA_Model [19]	66.49	76.15	92.29
VQA_Model [21]	61.19	70.84	90.61
VQA_Model [22]	55.09	67.9	89.52
VQA_Model [23]	67.51	76.7	92.41
VQA_Model [2]	61.6	71.6	90.9
VQA_Model [3]	65.4	75.1	92
VQA_Model [14]	70.98	78.35	92.87
Proposed Model	66.92	78.49	91.98

Table 3 Comparison of VQA models on DAQUAR. [27] dataset

Model	Acc	WUPS 0.9	WUPS 0.0
VQA_Model [24]	19.43	25.28	62
VQA_Model [18]	23.4	25.59	62.95
VQA_Model [20]	24.27	30.41	62.29
VQA_Model [13]	25.37	31.35	65.89
VQA_Model [17]	28.96	34.74	67.33
VQA_Model [21]	28.98	34.8	67.81
VQA_Model [15]	29.16	35.3	68.66
VQA_Model [2]	29.3	35.1	68.6
Proposed Model	29.2	34.34	67.69

References

1. Anderson P, He X, Buehler C, Teney D, Johnson M, Gould S, Zhang L (2018) Bottom-up and top-down attention for image captioning and visual question answering. In Proceedings of the IEEE conference on computer vision and pattern recognition, pp 6077–6086
2. Yang Z, He X, Gao J, Deng L, Smola A (2016) Stacked attention networks for image question answering. In Proceedings of the IEEE conference on computer vision and pattern recognition, pp 21–29
3. Lu J, Yang J, Batra D, Parikh D (2016) Hierarchical question-image co-attention for visual question answering. In Advances in neural information processing systems, pp 289–297
4. Ilievski I, Feng J (2017) Generative attention model with adversarial self-learning for visual question answering. In Proceedings of the on Thematic Workshops of ACM Multimedia 2017, pp 415–423
5. Kim JH, Jun J, Zhang BT (2018) Bilinear attention networks. In Advances in neural information processing systems, pp 1564–1574
6. Osman A, Samek W (2019) DRAU: dual recurrent attention units for visual question answering. Comput Vis Image Underst 185:24–30
7. Zhang D, Cao R, Wu S (2019) Information fusion in visual question answering: a survey. Inf Fusion 52:268–280
8. Nguyen DK, Okatani T (2018) Improved fusion of visual and language representations by dense symmetric co-attention for visual question answering. In Proceedings of the IEEE conference on computer vision and pattern recognition, pp 6087–6096
9. Su Z, Zhu C, Dong Y, Cai D, Chen Y, Li J (2018) Learning visual knowledge memory networks for visual question answering. In Proceedings of the IEEE conference on computer vision and pattern recognition, pp 7736–7745
10. Yu Z, Yu J, Xiang C, Fan J, Tao D (2018) Beyond bilinear: Generalized multimodal factorized high-order pooling for visual question answering. IEEE Trans Neural Netw Learn Syst 29(12):5947–5959
11. Singh J, Ying V, Nutkiewicz A (2018) Attention on attention: architectures for visual question answering (VQA). arXiv preprint arXiv:1803.07724
12. Teney D, Anderson P, He X, Van Den Hengel A (2018) Tips and tricks for visual question answering: Learnings from the 2017 challenge. In Proceedings of the IEEE conference on computer vision and pattern recognition, pp 4223–4232
13. Chen K, Wang J, Chen LC, Gao H, Xu W, Nevatia R (2015) Abc-cnn: an attention based convolutional neural network for visual question answering. arXiv preprint arXiv:1511.05960
14. Wu Q, Shen C, Wang P, Dick A, van den Hengel A (2017) Image captioning and visual question answering based on attributes and external knowledge. IEEE Trans Pattern Anal Mach Intell 40(6):1367–1381

15. Wu Q, Wang P, Shen C, Dick A, Van Den Hengel A (2016) Ask me anything: free-form visual question answering based on knowledge from external sources. In Proceedings of the IEEE conference on computer vision and pattern recognition, pp 4622–4630
16. Li R, Jia J (2016) Visual question answering with question representation update (qru). In Advances in neural information processing systems, pp 4655–4663
17. Kafle K, Kanan C (2016) Answer-type prediction for visual question answering. In Proceedings of the IEEE conference on computer vision and pattern recognition, pp 4976–4984
18. Ma L, Lu Z, Li H (2016) Learning to answer questions from image using convolutional neural network. In Thirtieth AAAI conference on artificial intelligence
19. Lu P, Li H, Zhang W, Wang J, Wang X (2018) Co-attending free-form regions and detections with multi-modal multiplicative feature embedding for visual question answering. In Thirty-Second AAAI conference on artificial intelligence
20. Wu Q, Shen C, Liu L, Dick A, Van Den Hengel A (2016) What value do explicit high level concepts have in vision to language problems? In Proceedings of the IEEE conference on computer vision and pattern recognition, pp 203–212
21. Noh H, Hongsuck Seo P, Han B (2016) Image question answering using convolutional neural network with dynamic parameter prediction. In Proceedings of the IEEE conference on computer vision and pattern recognition, pp 30–38
22. Ren M, Kiros R, Zemel R (2015) Exploring models and data for image question answering. In Advances in neural information processing systems, pp 2953–2961
23. Song J, Zeng P, Gao L, Shen HT (2018) From pixels to objects: cubic visual attention for visual question answering. In IJCAI, pp. 906–912
24. Malinowski M, Rohrbach M, Fritz M (2015) Ask your neurons: a neural-based approach to answering questions about images. In Proceedings of the IEEE international conference on computer vision, pp 1–9
25. Sharma H, Jalal AS (2021) Visual question answering model based on graph neural network and contextual attention. Image Vision Comput 110:104165
26. Sharma H, Jalal AS (2020) Incorporating external knowledge for image captioning using CNN and LSTM. Mod Phys Lett B 34(28):2050315
27. Malinowski M, Fritz M (2014) A multi-world approach to question answering about real-world scenes based on uncertain input. In Advances in neural information processing systems, pp 1682–1690
28. Antol S, Agrawal A, Lu J, Mitchell M, Batra D, Lawrence Zitnick C, Parikh D, (2015) Vqa: visual question answering. In Proceedings of the IEEE international conference on computer vision, pp 2425–2433

Signal Processing

A Deep Convolution Neural Networks Framework for Analyzing Electroencephalography Signals in Neuromarketing

Shrasti Vyas and Ayan Seal

Abstract Neuromarketing is the field of neuroscience that potentially infers to gain knowledge of customers' choice by analyzing the physiological signals. Here, the motive is to maintain the sustainability among traditional market research, which depends upon the preference of the consumer explicitly, on the other side neuromarketing, which shows the preference implicitly. Research work in this fastly emerging field is highly demanded due to its inherent potential. But, due to the paucity of sophisticated data-mining approaches for predicting and classifying consumers' choices, it is not yet touched a gruntled level. Therefore, in this work, a deep convolutional neural network (DCNN) is proposed to understand consumer choice using electroencephalogram (EEG) signals which were recorded while consumers were browsing through various E-commerce products. The arbitrary time-domain signals are transformed into two-dimensional images, which are then used for the classification of the EEG signals using DCNN. The score-level fusion is used to examine the model performance and got 81% accuracy which surpasses the existing approaches.

Keywords Electroencephalography · Neuromarketing · Score-level fusion

1 Introduction

Neuromarketing is the combination of three concepts—marketing, neuroscience, and psychology. It is the latest technique to understand consumers' preferences eventually led to a greater understanding of human cognition and integration of biological and social sciences. A wide range of traditional market strategies exists which include questionnaires, verbal, and face-to-face communication, etc. But, these are not as

S. Vyas (✉) · A. Seal
PDPM Indian Institute of Information Technology, Design and Manufacturing, Jabalpur, Madhya Pradesh 482005, India
e-mail: 1911015@iiitdmj.ac.in

A. Seal
e-mail: ayan@iiitdmj.ac.in

S. Basu et al. (eds.), *Proceedings of International Conference on Frontiers in Computing and Systems*, Lecture Notes in Networks and Systems 404,
https://doi.org/10.1007/978-981-19-0105-8_12

efficacious as psychological methods because these traditional approaches can be easily distracted by any of the environmental or emotional factors. So, to rely on effective results, neuromarketing provides an insight into what actually consumer feels and what he/she wants. This can also help to enhance the products before real set out in the market, which eventually increases the sale and gain to the production unit.

A recent study reveals that market strategies inclined toward different neurophysiological methods for analyzing the behavior and preference of consumers to study several aspects of marketing [1]. This paramount growth has been noted because of advancements in technological instruments like electroencephalography (EEG), functional magnetic resonance imaging (fMRI), magnetoencephalography (MEG), and some other neurotechnologies, which are often utilized in the field of neuromarketing. The fMRI detects the changes related to blood flow by measuring brain activity. But, due to its additional costs, some portability issues with apex maintenance restraint the use of fMRI in that applications. The MEG, a non-invasive method to investigate the brain's neuronal activity, also costs high with technical complexities limited reach to subcortical areas for brain imaging, eventually not an efficient approach to be preferred to build neuromarketing systems.

For restraining the limitations of existing approaches, here, EEG is used which is a non-invasive inexpensive method offering high temporal resolution and is cheaper when compared with techniques like fMRI and have fewer electrodes (say 14). It has easy handling with wireless connectivity, and the maintenance cost is also low. Inspired by the above research, here, an EEG authorized system is developed that can easily supersede costly setups of nowadays neuromarketing system. The major contributions of the proposed work are as follows-

- A novel deep convolution neural network (DCNN) framework is introduced for the classification of consumers' choices in terms of 'like' and 'dislike.' The EEG recordings were recorded for 4 s using 14 channels. The deployment of the proposed method in practical terms is efficient, due to the use of short EEG signal recordings.
- It involves 1045 samples and acquires 81% accuracy, which surpasses the result from other nearly similar DCNN-based approaches. Furthermore, because of the accuracy paradox, some other quantitative classification metrics like individual score, precision, recall, and f1-score are also taken into consideration while collating with the results of this proposed method with some of the state-of-the-art methods.
- Moreover, score-level fusion is also considered for validating the model performances. Late fusion at score level is employed to classify signal images in terms of two different classes.

The paper is organized as follows: An overview of the existing research is provided in Sect. 2. Section 3 illustrates the proposed DCNN. The results of the proposed method along with other state-of-the-art methods are provided in Section IV. Finally, the conclusion of the work is drawn in Section V.

2 Related Works

The recent study that includes the activity of EEG signals in predicting the consumers' preference and behavior is summarized in Table 1. Murugappan et al. [2] discussed the relationship between EEG signals and the consumer decision for foretelling the most liked automobile company in that four types with 12 subjects. In [3], Boksem et al. introduced a modeling framework involving 32 subjects and 56 movie trailers using high-frequency beta and gamma waves based on consumers' preference. Kawasaki et al. in 2012 [4] introduced the effects of color preferences by noticing the alterations in oscillations of EEG signals based on two colors which were shown on a digital screen to 19 subjects for 1 s. In [5], EEG and EMG were used for ranking of movie scenes. In [6], highly synched frontal and occipital areas of the brain with greater values for different waves are reported. Mahendra et al. in 2017 proposed a predictive modeling framework for predicting consumers' preferences using images of E-commerce products and recorded EEG signals at the same time, and 70.33% accuracy was noted with hidden Markov model classifier.

3 Methodology

The methodology used for analyzing the impact of EEG signals in neuromarketing for predicting consumers' preferences is discussed in this section. This is divided into three stages. Figure 1 shows the block diagram of the proposed system. Stage-1 includes preprocessing of the dataset, where transforming signals to images and reshaping them using bilinear interpolation are carried out. Figure 2 illustrates pre-

Table 1 Summary of the related works

Author and year	Method	Channel count	Dataset and subjects
Murugappan et al. [2]	EEG, Butterworth filter, FFT, using k-NN & PNN	14	4 brands,12 subjects
Boksem et al. [3]	EEG, beta and gamma waves	64	56 movie trailers,32 subjects
Kawasaki et al. [4]	EEG, theta waves	60	2 colors, 19 subjects
Soleymani et al. [5]	EEG, EMG, self assessed ratings, correlation	64	64 movie scenes,8 subjects
Stickel et al. [6]	EEG, alpha beta and gamma waves	1	26 subjects
Mahendra et al. [1]	EEG, DWT, hidden Markov model	14	25 male 15 female subjects

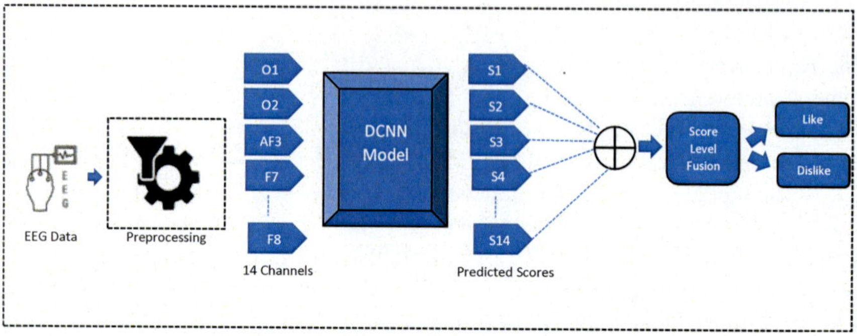

Fig. 1 The representative diagram of the proposed system

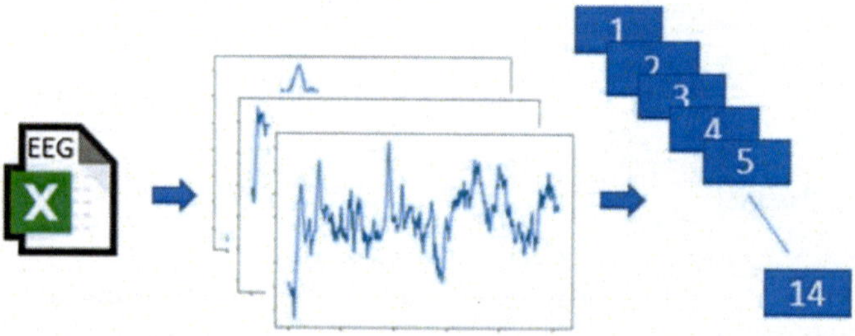

Fig. 2 Preprocessing: EEG Data-to-signal and channel-wise images

processing of dataset. In stage-2, channel-wise images are fed to the proposed DCNN for extracting features followed by computing scores. Finally, scores for all the channels are collected, and late score-level fusion is done in stage-3.

3.1 Dataset Description

The EEG dataset considered for this work was recorded using 14 channels from 25 persons when they are allowed to browse through different E-commerce products [1]. Though data is recorded 128 Hz sampling frequency, four seconds recording has 512 (4×128) time points for one sample. A total of 1045 samples were collected and stored. These time-domain EEG data is then transformed to two-dimensional images of size 224×224. 1045 samples are split into testing and training sets in an 80:20 ratio.

3.2 The Architecture of the Proposed DCNN Framework

The proposed model consists of six convolutional (Conv), batch normalization (BM), and max-pooling (MP) layers with three dense layers. A sigmoid activation function is used for a fully connected layer while initial layers use rectified linear unit (Relu) activation function and last two layers make use of LeakyRelu activation function [7]. The details of filters and parameters are mentioned in Fig. 3, and schematic diagram for the same is shown in Fig. 4, where each block represents one convolutional layer, one batch normalization layer and one max-pooling layer. 256 filters were chosen for the first convolution layer to extract most of the essential low-level features. These numbers are decreased with the network width for converting the data into a low dimensional space. Essential high-level features are contained in the last layers. Then batch normalization is applied to stabilize the neural network by normalizing the output of the previous layer. Binary cross-entropy loss is considered as loss function for the model as $J(\Theta)$ [7–9].

$$J(\Theta) = -\frac{1}{N}\sum_{i=1}^{N} y_i \log(\hat{y}_i) + (1 - y_i)\log(1 - \hat{y}_i), \tag{1}$$

where $\hat{y}_i$ and y_i are considered as predicted and actual class labels and N is the total number of samples in Eq. 1. DCNN is trained over many epochs to reduce binary cross-entropy loss. Combining multiple channels can significantly increase accuracy as compared to that of a single channel alone. The fusion is calculated by few functions on the predicted scores, s_i, typically from each of the 14 channels.

$$S_f = s_1 + s_2 + ...s_{14}, \tag{2}$$

where S_f is the fused score.

S.No	Layers Name	Layers Type	Filters Size	Filters & Neurons	Stride-x	Stride-y	Output size	Parameters
1	Conv1	Convolution 1	1×5	256 filters	1	2	224×220×256	4096
2	BN1	Batch normalization 1					224×220×256	1024
3	MP1	Max_pool 1	1×2		1	2	224×110×256	
4	Conv2	Convolution 2	1×5	128 filters	1	2	224×106×128	163968
5	BN2	Batch normalization 2					224×106×128	512
6	MP2	Max_pool 2	1×2		1	2	224×53×128	
7	Conv3	Convolution 3	1×5	64 filters	1	2	224×49×64	41024
8	BN3	Batch normalization 3					224×49×64	256
9	MP3	Max_pool 3	1×2		1	2	224×24×64	
10	Conv4	Convolution 4	1×5	64 filters	1	2	224×20×64	20544
11	BN4	Batch normalization 4					224×20×64	256
12	MP4	Max_pool4	1×2		1	2	224×10×64	
13	Conv5	Convolution 5	1×3	32 filters	1	2	224×8×32	6176
14	BN5	Batch normalization 5					224×8×32	128
15	MP5	Max_pool5	1×2		1	2	224×4×32	
16	Conv6	Convolution 6	1×2	32 filters	1	2	224×3×32	2080
17	BN6	Batch normalization 6					224×3×32	128
18	MP6	Max_pool6	1×2		1	2	224×1×32	
19	Den1	Dense 1		16 Neurons			16	114704
20	Den2	Dense 2		8 Neurons			8	136
21	Den3	Dense 3 (Sigmoid)		2 Neurons			2	18

Fig. 3 The architecture of the proposed DCNN framework

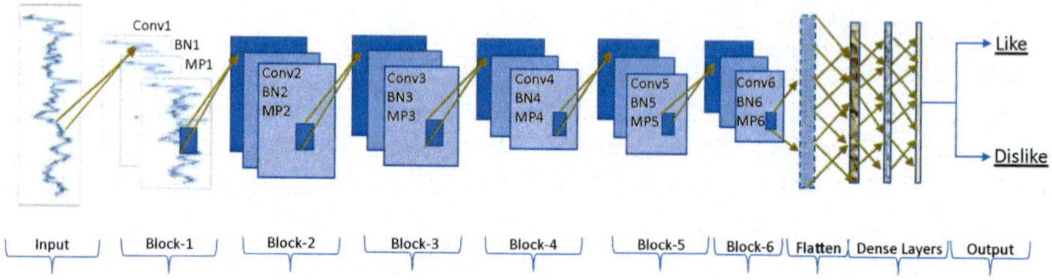

Fig. 4 Schematic architecture of the proposed DCNN

4 Experimental Results and Discussion

This section includes the machine details, evaluation metrics with results achieved after classification from DCNN.

4.1 Environment Settings

Here, Colab, the Google platform, is chosen for the training and testing of a DCNN with Keras framework. Python 3.8.1 is used because the development of many of the deep learning libraries is done through it. GPU RAM of 8GB, CPU Intel(R) i5-9300H @ 2.40GHz and graphic processor Intel(R) UHD graphics 630 are used.

4.2 Evaluation Metrics

Here, accuracy is taken as a measure of comparison among several approaches. Although because of the accuracy paradox, some other parameters like precision, recall, and others were also considered. The details of these data are beyond the scope of this work, as they are well-known methods. However, if interested, readers can refer to [10–12] for more information.

4.3 Results and Comparison

The classification result of the proposed DCNN is shown in Table 2. The architecture of the model is finalized by performing the ablation study [13].

- In Ablation-1, the layers are reduced to five, i.e., only the initial five layers from the proposed model are kept intact and the last one is dropped, to learn the effects

Table 2 Channel-wise accuracies for proposed DCNN in comparison with three different ablation study

Channels	Ablation-1	Ablation-2	Ablation-3	Proposed DCNN
1	0.64	0.61	0.65	0.78
2	0.57	0.595	0.6875	0.75
3	0.73	0.68	0.5937	0.875
4	0.67	0.633	0.625	0.78
5	0.56	0.568	0.6876	0.71
6	0.64	0.53	0.5937	0.718
7	0.75	0.62	0.718	0.81
8	0.78	0.64	0.76	0.78
9	0.51	0.58	0.65	0.683
10	0.64	0.51	0.52	0.723
11	0.55	0.59	0.68	0.76
12	0.57	0.61	0.75	0.72
13	0.67	0.63	0.71	0.71
14	0.69	0.61	0.68	0.716

of the spatial and temporal dimensions on signal images. The other functions like activation and loss are kept same as that used in proposed model.

- In Ablation-2, optimizer and activation functions of the proposed model were altered with RMSprop and softmax, and rest of the layers and parameters were kept as it is.
- In Ablation-3, the layer is added at the end to the proposed model. The optimizer was RMSprop, softmax as activation function and binary cross-entropy loss alike proposed model.

The proposed DCNN outperforms the other models for approximately all the channels. Channel F3, F4, F7, and F8 are frontal lobe channels and mainly responsible for voluntary movement and expressive language and thus giving us good results when compared to other lobes.

On comparing with other state-of-the-art approaches like AlexNet, VGG16, VGG19 and InceptionV3, the proposed DCNN performs better in terms of accuracy, precision, recall, etc. The comparative results are mentioned in Table 3 for both the classes.

Table 3 Comparative analysis of different CNN models using score-level fusion

Models	Classes	Precision	Recall	F1-score	Accuracy
Telpaz et al. [14]	–	–	–	–	0.65
	–	–	–	–	
Mahendra et al. [1]	–	–	–	–	0.7033
	–	–	–	–	
AlexNet	Like	0.79	0.55	0.65	0.74
	Dislike	0.71	0.89	0.79	
VGG16	Like	0.72	0.55	0.62	0.71
	Dislike	0.70	0.83	0.76	
VGG19	Like	0.74	0.59	0.63	0.73
	Dislike	0.71	0.81	0.77	
InceptionV3	Like	0.74	0.71	0.73	0.693
	Dislike	0.65	0.68	0.67	
Proposed DCNN	Like	0.87	0.67	0.76	0.81
	Dislike	0.78	0.92	0.84	

5 Conclusion

This work propitiously used DCNN models for analyzing the EEG signals and explaining the brain's electrical activities transformation in the neuromarketing field. Concludingly, the proposed DCNN, a CNN-based deep learning model with 81% accuracy, performs better when compared to the other baseline methods. For channel-wise comparisons, scores for individual channels were calculated then added up to form the overall predicted class matrix. This is then compared with actual class labels to get the overall accuracy and other evaluation metrics. The results obtained in this study are far better than the existing models, and it can be further extended by assuming multiple factors shortly for better performance [15, 16].

References

1. Yadava M, Kumar P, Saini R, Roy PP, Dogra DP (2017) Analysis of EEG signals and its application to neuromarketing. Multimedia Tools Appl 76(18):19087–19111
2. Murugappan M, Murugappan S, Gerard C (2014, March) Wireless EEG signals based neuromarketing system using Fast Fourier Transform (FFT). In: 2014 IEEE 10th international colloquium on signal processing and its applications. IEEE, pp 25–30

3. Boksem MA, Smidts A (2015) Brain responses to movie trailers predict individual preferences for movies and their population-wide commercial success. J Mark Res 52(4):482–492
4. Kawasaki M, Yamaguchi Y (2012) Effects of subjective preference of colors on attention-related occipital theta oscillations. NeuroImage 59(1):808–814
5. Soleymani M, Chanel G, Kierkels JJ, Pun T (2008, October) Affective ranking of movie scenes using physiological signals and content analysis. In: Proceedings of the 2nd ACM workshop on multimedia semantics, pp 32–39
6. Stickel C, Fink J, Holzinger A (2007, July) Enhancing universal access–EEG based learnability assessment. In: International conference on universal access in human-computer interaction. Springer, Berlin, Heidelberg, pp 813–822
7. Seal A, Bajpai R, Agnihotri J, Yazidi A, Herrera-Viedma E, Krejcar O (2021) DeprNet: a deep convolution neural network framework for detecting depression using EEG. IEEE Trans Instrume Measur 70:1–13
8. Mohan K, Seal A, Krejcar O, Yazidi A (2020) Facial expression recognition using local gravitational force descriptor-based deep convolution neural networks. IEEE Trans Instrum Measur 70:1–12
9. Mohan K, Seal A, Krejcar O, Yazidi A (2021) FER-net: facial expression recognition using deep neural net. Neural Comput Appl 1–12
10. Sharma KK, Seal A (2021) Multi-view spectral clustering for uncertain objects. Inf Sci 547:723–745
11. Sharma KK, Seal A (2019) Modeling uncertain data using Monte Carlo integration method for clustering. Expert Syst Appl 137:100–116
12. Sharma KK, Seal A (2020) Clustering analysis using an adaptive fused distance. Eng Appl Artif Intell 96
13. Gupta AK, Seal A, Khanna P, Yazidi A, Krejcar O (2021) Gated contextual features for salient object detection. IEEE Trans Instrum Measur 70:1–13
14. Telpaz A, Webb R, Levy DJ (2015) Using EEG to predict consumers' future choices. J Mark Res 52(4):511–529
15. Holzinger A, Stocker C, Bruschi M, Auinger A, Silva H, Gamboa H, Fred A (2012, December) On applying approximate entropy to ECG signals for knowledge discovery on the example of big sensor data. In: International conference on active media technology. Springer, Berlin, Heidelberg, pp 209–213
16. Vyas S, Seal A (2020, November) A comparative study of different feature extraction techniques for identifying COVID-19 patients using chest X-rays images. In: 2020 international conference on decision aid sciences and application (DASA). IEEE, pp 209–213

Semantic Segmentation of Road Scene Using Deep Learning

Ritesh Kumar and Khwairakpam Amitab

Abstract The semantic segmentation task is the process of predicting each pixel in the image to some class. Classes could be road, person, car, tree, etc. Road scene segmentation has a vast application in the autonomous driving system. Researchers emphasize autonomous vehicles to reduce the human effort of driving and reduce the chances of road accidents caused by human error. Much research has been done to segment and classify objects in the road scene. We humans can easily detect and understand the objects around the road scene, but it is quite a challenging task for any machine. This article presents some state-of-the-art deep learning-based semantic segmentation techniques, namely FCN, SegNet, and UNet. And a new architecture named PNet for road scene semantic segmentation is proposed. The proposed network is trained end-to-end and comparatively evaluated with the state-of-the-art by using segmentation performance matrices mean Intersection over Union (mIoU) and Dice Coefficient measure (DCM). The publicly available CamVid dataset is used to train and validate the proposed PNet. We have achieved 70.7% mIoU and 80.9% DCM, which is comparatively better than the state-of-the-art.

Keywords CNN · DCM · FCN · mIoU · PNet · SegNet · Semantic segmentation · UNet

1 Introduction

Semantic segmentation [1] is a pixel-wise labeling task. It assigns each pixel into a certain class label [2]. Labels could be a bus, road, tree, building, etc. For example, in an image with many persons, the semantic segmentation will label all the persons as one class of person. The key idea is to recognize and understand what is present in the image at the pixel level. It is interesting to note that the output image size must be the same as the input image, which is essential in semantic segmentation. A semantic segmentation task is a powerful technique to distinguish the

R. Kumar (✉) · K. Amitab
Department of Information Technology, NEHU, Shillong, Meghalaya, India

S. Basu et al. (eds.), *Proceedings of International Conference on Frontiers in Computing and Systems*, Lecture Notes in Networks and Systems 404,
https://doi.org/10.1007/978-981-19-0105-8_13

objects present in the images. Hence, it is used in various aspects of computer vision. Wang et al. [3] have stated that semantic segmentation's primary applications include autonomous driving systems, industrial inspection, medical image analysis, satellite image processing, precision agriculture, geo-sensing, or land usage analysis, facial segmentation, robotics, etc.

Convolution neural network (CNN)-based deep learning models are gaining popularity in image classification, text or facial recognition, and image segmentation. [2]. Due to their automatic feature extraction capabilities and high prediction accuracy, researchers have made extensive efforts to develop deep CNN for semantic segmentation of road scenes [4]. CNNs such as SegNet, fully convolutional network (FCN), and UNet have achieved remarkable successes in the field of semantic segmentation. Generally, the semantic image segmentation model consists of downsampling and upsampling followed by a softmax activation function [5]. Convolution and pooling operations downsample the image, which converts a high-resolution image into low resolution. Max pooling operation helps to understand what is present in the image by reducing the spatial size of an image, but it also reduces the number of parameters, resulting in losing the location information of the objects in the image. Therefore, image needs to be recovered to full size. In road scene semantic segmentation task, it is not only important to understand what is present in the image, but it is also equally important to know where the desired object is located in the image. Hence, upsampling is necessary. It helps to restore the image's actual size and provides information about the location of the desired object in the image.

This article presents a new semantic segmentation network named as PNet. It consists of a downsampling path, bottleneck, and upsampling path. The proposed model PNet is inspired by SegNet [6], FCN [7], and UNet [8]. We have trained and tested the PNet on the CamVid dataset and achieved segmentation accuracy of 70.7% mIoU and 80.9% Dice score.

The rest of the paper is organized as follows. Section 2 presents the literature review of state-of-the-art techniques for semantic segmentation. In Sect. 3, the detailed architecture of the proposed model PNet is presented. Section 4 discusses the details of the training process, results, and analysis. This article is concluded in Sect. 5.

2 Literature Survey

Several researchers have used deep CNN for the segmentation of the road scene. Alvarez et al. [9] have proposed a CNN-based semantic segmentation algorithm. The experiment was carried out on the CamVid dataset. The authors have used 367 and 233 images for training and testing, respectively, which includes 11 different classes of objects and achieved 56.6% mIoU. Full-resolution residual networks for semantic segmentation of street scene was proposed by Pohlen et al. [10]. The model is inspired by ResNet. They have used the Cityscape dataset consisting of 5000 images. The dataset was trained on 2975 images, validated on 500 images, and tested on 1525 images. The network is trained from scratch and achieved mIoU of 71.85% with 30 different classes.

2.1 State-of-the-Art

2.1.1 SegNet

The SegNet model developed by Badrinarayanan et al. [6] uses the CNN model for semantic segmentation. SegNet is specially designed for road scene segmentation because it can handle multiple classes of objects very efficiently. It is also called the encoder–decoder network. The encoder's weights are fetched from the VGG-16 [11] that is a pre-trained classification model with 16 convolutions and pooling layers to downsample the image. The decoder upsamples it using memorized pooling indices from the corresponding encoding path. And the last layer is a softmax classifier to classify each pixel and assign the class label. Each layer in the encoder side performs convolution with 3×3 filter mask and batch normalized, and then the feature maps are passed through ReLu, followed by 2×2 max pooling which reduces the number of parameters, and the max pooling indices are stored so that it could be used for upsampling the image. SegNet has significantly fewer parameters compared to FCN. FCN has 134 Million (M) parameters, whereas SegNet has 14.7 M parameters, which results in less network complexity.

2.1.2 FCN

FCN for semantic segmentation proposed by Long et al. [7] also uses VGG network architecture at the downsampling path. The FCN was trained end-to-end and pixel-to-pixel for semantic segmentation. The downsampling path uses convolution and pooling operations which reduces the size of the image. By downsampling, the model understands what is present in the image but loses the spatial information. Therefore, upsampling the image is required. Upsampling path uses transposed convolution instead of memorizing pooling indices, unlike SegNet. The input image gets converted from a low-resolution image vector to a high-resolution image by transposed convolution. FCN has three variants, namely FCN-32, FCN-16, and FCN-8. We have evaluated FCN-32 and FCN-8 for semantic segmentation on the CamVid dataset. In FCN-32, the segmented image is directly produced from the seventh layer of the convolution block and upsampled 32 times to get the semantically segmented image. It is observed that the output feature map of FCN-32 is rough and generates a coarse image. FCN-8 uses skip connections between the layers. Skip connection allows taking the output of activation function from one layer and feed it directly into another deeper layer in the same network. Using this approach, the network can be deeper by skipping many layers in between. In FCN-8, the seventh convolution layer is convolved four times. The fourth pooling layer is upsampled twice and summed with the third pooling layer using 1×1convolution. And after merging, it is upsampled eight times using transposed convolution to get the segmented image.

2.1.3 UNet

The UNet was proposed in the article written by Ronneberger et al. [8]. The author claimed that UNet is mostly suitable for biomedical image segmentation. The architecture consists of two paths, contraction and expanding path and bottleneck in the middle. The UNet is different from other architectures because downsampling and the upsampling vectors are concatenated instead of the sum operation. The contracting path consists of four blocks. Each block has two 3×3 convolution layers followed by ReLU and batch normalization followed by 2×2 max pooling. The main intention of concatenation operation is to capture the contextual information, and this information is passed to the upsampling blocks to understand where the desired objects are located in the image. The bottleneck in the middle consists of two convolutional layers with batch normalization. The number of expansion blocks is similar to the contraction block. In each layer, deconvolution is performed with stride 2, the output of the upsampled map is concatenated with a corresponding feature map from contraction blocks followed by two 3×3 convolution layers + ReLU and batch normalization.

3 Proposed PNet Model

The proposed network architecture is named PNet (Pyramid Network) which consists of a downsampling path at the left-hand side, Upsampling path on the right-hand side, and a bottleneck in the middle. The proposed model PNet is inspired by SegNet [6], FCN [7], and UNet [8]. Each blue box in Fig. 1, represents a multi-channel feature map, and the dimensions are shown below the blocks. The blue arrows represent concatenation operation. Red and yellow blocks represent pooling and transposed convolution, respectively. Each green-colored block consists of three convolution blocks of size 1×1, 3×3, and 5×5, and all the three blue blocks in the second part of the Fig. 1 are concatenated. These green blocks are repeated four more times at downsampling path.

Input of $224 \times 224 \times 3$ is fed to the network, where 224×224 is the dimensions of the image vector and 3 is the RGB channels. The first block consists of 3×3 convolutions + ReLu + batch normalization, which consist of "N" number of filters. "N" initialized with 64 filters. And in the second block, there are three convolution operations using 1×1, 3×3, 5×5 filters. All these three convolutions are performed on the first convolution layer with "N/2" number of filters; then, all the three convolutions block is merged using concatenation operation as shown in Fig. 1, each followed by ReLU. After the activation, a 2×2 max pooling operation with stride 2

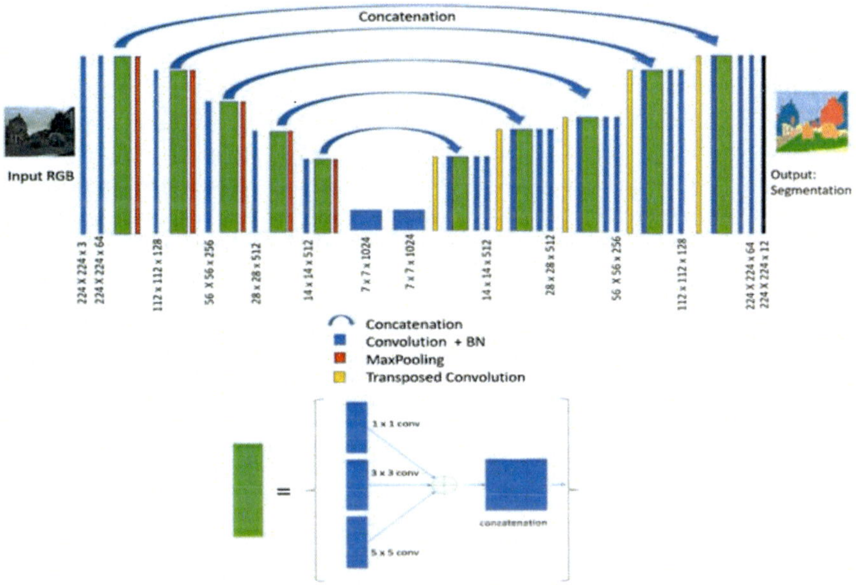

Fig. 1 PNet architecture

is performed for capturing contextual information during downsampling. The above operations is performed four more times with different filter numbers. These contextual pieces of information are used to transfer to the upsampling path for predicting the pixel-wise segmentation. At each downsampling path, we double the number of channels. Initially, 64 feature is used followed by 128, 256, 512, and 1024 features. At the bottleneck, two convolutions are performed without max pooling with channel 1024 each, followed by dropout layer. In the upsampling path, the input image is upsized to its original dimensions.

4 Result and Analysis

The experiment is performed on the publicly available Cambridge-driving Labeled Video database (CamVid image dataset) [12]. The CamVid dataset is a collection of videos and images consisting of 32 classes for semantic road scene segmentation labeling. Twelve classes of objects are selected for experimentation which include bicycle, tree, car, road, side walk, building, sign symbol, fence, pole, pedestrian, sky, and others. The original images in the dataset have a resolution of 360×480. Initially, 367 images are selected for training and validation of all the models (FCN-32, FCN- 8, SegNet, UNet, and PNet). To simplify the problem, all the 367 images are resized to the input size of VGG-16, i.e., 224×224. The existing and the proposed models are trained on 269 images and validated on 74 images. All the models are trained for 100 epochs with Adam optimizer. To avoid buffer overflow in training the networks, the buffer size was set to 32 images. But due to an insufficient number of training images, overfitting occurs, and we got low test accuracy. Therefore, to reduce overfitting, the number of images was increased to 701 and the proposed network was trained on 600 images and validated on 101 images.

It is observed from Table 1 that UNet achieved the lowest accuracy with 42.5% mIoU and comparatively consumes more training time. It has fewer parameters (14M) and failed to recognize multiple class labels. Researchers commonly use UNet for medical image segmentation. On the other hand, SegNet model consists of 32M parameters and uses memorized pooling indices from the encoder to the decoder. It is observed that it also fails to recognize very small objects in the images like "Pole" and "Bicycle" but provides better segmentation results compared to UNet. FCN has 138M trainable parameters. It uses fully connected layers at the last. The FCN has the highest accuracy, followed by SegNet and UNet. As FCN uses transposed convolution for upsampling, it gives a better result compared to other state-of-the-art. Moreover, FCN-8 uses skip connections and allows the network to take activation function from one layer and directly feed it to another deeper layer in the network. The main disadvantage of FCN is the computational complexity because it contains fully connected layers in which every output neuron in one layer is connected to every neuron in the previous layer. This network also fails to recognize very small objects like "Pole." The FCN-8 gives a better segmentation result compared to FCN-32 by reducing coarse in the image.

It can be seen in Fig. 2 that the proposed PNet performs efficiently in semantic road scene segmentation. PNet improves the existing models by introducing batch normalization and dropout layer at the bottleneck, which reduces the parameters to 56M. Initially, the proposed model was trained and validated on 367 images and achieved 58.9% mIoU and 73.4% Dice coefficient, though it has better accuracy in comparison to FCN, SegNet, and UNet. We observed that the PNet was overfitting due to fewer training images, which results in coarse images. Hence, the number of images was increased to 701. The proposed PNet was retrained and validated on 701 images and achieved 70.7% mIoU, and 80.9% Dice coefficient (Table 2). PNet model took approximately 1 h to train on 701 CamVid images, which is comparatively lesser than the other architectures (Table 3).

5 Conclusion

The deep learning model has set a benchmark in the field of computer vision. This article reviews some state-of-the-art deep learning-based semantic segmentation architecture, namely SegNet, FCN-32, FCN- 8, and UNet. A pyramid-based network named PNet is proposed for semantic segmentation of road scenes. The proposed and the state-of-the-art models were trained and validated on the publicly available CamVid image dataset and analyzed their segmentation performance and found that the proposed PNet requires less computation time and performs much better in pixel-

Table 1 Mean intersection over union (mIoU)

Architecture	Bicycle	Tree	Car	Road	Side walk	Building	Sign	Fence	Pole	Pedestrian	Sky	Other	Training	mIoU
SegNet	21.8	72.7	69.6	92.1	59.3	83.2	27.3	51.5	11.4	23.0	84.5	55.2	87.5	54.3
FCN-32	19.1	74.7	75.1	93.7	66.5	85.6	27.0	57.8	2.40	17.7	87.4	54.3	82.6	55.1
FCN-8	23.4	72.1	70.8	92.2	62.0	82.1	35.4	52.1	16.0	29.8	84.2	56.0	84.4	56.3
UNet	13.1	56.0	57.3	89.5	48.9	73.5	9.4	38.4	2.9	7.1	76.8	36.5	74.1	42.5
PNet	25.6	74.9	75.1	94.0	65.3	85.3	40.9	64.2	4.2	24	86.3	60.5	89.5	58.9
PNet 701 Images	57.2	85.5	84.4	94.9	77.9	89.6	51.1	70.5	14.8	37.3	90.9	67.9	91.2	70.7

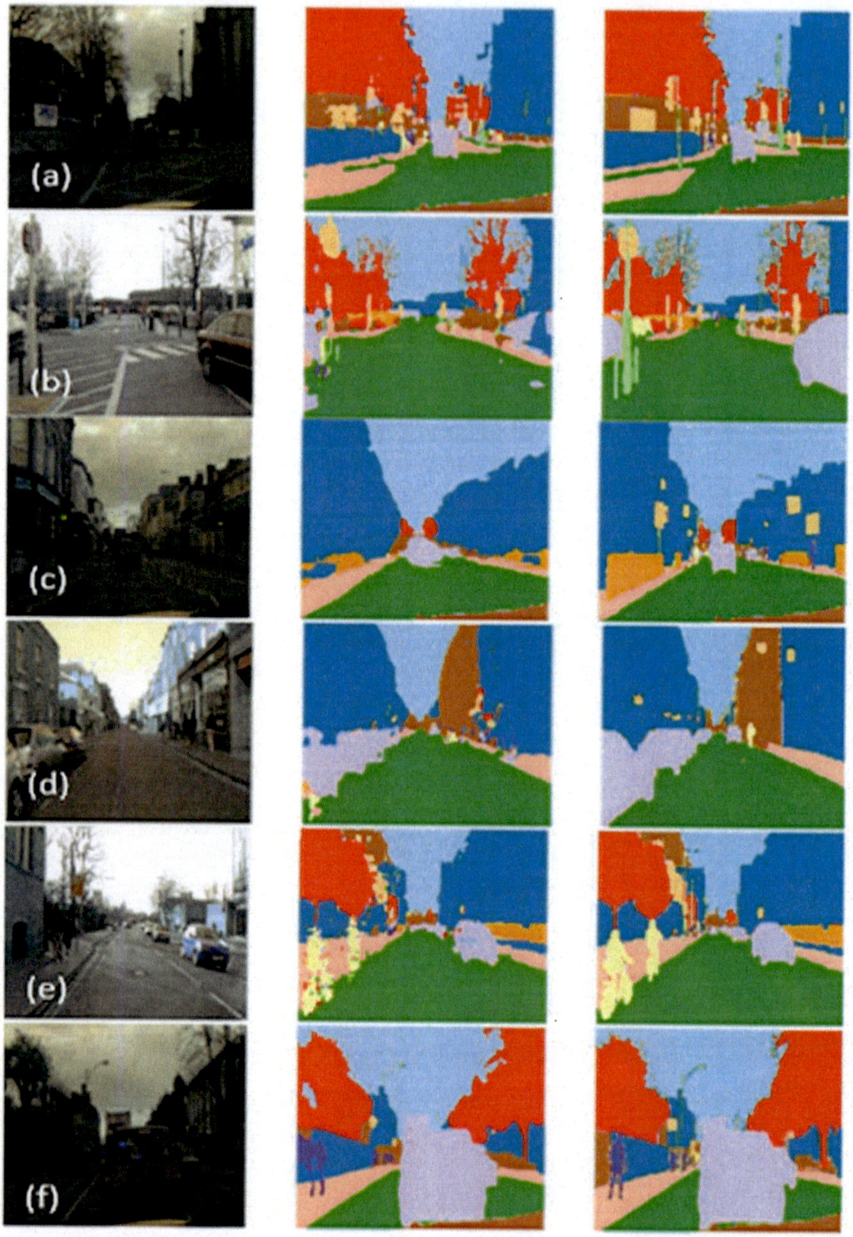

Fig. 2 **a** SegNet; **b** FCN-32; **c** FCN-8; **d** UNet; **e** PNet; **f** PNet (701 images). First, second, and third column represent original, predicted, and true class respectively

Table 2 Dice coefficient score

Architecture	Bicycle	Tree	Car	Road	Side walk	Building	Sign	Fence	Pole	Pedestrian	Sky	Other	Training	Dice Coeff.
SegNet	43.2	84.3	79.8	95.1	72.8	89.8	56.0	63.3	70.0	32.6	92.3	70.5	87.5	70.1
FCN-32	44.3	84.1	81.3	96.1	73.1	88.3	56.5	64.2	90.1	33.6	93.2	71.6	82.6	72.5
FCN-8	45.3	85.3	79.3	96.6	74.2	90.1	57.3	65.3	10.2	40.3	93.5	72.3	84.4	74.3
UNet	44.2	82.2	78.9	93.1	72.1	87.2	54.2	62.1	60.0	30.3	88.3	69.2	74.1	69.8
PNet	40.8	85.7	85.8	96.9	79	92.0	58.1	78.2	0.9	38.7	40.8	92.7	89.5	73.4
PNet 701 Images	54.2	85.5	84.4	94.9	77.9	89.6	51.1	70.5	14.8	37.3	90.9	67.9	91.2	80.9

Table 3 Parameters and training time

S.No.	Architecture	Images	Parameters in million	Training time in hour
1.	SegNet	367	32	1 h 15 min
2.	FCN- 32	367	134	1 h 30 min
3.	FCN- 8	367	134	1 h 25 min
4.	UNet	367	14	2 h
5.	PNet	367	56	50 min
6.	PNet	701	56	1 h

wise segmentation compared to the state-of-the-art. The proposed model was initially trained and validated on 367 images. To further improve the accuracy, the model was retrained on 701 images dataset and achieved 70.7% mIoU and 80.9% Dice score.

References

1. Swarnendu G, Nibaran D, Ishita D, Ujjwal M (2019) Understanding deep learning techniques for image segmentation. ACM Comput Surv 52(4):1–35. https://doi.org/10.1145/3329784
2. Albawi S, Mohammed TA, Al-Zawi S (2017) Understanding of a convolutional neural network. In: 2017 international conference on engineering and technology (ICET), Antalya, Turkey, pp 1–6. https://doi.org/10.1109/ICEngTechnol.2017.8308186
3. Wang C, Mauch L, Guo Z, Yang B (2016) On semantic image segmentation using deep convolutional neural network with shortcuts and easy class extension. In: 2016 sixth international conference on image processing theory, tools and applications (IPTA), Oulu, pp 1–6. https://doi.org/10.1109/IPTA.2016.7821005
4. Sevak JS, Kapadia AD, Chavda JB, Shah A, Rahevar M (2017) Survey on semantic image segmentation techniques. In: 2017 international conference on intelligent sustainable systems (ICISS), Palladam, India, pp 306–313. https://doi.org/10.1109/ISS1.2017.8389420
5. Pal A, Jaiswal S, Ghosh S, Das N, Nasipuri M (2018) SegFast: a faster SqueezeNet based semantic image segmentation technique using depth-wise separable convolutions. https://doi.org/10.1145/3293353.3293406
6. Badrinarayanan V, Kendall A, Cipolla R (2017) SegNet: a deep convolutional encoder-decoder architecture for image segmentation. IEEE Trans Pattern Anal Mach Intell 39(12):2481–2495. https://doi.org/10.1109/TPAMI.2016.2644615
7. Long J, Shelhamer E, Darrell T (2015) Fully convolutional networks for semantic segmentation. In: 2015 IEEE conference on computer vision and pattern recognition (CVPR), Boston, MA, USA, pp 3431–3440. https://doi.org/10.1109/CVPR.2015.7298965
8. Ronneberger O, Fischer P, Brox T (2015) U-Net: convolutional networks for biomedical image segmentation. Comput Vis Pattern Recogn. arXiv:1505.04597
9. Alvarez JM, LeCun Y, Gevers T, Lopez AM (2012) Semantic road segmentation via multi-scale ensembles of learned features. In: Fusiello A, Murino V, Cucchiara R (eds) Computer vision—ECCV 2012. Workshops and demonstrations. ECCV. Lecture notes in computer science, vol 7584. Springer, Berlin, Heidelberg
10. Pohlen T, Hermans A, Mathias M, Leibe B (2017) Full-resolution residual networks for semantic segmentation in street scenes. In: 2017 IEEE conference on computer vision and pattern recognition (CVPR), Honolulu, HI, pp 3309–3318. https://doi.org/10.1109/CVPR.2017.353.

11. Qassim H, Verma A, Feinzimer D (2018) Compressed residual-VGG16 CNN model for big data places image recognition. In: IEEE 8th annual computing and communication workshop and conference (CCWC). Las Vegas, NV, USA, pp 169–175. https://doi.org/10.1109/CCWC.2018.8301729
12. Brostow GJ, Fauqueur J, Cipolla R (2008) Semantic object classes in video: a high-definition ground truth database. Patt Recog Lett
13. Swarnendu G, Anisha P, Shourya J. Santosh KC, Nibaran D, Mita N (2019) SegFast-V2: semantic image segmentation with less parameters in deep learning for autonomous driving. Int J Mach Learn Cybern. https://doi.org/10.1007/s13042-019-01005-5

Edge Detection and Segmentation Type Responses in Primary Visual Cortex

Satyabrat Malla Bujar Baruah, Uddipan Hazarika, and Soumik Roy

Abstract The primary visual cortex consists of varied neuronal structures and connectivity links contributing to complex feature extraction such as edge detection, contour detection, depth and motion detection. But the functional basis of such complex computation is not yet realized. In the proposed work, an attempt has been made to model a single layer of retinal ganglion cell network incorporating morphologically detailed neurons with active dendritic integration to investigate the role of morphology, localized active membrane and connectome specificity. Two different connectome-specific instances of the layer are simulated with one network connected with excitatory connections with ON-bipolar cell behaved as scene segmenter, whereas another network connected with specific connections with ON- and OFF-bipolar cell with excitatory and inhibitory connections, respectively, behaves as an edge detector.

Keywords Spiking neural network · Bipolar cells · Retinal ganglion cells · Active dendritic integration · Izhekivich membrane model

1 Introduction

Primate vision is one of the interesting aspects of neuroscience where visual stimuli are processed incessantly to extract feature and contour information, detect directional features and forward these features to successive neuronal networks to shape learning and cognition [14, 18]. Over the last few decades, a lot about neurons, specifically corresponding to the visual cortex, has been explored to understand the role of neurons and neuronal networks in shaping visual cognition, learning and

S. M. Bujar Baruah (✉) · U. Hazarika · S. Roy
Department of Electronics and Communication Engineering, Tezpur University, Napam, Tezpur, Assam, India
e-mail: baruah.satyabrat@gmail.com

S. Roy
e-mail: xoumik@tezu.ernet.in

S. Basu et al. (eds.), *Proceedings of International Conference on Frontiers in Computing and Systems*, Lecture Notes in Networks and Systems 404,
https://doi.org/10.1007/978-981-19-0105-8_14

memory. Even though a lot has been explored in the field of primate vision, there exists a bridging gap between types of information extracted in the primary visual cortex and how are such features being extracted [17]. Connectome specificity of neurons with different layers, specifically with bipolar cells, is yet to be recognized along with the role of dendritic arbor in shaping complex responses such as direction selectivity and orientation selectivity. [9]. Recent findings reported the existence of a wide range of active membrane dynamics in specific dendritic sites [10, 11] and their probable role in active dendritic integration and shaping of complex neuronal functions [4, 13]. But details of the functional importance of such localized active membrane dynamics in active dendritic integration are yet to be realized. Even though connectome specificity has been addressed as a basis of connectomic plasticity [2, 16], very little is known about distinct morphologies and their probable link with function formation [8, 19] and plasticity.

In the proposed work, an attempt has been made to link the missing bridge between neuronal morphology, connectome specificity and the role of active membrane dynamics in shaping complex neuronal responses such as edge detection and scene segmentation in early layers of the primate visual cortex. Details of the spiking neuronal network (SNN) model have been discussed in Sect. 2 in detail along with connectome specificity, neuronal morphology and localized dendritic membrane dynamics to simulate active integration in dendritic arbor. Later in the results and discussion, Sect. 3 puts forward the model response generated incorporating connectome specificity, neuronal membrane dynamics, electro-physiological and physico-chemical behavior along with neuronal morphology specific to edge detection and scene segmentation.

2 Methods

In the proposed model, a spiking neural network inspired from Poggio et al. [14] and neural network arrangement inspired from works of Hubel and Wiesel [5] has been constructed equipped with an amalgamation of distinct neuronal morphologies of Retinal Ganglion Cells (RGC's) with localized active membrane dynamics and connectome specificity arranged in precise repetitive modular arrangements [3, 12]. Distinct neuronal morphologies are used to mimic the behavior of midget and parasol cells [19] and localized active membrane dynamics has been incorporated at specific sites of the neuron such as distal dendritic ends, branching junctions and cell body [10, 11, 13] to simulate active dendritic integration [4, 13, 15] Figs. 2, 3. Connectome specificity of neurons with bipolar cell is displayed in Fig. 2 and Fig. 3, where connections in Fig. 2 correspond to an RGC morphology of edge detection type network, whereas Fig. 3 corresponds to RGC morphology of scene segmentation type network. The neuronal layer organization of different types of cells in the primary visual cortex is shown in Fig. 1.

Connectivity of RGC morphology in Figs. 2, 3 with bipolar cells is described using connectome specificity matrix shown in Fig. 4 where a '−1' represents inhibitory

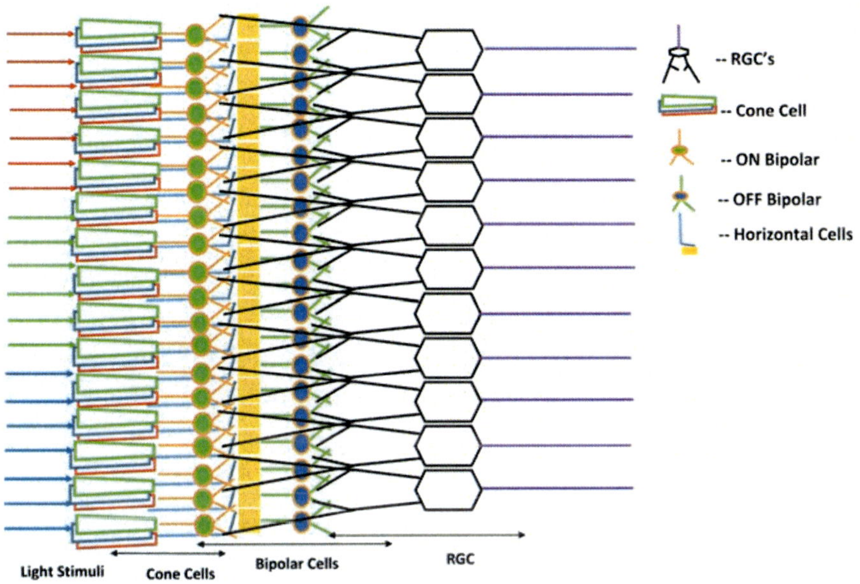

Fig. 1 Connectivity and arrangements of different types of cells in primary visual cortex with ON- and OFF-bipolar cells connectivity with cone cells

Fig. 2 Morphology of orientation selective RGC used in the proposed model, the light green and dark green synapses correspond to connectivity with ON-bipolar cell and OFF-bipolar cells, respectively

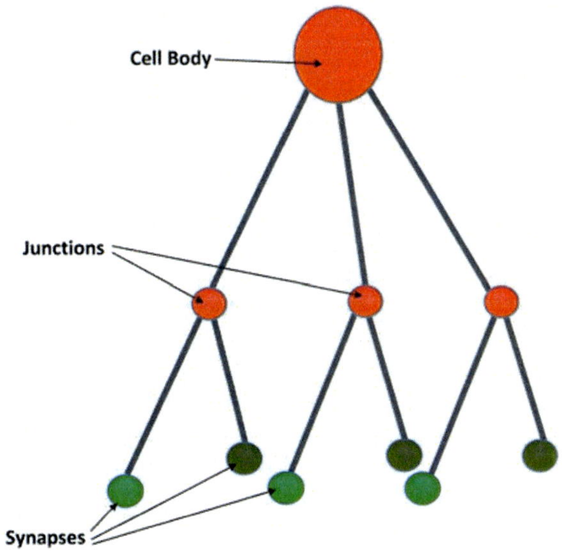

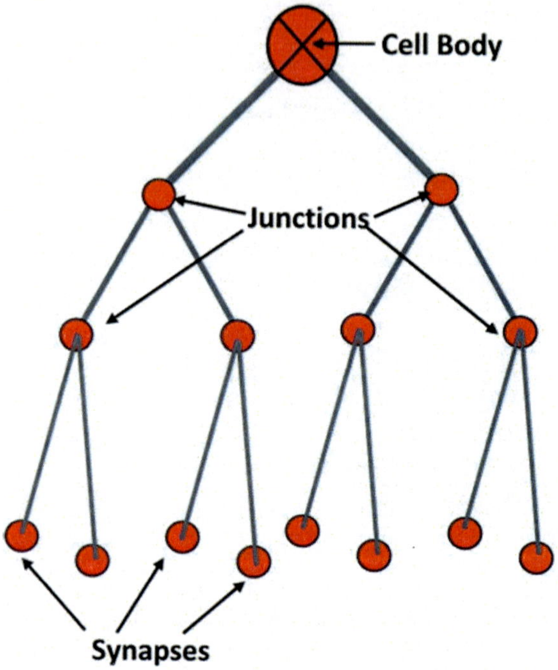

Fig. 3 Morphology of segmentation type RGC used in the proposed model where all synaptic connections are with ON-bipolar cells

connectivity with OFF-bipolar cell, '1' represents excitatory connectivity with ON-bipolar cell and '0' represents no connectivity.

Input to the model is 'tiff' and 'png' images that are considered as a stimulus to cone cell and rod cell arrays of photoreceptor cells. Photo receptor cells transduce the trapped light stimuli, ON-bipolar cells are directly connected to photoreceptor cells, whereas OFF-bipolar cells are connected to the photoreceptor cells via the horizontal cells and relay light and dark information from the scenes projected onto the photoreceptors. Responses from the ON- and OFF-bipolar cells are then fed to the respective RGC networks. Orientation selective network RGC are connected within a neighborhood of 3×3 bipolar cells, of which three excitatory synaptic inputs are

1	1	1
0	0	0
-1	-1	-1

-1	0	1
-1	0	1
-1	0	1

0	1	1
-1	0	1
-1	-1	0

1	1	0
1	0	-1
0	-1	-1

1	1	1
1	0	1
1	1	1

Fig. 4 Connectivity matrices for detection of four directional edges at 0°, 90°, 135° and 45° respectively from left to right for neuron morphology in Fig. 2 and segmentation type RGC connectivity for neuron morphology Fig. 3

connected to ON-bipolar cells and inhibitory synaptic inputs are connected to OFF-bipolar cell. On the other hand, segmentation type network RTCs are connected to only ON-bipolar cells with all excitatory connections as shown in Fig. 4. The RGCs then encode the incoming signals in terms of spiking activities. Spiking activities are simulated using Izhekivich 'bursting' and 'chattering' type membrane dynamics defined by the system of coupled Eq. 1 [6, 7].

$$C\frac{\mathrm{d}v}{\mathrm{d}t} = k\left((v - v_r)(v - v_t) - u + I\right),\ if\ v \geq v_t \tag{1}$$

$$\frac{\mathrm{d}u}{\mathrm{d}t} = a\left[b\,(v - v_r) - u\right], \qquad v \leftarrow c, u \leftarrow u + d \tag{2}$$

where a, b, c, d and k are the control parameters, I is the stimuli, v is the membrane potential, u is the recovery current, v_r resting potential and v_t, the threshold potential. Such encoding activities are seen in distal dendritic ends, dendritic junctions and the cell body. After signal encoding, the signal propagates through passive dendritic fiber incurring propagating losses to the signal. Junctions and cell bodies are taken as summing nodes, where the cumulative propagating signals are re-encoded and renewed for forward propagation. Propagating activity in passive dendritic fiber has been modeled using equations discussed in Eqs. 3 and 4 [1].

$$I_{\text{inTotal}} = I_t + I_{\text{out}},\ I_{\text{inTotal}} = \frac{(V_{\text{out}} - V_{\text{in}})}{R_{\text{lon}}} \tag{3}$$

$$I_t + C_m\frac{\mathrm{d}V_{out}}{\mathrm{d}t} + G_L\left(V_{\text{out}} - E_L\right) = 0 \tag{4}$$

Values associated with active membrane dynamics depicting 'chattering' and 'bursting' activities are given in Table 1 and values corresponding to modeling of propagation of the signal in passive fiber are summarized in Table 2.

3 Results and Discussion

Simulation of the proposed model has been conducted in 'Python 3' interpreter. Packages such as 'matplotlib,' 'numpy' and 'scipy' have been used for generating the responses and operating on the data respectively. Input to the model is 'tif' and 'png' images, where pixel intensities are taken as light stimuli trapped by the photoreceptor cell array and forwarded to the network of bipolar cells. In the orientation selective RGC network, intensity values are converted to grayscale and fed to the bipolar network whereas, in the segmentation type RGC network, RGB images are taken as stimuli to color selective RGC cone cells. ON- and OFF-bipolar cells convert this spatial information into temporal square waves of 200 ms with an offset of 20 ms and pulse width of 150 $mSec$ to replicate the behavior of bipolar cells without increasing the computational complexity of the model. Stimulus amplitudes are amplified within

Table 1 Izhikevich bursting and chattering membrane parameters [6, 7]

	Parameters								
Membrane type	a	b	c	d	C	k	V_r	V_t	V_{peak}
Bursting	0.01	5	56	130	150	1.2	−65	−35	40
Chattering	0.03	−2	−50	100	100	0.7	−60	−30	40

Table 2 Passive membrane parameters for facilitating signal transmission

C_m	R_{lon}	E_L	G_L
1 µF	2Ω	−65 mV	10^{-6} S

Fig. 5 Some of the test images used in the model taken from SIPI image database and BSD database

the range of −640 and 635 nA to mimic the current levels in bipolar cells due to darkness intensities and light intensities, respectively. In both the model network, when the temporal signals are fed to the local modular RGC's, stimulus strengths are encoded in terms of spiking activities and forward propagated through the dendritic fibers. Parameters for spiking activities are taken from Izhikevich model and given in Table 1. Similarly for simulating signal propagation behavior, passive membrane parameters used in the model are given in Table 2.

At the junction of both the networks, the signals are summed together, and the cumulative response is renewed and re-encoded until the signal reaches the cell body. The cell body finally gives the cumulative response of a cluster of signals acquired from different connections in terms of spiking frequency. In the proposed model, a combination of spiking activity such as 'chattering' along with 'bursting' type activity controls the bandwidth of the filter whereas dendritic spread controls the size of the receptive field. The model has been implemented and tested using the SIPI image database, which is an assembly of digitized images that aids in image processing and analysis, and Berkeley's BSD dataset consisting of ground truths for a set of natural scenes, and seems consistent. Some of the images considered for implementation of the model are shown in Fig. 5. Shown in Fig. 6 are the orientation selective response of RGC SNN, and Fig. 7 shows edge detection type behavior achieved by max-pooling of four cardinal direction responses.

Shown in Fig. 8 are responses generated by segmentation type RGC SNN that shows image segmentation type behavior by clustering the stimuli neighborhood.

Fig. 6 Orientation data at 0°, 90°, 135° and 45°, respectively, from left to right from orientation selective RGC SNN from house image from test images shown in Fig. 5 as discussed in [5, 14]

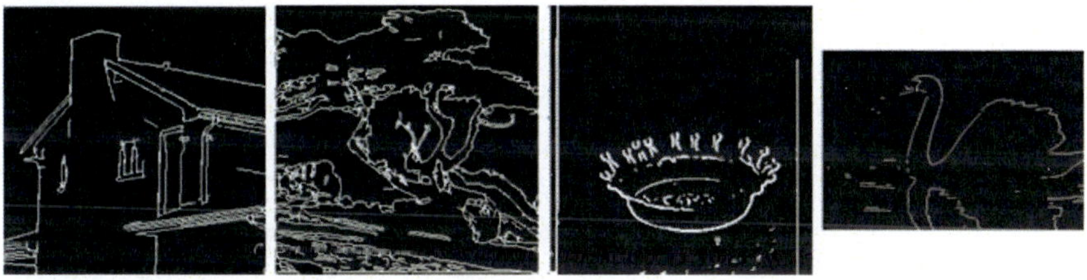

Fig. 7 Edge detection type RGC SNN response for test images shown in Fig. 5

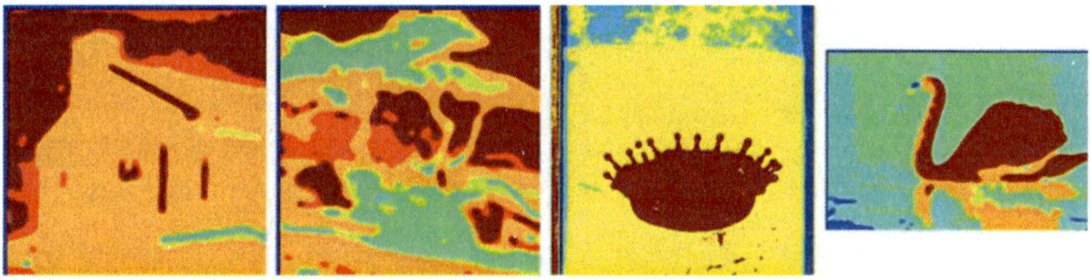

Fig. 8 Segmentation type RGC SNN response for test images shown in Fig. 5

Such clustering capability might be achieved due to the bandwidth controllability and data reprojection due to a combination of different active integration in the dendritic arbor.

4 Conclusion

The proposed model has been incorporated to gain insight into probable functions and computations performed in a single-layered neural network of RGC's in the primary visual cortex. RGC network of similar morphology but with different connectomics achieving different functionalities can be concluded from the proposed model. The edge detection model shows promising results by identifying sharp edges and non-maxima suppression type behavior whereas the segmentation model seems to be able to handle classification problems with ease. Preliminary quantitative comparison of the proposed model response with Berkeley Segmentation Dataset seems promising in terms of edge detection accuracy. Such a model can shed new understanding of types of features extracted by the primary visual cortex and their probable role in learning, plasticity and cognition. Further, this model can be extended in the direction of shape perception, contour detection, etc. incorporating machine learning.

Acknowledgements This publication is an outcome of the R&D work undertaken project under the Visvesvaraya Ph.D. Scheme of Ministry of Electronics & Information Technology, Government of India, being implemented by Digital India Corporation.

References

1. Bujar Baruah SM, Gogoi P, Roy S (2019) From cable equation to active and passive nerve membrane model. In: 2019 second international conference on advanced computational and communication paradigms (ICACCP). IEEE, pp 1–5
2. Bhattacharya A, Aghayeva U, Berghoff EG, Hobert O (2019) Plasticity of the electrical connectome of *c. elegans*. Cell 176(5):1174–1189
3. DeAngelis GC, Ghose GM, Ohzawa I, Freeman RD (1999) Functional micro-organization of primary visual cortex: receptive field analysis of nearby neurons. J Neurosci 19(10):4046–4064
4. Grienberger C, Chen X, Konnerth A (2015) Dendritic function in vivo. Trends Neurosci 38(1):45–54
5. Hubel DH, Wiesel TN (1968) Receptive fields and functional architecture of monkey striate cortex. J Physiol 195(1):215–243
6. Izhikevich EM (2003) Simple model of spiking neurons. IEEE Trans Neural Netw 14(6):1569–1572
7. Izhikevich EM (2007) Dynamical systems in neuroscience. MIT Press
8. Krichmar JL, Nasuto SJ, Scorcioni R, Washington SD, Ascoli GA (2002) Effects of dendritic morphology on ca3 pyramidal cell electrophysiology: a simulation study. Brain Res 941(1–2):11–28
9. Liu X, Robinson PA (2021) Analytic model for feature maps in the primary visual cortex. arXiv preprint arXiv:2103.09954
10. Misonou H (2018) Precise localizations of voltage-gated sodium and potassium channels in neurons. Dev Neurobiol 78(3):271–282
11. Nusser Z (2012) Differential subcellular distribution of ion channels and the diversity of neuronal function. Current Opin Neurobiol 22(3):366–371
12. Ohki K, Chung S, Ch'ng YH, Kara P, Reid RC (2005) Functional imaging with cellular resolution reveals precise micro-architecture in visual cortex. Nature 433(7026):597
13. Ranganathan GN, Apostolides PF, Harnett MT, Xu N-L, Druckmann S, Magee JC (2018) Active dendritic integration and mixed neocortical network representations during an adaptive sensing behavior. Nat Neurosci 21(11):1583–1590
14. Riesenhuber M, Poggio T (1999) Hierarchical models of object recognition in cortex. Nature Neurosci 2(11):1019–1025
15. Sheasby BW, Fohlmeister JF (1999) Impulse encoding across the dendritic morphologies of retinal ganglion cells. J Neurophysiol 81(4):1685–1698
16. Sporns O (2013) The human connectome: origins and challenges. Neuroimage 80:53–61
17. Tchinda BS, Tchiotsop D, Noubom M, Louis-Dorr V, Wolf D (2021) Retinal blood vessels segmentation using classical edge detection filters and the neural network. Inf Med Unlocked 23:100521
18. Ts'o DY, Frostig RD, Lieke EE, Grinvald A (1990) Functional organization of primate visual cortex revealed by high resolution optical imaging. Science 249(4967):417–420
19. Weber AJ, Harman CD (2005) Structure–function relations of parasol cells in the normal and glaucomatous primate retina. Invest Ophthalmol Visual Sci 46(9):3197–3207

Hand Gesture Based Computer Vision Mouse

Bibekananda Buragohain, Dhritiman Das, Bhabajyoti Baishya, and Minakshi Gogoi

Abstract The popularity of the Human-Computer Interaction (HCI) technology exponentially grew with the advancement of artificial intelligences. This work with the help of computer vision will enable the user to interact with the computer at their fingertips. The user would essentially be able to control the cursor and do associative functionalities such as left-click, right-click, move, and almost everything a mouse can do. A cross-platform application is developed which is the underlining vision of replacing a mouse.

Keywords HCI · Fingertips · Cursor control · Gesture control

1 Introduction

Over the last two decades, technologies driving the usage of computer has seen phenomenal changes. From past Thin Film Transistor LCD, Track and ball mouse to current Organic light-emitting diode (OLED), optical mouse has greatly enhanced the ability of the computer systems and human to do many powerful calculations and novel activities from solving complex mathematical equations to contribute to augmented reality (AR) and virtual reality (VR) activities. With growing usage of computers in people's daily activities has allowed computer systems has become a powerful and integral part of people's lives. Currently, for any activities to be performed by the user on the computer the interaction between the user and the computer is mainly carried by the mouse and keyboard. This usage of the traditional mouse is challenged constantly by academicians, researchers, and industries. These parties are constantly striving for a novel approach for human and computer interactions or replace the traditional mice and keyboard. The most natural and instinctive technique for HCI, which can support a viable replacement for the computer mouse is with the use of hand gestures. Therefore, this work intends to develop a cursor

B. Buragohain · D. Das · B. Baishya · M. Gogoi (✉)
Department of Computer Science and Engineering, Girijananda Chowdhury Institute of Management and Technology, Guwahati, Assam, India
e-mail: minakshi_cse@gimt-guwahati.ac.in

S. Basu et al. (eds.), *Proceedings of International Conference on Frontiers in Computing and Systems*, Lecture Notes in Networks and Systems 404,
https://doi.org/10.1007/978-981-19-0105-8_15

control system based on hand gestures and primarily an investigative to test the feasibility of a wireless hand gesture-based interaction between the computer system and humans. The proposed work makes use of a webcam that has been part of laptops for implementation of the system to replace the need and requirement of traditional computer mice.

This work is chosen with an aim of facilitating direct interaction between humans and consumer electronic devices. The dependence on age-old peripheral devices maybe phased out or minimized with the help of gesture control-based technologies.

In our day-to-day activities with our electronic devices such as computers, televisions, projectors, motion or gesture control technology would be immensely useful for tasks such as selecting, scrolling, and clicking. The required hardware for the implementation of gesture-controlled technology was not feasible due to time and monetary budget constraints. Hence, an introductory software implementation is designed for a computer to test the viability of replacing a mouse.

In this work, we have concentrated on building a computer vision mouse model fully controlled by fingers without any external devices using machine learning. The method exploits a hand tracking solution and implements a palm detector model. A hand landmark model is established to return hand 3D key points. The work also developed a key point classifier for hand gesture recognition. In the next section, we are describing the prior related work.

2 Prior Related Study

The resourcefulness of a webcam could be assessed or seen on varying applications such as in a database for sign language or fields of motion tracking or controller. Current products such as Microsoft Azure Kinect [1] and Nintendo Wii Remote [2] are the results due to significant advancements in gaming technologies that can be scalable in other human and computer interactions domain. These gaming technologies are developed to enhance the user's experiences while playing video games. These technologies curved to give experiences of real, natural, interactive, and intuitive. This could serve as one reason behind the growing gaming industry worth 160$ Billion. HCI using hand gestures is instinctive, promising, and effective for one-to-one interaction between computers and humans as it provides a Natural User Interface (NUI). In this competitive world, there has been a quite lot of research on cursor control systems based on hand gestures. Moreover, a hand gesture-based recognition system is used for the interpretation of signs language, security purposes, and many other fields.

Hand gesture recognition has been worked upon or investigated by many. To name a few are, Francis [3], Mitra [4] and Ghotkar [5]. Each of them submitted their findings separately in various IEEE journal papers. Francis and Mitra investigated or analyzed various approaches or methods for gesture-based recognition systems ranging from cars to Kinect-based devices and from Markov models to finite state machines. While Ghotkar read out a novel implementation for hand segmentation and gesture recognition with help of usage of various color spaces.

The methods or approaches proposed by Mitra and Francis are computationally heavy, hardware-centric, expensive, and high processing time. On the counter, we require a simpler and effective way to recognize the hand gestures and use them to control the functions of a computer mouse.

3 Background Study

Human-Computer Interaction is probably the current hotbed field in the IT sector. The aim of HCI is solely to improve the interaction between users and computers to user's comfort. Various techniques and methodologies are currently developing and many are developed to counter the increasing difficulties of HCI. Some of the predominant existing ways of interacting with a computer are through mouse and touch. With the rapid advancement of technologies in machine learning and artificial intelligence, many novels and alternative ways of interaction are being currently developed or are on the path of breakthrough. A few of them are speech-based interaction, hand-based interaction, head movement-based interaction, and eye movement-based interaction. Improvement in HCI technologies, apart from easing for a healthy human being. Also, will be of profound help to people with disabilities.

The underlying vision of our work is to create a hand gesture-based cursor control system (HGCCS). Although HGCCS is a well-practiced idea and has still on the path of breakthrough. Many image processing techniques are explored to create the same, to be named a few are skin detection, color detection, background subtraction, and hand contour extraction. The latency of the other techniques is usually high, which is slow and breaks the vision of HCI. In June 2019 at the conference of CVPR google published a novel real-time hand tracking technique [6]. This machine learning approach results or displays real-time 21 3D fingertips along with a quality hand. Presently, most of the available approaches are heavily dependent upon high-end computer systems with powerful processing power. The Google implemented approach is scalable to multiple hands and results in real-time performance even on a mobile phone.

Our work implements this Google provided ML pipeline solution to track the hand and do gesture recognition. The associated vision of controlling the cursor is with hand gestures is put into a framework with the help of python libraries and modules (see Fig. 1).

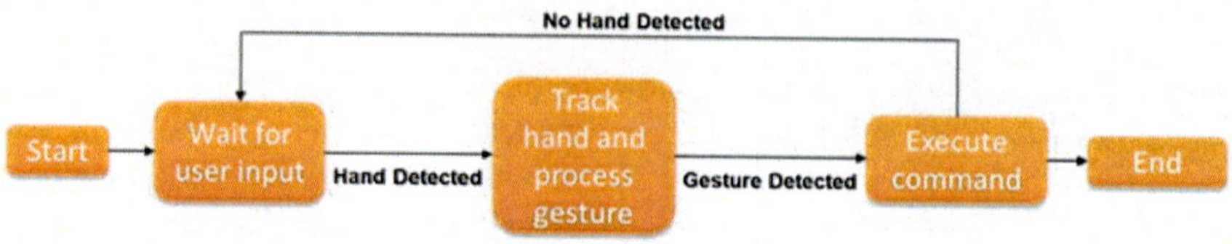

Fig. 1 System design of a generic computer vision mouse

3.1 Prerequisites

Python [7] Is a just-in-time high-level programming language. Its design prioritizes code readability with its use of indentation. It is a multi-paradigms programming language. One of its greatest strengths is its comprehensive standard library.

PyAutoGUI [8] and Flask [9] They are python-based libraries, both are used extensively to develop the work. To control the mouse and keyboard PyAutoGUI is used which is a cross-platform GUI automation python module. While Flask is a python-based web framework, used for creating local API.

JavaScript [10] JavaScript is a just-in-time or interpreted, heavily used programming language by developers around the world. It falls under the category of Scripting language and is used for developing applications, web pages, websites, games, etc.

React [11] It is an open-source JavaScript library for creating user interfaces and is also known as React.js or ReactJs and is a frequently used language in the development of single-page or mobile app development.

Visual Studio Code [12] It is an integrated development environment (IDE) made by Microsoft. It supports syntax highlighting, debugging, code refactoring, and embedded Git. FTP includes multiple extensions which have allowed software developers greatly to use as a substitute for various applications and web development.

TensorFlow.js [13] It is a library in JavaScript for machine learning. It is used to develop machine learning models in JavaScript and allows to use of ML directly in Node.jsor the browser.

4 Methodology and Proposed Method

4.1 Method Used

Google designed, developed, and implemented an ML solution on a publicly available cross-platform framework (MediaPipe) largely used to build pipelines to exercise

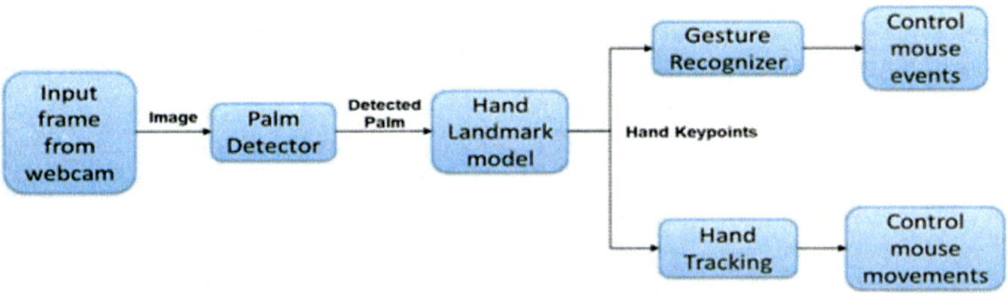

Fig. 2 Our proposed approach

perceptual data of different modalities (audio and video). This machine learning approach results or displays real-time 21 3D hand key points from just a single frame. They get an average precision of 95.7% in palm detection using the strategies described above. A typical cross-entropy loss with no decoder yields a baseline of just 86.22%.

Palm detection model This model uses a Single-shot detector (SSD) which is optimized for mobile real-time object detection or palm detection.

Hand Landmark Model It provides high accuracy hand and finger tracking which employs machine learning to give 21 numbers of 3D hand key points from just a single frame.

From the previously computed key points, we designed a hand gesture recognition model which classifies those key points into a distinct set of gestures (see Fig. 2).

4.2 *Palm Detector*

In the paper, the Palm Detector is implemented (see Fig. 3). The model used to detect hand locations is BlazePalm. The model type is convolutional neural network. The

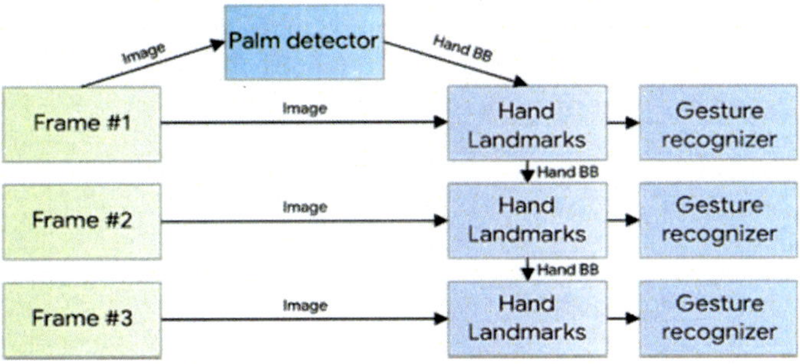

Fig. 3 System flow diagram of BlazePalm model

model architecture is a palm detector that uses a Single-shot detector (SSD) with a custom encoder and non-max suppression algorithm [14] to filter the palm from different objects identified by the SSD.

It takes input as an image (possibly a video frame) resized to 256 × 256 pixels and gives output as the palm bounding box coordinates with 7 Palm 2D key points, 21 3D hand key points, and detection of confidence score.

4.3 Gesture Recognizer

The following mentioned steps explain the working of the gesture recognizer:

1. Firstly, the gesture recognizer identifies all the hand landmarks from the video frame.
2. Secondly, the curl and direction of each finger are estimated.
3. Finally, the output is compared to a set of predefined gesture descriptions.

All the above specified steps of gesture recognizer are handled by TensorFlow's hand pose model. The curl and direction of each finger together define a gesture. An example of a gesture would be—Thumbs Up. While forming a 'thumbs up' gesture, all the fingers other than the thumb are curled and the thumb is straightened up pointing to either left or right.

5 Proposed Work

The workflow diagram of our proposed model is given below (see Fig. 4). The source code of our proposed work is available on GitHub [15].

To do cursor control, a method is proposed with the help of BlazePalm and the gesture control model. The method for left-click and right-click are:

Left-Click. This is implemented with the gesture sign involving the index and middle finger making the letter '*V*' of the English alphabet while all other fingers stay closed under the thumb.

Right-Click. This is implemented with the gesture sign involving the thumb and the index finger making the letter '*O*' of the English alphabet and the other three fingers stay upright.

6 Results

A vision-based cursor control system using hand gestures was developed under python and used a publicly available TensorFlow hand tracking model named hand pose model. The system was successfully able to do functionalities as the traditional

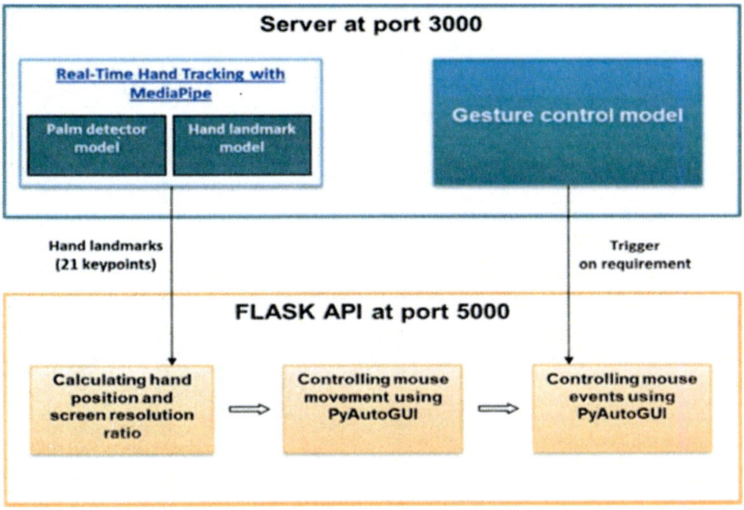

Fig. 4 Workflow for proposed work

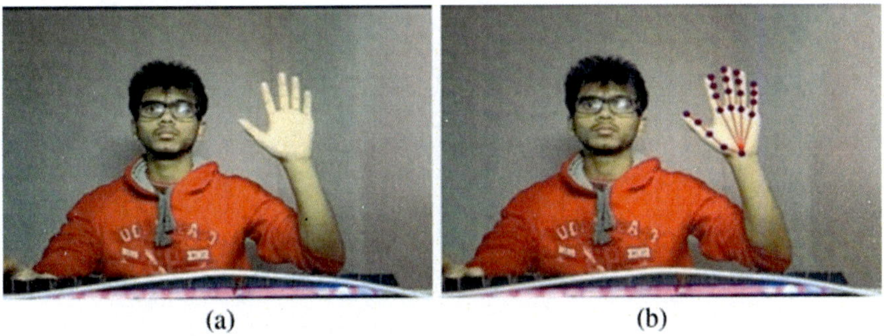

Fig. 5 **a** Input frame and, **b** output frame with 21 3D key points

mouse such as control of the movement of cursor by hand and cursor functions by a set of different hand gestures. An input frame (see Fig. 5a) and the output frame (see Fig. 5b) with 21 3D hand key points. The left-click (see Fig. 6) and right-click (see Fig. 7) demonstration are shown.

7 Conclusion

We can consider our implementation of the system was largely successful at the investigative stage, but there is significant room for improvement. The system implements a high-precision ML solution capable of running in real-time on various platforms

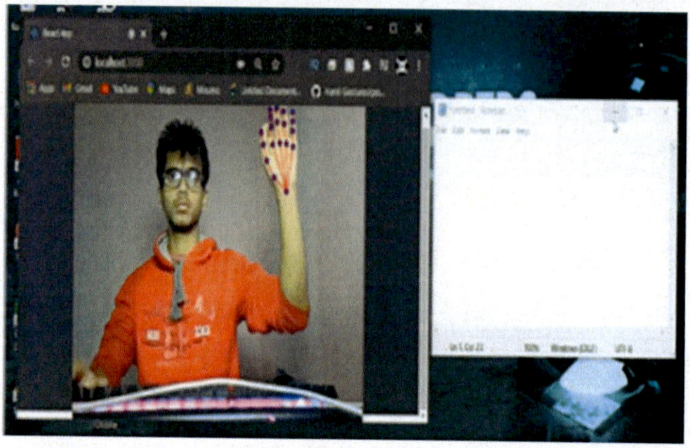

Fig. 6 Left-click demonstration

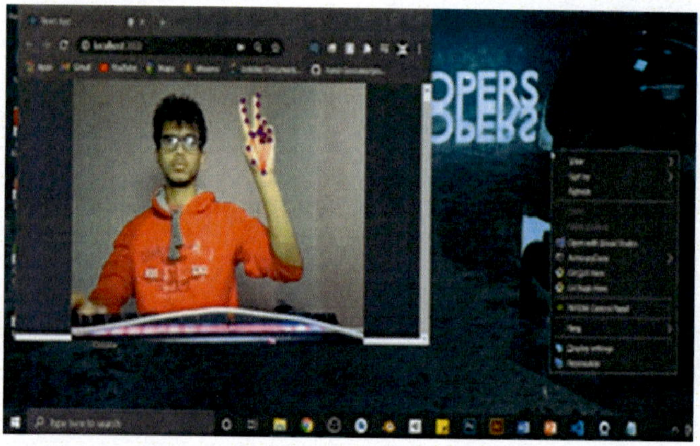

Fig. 7 Right-click demonstration

(cross-platform). There are certain limitations to be mentioned while working with a single view camera, the primary problem being the lightning, i.e., poor illumination and very high illumination. Our hand gesture-based mouse has many advantages that is it scalable, modifiable, wireless, and user-friendly. It is scalable as ample gestures could be added or dissolved as per wishes and the comfortability of users. It can also allow multiple user interactions. Depending on the choice of users, it is possible to do modifications to any gestures. The developed model being software-based can replace the generic mouse and can give a greater distance between the user and the screen. The software being gesture-based gives a great amount of leisure for hand and cursor coordination.

Future work
The technology has the potential to be a viable substitute for the computer mouse by extending with more robust and consistent tracking, increasing the number of gestures we can detect reliably, and supporting dynamic movements unfolding in time.

References

1. Microsoft Kinect Wikipedia Homepage. https://en.wikipedia.org/wiki/Kinect
2. Nintendo Wii Remote Wikipedia Homepage. https://en.wikipedia.org/wiki/Wii_Remote
3. Francis J, Anoop BK (2014) Significance of hand gesture recognition systems in vehicular automation-a survey. Int J Comput Appl 99(7):50–55
4. Mitra S, Acharya T (2007) Gesture recognition: a survey. IEEE Trans Syst Man Cybern part C, 37(3):311324
5. Ghotkar AS, Kharate GK (2012) Hand segmentation techniques to hand gesture recognition for natural human-computer interaction. Int J Hum Comput Interact 3(1):15–25
6. On-device, real-time hand tracking with media pipe by Valentin Bazarevsky and Fan Zhang, Research Engineers, Google Research. https://ai.googleblog.com/2019/08/on-device-real-time-hand-tracking-with:html
7. Python Wikipedia Homepage. https://en.wikipedia.org/wiki/Python(programming_language)
8. PyAutoGUI Homepage. https://pypi.org/project/PyAutoGUI/
9. Flask Wikipedia Homepage. https://en.wikipedia.org/wiki/Flask_(web_framework)
10. JavaScript Wikipedia Homepage. https://en.wikipedia.org/wiki/JavaScript
11. ReactJs Wikipedia Homepage. https://en.wikipedia.org/wiki/React_(JavaScript_library)
12. Visual Studio Code Wikipedia Homepage. https://en.wikipedia.org/wiki/Visual_Studio_Code
13. Tensorflow.js Homepage. https://www.tensorflow.org/js
14. Non-max suppression. https://learnopencv.com/non-maximum-suppression-theory-and-implementation-in-pytorch/
15. Project Source Code. https://github.com/bibekagohain/computer-vision-mouse

A Novel Time-Stamp-Based Audio Encryption Scheme Using Sudoku Puzzle

Sunanda Jana, Neha Dutta, Arnab Kumar Maji, and Rajat Kumar Pal

Abstract An innovative alternative method in the field of security that can enforce safety and security of information has been proposed here by performing audio encryption. Cryptography, a way to secure communications, allows only the sender and the intended recipient of a message to access its contents. Cryptography is derived from the Greek word "*Kryptos*," meaning hidden, which is associated with content hiding by encryption. In this paper, we develop a novel time-stamp-based audio encryption method using the mind-cracking number place challenger Sudoku puzzle of higher dimension, i.e., 16 × 16 (popularly known as Super Sudoku) to add a new era in the field of security. In this work, we have developed a Super Sudoku cipher by applying time-stamp in it. Immediately after that, the extracted audio frames of the input audio files are XORed with our developed innovative Cipher to generate encrypted audio frames. Then these encoded audio frames are converted to an encrypted audio file to transmit with safety and security. At the receiver end, by applying the same cipher key, we can extract our original audio file.

Keywords Audio encryption · Audio steganography · Super Sudoku · Time-stamp

1 Introduction

The explosion of information in the form of digital transmission has fascinated special importance from researchers regarding safety and security. Techniques like cryptography [1], steganography [2, 3], and watermarking have extensive use in this field.

S. Jana (✉)
Department of Computer Science and Engineering, Haldia Institute of Technology, Haldia Purba Medinipur, West Bengal 721657, India
e-mail: Sunanda.only1@gmail.com

N. Dutta · R. K. Pal
Department of Computer Science and Engineering, University of Calcutta, JD-2, Sector-III, Salt Lake, Kolkata, West Bengal 700106, India

A. K. Maji
Department of Information Technology, North-Eastern Hill University, Umshing Mawkynroh, Shillong, Meghalaya 793022, India

S. Basu et al. (eds.), *Proceedings of International Conference on Frontiers in Computing and Systems*, Lecture Notes in Networks and Systems 404,
https://doi.org/10.1007/978-981-19-0105-8_16

To protect data, cryptography uses encryption techniques, where the sender uses a key to generate an encrypted secret data. Now it may be quite difficult to extract the encrypted secret message [1] without knowing the key. Cipher is used for hiding data with encryption by replacing original data [4]. Different types of cipher have been developed to ensure data security. In the proposed work, we have generated a Super Sudoku cipher as a key to encrypt the audio file. This method has used one of the solutions of a number place challenger Sudoku of size 16×16 and the Unix time-stamp concept to enlighten the field of audio encryption. Sudoku, the logic-based puzzle, have different variants based on its size like 4×4 grids with 2×2 regions [5], "Logi-5," a 5×5 grid with 5-omino regions [5], "Mini Sudoku," a 6×6 variant with 3×2 regions [5], 7×7 grid with 6-omino regions and a disjoint region, "Number Place Challenger or Super Sudoku," 16×16 puzzles having 4×4 minigrids [6].

Sudoku is indicated by $n \times n$, where n must be a perfect square integer. The logic for standard Sudoku is such that, each row, each column, and each minigrid contains integers 1 through n without repetition. Sudoku is not just a game. It has a vast area of applications starting from steganography, image authentication to data security using encryption. We have proposed a simple and robust audio encryption because nowadays securing audio, video in today's digital world has become especially important.

In our proposed algorithm, we have applied the UNIX time-stamp [7] concept because they can represent entire time zones globally. For example, 1296001734 UNIX time-stamp can represent 6.28 pm on the 25th January 2011 in London, 11.28 am on the 25th January 2011 in New York, and 5.28 am on 26th January in Wellington, New Zealand. It has wide application in website making where everything will be in the form of real-time. We have used the concept of the time-stamp to generate our Sudoku cipher because, as it is changing every moment, so it is quite difficult to track that time-stamp which sender has embedded to generate cipher. As well as in the receiver site, decryption is possible by using the key cipher with that time-stamp. For that reason, the sender sends both cipher and time-stamp so that that receiver can extract the original audio file. So, it is difficult for an attacker to predict the time-stamp, which enhances the security of our proposed method by double.

1.1 Two-Dimensional Sudoku

The Sudoku structure we are familiar with is a 2D grid structure in which each constraint of Sudoku is applied in each row, column, and minigrid (each $\sqrt{n} \times \sqrt{n}$ square represented by bold outline). Generally, we find a 9×9 2D Sudoku as a standard puzzle in various newspapers. Sudoku is not just a game. It has a vast area of applications starting from steganography, image authentication to data security using encryption. As in our work, we are dealing with bigger Sudoku instances, i.e., number place challenger (16×16 Sudoku having 4×4 minigrids) which is also

2	11	9	5	8	16	13	4	12	3	14	7	10	6	15	1
4	12	15	10	3	6	9	11	13	5	8	1	16	7	14	2
1	14	6	7	15	2	5	12	11	9	10	16	3	13	8	4
16	13	8	3	14	1	10	7	4	6	2	15	9	11	5	12
12	2	16	9	10	14	15	13	8	1	5	3	6	4	11	7
6	7	1	11	5	12	8	16	9	15	4	2	14	10	3	13
14	5	4	13	6	11	1	3	16	12	7	10	8	9	2	15
3	8	10	15	4	7	2	9	6	14	13	11	1	12	6	5
13	9	2	16	7	8	14	10	3	4	15	6	12	5	1	11
5	4	14	6	2	13	12	1	10	16	11	8	15	3	7	9
7	1	11	12	16	4	3	15	5	13	9	14	2	8	10	6
10	15	3	8	9	5	11	6	2	7	1	12	4	14	13	16
11	10	13	14	1	9	7	8	15	2	6	4	5	16	12	3
15	3	7	4	12	10	6	5	1	8	16	13	11	2	9	14
8	6	5	1	13	3	6	2	14	11	12	9	7	15	4	10
9	16	12	2	11	15	4	14	7	10	3	5	13	1	6	8

Fig. 1 One of the solutions of a 2D number place generator (16×16) Sudoku instance

known Super Sudoku, so an example of such 16×16 Sudoku solution is given in Fig. 1.

1.2 A Brief Description of Audio Encryption and Audio Decryption

Encryption is the process of encoding information into encrypted cipher form which is indecipherable without a decoding key to prevent anyone except the intended recipient from accessing that data. Decryption is the reverse process of extracting the original file or data from its encoded form [8]. When a user encodes a file, another user cannot decode and read the file without the decryption key. In this process, a key [8] is used to decode the information, which may be in any form like text, image, audio, or video. Get the original plain text. An encryption algorithm takes the original message and a key, and alters the original message based on the key to a ciphertext and creates a new encrypted message. There are two encryption methods available:

- Symmetric key encryption, and
- Public-key encryption.

In symmetric encryption, the same key is used for both encoding and decoding the data. In our work, we have incorporated the concept of symmetric key. Here we have generated a key by using two import concepts to maximize security by constructing the key in such a way that it is very tough to access. Firstly, we have considered time-stamp of a particular moment, and secondly, we have applied that time-stamp

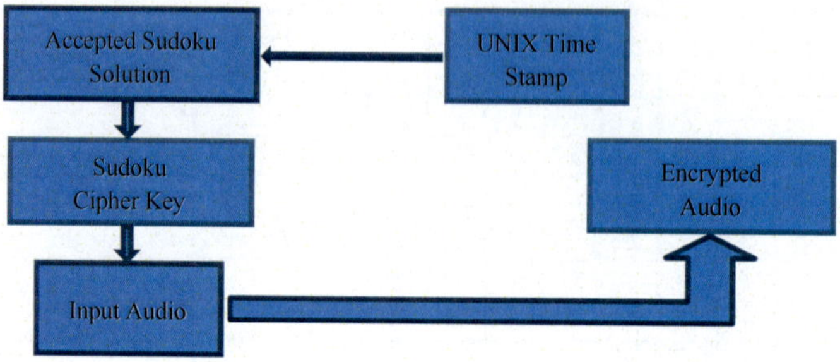

Fig. 2 An audio encryption process

into that particular Sudoku to generate cipher array. Then this Super Sudoku cipher and that specific time-stamp will be used for both encryption and decryption (Fig. 2).

In public-key encryption two dissimilar keys, a public key and a private key or a secret key, are applied. One key encodes the data, and the other performs decoding. In our proposed method, our input data is an audio file. We are performing audio encryption using symmetric key encryption.

1.3 Existing Literature on Audio Encryption

We know that the concept of encryption is to perform secure communication in an untraceable way [9]. Till now no such extensive works have been done in the field of audio encryption using Sudoku. In 2011, Y. Bassil developed an audio steganographic technique [10] based on two intermediates for secret data for transmission. Two intermediates are an audio file and its grammatically correct English text. They use location encoding of the randomly taken samples audio of the audio file as their audio steganographic method. In 2020, A. H. Ali proposes an audio steganography method [11] having a capacity of higher level, which uses fractal coding and uniform coefficient modulation, keeping in mind about the transparency and robustness. Various method of audio steganography adds safety and security in the field of digital transmission. Gupta et al. [12] have come up with an information hiding scheme which combines image steganography with a cryptographic method where encrypted data is embedded into the image. Khare et al. [13] developed a technique that hides the text messages into a digital image. However, implementing audio steganography by using Super Sudoku cipher, which is again achieved by applying the unique time-stamp concept is completely an innovative work in this field of information security.

2 Proposed Scheme for Encrypting Audio Using Bigger Sudoku Instance

In our proposed Audio encryption Scheme, we have used one of the solutions of number place challenger Sudoku, 16 × 16. Then we have used the UNIX time-stamp (which is sent along with the file containing one of the Sudoku solutions), the unique number whose value is changing every moment globally, to create Super Sudoku cipher as our key. By applying that key, we encrypt our audio frames of our input audio file to generate encrypted audio, which is transmitted securely, and the receiver decrypts the original frame's bits by again XORing the encrypted frame bits with that Super Sudoku cipher key. Every time the file is sent, the UNIX time-stamp changes according to the current file creation time.

So, our proposed innovative method uses 16 × 16 number place challenger Sudoku matrix-based Super Sudoku cipher for encrypting and decrypting the audio file. Many algorithms in the field of image encryption [14], image steganography and watermarking [15] have been proposed until today applying Sudoku puzzle. Till date, no such effort has been made to perform audio encryption using Sudoku-based cipher. The main reason for choosing number place challenger Sudoku is that a blank Sudoku puzzle produces a huge number, i.e., about 5.96×10^{98} different Sudoku matrices as solutions [16]. We can choose any solution matrix among this huge number of solutions for the cipher generation for our proposed method. Because of this reason, if we pick Sudoku for cipher generation, there will be an extremely limited chance for the attacker to extract the key. Another important fact is Sudoku's uniqueness property which means digits from 1 to 16 occur exactly once for each row, column, and minigrid. So, any ciphertext which is encrypted using the Sudoku matrix is less susceptible to modify by the attacker as the uniqueness constraint of this Sudoku matrix may be debased.

2.1 Time-Stamp in Proposed Model

As our proposed concept is working by considering time-stamp, so first we will discuss the concept behind time-stamp. Trailing time in the form of running total of seconds is considered, in the UNIX time-stamp [17]. It is kept as a 32-bit signed binary integer. It is a way to track time as a running total of seconds. This time count begins at the Unix Epoch on the first day of January 1970 at UTC. This is the number of seconds between a specific date and the Unix Epoch. This time does not vary depending on location throughout the globe.

By using the time-stamp, our proposed algorithm is calculating Super Sudoku cipher, so that an attacker cannot extract the key. It is so because every moment time-stamp is changing globally. For that reason, it is quite impossible to track which time-stamp a sender has used to generate Super Sudoku cipher, and at the receiver end, this key and the time-stamp are shared to extract the original audio file. In our proposed method, we have used time-stamp value 1,584,029,637, which is corresponding to Thursday, 12th March 2020 16:13:57 +0000, and from this time-stamp we extract a code by summing all its digits and by performing modulus operation with this we choose which particular Sudoku solution we have to use among multiple solutions [16, 18] as well as we can generate Sudoku cipher.

2.2 *Proposed Algorithm for Audio Encryption*

Here we elaborate our proposed novel idea where we have combined the concept of the time-stamp with 16 × 16 Sudoku, i.e., a number place challenger to generate an untraceable symmetric key. In this way, we have enlightened the field of audio encryption by developing a unique encrypted audio code which is quite impossible to extract by an attacker.

Algorithm

Let the Sudoku size be $m \times m$. We have used here a number place challenger of size 16 × 16.

Input: An audio file (max 30 s and in .wav format) and a stream of solved number place challenger (say, $n = 100$) in a file.

Output: An encrypted audio file.

STEP 1: Read the frames of the input audio file and convert it into its binary format and store each frame in each line in a file (audtobin.txt). Then calculate the size of this file and store it in a variable (say n).

STEP 2: Take the 32-bit UNIX time-stamp and sum up all its digits and store the summation in a long int type variable (say p). Modify the number p as follows

$t = p\%16.$
$u = p\%100.$

STEP 3: Break the u-th number place challenger (as $n = 100$) Sudoku, which is shown in

Fig. 1

into all possible 2 × 2, 3 × 3, ..., $m \times m$ in both row-major and column-major order.
Multiplied with the number t and converted to its binary using a function and then stored this Super Sudoku cipher key in an array.

Then check–if the end of the Sudoku has reached or not.

If yes, then check–still reached n or not.

If not reached n, then start repeating the values of the array from the beginning until we reach n.

STEP 4: Then XORed the values of the Super Sudoku cipher key array with the audio bits in the audtobin.txt file to generate an encrypted frame.

STEP 5: The encrypted audio frames are then converted to an encrypted audio file.

Decryption

We can easily get back the original frame bits by again XORing the encrypted frame bits with the Sudoku cipher key array. This will be the procedure of decryption. For simplification of understanding, a stepwise explanation is given below.

STEP 1: We have taken an audio file (a voice recording) of 27 seconds length.

As it is quite difficult to show the entire 2.5 lac frames, so we have explained the working methodology of the algorithm using the following two frames–

['0b1000' '0b1000']
['0b10100' '0b10100']
Since all the audio records are recorded nowadays through most of the electronic gadgets and they are using dual-channel, so the repetition of the bits in each frame can be observed.

STEP 2: For simplification of understanding, we have considered a sample time-stamp.
time-stamp = 1584029637 corresponding to Thursday, 12 March 2020 16:13:57 +0000.
p = 1+5+8+4+0+2+9+6+3+7 = 45.
t = 45%16 = 13.
u = 45%100 = 45.

STEP 3: Break the u-th Sudoku into all possible 2×2, 3×3, …, $m \times m$ in both row-major and column-major order which is shown in

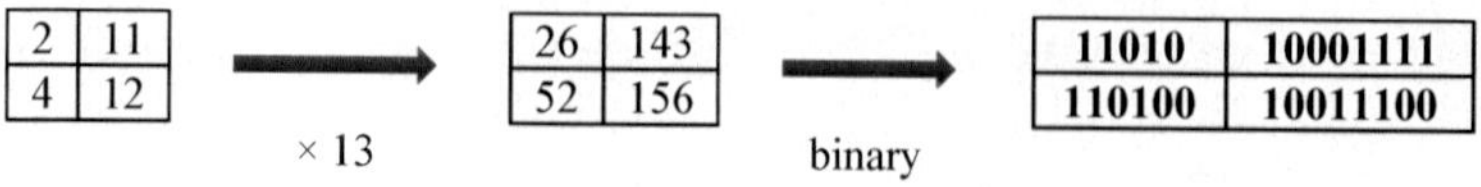

Fig. 3 Generating Sudoku cipher key

Fig. 3.

Then multiply it with the number, t and converted to its binary form using a function and stored that Super Sudoku cipher in an array.

Thus, the Sudoku cipher array becomes
1,1,0,1,0,1,0,0,0,1,1,1,1,1,1,0,1,0,0,1,0,0, ...
We calculate this up to the size, n.
The first Super Sudoku cipher bit is used to encrypt both the bits in the first valid frame, then we move on to the second frame, and so on. Hence, we get all the encrypted audio frames. Here, we have encrypted all the bits of all the valid frames in the audtobin.txt file. There are few frames with '0' value; we have not encrypted those frames.

STEP 4: The frames were–

['0b1000' '0b1000']
['0b10110' '0b10110']
For the first frame, the digits are XORed with 1.
For the second frame, the digits are XORed with 0.

The encrypted frames are–

['0b0111' '0b0111']
['0b10110' '0b10110']
Then, convert these encrypted frames into an audio file which we have to transmit.

Decryption
We already have the Super Sudoku cipher array with us which we got from Step 3. We again XORed it with the last two bits of the encrypted frame.

The encrypted frames were—
['0b0111' '0b0111']
['0b10110' '0b10110']
For the first frame, the digits are XORed with 1.
For the second frame, the digits are XORed with 0.
Thus, the decrypted frames are—
['0b1000' '0b1000']
['0b10110' '0b10110']

We can observe that these are the original frames.

3 Experimental Results Analysis

The system configuration of the machine on which the program was executed, and the platform used is—

- Processor: AMD Quad-core processor.
- RAM: 4 GB.
- Windows 10 64-bit.
- Python 3.8.

We have executed the algorithm for audio songs, human voice, animal noises (within 30 s). The execution time varies from 30 s to 2 min maximum.

Considering the solution of a sample Super Sudoku given in Fig. 1 we will perform our implementation.

First few frames are blank and with smaller values. Therefore, skipping these frames we have taken the ten frames from the beginning of the file so that the working of the algorithm could be better understandable. Out of the total 227,360 frames (approximately), the first five frames are shown in Table 1.

time-stamp = 1,584,029,637 corresponding to Thursday, 12 March 2020 16:13:57 + 0000.

$p = 1 + 5 + 8 + 4 + 0 + 2 + 9 + 6 + 3 + 7 = 45.$

$t = 45\%16 = 13.$

$u = 45\%100 = 45.$

Sudoku cipher (first ten bits, the original size is 2.5 lac bits approximately) 1101010001 … …

Table 1 First 5 frames of original audio and encrypted audio

Frames of original audio	Frames of encrypted audio
['-0b111' '-0b111']	['-0b000' '-0b000']
['-0b110' '-0b110']	['-0b001' '-0b001']
['-0b101' '-0b101']	['-0b101' '-0b101']
['-0b100' '-0b100']	['-0b011' '-0b011']
['-0b101' '-0b101']	['-0b101' '-0b101']

Table 2 Comparison between our proposed method and Agarwal et al. [7]

Encryption technique	Complexity	Key type	Possible key combinations
AES-128	Complex	Private	2128
Proposed algorithm	Complex	Private	5.96×10^{98}

The data frames (out of the total 227,360 frames, the first five are shown) of the encrypted audio file are shown in Table 1 and in Table 2 our proposed work has been compared with Agarwal et al. [7] where they have used AES for encryption. The decrypted file is the same as that of the original data frame file. So, we have not shown it again.

We can clearly observe that security of our proposed encryption algorithm is much higher compared to AES algorithm. In this algorithm not only more possible key combination but also unique time-stamp concept makes it unbreakable and robust against different attacks.

4 Computational Complexity

This part focuses on complexity issues of the proposed algorithm. In this work, firstly the time-stamp calculation part is a one-time job which can, therefore, be assumed to be in constant time, say k. Then, Sudoku array cipher is generated up to the size of the total number of frames in the original audio file. Considering the size of the Sudoku to be $m \times m$, and the number of frames in the audio file to be n, the generation of the cipher array can take maximum n time. Therefore, we can say that n frames are modified using XOR operation with the n bits cipher array, again this will take at most n time. So, the complexity is $O\ (k + n + n) = O(n)$, since k is a constant and the value of k is much less than that of n.

5 Conclusion and Future Scope

In this paper, we have attempted to explore how, by using a number place challenger Sudoku (i.e., 16×16) solution, we can enhance the hiding power of audio cryptography. Since a Sudoku puzzle may have multiple solutions, so we have used it to encrypt our message for security, and an intruder is expected not to know which solution is used to encrypt the message. Following such an approach, we brought this novel algorithm to encode and decode our audio data. Time-stamp added enhanced security in our proposed work by applying its global-wise uniqueness concept. This approach is just an initial step toward the applications of larger Sudoku instances, which if used and implemented wisely and efficiently, can be of great help in the field of the message and information security.

References

1. Sharma N, Prabhjot, Kaur H (2017) A review of information security using cryptography technique. Int J Adv Res Comput Sci 8(no. Special Issue):323–326
2. Jana S, Maji AK, Pal RK (2019) A novel SPN-based video steganographic scheme using Sudoku puzzle for secured data hiding. Innov Syst Softw Eng 15(1):65–73
3. Maji AK, Roy S, Pal RK (2014) A novel steganographic scheme using Sudoku. In: IEEE international conference on electrical information and communication technology, KUET, Khulna, Bangladesh, pp 116–121
4. https://pingler.com/blog/what-are-unix-timestamps-and-why-are-they-useful/
5. https://en.wikipedia.org/wiki/Sudoku
6. http://www.en.wikipedia.org/wiki/SuperSudoku
7. Agarwal A, Singh PR, Katiyar S (2019) Secured audio encryption using AES algorithm. Int J Comput Appl (0975–8887) 178(22):29–33
8. Kahate A (2003) Cryptography and network security. Tata McGraw-Hill Education
9. Cvejic N, Seppben T (2002) Increasing the capacity of LSB-based audio steganography. FIN90014 University of Oulu, Finland
10. Bassil Y (2012) A two intermediates audio steganography technique. J Emerg Trends Comput Inform Sci 3(11)
11. Ali AH, Georg LE, Mokhtar MR (2020) An adaptive high capacity model for secure audio communication based on fractal coding and uniform coefficient modulation. Circ Syst Signal Process 39:5198–5225
12. Gupta S, Goyal A, Bhushan B (2012) Information hiding using least significant bit steganography and cryptography. Int J Mod Educ Comput Sci 6:27–34
13. Saradha P, Swamy B (2012) Improving image data hiding capacity scheme using Sudoku puzzle in colour images. Int J Eng Res Appl 2(3):2741–2744
14. Wu Y, Noonan JP, Agaian S (2011) Image encryption using the rectangular Sudoku Cipher. In: International conference on system science and engineering proceedings, pp 704–709
15. Felgenhauer B, Jarvis F (2005) Enumerating possible Sudoku grids. Math Spect (Corpus ID: 16859520), pp 1–7
16. https://www.techopedia.com/definition/6472/cipher. Last Accessed 29th Oct 2018
17. https://www.unixtimestamp.com/index.php
18. www.afjarvis.staff.shef.ac.uk/sudoku

MathIRs: A One-Stop Solution to Several Mathematical Information Retrieval Needs

Amarnath Pathak, Partha Pakray, and Ranjita Das

Abstract Recent few years have witnessed a huge surge in the scientific documents on the Web, and consequently, the systems capable of retrieving mathematical information from such documents have also come into being. All the existing systems adopt a suitable strategy to handle a particular challenge of mathematical information retrieval (MIR). However, it is quite uncommon to find a single system which alone handles multiple unrelated challenges of MIR. This paper, therefore, discusses implementation of a mathematical information retrieval system (MathIRs) which comprises different modules to account for several different unrelated challenges and serves as one-stop for several mathematical information retrieval needs. Evaluation of MathIRs using standard scientific document set and query set and its comparison with other competent systems proves its competence in the field of MIR.

Keywords Math Information Retrieval (MIR) · Long Short-Term Memory (LSTM) · Context guided retrieval · Formula context extraction · Formula embedding · MathIRs · NTCIR

1 Introduction

Mathematical information retrieval (MIR) concerns retrieval of mathematical contents, such as formula, equation, and symbols, contained inside scientific documents. Mathematical contents are significantly different from plain text and, hence, need to be handled differently. For instance, the conventional text retrieval techniques of

A. Pathak (✉) · R. Das
National Institute of Technology Mizoram, Aizawl, India
e-mail: amar4gate@gmail.com

R. Das
e-mail: rdas@nitmz.ac.in

P. Pakray
National Institute of Technology Silchar, Silchar, Assam, India
e-mail: partha@cse.nits.ac.in

S. Basu et al. (eds.), *Proceedings of International Conference on Frontiers in Computing and Systems*, Lecture Notes in Networks and Systems 404,
https://doi.org/10.1007/978-981-19-0105-8_17

lexical matching of query with indexed contents and generating inverted index for contents inside documents fail to retrieve mathematical contents. The following have been identified as some of the noteworthy challenges of MIR:

1. *Retrieving similar formula, sub-formula, parent formula, and alternate representations alongside exact formula*:
   A text retrieval system retrieves synonyms, heteronyms etc., of the text query alongside exact text. Similarly, for a given formula query (say $x^2 = y^2 + z^2$), a Math Search Engine (MSE) should retrieve exact formula, similar formula (say $a^2 = b^2 + c^2$), sub-formula (say $y^2 + z^2$), parent formula (say $x^2 = y^2 + z^2 + a^2 + b^2$), and different possible alternate representations.
2. *Recognizing entailment relationship between pairs of formula*:
   An MSE should also account for entailment relationship, if any, between pairs of formulae. For instance, the formula $e^{\ln x}$ entails (evaluates to) x, although the two formulae look syntactically dissimilar.
3. *Context-guided retrieval of math formula*:
   The textual description of a formula is often referred to as its context, and it often features in the line or the paragraph in which the formula is present. For instance, in the sentence "*Sum of the squares of length and breadth of a right-angled triangle equals square of hypotenuse is popularly known as Pythagorean theorem, and is mathematically described as* $l^2 + b^2 = h^2$," the appropriate context associated with the formula $l^2 + b^2 = h^2$ is "*Pythagorean theorem.*" An MSE should facilitate searching a formula using a formula query as well as its context information.
4. *Extracting context of math formula*:
   An end user may also be interested in retrieving context information associated with the formula. Thus, given a formula query, an MSE should also be able to retrieve all possible associated context information from inside the indexed documents.
5. *Use of state-of-the-art techniques for MIR*:
   Use of state-of-the-art techniques, specifically long short-term memory (LSTM) neural networks, has showcased promising performance in several application domains of natural language processing (NLP), such as text summarization and machine translation. However, use of such networks for MIR is obscure and less known.

Many of the aforesaid challenges have been comprehensively addressed in past. Moreover, the math tasks [1, 2, 11] of NII Testbeds and Community for Information access Research (NTCIR) workshops have contributed a number of efficient systems, which address several challenges of MIR domain.

However, it is rare to see a single system, which can successfully account for all or many such unrelated challenges. In order to alleviate this problem and to qualitatively boost the retrieval performance, this research discusses implementation of a Math Information Retrieval system (MathIRs), which handles majority of the above-mentioned challenges and serves as one-stop for several MIR needs. More specifically, the implementation of MathIRs covers the following modules: (a)

Binary Vector Transformation of Math Formula (BVTMF) approach to retrieve similar formula, sub-formula and parent formula (b) an editor to input text and formula query in MathML notation (c) a long short-term memory (LSTM) neural network to recognize entailment between pairs of formula and to perform MIR (d) use of a knowledge base comprising of context-formula pairs to retrieve formula using context information, and (e) use of a heuristic-based approach followed by weight tuning to retrieve context information associated with a query formula. The performance of MathIRs has been evaluated using Wikipedia corpus and Wikipedia Main Task query set of NTCIR-12 MathIR task [11]. It outperforms the other NTCIR systems for relevant as well as partially relevant result categories on the grounds of Precision score.

Rest of the paper is structured as follows: Sect. 2 describes architecture and different components of the systems. Section 3 describes experimental results and their analysis. Section 4 concludes the paper and points directions for future research.

2 System Description

2.1 Corpus, Query Set, and Gold Dataset Description

In this paper, the MathTagArticles inside Wikipedia corpus of NTICR-12 MathIR task [11] have been used for experimentation. This corpus is composed of 31,839 articles, which contain around 579,268 formulae alongside scientific text. The formulae inside the corpus are encoded using three different languages, namely Presentation MathML, Content MathML, and LaTeX.

The query set provided by NTCIR-12 MathIR task organizers contain 30 queries. Majority of these queries contain text keywords alongside formula. Ten out of 30 queries contain query variable, which act as wildcard and can match to a single symbol or even a more complex mathematical sub-expression. These wildcards in the query are represented using a number surrounded by asterisk, e.g., $*1*$.

The Gold dataset of Wikipedia main task comprises relevant as well as partially documents corresponding to each query in the query set. The gold dataset provided by the organizers follows trec_qrel[1] format.

2.2 System Architecture

Figure 1 shows the architecture of MathIRs. It comprises of number of modules which work in coordination to furnish search results corresponding to the user query.

---

[1] https://trec.nist.gov/.

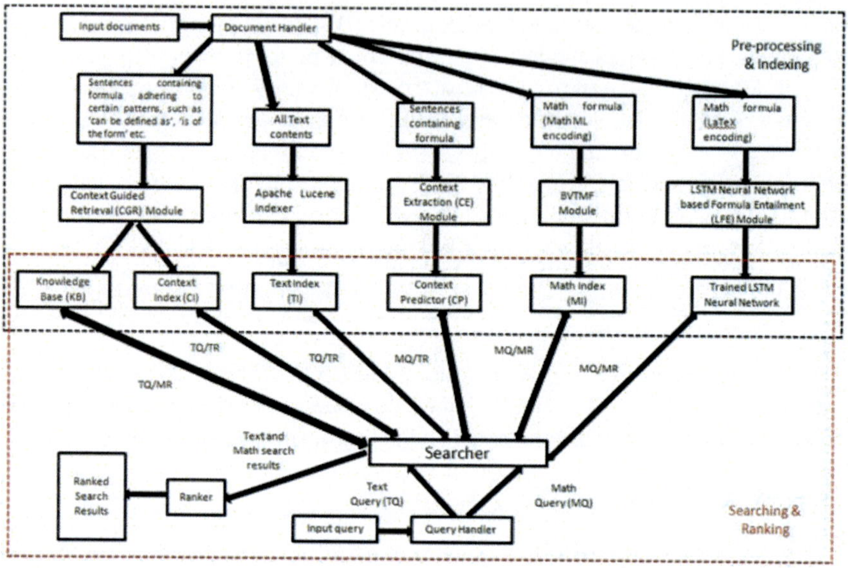

Fig. 1 Architecture of Math Information Retrieval system (MathIRs)

Document Handler takes each HTML document of the corpus as input and extracts following information in isolated manner: (a) all sentences which contain formula (b) all sentences which contain formula and the formula and context are linked through 09 characteristic patterns, namely "*is of the form*," "***can be defined as***," "***is defined to be***," "*is given by*," "*is then described by*," "*can be expressed as*," "*can be written as*", "***is defined as***" and "*has the following form*" (c) all text contents inside document (d) all MathML encoded formula, and (e) all LaTeX encoded formula. All such information serve as inputs to different modules of MathIRs.

Similar to Document Handler, the Query Handler separates an input query into text query (TQ) and math query (MQ), which are fed to the Searcher for further processing. Separating a query into text and math parts is essential as different components of MathIRs operate on text and math parts in isolated fashion.

The context-guided retrieval (CGR) module first generates a knowledge base (KB) comprising of context–formula pairs. All sentences containing formula and any of the 09 characteristic patterns are examined by CGR for this purpose. For instance, given the sentence "*Pythagorean theorem is given by* $h^2 = l^2 + b^2$," the CGR extracts the context (i.e., "*Pythagorean theorem*")—formula (i.e., $h^2 = l^2 + b^2$) pair and stores it into KB. Thereafter, the contexts inside KB are indexed using Apache Lucene[2]. indexer to expedite retrieval of formula search results for a given TQ. Searcher searches a TQ in the Context Index (CI), which in turn provides two most relevant contexts. Subsequently, the retrieved contexts for TQ are searched in KB, and the

[2] https://lucene.apache.org/

corresponding Math Result (MR) and the documents containing them are retrieved. Further details related to CGR module can be found in [6].

To allow searching of text contents, an inverted Text Index (TI) of all the text contents present inside documents is generated by Apache Lucene Indexer available with Apache Nutch[3] platform. The indexer uses different sub-components, such as injector, generator, updater, segment directory, fetcher, and parser for this purpose. The Apache Lucene Indexer and Searcher (ALIS) module retrieves text results (TRs) corresponding to the text query. Also, the default scoring mechanism based on page rank algorithm is used by ALIS to assign score to the retrieved documents.

Context extraction (CE) module takes all sentences containing formula as input and generates a context predictor (CP), which predicts relevant contexts for an MQ. The foremost step is to parse all such sentences using Stanford Shift-Reduce Constituency Parser[4] and extract all noun phrases (NPs) from the parse tree of sentences. It is rightly assumed that the context corresponding to formula in a sentence is one of the NPs present in the sentence. However, to select one of the NPs, weights are assigned to NPs based on certain assumptions. Subsequently, the weights are tuned using a development set comprising of formulae and their respective gold contexts. The best set of weights, i.e., the best configuration is termed as CP, which is used to predict context for a given MQ. Further details related to CE module can be found in [5].

Using a Bit Position Information Table (BPIT), the BVTMF module transforms formulae into binary vectors and generates a Math Index (MI) containing vectors and their respective document information. BPIT maps each mathematical entity to its position in a binary vector of length 202. An MQ is also transformed into binary vector in similar fashion, and its similarity is computed with each formula in the index. Subsequently, the relevant documents containing MQ are retrieved from the MI. Further details can be found in [8].

An LSTM neural network-based formula entailment (LFE) module, which checks for entailment between MQ and document formulae, has been implemented and used in MathIRs. The MATHENTAIL dataset containing annotated symbol pairs, wherein each symbol pair is either annotated either "*entailment*" or "*neutral*," is used to train a two-layer LSTM neural network. A formula query is also tokenized and its entailment is computed with each formula inside a document. The document is considered relevant if the fraction of formulae inside document (*efrac*), which entail the user query, exceed a threshold 0.95. Further details related to LSTM neural network-based MIR and creation of MATHENTAIL dataset can be found in [7].

The Searcher module employs BVTMF, CGR, LFE, ALIS, and CE modules to predict relevant documents corresponding to a query. Further, the documents retrieved by Searcher module are integrated by the Ranker module using a suitable ranking mechanism. The Ranker module assigns more preference to the documents retrieved by BVTMF followed by LFE, CGR, ALIS, and CE modules. The documents con-

[3] http://nutch.apache.org/.

[4] https://nlp.stanford.edu/software/srparser.html.

taining both text as well as math are preferred over those containing only math or only text.

3 Experimental Results and Analysis

3.1 Evaluation Parameters

In order to enable performance comparison of MathIRs with other systems, the same Precision scores, i.e., P@5, P@10, P@15, and P@20, as used by the NTCIR MathIR task organizers, have been used to evaluate the performance. These evaluation parameters are chosen because of their simplistic nature and their ability to characterize retrieval behavior. P@k score is indicative of the number of relevant documents out of first k retrieved documents. The trec_eval tool operates upon Gold dataset and ResultSet to output values of these parameters.

3.2 System Results

Table 1 shows performance comparison of MathIRs with other NTCIR systems implemented for Wikipedia Main Task. As can be seen from the table, MathIRs outperforms other systems for both relevant as well as partially relevant result categories. This is attributed to the competence of MathIRs and shortcomings of the NTCIR systems, some of which are as follow:

Table 1 Comparison of MathIRs with other NTCIR systems implemented for Wikipedia main task

NTCIR systems	Relevant				Partially relevant			
	P@5	P@10	P@15	P@20	P@5	P@10	P@15	P@20
ICST	0.4733	0.3767	0.2978	0.2617	0.8533	0.7900	0.7133	0.6600
MathIRs (Implemented system)	**0.5067**	**0.4233**	**0.3333**	**0.2867**	**0.8733**	**0.8133**	**0.7311**	**0.6617**
MCAT	0.3600	0.3233	0.2689	0.2433	0.7667	0.7167	0.6867	0.6533
SMSG5	0.3700	0.2900	0.2400	0.2200	0.7500	0.7000	0.6700	0.6400
TANGENT	0.2533	0.2467	0.2156	0.2017	0.4933	0.4900	0.4822	0.4700

The row highlighted in bold refers to the precision scores of implemented MathIRs system for relevant as well as partially relevant result categories.

Table 2 Sample search results retrieved by MathIRs in the form of documents and their excerpts for some of the Wikipedia Main task queries; *R*: relevant, PR: partially relevant

NTCIR query	Excerpt	Document	Annotation
$F = ma$	$\mathbf{F} = m\mathbf{a}$	Cloth_modeling	R
Proof $x^2 + y^2 = z^2$	$z^2 = x^2 + y^2$	Pseudomanifold	PR
Legendre $(\frac{a}{p})$	Legendre symbol ...$(\frac{p}{q})$..	Reciprocity_law	R

First, MathIRs integrates results from different modules, which are designed to handle different set of challenges. The search results containing text and math, which are highly desirable, are assigned highest preference. Some of the sample search results retrieved by MathIRs are shown in Table 2. These sample results depict that MathIRs successfully retrieves search results containing similar formula, sub-formula, parent formula, and those containing text as well as formula. *Second,* the best performing ICST math retrieval system [3] uses semantic enrichment technique to extract structural and semantic information from layout presentations of formulae. However, the ICST system only uses few of the Wikipedia files for experimentation. Moreover, it fails to handle correlation between text query and indexed mathematical contents. *Third*, the MCAT math retrieval system [4] uses dependency graph method to enrich the textual information and perform retrieval. However, the failure of dependency graph method by MCAT owes to the fact that majority of the text keywords present in the query resemble stopwords, and hence, their enrichment is not beneficial. Only 5 out of 30 queries contain keywords which are uncommon and specific to the formula. *Fourth*, the TANGENT system [9] also fails to account for relationships between text and math and does not use text features or contexts associated with the formula. *Fifth*, the Borda count-based ranking mechanism used by SMSG5 system [10] takes into consideration different knowledge bases, but assigns same weightage to the results from these knowledge bases, which does not work well for majority of the queries.

Furthermore, to assess the contribution of different modules of MathIRs, precision scores are computed for different combination of constituent modules. More specifically, the search results retrieved by different constituent modules are removed one by one from the complete search result, and the effect on precision score is observed. The precision scores obtained for different variants of MathIRs for relevant and partially relevant search results are shown in Tables 3 and 4, respectively. The performance of MathIRs is chiefly affected by the removal of ALIS module search results. This is because of the fact that the Ranker module restricts the ALIS module to retrieve only those documents which contain text as well as math. Since such documents are highly desirable and relevant, their removal from the result list degrades the performance of MathIRs.

Table 3 Precision scores of different variants of MathIRs for relevant search results

	P@5	P@10	P@15	P@20
MathIRs (BVTMF+LFE+CGR+ALIS+CE)	0.5067	0.4233	0.3333	0.2867
BVTMF+CGR+ALIS+CE	0.5000	0.4133	0.3222	0.2800
BVTMF+ALIS	0.5000	0.4133	0.3222	0.2617
BVTMF	0.4467	0.3533	0.2689	0.2350

Table 4 Precision scores of different variants of MathIRs for partially relevant search results

	P@5	P@10	P@15	P@20
MathIRs (BVTMF+LFE+CGR+ALIS+CE)	0.8733	0.8133	0.7311	0.6617
BVTMF+CGR+ALIS+CE	0.8467	0.7933	0.7111	0.6450
BVTMF+ALIS	0.8467	0.7933	0.7111	0.6417
BVTMF	0.7800	0.7100	0.6311	0.5650

4 Conclusion and Future Directions

This paper describes implementation of Math Information Retrieval system (MathIRs), an application which comprises five distinct modules to retrieve relevant search results for queries containing text as well as math. The three of the five modules, namely binary vector transformation module, LSTM-based formula entailment module and context-guided retrieval module, retrieve formula search results, whereas the other two modules, namely Apache Lucene Indexer and Searcher Module and Context Extraction Module, retrieve text results. The search results retrieved by the above-mentioned five modules are ranked using a suitable ranking mechanism, which gives preference to the search results containing text as well as math. Incorporating a suitable mechanism to handle wildcards in the query and evaluating performance of other ranking mechanisms are some of the future directions worth exploring.

References

1. Aizawa A, Kohlhase M, Ounis I (2013) Ntcir-10 math pilot task overview. In: Proceedings of the 10th NTCIR Conference. Tokyo, Japan, pp 654–661
2. Aizawa A, Kohlhase M, Ounis I, Schubotz M (2014) Ntcir-11 math-2 task overview. In: Proceedings of the 11th NTCIR conference on evaluation of information access technologies, vol. 11. Citeseer, Tokyo, Japan, pp 88–98
3. Gao L, Wang Y, Hao L, Tang Z (2014) Icst math retrieval system for ntcir-11 math-2 task. In: Proceedings of the 11th NTCIR conference. Citeseer, Tokyo, Japan, pp 99–102
4. Kristianto GY, Topic G, Ho F, Aizawa A (2014) The mcat math retrieval system for ntcir-10 math track. In: Proceedings of the 10th NTCIR conference. Citeseer, Tokyo, Japan, pp 680–685

5. Pathak A, Das R, Pakray P, Gelbukh A (2019) Extracting context of math formulae contained inside scientific documents. Computación y Sistemas 23(3)
6. Pathak A, Pakray P, Das R (2019) Context guided retrieval of math formulae from scientific documents. J Inf Optim Sci 40(8):1559–1574. https://doi.org/10.1080/02522667.2019.1703255
7. Pathak A, Pakray P, Das R (2019) LSTM neural network based math information retrieval. In: Second international conference on advanced computational and communication paradigms (ICACCP). IEEE, pp 1–6
8. Pathak A, Pakray P, Gelbukh A (2019) Binary vector transformation of math formula for mathematical information retrieval. J Intell Fuzzy Syst 36(5):4685–4695
9. Pattaniyil N, Zanibbi R (2014) Combining tf-idf text retrieval with an inverted index over symbol pairs in math expressions: the tangent math search engine at ntcir 2014. In: Proceedings of the 11th NTCIR conference. Tokyo, Japan
10. Thanda A, Agarwal A, Singla K, Prakash A, Gupta A (2016) A document retrieval system for math queries. In: Proceedings of the 12th NTCIR conference on evaluation of information access technologies. Tokyo, Japan, pp 346–353
11. Zanibbi R, Aizawa A, Kohlhase M, Ounis I, Topic G, Davila K (2016) Ntcir-12 mathir task overview. In: Proceedings of the 12th NTCIR conference on evaluation of information access technologies. Tokyo, Japan, pp 299–308

Model Structure from Laser Scanner Point Clouds

Bara' W. Al-Mistarehi, Ahmad H. Alomari, Maad M. Mijwil, Taiser S. Khedaywi, and Momen Ayasrah

Abstract The essential target of this research work is generating a three-dimensional (3D) design of cloud point laser scanner structure that performs computer vision system (CVS) processes intentionally fully capabilities of invest to estimate model structure resulting from laser scanner point clouds and recognizing main areas for further studies and development. This study is significant processes were displayed, begins from collecting the clouds of laser point in static mode, and then analyzing the existing data, an iterative closest point algorithm (ICP) used in point cloud registration and matching, noise reduction, synthesis scheme in laser scanner point clouds uses a random sample consensus algorithm (RANSAC), data generating 3D laser scanner, importing a structure model from laser scanner point clouds and fit in using CVS programs. The Poisson surface algorithm is utilized to pick up point clouds and mesh surfacing. The results displaced some of the structural troubles such as deformation and cracks over two places of the wall. The results of the proposed algorithm matched the original point clouds very well. This differential proved that the algorithm advanced in this research work is efficient and more workable for visualization, monitoring, and future data processing.

B. W. Al-Mistarehi (✉) · T. S. Khedaywi
Civil Engineering Department, Faculty of Engineering, Jordan University of Science and Technology, P.O. Box 3030, Irbid 22110, Jordan
e-mail: bwmistarehi@just.edu.jo

T. S. Khedaywi
e-mail: khedaywi@just.edu.jo

A. H. Alomari
Civil Engineering Department, Faculty of Engineering, Yarmouk University, Irbid, Jordan
e-mail: alomarish@yu.edu.jo

M. M. Mijwil
Computer Techniques Engineering Department, Baghdad College of Economic Sciences University, Baghdad, Iraq
e-mail: mr.maad.alnaimiy@baghdadcollege.edu.iq

M. Ayasrah
Master Civil Engineer & Research Assistant, Department of Civil Engineering, Jordan University of Science and Technology, Irbid, Jordan

S. Basu et al. (eds.), *Proceedings of International Conference on Frontiers in Computing and Systems*, Lecture Notes in Networks and Systems 404,
https://doi.org/10.1007/978-981-19-0105-8_18

Keywords Laser scanner · Point clouds · Computer vision system

1 Introduction and Literature Review

Three-dimensional (3D) scanning determines the shape of the surface or the volume of an object in 3D space. Laser scanner machine is a device for creating high-resolution, exact digital 3D models from real-world objects. The scanner is managed by 3D scanning software that works on a computer [11]. A point cloud is a collection of data points in a coordinate system. In a 3D coordinate system, these points are ordinarily determined by *X*, *Y*, and *Z* coordinates and are frequently aimed to draw the outer surface of an object [10].

The main advantage of laser scanning is that versatile scanning technology which allows a large amount of 3D information to be captured quickly and accurately in a way completely unobtrusive. It can be applied to a multitude of cases; the survey allows us to complete projects with previously accuracy and detail levels cannot be obtained with conventional survey methods [6]. Different methodologies have been traditionally used to quantify and monitor objects that use gypsum strips, geodetic methods, mechanical stretch gages, and electrical sensors. These procedures have the drawback that they contact tools are required, which must be placed manually on the structure like geodetic targets, scaled bar, or metric sensors. These devices are accurate, but their application depends on accessibility. Generally, they are undesirable for both esthetic and functional reasons since they are needed to, they are installed on the roof of buildings for a long time. Also, these methodologies tool up individual measurements point of the damaged area rather than a complete and continuous record of the extent of damage across the affected area. Therefore, these are usually not suitable for tasks that want many measurements points spread on the surface of the object [3].

Jamhawi et al. [7] studied the potential benefits of combine terrestrial laser scanner (TSR) and dense stereo matching (DSM) techniques to form a complete and detailed representation of Dar Es-Saraya Museum. They discussed the processing and registration of the terrestrial laser scanner and DSM data in creating a 3D point cloud of the museum to be used in a more accurate CAD model. They found that the combined approach optimized the geometric accuracy, the visual quality to have a high-resolution textured 3D model of the Dar Es-Saraya museum. The stereo matching process offers a fast solution to get 3D point clouds, flexible, and cheap. By combining all the data sources, the detailed 3D format and accurate features can be reconstructed accurately, because the point clouds interpretation improved. These functions can be added to the laser scanning data to provide a more realistic visualization of the entire scene.

Wang and Wang [12] investigated how a point cloud can be quickly registered with large amounts of data. The most distinguished offering is the proposed algorithm called iterative closest points (ICPs). In this way, the transformation parameters of the two-point sets are determined via the connection among the corresponding matching points of the two-point sets to meet the defined affinity accuracy. Lastly, each rotation and translation parameters are taken between the two points to complete the registration process. Nevertheless, there are several difficulties with the classical ICP algorithm. Where the primary value of the iteration needs to be determined when the primary step of the ICP algorithm is executed. The chosen primary value will have a significant impact on the final registration result. If the choice of the primary value is not suitable, the algorithm may direct to a local best, so that the iteration cannot converge to the correct registration result.

Therefore, this research work will build a 3D frame model from laser scanner point clouds and utilizing computer vision system (CVS) operations for better-resulting model visualization and overcoming some of the problems raised before. The developed algorithm was anticipated to building a 3D structure model safely, perform real-time operations, extract highly accurate data, and present the collected data spatially. Detection of the deformation from the 3D structure model of the wall was also performed.

2 Research Objective

This study will be a significant endeavor if it achieves the following goals for the area studied:

1 Building a 3D model frame from laser scanner point clouds.
2 Comparing the results from the detection and modeling algorithm in this research work with the original point cloud for validation purposes.
3 Identification of critical areas for further studies and development vjgy.

3 Methodology

This section depicts the overall protocol for 3D model structure and uses advances in technologies, including laser scanner point clouds and CVS. The model architecture is shown in Fig. 1.

3.1 Field Data Collection

For this study, laser scanner point clouds for parts of a wall in Germany were taken as a case study. These laser scanner data were surveyed and collected using a laser scanner for those measurements: HDS7000 from Leica geosystems (static mode), as shown in Fig. 2. The number of point clouds is 41,674 points. The point cloud was captured indoors, and the laser scanner was placed inside the room. The procedure for

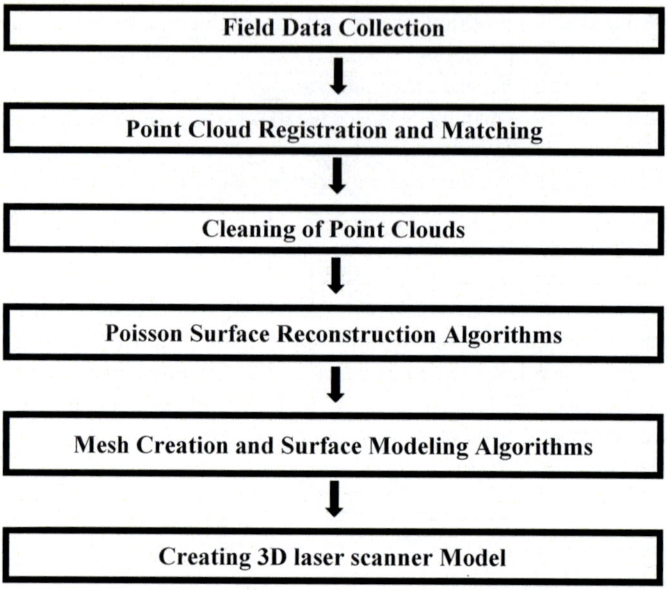

Fig. 1 Flowchart showing the study methodology

Fig. 2 Laser scanner HDS7000 [9]

collecting data was out the scope of this research work. This research work concentrates on the usage of the laser scanner point clouds for building model structure. The coordinates (x, y, z) for each laser scanner point and the intensity (brightness) value were identified as the independent variables.

The point cloud was captured indoors, so the laser scanner was placed inside the room. The distance between the laser scanner and targets is as follows: The first station of the laser scanner is approximately 1.51 m away from this wall, the second approx. 1.85 m, the third station approx. 2.75 m, and the fourth one approx. 2.82 m. This arrangement might look a little bit strange; a small object had been scanned on a desk, and the scanner was placed around this desk.

The scan time always depends on the chosen resolution, to choose between several resolution levels (angular resolution), which affects the distance of two measurements on a wall. For the lowest resolution, a scan (Faro scanner as well as the Leica HDS7000) takes one minute or less, for modeling structure elements point clouds cannot use. In the highest resolution, it takes several hours to finish a 360° scan. However, the amount of data would be so large that it cannot be saved on the inserted storage media. This point cloud is a combination of multiple scans in different resolutions.

The data are stored on USB sticks (HDS7000) or an SD card (Faro). After the scan process is finished, the data can be copied to a computer. The transfer durancy depends on the amount of data. Then, the data must be imported into suitable software. This might also take a few minutes to several hours, depending on the data amount, the computer itself, and the chosen software. At this moment, the single point clouds can be pinpoint (from each station). Four targets had been used to register the individual scans, the coordinates are (−1.393, −0.142, −0.616) m, (−1.409, 0.958, −0.766) m, (−0.420, 1.080, −0.681) m, (−0.791, 0.982, 0.075) m. The center of this coordinate system is the center of the laser scanner in the first scan. Further information that was used for the registration is that the scanner was leveled, so the z-axis is a plumb line.

3.2 *Point Cloud Registration and Matching*

The first step in generating a 3D model is registering individual scans obtained in the field. Scanning must be entered in to a common reference system to create a contiguous point cloud. This involves converting all surveys into one unified format. Registration is the method that requires the alignment of two or more-point datasets in various coordinate systems to several others, resulting in one coordinate system for both. In laser scanning systems, the method is performed by transforming one coordinate system of a point set to the coordinate system of another point set by applying specific transformations and rotations using rigid transformations (6 degrees of freedom), where the size also shapes are saved (scale factor is unity—orthogonal transformation). Therefore, the registration is, in principle, considered to be a coordinate transformation problem involving rotations and translations operations and the problem space is a subset of the overall affine coordinate transformation that is solved in the rigid body transformation or by the manner of similarity. The transformation problem of this coordinate can be described as a least-squares problem solved by minimizing the least-squares error [11].

Fig. 3 Points cloud after registration

The critical component in the processing of data is the registration, because the accuracy achieved at this stage affects every subsequent processing steps [1]. Leica geosystems are designed for the 3D point cloud processing software and are generally applied to survey terrestrial applications. The model presents two methods of integrating individual scanning into a single format system. The first method is based on goal-based registration. This method uses goals as common points for recording. The other depends on selecting corresponding cloud point features using the ICP algorithm [11]. The efficiency of the resulting shift relies on the accuracy of the scanned targets. In this study, target-based registration into cloud compare software [12] is used. Figure 3 shows the final registered point cloud model.

3.3 Cleaning of Point Clouds

Clearing data, i.e., removing objects that are not related to documents, are an essential step in 3D modeling. Cleaning improves the results of 3D structure modeling, particularly on the sites full of debris. During this research work, scan cleaning can be implemented on individual scans or a full-point cloud [7]. However, cleaning was done on the whole point clouds using the random sample consensus algorithm (RANSAC) [5] for cleaning point clouds. It is a non-deterministic method for robustly providing a model to a dataset contaminated with outliers. This algorithm is widely used for matching, visual motion estimation, and robust image feature detection, especially in the field of computer vision. Figure 4 shows the RANSAC algorithm for cleaning is implemented using cloud compare software to reduce the time of cleaning objects appearing in multiple scans that were removed in one operation.

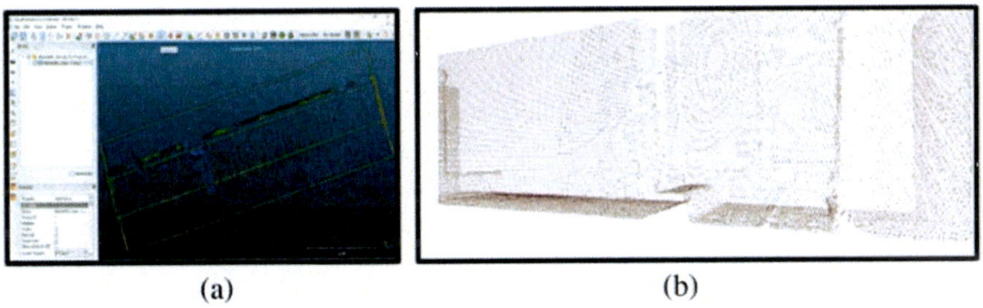

Fig. 4 After removing irrelevant cloud points; **a** during the cloud compare software processing using (RANSAC); **b** the final result of cleaning

3.4 Poisson Surface Reconstruction Algorithms

Reconstructing 3D surfaces from a point cloud is a common obstacle with computer graphics. It allows the creation of filling surface holes, scanned data, and accomplishing of current models. A new method is provided that reflects surface reconstruction as a solution to the Poisson equation. The reconstruction of the surfaces of the points faced several obstacles in practice. Point sampling is usually not standardized. The positions and standards are usually noisy due to inaccurate samples and bad registry scanning [6].

Accessibility restrictions through scanning may leave some surface areas free of data. Given these limitations, reconstruction methods try to infer the topology of the unknown surface, fill the holes reasonably, and also accurately match the noisy data [10].

The tremendous practical motivation for surface reconstruction is derived from the 3D survey, so this research work focused on reconstructing 3D models of real-world data. Surface reconstruction can be expressed using Poisson surface reconstruction (PSR) algorithms [11]. Figure 5 shows the results after applying the PSR algorithm using cloud compare software.

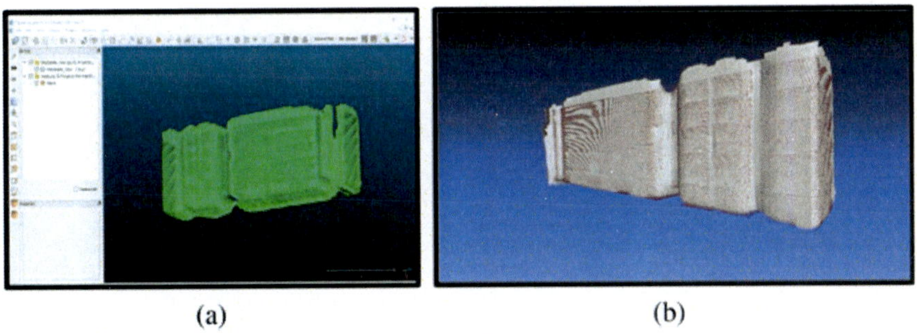

Fig. 5 Results after applying PSR; **a** during setup the parameters of Poisson; **b** the final result

3.5 Mesh Creation and Surface Modeling

Surfaces can be created either by connecting points directly and then becoming parts of the surface [2, 4] or by making the preferable surface apply through the points [8]. The first method lets the creation of the model form for a whole and unorganized point but has the problem of being quite prone to tool noise also scan alignment errors, while the best surface generation algorithms fit the noise reduction performance automatically by rounding the surface and as a result eliminating the noise. Besides, the risk in all smoothing purposes is the possible lack of relevant possible features. In this research work, the Poisson algorithm [11] was used for converting 3D points to the mesh surface for better visualization of deformation. The final 3D meshed model for visualization using cloud compare software is depicted in Fig. 6.

For better visualization of the results, 3D builder program [4] was used to create a 3D surface model after texturing, as shown in Fig. 7.

The realistic appearance of the model is checked against the original point cloud for validation purposes. The comparison showed that there is deformation in two places on the wall surface (Figs. 6 and 7; respectively). These scores match the original draw points excellent (Fig. 8). The simile is made manually by visual obser-

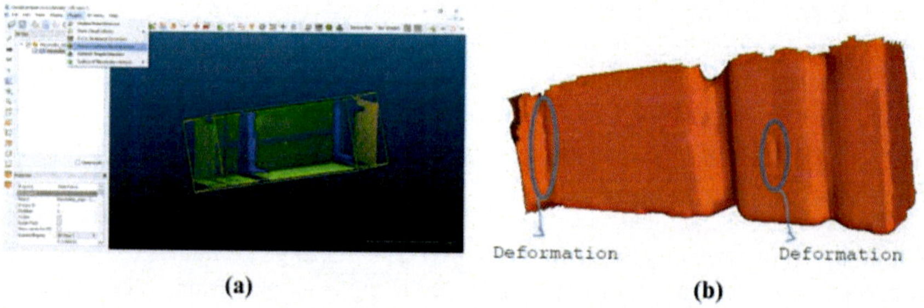

Fig. 6 Mesh surface of 3D points cloud surface using Poisson algorithm; **a** during the process; **b** final result

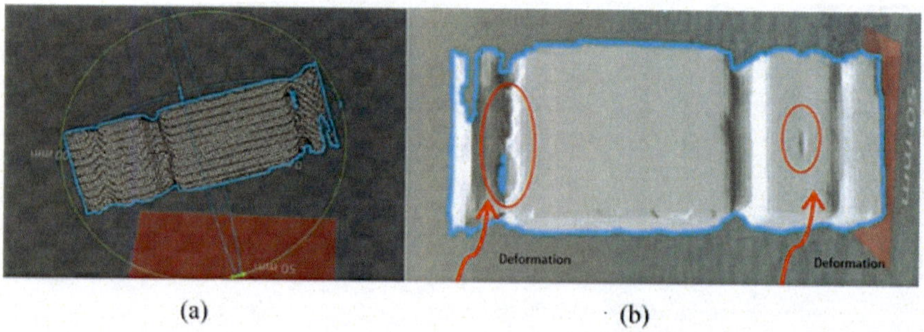

Fig. 7. 3D surface model from 3D builder after texturing; **a** during the process; **b** the final result

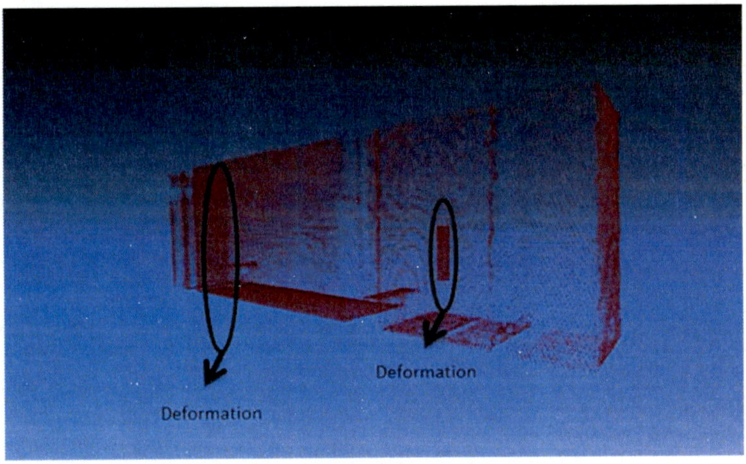

Fig. 8 Original point cloud for validation purpose

vations. This comparison proved that the algorithm aforementioned in this study is very useful and practical for future data processing, visualization, and observation.

4 Conclusions and Future Works

Processing the data to get a fully functional 3D model is a complex mission. It covers a wide range of planning, following the latest trends in the field, learning new technologies, and finally producing the model. At each step of the process, there can be seen some obstacles. The main benefit of using laser scanning is that it can collect many surface points very fast and directly. Detecting deformation and deflection of the structure are considered as a challenge, which requires unique acquisition methods and special processing algorithms. Building a 3D model frame from laser scanner point clouds, using a CVS operation, was presented in this work. It can be shown that the wall structure contained deformation in two places of the wall surface. The results verified that the algorithm mentioned in this study is effective and highly practical for future data processing, visualization, and observing. Future work should focus on the detection deformation of the wall and extract its characteristics like its length and depth automatically, not by visual observations.

References

1. Abdelmajid Y (2012) Investigation and comparison of 3D laser scanning software packages. Master of science thesis in Geodesy. School of Architecture and the Built Environment Royal, Institute of Technology (KTH), Stockholm, Sweden
2. Ahmadabadian AH, Robson S, Boehm J, Shortis M, Wenzel K, Fritsch D (2013) A comparison of dense matching algorithms for scaled surface reconstruction using stereo camera rigs. ISPRS J Photogramm Remote Sens 78:157–167
3. Alshawabkeh Y, El-Khalili M (2013) Detection and quantification of material displacements at historical structures using photogrammetry and laser scanning techniques. Mediter Archaeol Archaeom 13(2):57–67
4. Bernardini F, Mittleman J, Rushmeier H, Silva C, Taubin G (1999) The ball-pivoting algorithm for surface reconstruction. IEEE Trans Visual Comput Graph 5(4):349–359
5. Fischler MA, Bolles RC (1981) Random sample consensus: a paradigm for model fitting with applications to image analysis and automated cartography. Commun ACM 24(6):381–395
6. Ioannides M, Arnold D, Niccolucci F, Mania K (2006) The processing of laser scan data for the analysis of historic structures in Ireland. In: The 7th international symposium on virtual reality, archaeology and cultural heritage, VAST-2006, Nicosia, Cyprus
7. Jamhawi M, Alshawabkeh Y, Freewan A, Al-Gharaibeh R (2016) Combined laser scanner and dense stereo matching techniques for 3D modelling of heritage sites: Dar Es-Saraya Museum. Mediterr Archaeol Archaeometry 16(3)
8. Kazhdan M (2005) Reconstruction of solid models from oriented point sets. In: Proceedings of the third Eurographics symposium on geometry processing, Vienna, Austria, July 4–6, 2005, pp 73–82. ACM international conference proceeding series 255, Eurographics Association 2005, ISBN 3-905673-24-X
9. Leica Geosystems AG—Part of Hexagon (2020) HDS7000 user manual. http://w3.leica-geosystems.com/downloads123/hds/general/HDS7000/manuals/HDS7000_UserManual_en.pdf. Accessed Feb 2020
10. Levoy M, Whitted T (1985) The use of points as a display primitive. University of North Carolina, Department of Computer Science, Chapel Hill, NC 27514, pp 829–843
11. Rakitina E, Rakitin I, Staleva V, Arnaoutoglou F, Koutsoudis A, Pavlidis G (2008) An overview of 3D laser scanning technology. In: Proceedings of the international scientific conference, Varna, Bulgaria
12. Wang JW, Wang M (2009) 3D Reconstruction of shoe-last based on binocular stereo vision. Comput Technol Dev 19:224–230

Feature Extraction Techniques for Gender Classification Based on Handwritten Text: A Critical Review

Monika Sethi, M. K. Jindal, and Munish Kumar

Abstract Features of the handwritten text play a vital role in the area of handwriting identification. It became more challenging when one has to identify gender, age, and handedness of the person through handwriting. In last two decade, the use of various feature extraction techniques immerged having advantages one on the other. Ample research is done on writer identification systems by implementing various feature extraction techniques. In this paper, we have shifted the concern toward various features and features extraction techniques implemented on gender identification through handwriting. The objective of this survey is to present the critical review of work done in area feature extraction in gender identification taking only handwriting into consideration. We have categorized all the feature extraction techniques used by the researchers for gender classification into four broad categories: statistical-, transform-, gradient-, and model-based techniques. From the survey, we have identified few techniques that deserve future attention of the researchers for optimal results.

Keywords Gender classification · Handwriting · Text recognition · Feature extraction · Classification

1 Introduction

Gender identification system through handwriting has been divided into four phases: (i) acquisition of images, (ii) preprocessing, (iii) extraction of features, and (iv)

M. Sethi
Department of Information Technology, Goswami Ganesh Dutta Sanatan Dharma College, Chandigarh, India
e-mail: monika.sethi@ggdsd.ac.in

M. K. Jindal
Department of Computer Science and Applications, Panjab University Regional Centre, Muktsar, India

M. Kumar (✉)
Department of Computational Sciences, Maharaja Ranjit Singh Punjab Technical University, Bathinda, Punjab, India

S. Basu et al. (eds.), *Proceedings of International Conference on Frontiers in Computing and Systems*, Lecture Notes in Networks and Systems 404,
https://doi.org/10.1007/978-981-19-0105-8_19

classification. Two most important phases are feature extraction and classification. Poorly extracted features lead to poor recognition by the classifier and, hence, less accuracy in gender identification. Extraction of features refers to the technique of converting the image into numerical vector form. The converted image features are known as descriptors. More effective the descriptors, less the time are used for training the machine for the classifier model. According to our survey, the large number of techniques has been used by the researchers to study gender classification through handwriting. Figure 1 the Sankey diagram shows the association between the year of publication and feature extraction techniques used by the researchers across the world from 2012 to 2021 in gender classification. According to our survey, the large number of techniques has been used by the researchers to study gender classification through handwriting.

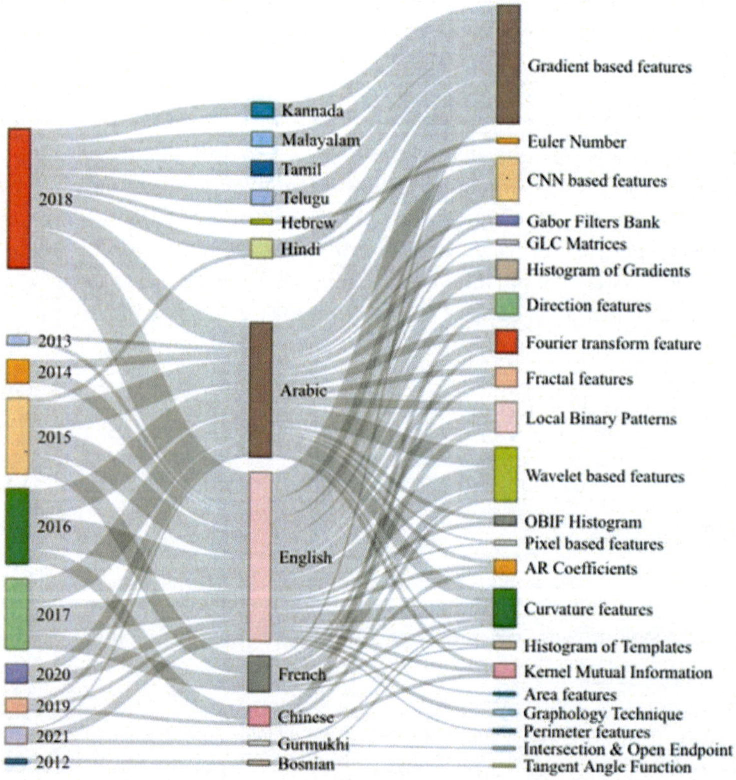

Fig. 1 Yearwise feature extraction techniques used by various researchers

2 Feature Extraction Techniques in Gender Identification

In this section, we will discuss the features extraction techniques used in gender identification through handwriting under four broad categories. Table 1 presents these four categories of feature extraction techniques and different features used by researchers for gender identification.

2.1 *Statistical Techniques*

Statistical features are geometric formulas and measurement for the analysis of the data. These are group of arithmetic parameters used to measure points, lines, and surface, etc. There are certain gender identification systems that worked on area and perimeter features. These are statistical features where perimeter deals with the circumference of the given characters and area determines the pixels contained in each character. Maken and Gupta [1] considered physical appearance of male and female handwriting and proved that perimeter of male handwriting is always greater. Bouadjenek et al. [2] used two pixel-based features: pixel density and pixel distribution. Density is grid-oriented features used in handwritten script. It depends on width of script generated due to pressure. Distribution means that available pixels to be distributed in the predefined space for processing.

Maadeed et al. [3] and Siddiqi et al. [4] used various curvature features that are based on angle of diversion of the straight line. Sharper turns lead to high curvature. Positive, negative, and zero curvature can be best depicted by Fig. 2. Tortuosity defines the shape property to have many turns and twists. Figure 3 is presenting the computation of tortuosity, done by (L/L_0) where actual length of the curve is depicted as (L) and (L_0) is the distance between the endpoints of the curve.

Table 1 Categories of feature extraction techniques and features

Feature extraction techniques	Features
Statistical techniques	Euler number; area features; perimeter features; AR coefficients; curvature features; pixel-based features; tangent angle function; direction-based features; intersection and open endpoint; gray-level co-occurrence matrices
Gradient techniques	Gradient-based features; histogram of gradients; OBIF histogram; histogram of templates; local binary patterns
Transform techniques	Fourier transform features; Gabor filters bank; wavelet-based features
Model/automatic techniques	Convolutional neural network; fractal-based features

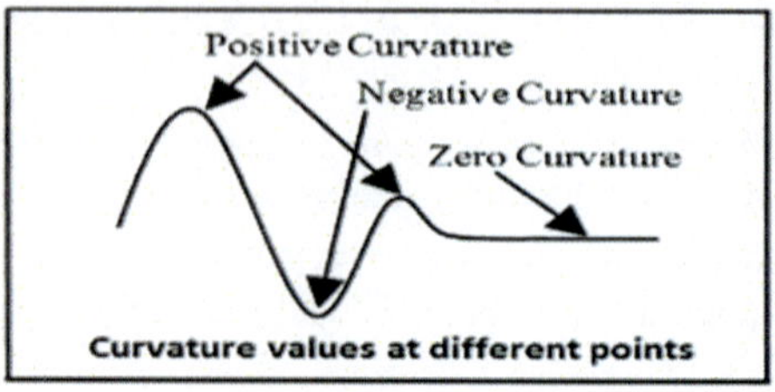

Fig. 2 Curvature feature

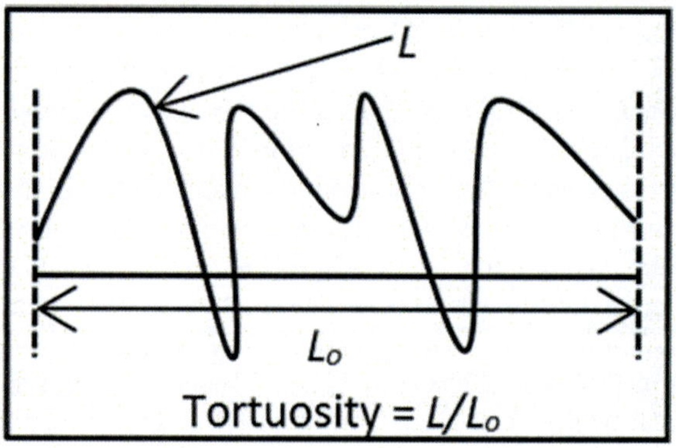

Fig. 3 Tortuosity feature

Bi et al. [5] research was based on KMI—kernel mutual information. In their research, they used three parameters kernel learning, min redundancy, and max relevance based on orientation and curvatures of handwritten texts. Cloud of line distribution and Hinge features also comes under the principle of curvature features used by Gattal et al. [6]. Ahmed et al. [7] used gray-level co-occurrence matrices along with LBP and HOG features to check co-occurrence of gray level in adjacent pixels in their research. The two pixels with (i, j)th element are separated by distance d and angle θ, then homogeneity and dissimilarity are identified along with probability also. Maji et al. [8] extracted features from handwritten text by including and excluding Euler number. It describes scalar value of difference, between total objects and total holes in image used in signature. Siddiqi et al. [4] worked with autoregressive (AR) coefficients, where features are extracted from binary data of written text and linear combination is used for predicted neighboring pixels. Direction features are one of the most used categories of statistical technique. Maadeed et al. [3] and Maken and Gupta [1] worked with the concept of rows and columns orientation of direction features. They used Sobel edge detection technique. The combined gradient magnitude is calculated as G in Eq. (1) and phase θ in Eq. (2).

$$G = \sqrt{G_x^2 + G_y^2} \tag{1}$$

$$\theta = \arctan\left[\frac{G_y}{G_x}\right] \tag{2}$$

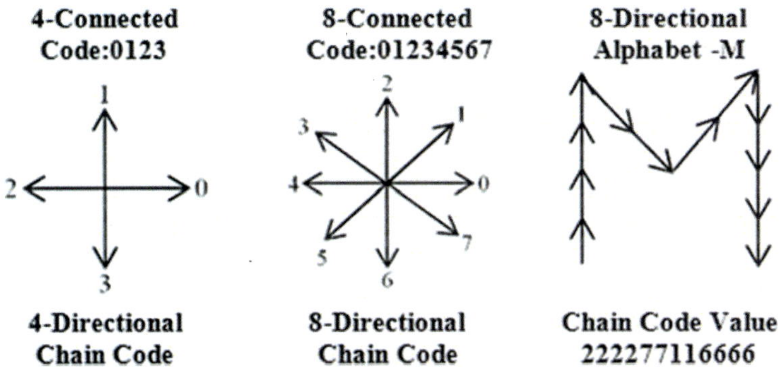

Fig. 4 Direction chain code, 8- direction chain code and example of chain code

In both directions, G_x and G_y are identified for the component of image. Intersection and open-end points are used by Dargan and Kumar [9]. They identified overlapping, density, direction, and pixel data using these features. Another directional feature known as tangent angle function is used by Sokic et al. [10] to identify directional movement of the pen nib. This function checks the continuity or discontinuity in the tangent angle when spike or circle is pointed out. Riza et al. [11] used graphology techniques to define features for their research. This technique used pressure created on pen while writing, height of every letter, margins between the letters and words.

Siddiqi et al. [4] extracted the features on basis of chain code technique. This feature is used to represent the boundary. Figure 4 represents the 4 directions and 8 directions of chain code. The boundary is composed of pixels and works on straight line segments. The length and direction of the line segment are specified and generally fixed with clockwise traversal.

2.2 *Gradient Techniques*

Gradient features represent vectors for change in intensity of color and direction of image.

These are the most used features for gender classification by the various researchers. Local binary patterns (LBP) are used by Siddiqi et al. [4], Ahmed et al. [7], and Bouadjenek et al. [12, 13] in their work for calculation and the comparison of gray value of the respective pixel at central position, with the neighboring gray levels in the circle as mentioned in Fig. 5. The value for neighboring position is considered 1 for lower gray level at central pixel, else 0. LBP code for P neighbors with radius R is depicted in Eqs. (3–5) where I_c = value of gray for central pixel and I_p = value of gray for neighbor at pth position.

Fig. 5 **a** Sample template, **b** Resulting LBP template

25	150	90
85	60	55
35	80	45

(a)

0	1	1
1		0
0	1	0

(b)

$$LBP_{P,R}(x, y) = \sum_{P=0}^{P} S(I_P - I_c)2^P \tag{3}$$

$$S(m) = \begin{cases} 1, |m \geq 0 \\ 0, |m < 0 \end{cases} \tag{4}$$

$$I_P = I\left\{\left(x + R\sin\frac{2\pi p}{P}\right), \left(y - R\cos\frac{2\pi p}{P}\right)\right\} \tag{5}$$

Ahmed et al. [7] and Bouadjenek et al. [2, 12, 14] worked with gradient local binary patterns (GLBP) feature by combining information of gradient data. In this, 8-neighboring pixels value is compared with central pixel, for greater value 1, else 0 is allotted in 8-binary code (0 to 7), and gradient magnitude is calculated as in Eq. (6).

$$M(x, y) = \sqrt{g_x(x, y)^2 + g_y(x, y)^2} \tag{6}$$

LBP is then calculated as given in Eqs. (3–5). Bouadjenek et al. [12–14] used histogram of oriented gradient (HOG) to analyze the distribution of text curvatures with histogram. The handwritten text images are partitioned into grid form to obtain several local regions, and then gradient orientation is identified using HOG calculation. For every pixel in the local grid region, gradient is calculated horizontally and vertically as depicted in Equation numbers (7, 8).

$$g_x(x, y) = I(x + 1, y) - I(\mathrm{x} - 1,\ \mathrm{y}) \tag{7}$$

$$g_y(x, y) = I(x, y + 1) - I(x, y - 1) \tag{8}$$

Then, phase of gradient is computed as given in Eq. (9).

$$\varphi(x, y) = \arctan\left[\frac{g_x(x, y)}{g_y(x, y)}\right] \tag{9}$$

Navya et al. [15, 16] worked on gradient of image and its complement image when they are in opposite direction using adaptive multi-gradient features (AMGF). However, it is not true for gradient direction at intersections and corners. If pixel shows symmetry with complement, then it is edge, otherwise considered as intersection or corner in the image. The oriented basic image features (OBIFs) are an identity in which every location is defined as flat, rotational, slope, line, light, or dark. Gattal et al. [17] used OBIFs with different scale factor to increase its performance under OBIFs column histogram scheme.

2.3 *Transform Techniques*

Transform techniques cover gaps between the methods which deal with variables and numerical variations for solving differential equations of linear order. Wavelet transforms are a framework for developing discrete division of frequencies called wavelet sub-band which is used by Akbari et al. [18]. Ibrahim et al. [19] and Youssef et al. [20] worked on identification wavelet-domain local binary pattern (WD-LBP) in gender handwriting identification. WD-LBP is used for decomposition of the image into wavelet sub-bands, so that for each sub-band, LBP can be calculated as shown in Eqs. (3–5). Curvelet transform can be used as extension of wavelet transformation. Guerbai et al. [21] and Tan et al. [22] used curvelets for the directional transformation which helps in representation of text images having edges (Table 2).

Fourier transform is a technique based on function (F), i.e., Fourier transform. In their works, Tan et al. [22] and Mirza et al. [23] used Fourier transform which divides the image (I) into sizes ($L \times W$) as in Eq. (10).

$$F(u, v) = \sum_{n=1}^{W}\sum_{m=1}^{H} I(n, m)e^{\left[-j2\pi\left(\frac{un}{L}+\frac{vm}{W}\right)\right]} \tag{10}$$

where u and v are horizontal and vertical frequencies of image component. It finds the average value of image color which further helps in textural analysis. Fourier transform illustrates the textural image with weighted combination to both direction sinusoids, i.e., horizontal and vertical. Gabor filter is used by Mirza et al. [23], Gaussian function to analyze the textural images. This linear filter technique uses the frequency content and orientation in the textural images for particular direction in the nearby region around the point of image analysis. For texture analysis, two-dimensional Gabor filter, plane wave multiplication with Gaussian function is used as Eq. (11).

$$g(x, y; \lambda, \theta, \psi, \sigma, \gamma) = \exp\left(-\frac{x'^2 + \gamma y'^2}{2\sigma^2}\right)\sin\left(2\pi\frac{x'}{\lambda} + \psi\right) \tag{11}$$

$$x' = x\cos\theta + y\sin\theta \tag{12}$$

$$y' = -x\sin\theta + y\cos\theta \tag{13}$$

Table 2 Review of feature extraction techniques

Year	References	Features
2012	Sokic et al. [10]	Tangent angle function; contour curvature
2013	Youssef et al. [20]	Wavelet-domain LBP; gradient-based features
2014	Bouadjenek et al. [13]	Local binary patterns; histogram of gradients
	Ibrahim et al. [19]	Gradient-based features; wavelet LBP
	Maadeed et al. [3]	Direction-based features; curvature features
2015	Bouadjenek et al.[14]	Histogram of gradients; gradient LBP
	Maji et al. [8]	Euler number
	Siddiqi et al. [4]	Chain code; fractal features; LBP; AR coefficients
2016	Bouadjenek et al. [2]	Pixel-based features; gradient LBP
	Mirza et al. [23]	Fourier transform; Gabor filters bank
	Tan et al. [22]	Curvelet features; direction features; Fourier coefficient
2017	Ahmed et al. [7]	SFTA; LBP; histogram of gradients; GCLM
	Akbari et al. [18]	Wavelet sub-band
	Bouadjenek et al. [12]	Histogram of templates; LBP; gradient LBP
	Guerbai et al. [21]	Curvelet transform
2018	Gattal et al. [17]	OBIF histogram; OBIF column histogram
	Illouz et al. [25]	CNN-based features
	Moetesum et al. [26]	CNN-based features
	Morera et al. [24]	CNN-based features
	Navya et al. [15, 16]	Multi-gradient directional features; AMGF
	Riza et al. [11]	Graphology technique
2019	Bi et al. [5]	Kernel mutual information
2020	Gattal et al. [6]	Cloud of line distribution; Hinge features
	Xue et al. [27]	ATP-DenseNet
2021	Maken and Gupta. [1]	Area features; perimeter features; pixel-based features
	Dargan and Kumar[9]	Intersection and open-end points; curve fitting
	Rahmanian and Shayegan [28]	DenseNet201; InceptionV3; Xception

Local binary patterns (LBP); cloud of line distribution (COLD); convolutional neural network (CNN); gradient LBP (GLBP); gray-level co-occurrence matrices (GLCM); histogram of gradients (HOG); Kernel mutual information (KMI); oriented basic image feature (OBIF); segmentation-based fractal texture analysis (SFTA); wavelet-domain LBP (WD-LBP); adaptive multi-gradient features (AMGF)

where γ is spatial aspect ratio, σ is standard deviation, $\theta =$ orientation for Gabor function, λ is sinusoidal factor wavelength, ψ is phase offset. 2D Gabor filter in the area of spatial domain is represented with $h(x, y)$ in Eq. (14).

$$h(x, y) = g(x, y) \exp(-2\pi j (u_o x + v_o y)) \tag{14}$$

$$g(x, y) = \exp\left(\frac{x^2 + y^2}{2\sigma^2}\right) \quad (15)$$

where $g(x, y)$ is Gaussian function in Eq. (15). Gabor filter space is also useful in fingerprint and optical character recognition with greater flexibility from degrees of freedom. Gabor filter bank works on frequency domain and spatial domain for multi-resolution decomposition. It is a non-orthogonal and does not changes with change in scale and rotation in the image.

2.4 *Model/Automatic Techniques*

Convolutional neural network (CNN) based on deep learning model combined with the series of features with input images of handwriting is used as single architecture for feature extraction and classification. This architecture works efficiently with 2D data. A convolution neural network has multiple layers architecture that is used to extract features. The layers are: (1) convolution, (2) rectified linear unit (ReLU), (3) pooling, and (4) fully connected layer. This deep learning technique is used by Morera et al. [24]. They worked on all above four layers, and for getting the output, they used softmax activation function. Illouz et al. [25] and Moetesum et al. [26] also worked on CNN techniques to extract features and classification results. Using the concept of CNN, Xue et al. [27] worked with dense blocks called ATP-DenseNet, for visual object recognition. Dense blocks are densely connected layers. Another model-based technique is fractal features used by Siddiqi et al. [4] in their research. Fractals work on the principle of series of patterns, that is, repeating itself many times on same or different scales. Rahmanian and Shayegan [28] used three different features for the extraction and classification based on CNN, DenseNet20, Xception, and InceptionV3. A convolutional neural network having 201 layers is known as DenseNet201, and Xception is having 71 layers, whereas InceptionV3 works with connections of 48 layers.

3 Conclusion

We have presented a comprehensive survey for last one decade, work done by various researchers, with languages and the feature extraction techniques used by them for their works on gender classification. We have categorized the feature extraction techniques in four categories statistical techniques, gradient technique, transform techniques, and model/automatic techniques. Among the four categories, statistical, gradient, and transform techniques are now well known. And, some of the features like LBP and its different variants are first choice of the researchers. Curvature features come at number two when we talked about gender classification through handwriting. In this survey, we have observed that much of the work is done using statistical and gradient features. For last two years, researchers have started using model/automatic techniques for feature extraction. However, with the advent of deep learning methods which we discussed in model/automatic techniques, other categories are not concurrent with the performance of new techniques. Gender classification through handwriting is in its growing phase, so it is very difficult to suggest the best techniques for feature extraction. Though very less work is done on model-based features, but it is a promising area and deserves attention for future research.

References

1. Maken P, Gupta A (2021) A method for automatic classification of gender based on text-independent handwriting. Multimed Tools Appl. https://doi.org/10.1007/s11042-021-10837-9
2. Bouadjenek N, Nemmour H, Chibani Y (2016) Robust soft-biometrics prediction from off-line handwriting analysis. Appl Soft Comput 46:980–990
3. Al Maadeed S, Hassaine A (2014) Automatic prediction of age, gender, and nationality in offline handwriting. Eurasip J Image Video Process, 1–10
4. Siddiqi I, Djeddi C, Raza A, Souici-meslati L (2015) Automatic analysis of handwriting for gender classification. Pattern Anal Appl 18(4):887–899. https://doi.org/10.1007/s10044-014-0371-0
5. Bi N, Suen CY, Nobile N, Tan J (2019) A multi-feature selection approach for gender identification of handwriting based on kernel mutual information. Pattern Recogn Lett 121:123–132. https://doi.org/10.1016/j.patrec.2018.05.005
6. Gattal A, Djeddi C, Bensefia A, Ennaji A (2020) Handwriting based gender classification using COLD and Hinge features. In: Image and signal processing: 9th international conference, ICISP 2020, Marrakesh, Morocco, June 4–6, 2020, proceedings, 12119, 233–242. https://doi.org/10.1007/978-3-030-51935-3_25
7. Ahmed M, Rasool AG, Afzal H, Siddiqi I (2017) Improving handwriting based gender classification using ensemble classifiers. Expert Syst Appl 85:158–168. https://doi.org/10.1016/j.eswa.2017.05.033
8. Maji P, Chakraborty S, Samanta S, Chatterjee S, Kausar N, Dey N (2015) Effect of Euler number as a feature in gender recognition system from offline handwritten signature using neural networks. In: 2015 international conference on computing for sustainable global development, INDIA Com, 1869–1873
9. Dargan S, Kumar M (2021) Gender classification and writer identification system based on handwriting in Gurumukhi Script. In: 2021 international conference on computing, communication, and intelligent systems (ICCCIS). (ICCCIS), Greater Noida, India, pp 388–393. https://doi.org/10.1109/ICCCIS51004.2021.9397201
10. Sokic E, Salihbegovic A, Ahic-Djokic M (2012) Analysis of off-line handwritten text samples of different gender using shape descriptors. In: 2012 9th international symposium on telecommunications, BIHTEL 2012 proceedings
11. Riza LS, Zainafif A, Rasim, & Nazir, S. (2018) Fuzzy rule-based classification systems for the gender prediction from handwriting. Telkomnika (Telecommun Comput Electron Control) 16(6):2725–2732
12. Bouadjenek N, Nemmour H, Chibani Y (2017) Fuzzy integrals for combining multiple SVM and histogram features for writer's gender prediction. IET Biometrics 6(6):429–437. https://doi.org/10.1049/iet-bmt.2016.0140
13. Bouadjenek N, Nemmour H, Chibani Y (2014) Local descriptors to improve off-line handwriting-based gender prediction. In: 6th international conference on soft computing and pattern recognition, SoCPaR, 43–47

14. Bouadjenek N, Nemmour H, Chibani Y (2015) Age, gender and handedness prediction from handwriting using gradient features. In: 2015 13th international conference on document analysis and recognition (ICDAR), 1116–1120
15. Navya BJ, Shivakumara P, Shwetha GC, Roy S, Guru DS, Pal U, Lu T (2018) Adaptive multi-gradient kernels for handwritting based gender identification. In: Proceedings of international conference on frontiers in handwriting recognition, ICFHR, 392–397. https://doi.org/10.1109/ICFHR-2018.2018.00075
16. Navya BJ, Swetha GC, Shivakumara P, Roy S, Guru DS, Pal U, Lu T (2018) Multi-gradient directional features for gender identification. In: Proceedings—international conference on pattern recognition, 3657–3662
17. Gattal A, Djeddi C, Siddiqi I, Chibani Y (2018) Gender classification from offline multi-script handwriting images using oriented basic image features (oBIFs). Expert Syst Appl 99:155–167
18. Akbari Y, Nouri K, Sadri J, Djeddi C, Siddiqi I (2017) Wavelet-based gender detection on off-line handwritten documents using probabilistic finite state automata. Image Vis Comput 59:17–30.https://doi.org/10.1016/j.imavis.2016.11.017
19. Ibrahim AS, Youssef AE, Abbott AL (2015) Global vs. local features for gender identification using Arabic and English handwriting. In: 2014 IEEE international symposium on signal processing and information technology, ISSPIT, 155–160
20. Youssef AE, Ibrahim AS, Lynn Abbott A (2013) Automated gender identification for Arabic and English handwriting. In: 5th international conference on imaging for crime detection and prevention, ICDP. https://doi.org/10.1049/ic.2013.0274
21. Guerbai Y, Chibani Y, Hadjadji B (2018) Handwriting gender recognition system based on the one-class support vector machines. In: Proceedings of the 7th international conference on image processing theory, tools and applications, IPTA 2017, 2018-Janua(4), 1–5. https://doi.org/10.1109/IPTA.2017.8310136
22. Tan J, Bi N, Suen CY, Nobile N (2016) Multi-feature selection of handwriting for gender identification using mutual information. In: Proceedings of international conference on frontiers in handwriting recognition, ICFHR, 0(Mi), 578–583
23. Mirza A, Moetesum M, Siddiqi I, Djeddi C (2016) Gender classification from offline handwriting images using textural features. In: Proceedings of international conference on frontiers in handwriting recognition, ICFHR, pp 395–398. https://doi.org/10.1109/ICFHR.2016.0080
24. Morera A, Sánchez v, Vélez JF, Moreno AB (2018) Gender and handedness prediction from offline handwriting using convolutional neural networks. Complexity https://doi.org/10.1155/2018/3891624
25. Illouz E, David E, Netanyahu NS (2019) Handwriting-based gender classification using end-to-end deep neural networks. In arXiv (Vol. 1). Springer International Publishing. https://doi.org/10.1007/978-3-030-01424-7
26. Moetesum M, Siddiqi I, Djeddi C, Hannad Y, Al-Maadeed S (2018) Data driven feature extraction for gender classification using multi-script handwritten texts. In: Proceedings of international conference on frontiers in handwriting recognition, ICFHR, 564–569.https://doi.org/10.1109/ICFHR-2018.2018.00104
27. Xue G, Liu S, Gong D, Ma Y (2020) ATP-DenseNet: a hybrid deep learning-based gender identification of handwriting. Neural Comput Appl 3. https://doi.org/10.1007/s00521-020-05237-3
28. Rahmanian M, Shayegan MA (2021) Handwriting-based gender and handedness classification using convolutional neural networks. Multimedia Tools Appl. https://doi.org/10.1007/s11042-020-10170-7

Rice Disease Identification Using Deep Learning Models

Sk Mahmudul Hassan and Arnab Kumar Maji

Abstract Automatic identification of rice disease is much important in the agricultural sector as a large number of people in the world lives on rice. Symptoms in rice diseases occur in the leaves, and identifying the diseases by an experienced pathologist through physical observation of infected leaves is a slow process and expensive. Instead, this machine learning approach, especially deep learning-based technique that does not need any pre-processing and feature extraction, can classify the diseases efficiently. Many researchers worked on identifying rice diseases and achieved different results according to the implemented methods. In this paper, we study different deep learning models and transfer learning of the deep learning models in the identification of rice diseases. The rice dataset used in this study consists of 5932 images of four different classes. In addition, the performance of the different deep learning models is evaluated in terms of different parameters such as accuracy, loss, precision, recall, f1-score, and training time.

Keyword Plant disease · Machine learning · Convolutional neural network · Transfer learning

1 Introduction

Rice is considered as one of the primary food in India and China and especially in the north-eastern region of India. Occurrences of diseases in rice plant effects on growth and quality of the production. Generally, crops such as rice, wheat, and corn are essential for guaranteed the supply of foods, and quality loss of these crops leads thread to food security [12]. Therefore, early identification and diagnosis of these crop diseases play an important role in agriculture which ensures high quality of crops. However, identifying the diseases by physically observing the disease symptoms requires continuous monitoring by the expert which is expensive [1]. Moreover in

S. Mahmudul Hassan (✉) · A. Kumar Maji
North Eastern Hill University, Shillong, Megahalya 793022, India
e-mail: hassanmahmudul89@gmail.com

S. Basu et al. (eds.), *Proceedings of International Conference on Frontiers in Computing and Systems*, Lecture Notes in Networks and Systems 404,
https://doi.org/10.1007/978-981-19-0105-8_20

rural areas, the farmer may travel a long distance to contact the expertise which is time-consuming and too expensive. Therefore, automatic identification of plant diseases is necessary which benefits the farmers as the farmers having less knowledge about agriculture and often failed to identify the diseases in crops.

Several works are carried out by the researcher to identify the diseases in the plant. Among these approaches, identification of diseases is broadly classified into traditional feature-based identification and deep learning-based identification. In the past decades, the traditional feature-based approach was used to identify the diseases in the plant which relies on different image segmentation and feature extraction techniques. The different feature sets that were used in plant disease identifications are a color feature, shape feature, and texture features using different algorithms such as GLCM, SHIFT, LBP, and GLLRM. After that, different classification techniques such as KNN, SVM, ANN, and RF are used to classify the disease categories. However, segment the region and extraction of features is time-consuming, and finding the minimal set of features to represent the diseases is not an easy task.

In recent times, deep learning-based achieves much attention in several areas such as object detection, image classification, human activity recognition, and video analysis. Application of deep learning extends in agricultural field such as disease detection, plant identification, weeds detection, pest recognition. In deep learning techniques, the features are learnt automatically from the model, and hence, the segmentation and extraction of features is not required. Several works are there in identification of diseases on several crops such as corn [13], wheat [10], potato [9], tomato [4], cucumber [28], cotton [14], cassava [18, 20] etc.

In this paper, we study several deep learning models and used these models in the identification of diseases on rice plants. Transfer learning-based approach is used to train the deep learning models and compare the performances of each model in terms of accuracy, precision, recall, f1-score, training loss, and training time.

The rest of this paper organized as follows: Sect. 2 gives the literature survey. Section 3 gives the dataset description and the deep learning architecture used in this work. Experimental results are discussed in Sect. 4. Finally, the paper concluded with Sect. 5.

2 Literature Survey

Designing a reliable system in the identification of plant diseases is a very complex and challenging task. Researchers have come up with different ideas to identify the diseases in the plant. In this section, we are going to discuss some of the research work on the identification of rice diseases using machine learning. In [19, 27], author used shape and texture feature to identify different rice diseases. In their paper, they used genetic algorithm (GA) and SVM classifier, respectively, to identify the diseases. Different color features, shape, and texture features are used in [2] to identify different paddy diseases in Srilanka.

Convolutional neural network (CNN)-based technique is used in [17] to identify different rice pest and their diseases and this model achieved training accuracy of 90.9% . Bacterial blight and blast diseases in rice plant is identified in [rice5] using CNN. AlexNet-based CNN is used in [16] which used to identify three different rice leaf diseases that occur in Bangladesh, and they recorded an accuracy rate of 99.42%. Rice blast disease is identified in [15] using deep CNN, and in their paper, they have shown that deep CNN techniques extracts more discriminative and effective features than traditional hand crafted-based features as well as some of texture features which were extracted using GLCM, LBP, and wavelet transform.

VGG with inception connection used by Chen et al. [3] to identify maize and rice leaf diseases. They have used transfer learning-based approach in which the model is pre-trained on large-scale ImageNet dataset and recorded an accuracy rate of 82% and 92% on corn and rice, respectively. Self-attention convolutional neural network (SACNN) which uses the residual connection between the layer is used by Zeng et al. [28] to identify cucumber and rice leaf diseases. In this paper, they showed that CNN with self-attention mainly focuses on the diseased region which improves the performances of the model. The obtained recognition accuracy is 98% and 95.33% on MK-D2 and AES-CD9214 dataset, respectively. Deep feature-based technique is used in [22] to identify rice leaf diseases. In their paper, the author used several deep learning models for feature extraction and SVM classifier used to classify the diseases. Features from ResNet50 along with SVM classifier give the maximum accuracy rate of 98.33% .

3 Materials and Methods

3.1 Dataset

Rice leaf disease dataset is used in this paper to evaluate the performances. The dataset [21] contains 5932 images of four different categories of rice diseases, namely bacterial blight, blast, brown spot, and tungro. The images were captured from different rice fields from western Odisha. The images in the dataset are of size 300×300. The dataset contains 1584 bacterial blight, 1440 blast, 1600 brown spot, and 1308 tungro images. Table 1 gives the description of the dataset including the number of images used in training and testing. Figure 1 shows the sample images of the dataset. The dataset splits randomly into 80% training and 20% testing set. The images are resized accordingly in different deep learning models.

(a) Bacterial Blight (b) Blast (c) Brown Spot (d) Tungro

Fig. 1 Sample images of rice plant diseases

3.2 *State of Art Deep Learning Models*

Different CNN-based deep learning models are used in this paper to identify rice plant diseases. The implemented deep learning models are AlexNet, VGG16, VGG19, ResNet50, InceptionV3, InceptionV4, DenseNet121, and DenseNet201. Analyze the performances of each model and find the model which performs better in identification. Figure 2 shows the basic CNN architecture.

Table 1 Dataset details of rice plant diseases

Rice disease	No of train images	No of test images	Total no of images
Bacterial blight	1267	317	1584
Blast	1152	288	1440
Brown spot	1280	320	1600
Tungro	1046	262	1308

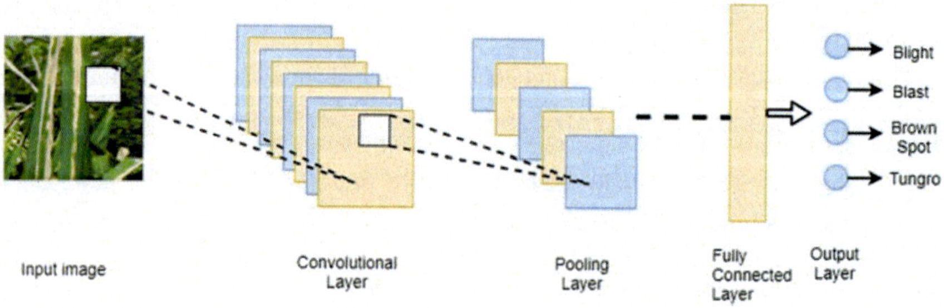

Fig. 2 Basic CNN model

3.2.1 AlexNet

AlexNet architecture was proposed by Krizhevsky et al. [11] which won the ImageNet Large-Scale Visual Recognition Challenge (ILSVRC-2012) in 2012. The design of AlexNet architecture is similar as LeNet-5 with more number of depth in the network. The AlexNet architecture is of eight-layer CNN model which generates nearly 61 million parameter. Among the eight layers, five layers are convolution layer with max pooling and three fully connected layers and a classification layer called softmax layer. It uses ReLu activation function to speed up the training process and dropout to prevent the model overfitting. Input used in this model is of dimension 227 $\times$ 227

3.2.2 VGG16

VGG16 is a deep learning architecture proposed by Simonyan et al. [23]. VGG16 model has 13 convolution layers, five maxpool layers, and two fully connected layers with one softmax layer. In each convolution layer, it uses 3 $\times$ 3 filter with zero padding and an stride of 1. Max-pooling layer uses 2 $\times$ 2 filter in dimension reduction with an stride of 2. Finally, the fully connected layers with ReLu as a nonlinearity to improve the model performance. VGG16 architecture takes the input image of dimension 224 $\times$ 224.

3.2.3 VGG19

VGG19 is has almost similar architecture with VGG16 with more number of layers. VGG19 architecture has 16 convolution layer, five maxpooling layer with three fully connected layers.

3.2.4 ResNet50

ResNet is a powerful deep learning model which won ILSVRC 2015 classification challenge in 2015 on ImageNet detection, ImageNet localization. Increasing the depth of the network degrades the accuracy and adding more number of layers cause higher training error. To address this performance degradation, a deep residual learning model introduced by He et al. [6]. Instead of stacked convolution layer, it fits the layer in a residual mapping order. Different variant of ResNet are there as ResNet34, ResNet50, ResNet101, RsNet152. In this paper, we compare the results with ResNet50 as it is one of the most dynamic network model.

3.2.5 Inception

The "Inception" is introduced first time in the GoogLeNet architecture by Szegedy et al. [25] and was called Inception-v1. Later on, batch normalization was introduced in the inception-v1 by Ioffe and Szegedy [8] and called as Inception-v2. Later, improved the architecture by additional factorization ideas which referred as Inception-v3 by Szegedy et al. [26]. More uniform simplified architecture and more inception modules added and termed the architecture as Inception-v4 [24]. Inception-v4 is the combination of inception and resnet architecture. The architecture used in this paper is Inception V3 and Inception V4.

3.2.6 DenseNet

DenseNet is a new architecture, and it is an extension of ResNet architecture with some fundamental differences [7]. To merge the previous layer output with next layer output, ResNet uses additive method(+), whereas DenseNet uses concatenation(.) method. In DenseNet, for each layer, all previous layers feature are used as inputs to the current layer. For next layer, the outputs of all previous layer along with the current layer output used as input to the next sub-sequent layer. DenseNet has several advantages as strengthen feature propagation, encourage feature reuse, and reduce number of parameter. Different DenseNet architectures are there as DenseNet121, DenseNet169, DenseNet201, and DenseNet264. In this paper, we have used DenseNet121 and DenseNet 201 architectures.

3.2.7 Transfer Learning

Transfer learning is the most popular approach in machine learning. Transfer learning is the reuse of the pre-trained model [5] for a similar kind of task in machine learning. In transfer learning, some of the layers in trained model are replaced with new layers and retrained the network with the new layers for a new task. Transfer learning uses the knowledge of the pre-trained model which was previously trained on large datasets such as the ImageNet dataset, coco dataset which consist of large number of images. The advantage of using transfer learning is that, to train the network, it requires less training time and gives higher performance accuracy as compared to the learning from scratch. Moreover, deep learning requires a large amount of data to achieve high-performance accuracy but using transfer learning, we can get the same performance on a small dataset.

4 Result

4.1 Experiment Setup

The experiment is performed on CPU mode, and the machine used is of memory 8 GB, i7 processor, and operating system is Windows10. The experiment is performed in Python 3.7, Jupyter Notebook, and Keras as neural network library.

4.2 Experimental Result

To evaluate the performances of the deep learning models, we have used different performance measure metrics such as accuracy, precision, recall, f1-score. To evaluate this metrics, we need several indices such as true positive (TP), true negative (TN), false positive (FP), and false negative (FN). TP is the correctly classified images in a particular category. FP is the number of wrongly classified image. Whereas TN is the sum of the correctly classified images in all other categories, FN is the number of misclassified images from the relevant category. The performance metrics are given by the following equations

$$\text{Accuracy} = \frac{\text{TP} + \text{TN}}{\text{TP} + \text{FP} + \text{TN} + \text{FN}} \tag{1}$$

$$\text{Precision} = \frac{\text{TP}}{\text{TP} + \text{FP}} \tag{2}$$

$$\text{Recall} = \frac{\text{TP}}{\text{TP} + \text{FN}} \tag{3}$$

$$f1\text{-score} = \frac{2 \times \text{Precision} \times \text{Recall}}{\text{Precision} + \text{Recall}} \tag{4}$$

From Table 2, it is seen that VGG16 and VGG19 give highest training accuracy of 99.89% and 99.96%, respectively. Among the deep learning models, VGG19 gives less training loss of 0.0015. To train the models, AlexNet requires much less time in comparison with other deep learning models as the parameter used in AlexNet is much less.

Table 3 gives the testing performances of different deep learning models. From the table, it is seen that VGG19 and VGG16 give higher performance accuracy than other implemented deep learning models in terms of all the evaluated parameter. AlexNet, VGG16, and VGG19 consist of simple convolutional layers which mainly extract the primary features. Other implemented deep learning models such that DenseNet121, DenseNet201, InceptionV3, and InceptionV4 consist of more convolutional layers with residual connection and concatenation which extracts more abstract features which is the reason for the lower performance accuracy in rice diseases. Figure 3 gives the testing accuracies of implemented deep learning models.

5 Conclusion

Timely and accurate identification of rice diseases is useful in agriculture to protect the quality and quantity of the crops. Deep learning is one of the most popular approaches in image processing and pattern recognition task. In this paper, we have studied and implemented several deep learning models in transfer learning to identify different rice leaf diseases. We have compared the performances of the deep learning models. Among all the implemented models VGG19 gives better performance in training with an accuracy rate of 99.96% and loss is 0.0015 after 50 epoch. AlexNet, VGG16, and VGG19 perform better than other implemented deep learning models. Due to the smaller datasize performance in DenseNet, Inception and ResNet model decreases. Furthermore, the future work of this paper includes identifying more varieties of rice diseases with different deep learning architectures and evaluate the

Table 2 Performance evaluation on training the deep learning models

Deep learning model	Training accuracy	Training loss	Training time (s) (Per epoch)
AlexNet	0.9921	0.0079	227
VGG16	0.9989	0.0021	490
VGG19	0.9996	0.0015	608
ResNet50	0.9977	0.0080	1434
DenseNet121	0.9623	0.1268	1310
DenseNet201	0.9687	0.1143	2582
InceptionV3	0.9292	0.2157	697
InceptionV4	0.9334	0.1428	690

Table 3 Performance analysis of different deep learning models

Deep learning model	Accuracy	Precision	Recall	F1-score	Time required (in s)
AlexNet	0.9645	0.9642	0.9646	0.9643	112
VGG16	0.9931	0.9923	0.9935	0.9928	123
VGG19	0.9956	0.9951	0.9958	0.9954	153
ResNet50	0.9337	0.9329	0.9336	0.9332	406
DenseNet121	0.8259	0.8256	0.8261	0.8258	536
DenseNet201	0.8481	0.8480	0.8483	0.8481	556
InceptionV3	0.7924	0.7921	0.7923	0.7922	206
InceptionV4	0.8346	0.8341	0.8344	0.8342	223

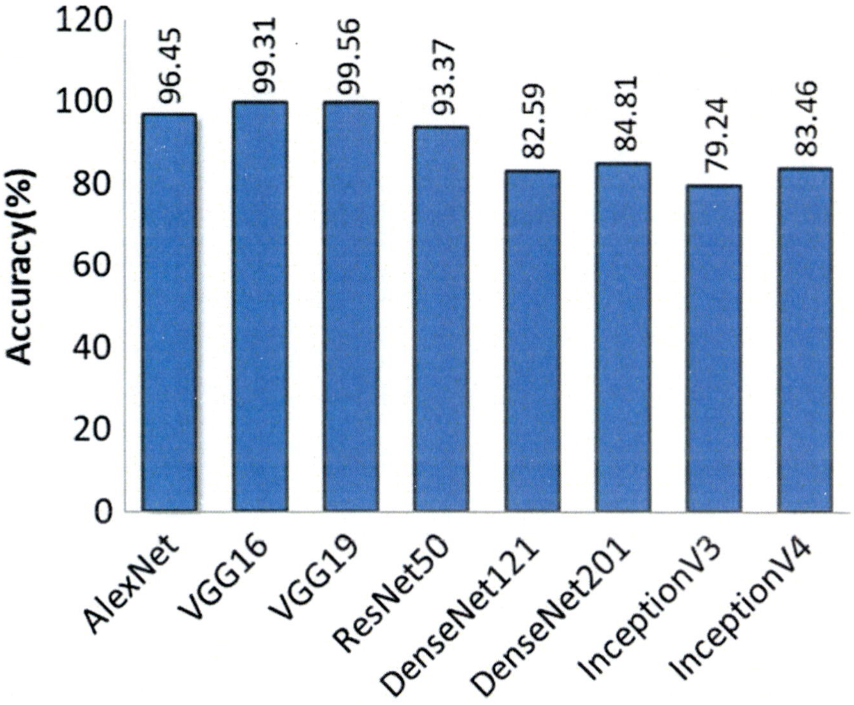

Fig. 3 Training accuracy of different deep learning models

model which performs better. Developing a model for low-powered devices is also an important area to be focus.

References

1. Al-Hiary H, Bani-Ahmad S, Reyalat M, Braik M, Alrahamneh Z (2011) Fast and accurate detection and classification of plant diseases. Int J Comput Appl 17(1):31–38
2. Anthonys G, Wickramarachchi N (2009) An image recognition system for crop disease identification of paddy fields in Sri Lanka. In: 2009 international conference on industrial and information systems (ICIIS). IEEE, pp 403–407
3. Chen J, Chen J, Zhang D, Sun Y, Nanehkaran YA (2020) Using deep transfer learning for image-based plant disease identification. Comput Electron Agric 173:105393
4. Fuentes A, Yoon S, Kim SC, Park DS (2017) A robust deep-learning-based detector for real-time tomato plant diseases and pests recognition. Sensors 17(9):2022
5. Hassan SM, Maji AK, Jasiński M, Leonowicz Z, Jasińska E (2021) Identification of plant-leaf diseases using CNN and transfer-learning approach. Electronics 10(12):1388
6. He K, Zhang X, Ren S, Sun J (2016) Deep residual learning for image recognition. In: Proceedings of the IEEE conference on computer vision and pattern recognition, pp 770–778
7. Huang G, Liu Z, Van Der Maaten L, Weinberger KQ (2017) Densely connected convolutional networks. In: Proceedings of the IEEE conference on computer vision and pattern recognition, pp 4700–4708

8. Ioffe S, Szegedy C (2015) Batch normalization: accelerating deep network training by reducing internal covariate shift. In: International conference on machine learning. PMLR, pp 448–456
9. Islam M, Dinh A, Wahid K, Bhowmik P (2017) Detection of potato diseases using image segmentation and multiclass support vector machine. In: 2017 IEEE 30th Canadian conference on electrical and computer engineering (CCECE). IEEE, pp 1–4
10. Johannes A, Picon A, Alvarez-Gila A, Echazarra J, Rodriguez-Vaamonde S, Navajas AD, Ortiz-Barredo A (2017) Automatic plant disease diagnosis using mobile capture devices, applied on a wheat use case. Comput Electron Agric 138:200–209
11. Krizhevsky A, Sutskever I, Hinton GE (2012) Imagenet classification with deep convolutional neural networks. Adv Neural Inf Process Syst 25:1097–1105
12. Lee SH, Goëau H, Bonnet P, Joly A (2020) New perspectives on plant disease characterization based on deep learning. Comput Electron Agric 170:105220
13. Li Y, Nie J, Chao X (2020) Do we really need deep cnn for plant diseases identification? Comput Electron Agric 178:105803
14. Li Y, Yang J (2020) Few-shot cotton pest recognition and terminal realization. Comput Electron Agric 169:105240
15. Liang WJ, Zhang H, Zhang GF, Cao HX (2019) Rice blast disease recognition using a deep convolutional neural network. Sci Rep 9(1):1–10
16. Matin MMH, Khatun A, Moazzam MG, Uddin MS et al (2020) An efficient disease detection technique of rice leaf using alexnet. J Comput Commun 8(12):49
17. Mique Jr EL, Palaoag TD (2018) Rice pest and disease detection using convolutional neural network. In: Proceedings of the 2018 international conference on information science and system, pp 147–151
18. Oyewola DO, Dada EG, Misra S, Damaševičius R (2021) Detecting cassava mosaic disease using a deep residual convolutional neural network with distinct block processing. Peer J Comput Sci 7:e352
19. Phadikar S, Sil J, Das AK (2013) Rice diseases classification using feature selection and rule generation techniques. Comput Electron Agric 90:76–85
20. Sambasivam G, Opiyo GD (2021) A predictive machine learning application in agriculture: Cassava disease detection and classification with imbalanced dataset using convolutional neural networks. Egyptian Informatics Journal 22(1):27–34
21. Sethy PK (2020) Rice leaf disease image samples. Mendeley Data VI. 10.17632/fwcj7stb8r.1
22. Sethy PK, Barpanda NK, Rath AK, Behera SK (2020) Deep feature based rice leaf disease identification using support vector machine. Comput Electron Agric 175:105527
23. Simonyan K, Zisserman A (2014) Very deep convolutional networks for large-scale image recognition. arXiv preprint arXiv:1409.1556
24. Szegedy C, Ioffe S, Vanhoucke V, Alemi A (2017) Inception-v4, inception-resnet and the impact of residual connections on learning. In: Proceedings of the AAAI conference on artificial intelligence, vol 31
25. Szegedy C, Liu W, Jia Y, Sermanet P, Reed S, Anguelov D, Erhan D, Vanhoucke V, Rabinovich A (2015) Going deeper with convolutions. In: Proceedings of the IEEE conference on computer vision and pattern recognition, pp 1–9
26. Szegedy C, Vanhoucke V, Ioffe S, Shlens J, Wojna Z (216) Rethinking the inception architecture for computer vision. In: Proceedings of the IEEE conference on computer vision and pattern recognition, pp 2818–2826
27. Yao Q, Guan Z, Zhou Y, Tang J, Hu Y, Yang B (2009) Application of support vector machine for detecting rice diseases using shape and color texture features. In: 2009 international conference on engineering computation. IEEE, pp 79–83
28. Zeng W, Li M (2020) Crop leaf disease recognition based on self-attention convolutional neural network. Comput Electron Agric 172:105341

Convexity Defects-Based Fingertip Detection and Hand Gesture Recognition

Soumi Paul, Shrouti Gangopadhyay, Ayatullah Faruk Mollah, Subhadip Basu, and Mita Nasipuri

Abstract Vision-based hand gesture recognition is perhaps the most widely used technique for human–computer interaction technology with special application toward sign languages used by differently abled people. Perhaps the most important step in such a gesture recognition system is feature extraction. In this paper, we use a novel fingertip detection mechanism based on convexity defects and use nine geometrical features that are translation and rotation invariant. We use seven different classifiers on two different public hand digit datasets (NTU hand digit dataset and SP-EMD color-depth hand gesture dataset) and find that for both the datasets, the random forest classifier gives the best classification accuracy (94.2% and 90.1%, respectively).

Keywords Convexity defect · Fingertip detection · Hand gesture recognition · Human–computer interaction

1 Introduction

Among different forms of inputs for human–computer interaction, hand gesture is perhaps the most natural, simple, and yet, rich in expressing varied physical behaviors or emotions. Automated hand gesture recognition [1] is very useful for speech and hearing-impaired people for their communications using sign languages [2] with the rest of the world.

S. Paul (✉) · S. Gangopadhyay · S. Basu · M. Nasipuri
Department of Computer Science and Engineering, Jadavpur University, Kolkata 700032, India
e-mail: spaul.cse.rs@jadavpuruniversity.in

S. Basu
e-mail: subhadip.basu@jadavpuruniversity.in

M. Nasipuri
e-mail: mita.nasipuri@jadavpuruniversity.in

A. F. Mollah
Department of Computer Science and Engineering, Aliah University, Kolkata 700160, India
e-mail: afmollah@aliah.ac.in

S. Basu et al. (eds.), *Proceedings of International Conference on Frontiers in Computing and Systems*, Lecture Notes in Networks and Systems 404,
https://doi.org/10.1007/978-981-19-0105-8_21

There are two ways in which image acquisition for hand gesture recognition happens. The first one is acceleration hardware-based [3], and it uses complex data gloves and position tracker and, hence, is expensive and less popular. The second one, on the other hand, is vision-based [4, 5] and is simple and easy to operate without complex devices or interfaces. There are four stages of vision-based hand gesture recognition: (i) sample images capturing, (ii) preprocessing of raw image, (iii) feature or descriptor extraction, and (iv) recognition (classification).

The most common features of hand gestures of static type comprise of gradient histogram [6], image subspace projection [7], shape features [8] etc. The gradient histogram is easy to compute, and it is invariant to translation but is affected by rotation. Image subspace projection removes higher-order correlations and creates an efficient representation of local descriptors of training images. However, it is dependent on position, scale, and rotation. Features based on shape, e.g., contour, silhouette, fingertips, are now the popularly used ones, as they are independent of translation, size, and rotations. In this paper, we introduce a series of fingertip-related novel features, exploiting convexity defects [9], and related geometrical shapes.

1.1 Related Work

There are a few existing works on gesture recognition based on first fingertip detection and then extraction of suitable features. For instance, in [10] the authors apply convexity hull defects algorithm to determine the finger count. Maisto et al. [11] recognize fingertips and identify their position. In another work [12], the authors apply K-curvature algorithm to locate fingertips on the contour and then use dynamic time warping (DTW) to compare an observed gesture with a set of reference gestures, whereas in [13], the authors first detect a convex hull of the hand gesture so that the fingertips can be identified and located accurately. In [14], the authors extract a series of features based on convex defect detection, taking advantage of the close relationship of convexity defect and fingertips. Xu et al. [15] worked with convexity defect detection with density distribution to analyze the characteristics of hand gesture. Some other recent relative works are [16–19], where fingertip has been detected from convex hull and convexity defects for gesture classification.

1.2 Contribution

We have extracted nine descriptors, namely minimum, maximum, average, and sum from center of gravity to fingertip distance, minimum, maximum, average, and sum from distance among fingertips in a particular direction and count of the fingertips. This helps to identify the fingertips, from two publicly available datasets, namely NTU hand digit dataset [20, 21] and SP-EMD color-depth hand gesture dataset [22, 23]. We have used seven classifiers such as, naïve Bayes (NB), J48, bagging, multi-layer perceptron (MLP), support vector machine (SVM) with linear kernel, SVM with Pearson VII Universal Kernel (PUK), and random forest (RF) to study the comparative performance on the datasets.

The remainder of this paper is organized as follows. Section 2 presents the materials and methods, then fingertip detection is illustrated in Sect. 3. Feature extraction explained in Sect. 4. Results and discussion are presented in Sect. 5 and with conclusion in Sect. 6.

2 Materials and Methods

In this section, we describe the datasets and how they were collected, and then discuss the preprocessing and finally present the key ingredient of our fingertip detection mechanism, namely the convexity approach.

2.1 DataSet

Kinect is a depth-sensing sensors which was released in the market primarily for games. Then, slowly, microsoft released different versions of it in the market, like Kinect for Xbox 360, Kinect for Windows, Kinect for Xbox One which was released with console and Kinect 2 for Windows. Developers in the field of gesture recognition using depth-sensing device welcomed it and started capturing gestures in RGB-D. Hardware or functional differences between these different released versions are explained by Paul et al. [24].

In this work, we have used two public image datasets for experimentation which were captured using Kinect sensor. The first one is the NTU hand digit dataset [20, 21], where 10 subjects have contributed 10 gestures from numbers 0 to 9 with 10 different poses performed for the same gesture by each subject. This makes the dataset consisting of 1000 images in total. As a second dataset, we use the one from the SP-EMD project, named SP-EMD color-depth hand gesture dataset [22, 23] that contains 10 gestures with 20 different poses from 5 subjects, giving a total of 1000 images. Sample images of these datasets are shown in Fig. 1.

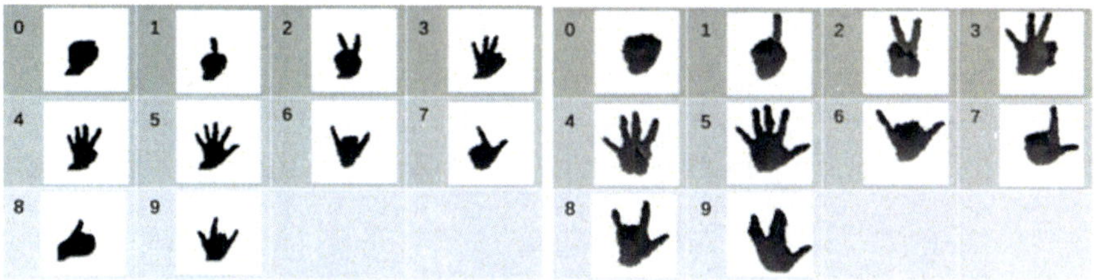

Fig. 1 Left: NTU hand digit dataset showing gestures from 0 to 9; right: SP-EMD color-depth hand gesture dataset showing gestures from 0 to 9

2.2 Preprocessing

For the NTU hand digit dataset, the region of interest (ROI) in RGB images has been annotated manually. After performing depth thresholding, the exact ROI regions from the depth images are cropped, and this is done by comparing it with RGB images. For the SP-EMD color-depth hand gesture dataset, the mean and standard deviation of the depth values are calculated instead of depth thresholding following the work by Paul et al. [25]. This helps the hand shapes to be correctly segmented.

2.3 Convexity Approach

For gesture recognition, fingertip is a vital part to detect. The number of fingertips, in most of the cases, denotes the number of fingers. Also, their directions can be useful in conveying the information about which gesture is being shown. The most commonly used fingertip detection method is contour analysis. In this work, we use the method of convexity defect, based on the contour shapes, to detect the fingertips.

Contour. Contour is mainly the series of boundary pixels of the hand area. As contour can be considered as the collection of edges, various edge detection algorithms can be used to detect the contour, as for example, fuzzy, Prewitt, Robert, Sobel, Canny. Here, we use Canny edge detection method to extract the contour of hand gesture, the green curve in left image of Fig. 2 shows the contour of a hand gesture.

Convex Hull. The smallest convex set, containing the contour, is called the convex hull. It can also be represented as the smallest convex polygon being surrounded by every convex vertices in the contour. The middle image of Fig. 2 shows the convex hull of a contour.

Convexity Defect. Convexity defect is the difference between the convex hull and the contour. The object's (whose contour or convex hull is determined) area is not included in the convexity defect calculation. A convexity defect is specified by three

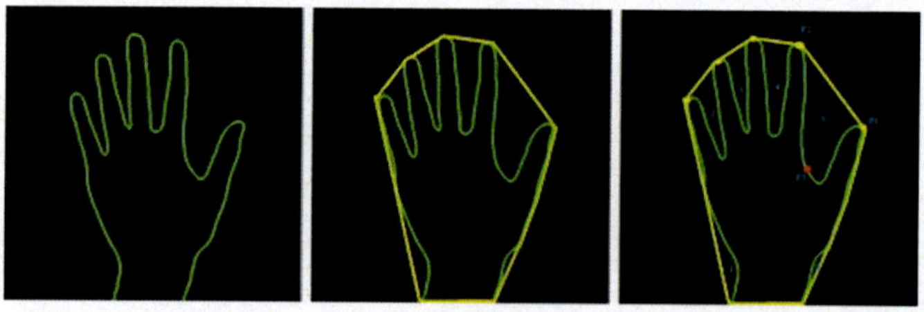

Fig. 2 Left: the extracted contour of depth thresholded image; middle: convex hull on extracted contour; right: the convexity defects

points—the start contour point, the end contour point, and the far (concave part) contour point. An example is shown in the right most picture of Fig. 2. $P1$ is the starting point of the convexity defect, $P2$ is the ending point, and $P3$ is the farthest concave point from the convex hull. The detection of fingertip is very closely related to the convexity defect, as the potential fingertips reside close to the start and end points of the convexity defect.

3 Fingertip Detection

After the convexity defects are detected, the start point and the nearest end point together define a potential fingertip. However, the start and end points of every convexity defect do not indicate a potential fingertip. To eliminate false positives, the angles between the straight lines connecting the far concave point with both the start and the end convexity defect points are considered. An angle not exceeding 120 °C is considered as a potential convexity defect generated by two fingers on its both sides.

As it can be seen in Fig. 3, the angle between d2Start and d2End with d2Far is less than 120 and that is taken as potential fingertips, but the angle between d1Start and d1End (which is also d2Start) is greater than 120 °C and is not taken into consideration.

Thus, d2Start and d2End can be considered as fingertips. To avoid redundancy in counting fingers, each time the start position of the convexity defect is considered as a fingertip. Only for the last convexity defect in any particular orientation, the end point is also considered as a fingertip.

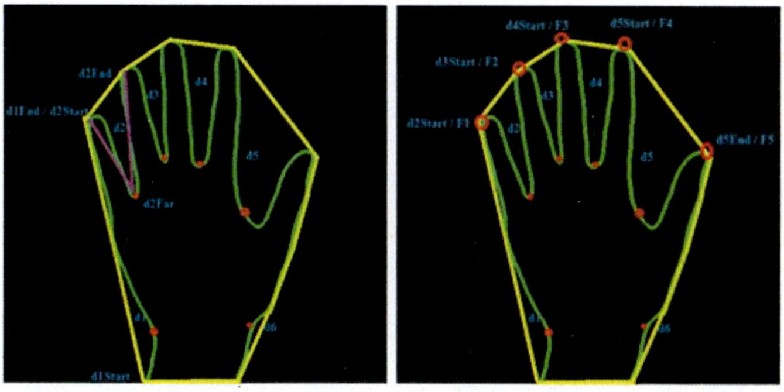

Fig. 3 Left: the defect areas for possible fingertips; right: the accurate fingertips after angle thresholding

4 Feature Extraction

After the fingertips are detected, we proceed to the feature extraction stage to identify the gesture of the hand. First, the moments are calculated, from which we get the center of gravity of the hand palm.

Let $C = (C_x, C_y)$ be the center of gravity (COG) of the hand and $F1$, $F2$, $F3$, $F4$, $F5$ be the tips of the five fingers. For each Fi, $1 \leq i \leq 5$, let Fi_x and Fi_y be the X and Y coordinates of Fi.

Center of Gravity to Fingertip Distance. The Euclidean distances from the center of gravity to the i-th fingertip point is given by

$$D_i = \sqrt{(C_x - Fi_x)^2 + \left(C_y - Fi_y\right)^2},$$

for $i = 1$ to 5. Then we find the following features by calculating minimum, maximum, average, and sum of all the distances.

$$D_{\min} = \min_{i=1}^{5}\{D_i\}, \qquad D_{\max} = \max_{i=1}^{5}\{D_i\},$$

$$D_{\text{sum}} = \sum_{i=1}^{5} D_i, \qquad D_{\text{avg}} = \frac{1}{5}\sum_{i=1}^{5} D_i.$$

Distance Among Fingertips in a Particular Direction. Next, we find the distances between consecutive fingertips as follows

$$\text{Dcon}_i = \sqrt{(F(i+1)_x - Fi_x)^2 + \left(F(i+1)_y - Fi_y\right)^2},$$

for $i = 1$ to 4. Then, we find the following features by calculating minimum, maximum, average, and sum of all the consecutive distances.

$$\text{Dcon}_{\min} = \min_{i=1}^{4}\{\text{Dcon}_i\}, \qquad \text{Dcon}_{\max} = \max_{i=1}^{4}\{\text{Dcon}_i\},$$

$$\text{Dcon}_{\text{sum}} = \sum_{i=1}^{4} \text{Dcon}_i, \qquad \text{Dcon}_{\text{avg}} = \frac{1}{4}\sum_{i=1}^{4} \text{Dcon}_i.$$

Thus, we have 4 + 4 = 8 features from above. Adding to this the number of fingertips, we have a total of 9 features.

Note that, a certain gesture may not have all the five fingertips. There are at most 5 fingertips, and therefore, 5 distances from the center of gravity. Also, there are at most 4 consecutive distances. But the number of features is fixed to 9, as instead of considering the actual distances, we are calculating four functions of them.

Table 1 Accuracy comparison using different classifiers on NTU hand digit dataset and SP-EMD color-depth hand gesture dataset (Highest Accuracy in each dataset is marked in Bold)

Dataset	Naïve Bayes	J48	Bagging	MLP	SVM (Linear)	SVM (PUK)	Random Forest
NTU (5 fold CV)	86.1	86.7	90.2	91.9	87.9	92.5	93.7
NTU (10 fold CV)	86.1	88.8	89.6	93.2	88.3	93.0	**94.2**
SP-EMD (5 fold CV)	75.5	82.7	84.2	87.8	81.7	87.6	89.7
SP-EMD (10 fold CV)	75.7	84.3	84.9	87.5	82.6	88.3	**90.1**

5 Results and Discussion

Initially, the experiments have been carried out on the two public datasets, namely the NTU hand digit dataset and the SP-EMD color-depth hand gesture dataset, using many conventional supervised classifiers, using both five and tenfold cross-validations. Out of them, seven classifiers, namely NB, J48, Bagging, MLP, SVM with Linear Kernel, SVM with PUK, and RF yielded comparatively better accuracy results which we report in Table 1.

Note that for both the datasets, RF classifier gives the highest accuracy. It generates a set of decision trees from random subsets of the training set and combines the votes from them to select the final class of the test object. A single decision tree may be subject to noise, but an aggregate of many is likely to reduce the effect of noise giving better accuracy. RF classifier also reduces overfitting and is robust to outliers. Since it uses rule-based approach instead of distance, no scaling such as standardization, normalization of features is necessary.

Among the two cross-validations, always the tenfold one gives better result. The best results on the NTU hand digit dataset and the SP-EMD color-depth hand gesture dataset are 94.2% and 90.1%, respectively, shown in bold in Table 1.

6 Conclusion

In this work, we have used convexity defects to detect fingertips and then extracted statistical features from them for subsequent hand gesture recognition using several machine learning algorithms. The proposed method is tested on NTU hand digit dataset [20, 21] and SP-EMD color-depth hand gesture dataset [22, 23], and for both of these, the random forest classifier yields the highest accuracy.

This work also has some limitations, like detecting false fingertips, counting a single finger multiple times, merging two fingers, etc., though for only a few cases. In our future work, we plan to improve our fingertip detection algorithm to mitigate these issues. We also aim to incorporate more public datasets for an extensive comparative study.

References

1. Mitra S, Acharya T (2007) Gesture recognition: a survey. IEEE Trans Syst Man Cybern Part C (Appl Rev) 37(3):311–324
2. Shi B, Brentari D, Shakhnarovich G, Livescu K (2021) Fingerspelling detection in American sign language. arXiv preprint arXiv:2104.01291
3. Hussain SMA, Rashid AH (2012) User independent hand gesture recognition by accelerated DTW. In: 2012 international conference on informatics, electronics and vision (ICIEV). IEEE, 1033–1037
4. Choi J, Seo BK, Lee D, Park H, Park JI (2013) Rgb-d camera-based hand shape recognition for human-robot interaction. In: IEEE ISR 2013. IEEE, 1–2
5. Palacios JM, Sagüés C, Montijano E, Llorente S (2013) Human-computer interaction based on hand gestures using rgb-d sensors. Sensors 13(9):11842–11860
6. Tiantian L, Jinyuan S, Runjie L, Yingying G (2015) Hand gesture recognition based on improved histograms of oriented gradients. In: The 27th Chinese control and decision conference (2015 CCDC), 4211–4215
7. Jiang F, Wang C, Gao Y, Wu S, Zhao D (2015) Discriminating features learning in hand gesture classification. IET Comput Vis 9(5):673–680
8. Panwar M (2012) Hand gesture recognition based on shape parameters. In: 2012 international conference on computing, communication and applications, 1–6
9. Mesbahi SC, Mahraz MA, Riffi J, Tairi H (2018) Hand gesture recognition based on convexity approach and background subtraction. In: 2018 international conference on intelligent systems and computer vision (ISCV). IEEE, 1–5
10. Ganapathyraju S (2013) Hand gesture recognition using convexity hull defects to control an industrial robot. In: 2013 3rd international conference on instrumentation control and automation (ICA). IEEE, 63–67
11. Maisto M, Panella M, Liparulo L, Proietti A (2013) An accurate algorithm for the identification of fingertips using an rgb-d camera. IEEE J Emerg Sel Top Circ Syst 3(2):272–283
12. Plouffe G, Cretu AM (2015) Static and dynamic hand gesture recognition in depth data using dynamic time warping. IEEE Trans Instrum Measur 65(2):305–316
13. Ma X, Peng J (2018) Kinect sensor-based long-distance hand gesture recognition and fingertip detection with depth information. J Sens
14. Xu Y, Park DW, Pok G (2017) Hand gesture recognition based on convex defect detection. Int J Appl Eng Res 12(18):7075–7079
15. Xu Y, Dai Y (2018) Convex defect detection and density distribution based hand gesture recognition. J Comput 29(5):29–40
16. Kakkoth SS, Gharge, S (2018) Visual descriptors based real time hand gesture recognition. In: International conference on advances in communication and computing technology (ICACCT). IEEE, 361–367
17. Hartanto R, Kartikasari A (2016) Android based real-time static Indonesian sign language recognition system prototype. In: 2016 8th international conference on information technology and electrical engineering (ICITEE). IEEE, 1–6
18. Li Y (2012) Multi-scenario gesture recognition using kinect. In: 2012 17th international conference on computer games (CGAMES). IEEE, 126–130
19. Saha HN, Tapadar S, Ray S, Chatterjee SK, Saha S (2018) A machine learning based approach for hand gesture recognition using distinctive feature extraction. In: IEEE 8th annual computing and communication workshop and conference (CCWC). IEEE, 91–98

20. Ren Z, Meng J, Yuan J (2011) Depth camera based hand gesture recognition and its applications in human-computer-interaction. In: 2011 8th international conference on information, communications and signal processing (ICICS). IEEE, 1–5
21. Ren Z, Yuan J, Meng J, Zhang Z (2013) Robust part-based hand gesture recognition using kinect sensor. IEEE Trans Multimedia 15(5):1110–1120
22. Wang C, Chan S (2014) A new hand gesture recognition algorithm based on joint color-depth superpixel earth mover's distance. In: 4th international workshop on cognitive information processing (CIP). IEEE, 1–6
23. Wang C, Liu Z, Chan SC (2014) Superpixel-based hand gesture recognition with kinect depth camera. IEEE Trans Multimedia 17(1):29–39
24. Paul S, Basu S, Nasipuri M (2015) Microsoft kinect in gesture recognition: a short review. Int J Control Theory Appl 8(5):20712076
25. Paul S, Bhattacharyya A, Mollah AF, Basu S, Nasipuri M (2019) Hand segmentation from complex background for gesture recognition. In: Emerging technology in modelling and graphics. Springer Singapore, 775–782

Variable-Length Genetic Algorithm and Multiple Entropic Functions-Based Satellite Image Segmentation

Ramen Pal, Somnath Mukhopadhyay, and Debasish Chakraborty

Abstract The genetic algorithm (GA) was invented to mimic the realistic biological evaluation for optimization. But, in literature, we can observe that the genetic operators are purely based on either mathematical or random operations. Our proposed genetic operators are designed to mimic realistic behavior. For this purpose, we have introduced the concept of gene casements in crossover and Levy flight function in mutation. We have also incorporated solution combination operation to give it an elitist nature. Otsu's, Tsallis, and Reyni's entropic functions were validated as efficient thresholding techniques. Researchers weren't able to conclude which method is superior among them. Their performance is highly dependent on the dataset. So, we have considered all of these methods to frame our methodology. We have considered the dataset from the WorldView-2 satellite sensor for the experimental study. The performance of the proposed method is validated with three different state-of-the-art methods.

Keywords VHR image · LULC segmentation · Variable-length GA · Lèvy flight · Entropic function

1 Introduction

Multi-spectral image segmentation is the procedure of distinguishing different objects or land-use land-cover (LULC) classes in a satellite image. The success of remote sensing applications like, LULC change detection [1], prediction [2], identification [1], etc., solely depend on the efficiency of the backhand segmentation algorithm [3]. In literature, we can observe that satellite image segmentation algorithms are framed by using: (i) edge, (ii) region, (iii) cluster, (iv) artificial neural network,

R. Pal · S. Mukhopadhyay (✉)
Department of Computer Science and Engineering, Assam University, Silchar, India
e-mail: som.cse@live.com

D. Chakraborty
RRSC-East, ISRO, Kolkata, India

S. Basu et al. (eds.), *Proceedings of International Conference on Frontiers in Computing and Systems*, Lecture Notes in Networks and Systems 404,
https://doi.org/10.1007/978-981-19-0105-8_22

(v) thresholding, etc. [4–9]. Among these approaches, cluster- and threshold-based segmentation techniques are proven as efficient [10, 11]. Clustering is a method to separate a satellite image into distinct classes. All the elements within a class should be highly cohesive, and between the classes are highly coupled in nature [12]. Clustering-based techniques generally suffer with problems like, (1) stacking into the regional optima, (2) highly dependent on initialization. These issues can be solved by using metaheuristic-based techniques. It can produce high-quality segmented images. Metaheuristic algorithms like, genetic algorithm, particle swarm optimization, ant colony optimization, cuckoo search, and multi-objective optimization(MOO) algorithms are useful for these purposes. Whereas, thresholding techniques find out optimal ranges with respect to color information. A range is defined as a LULC class. This is done by using the help of entropic functions. In last few decades, entropic functions like, Tsalis, Otsu's, Kapur, Rènyi, etc., are extensively used to segment an image with bi-modal or multi-modal histogram. These functions did not have any optimization tool to optimize these threshold values. This issue can be resolved by hybridizing a metaheuristic algorithm with entropic function. In literature, all the previous researches in this regard are suffering with issues like: (i) automatic identification of the total number distinct LULC classes within a satellite image and (ii) identification of proper entropic function as objective. Justification of the efficiency of an entropic function will not be proper. Since the performance of a function depends on the types of the satellite image. It is proven that Tsallis, Otsu's, and Rènyi's entropic functions are efficient in two different aspect like, types of the image and modal of the histogram. But, researchers weren't able to conclude that a particular entropic function is highly efficient compare to others in general. So, we should formulate the objective function of an optimization method by considering these aforementioned functions. In a satellite image, total number of LULC classes varies with the image. So, there is a strong need to create a variable-length nature for an algorithm that can cope-up with this situation. Genetic algorithm (GA) is one of the prominent metaheuristic approaches that solve optimization problems efficiently. But, this algorithm suffers with the following issues: (i) In mutation, this algorithm states that some weakest genes should be updated to make the chromosome more efficient. This should be done with respect to the fittest solution. But, in reality, some random changes are applied. (ii) Elitism is optional, and (iii) In crossover, dissimilar portion of both of the parent chromosomes should be considered. But, in reality, similar portions of gene sequences are also considered. So, it motivated us to propose an image segmentation method that is free from all of these dependencies or issues. Contributions of this paper are as follows:

1. A new approach toward designing a variable length and elitist satellite image thresholding algorithm.
2. Considering Tsallis, Otsu's, and Rènyi's entropic function collectively to formulate the objective function.
3. Automatic identification of the total number distinct LULC classes within a satellite image.

4. Proposing a new realistic technique for mutation by using Levy flight generation function.
5. Proposing a new realistic technique for crossover by using gene casements.

In the rest of the paper, Sect. 2 considered the discussion on the proposed methodology, Result and analysis are presented in Sect. 3, and conclusion is presented in Sect. 4.

2 Proposed Methodology

It is a two-step process. In the first step, threshold values for the satellite image are optimized, and corresponding segmentation is done. Then, in the second step, classes obtained in the segmented images are mapped with the proper LULC class by using ground truthing information. The proposed algorithm is presented in 1.

First, initial encoding of c number of chromosomes are done. Here, a chromosome is represented as a vector. It represents a combination of b number of solutions. Here, b represents the total number of bands in an input image. Each solution represents a probable set of threshold values for a band of the input image. All solutions will have equal length (l). Value for l is chosen randomly. Here, b represents the total number of bands in an image. Since the input image can have bi-modal or multi-modal histogram. So, we have limited the value for l. It should be ≥ 2. By iterating the above-mentioned procedure, we have encoded c number of chromosomes. After that, genetic operators and corresponding operations are applied iteratively based on a condition. In this process, first, the fitness value (F) of each chromosome is evaluated by using Eq. 1. Here, O_E represents the average Otsu's Entropy for ith chromosome.

$$F_i = \left(\frac{\sum_{i=0}^{b} O_E}{b} + \frac{\sum_{i=0}^{b} T_E}{b} + \frac{\sum_{i=0}^{b} R_E}{b} \right) \Big/ 3 \tag{1}$$

It is a non-parametric and between class variance function. It maximizes the inter-cluster distance. It is presented in Eq. 2.

$$O_E = \sum_{i=0}^{m} \sigma_i \tag{2}$$

Here, m represents the total number of threshold values. σ_i represents the between class variance of the ith class. It is calculated by using Eq. 3.

$$\sigma_i = \omega_i \times (\mu_T - \mu_i)^2 \text{ Where,} 0 \leq i \leq n \tag{3}$$

Here, ω represents the summation of probability of intensity values. μ_T represents the global mean of intensity values, and μ_i represents the mean intensity value for ith

Algorithm 1: Variable-length and Elitist Genetic Algorithm Based Satellite Image Segmentation and Change Detection Method.

Input: MS imagery (img), Number of iterations (z), Size of the population(c)
Output: Segmented or Thresholded Image
Begin;
b= Total number of bands in the input image;
Initial population is randomly encoded with c number of chromosomes. Here, the length of each gene is b ; `// 2≤ chromosome length ≤ 20`
for $g = 1$ *to* z **do**
  if $g = 1$ **then**
    Fitness of each chromosome is evaluated by using Equation 1;
    Store the fittest chromosome in G_{best};
    Roulette wheel or fitness proportionate selection is done to select the mating pool m of size c;
    Real coded crossover function is applied on m;
    Real coded mutation function is applied on m by using G_{best};
  end
  else
    Combined mating pools of current and previous generations to create a pool with $2 \times c$ number of chromosomes;
    Fitness of each chromosome is evaluated by using Equation 1;
    Select the fittest chromosome (P);
    if P *is fittest than from* G_best **then**
      $G_{best} = P$.
    end
    Roulette wheel or fitness proportionate selection is done to select the mating pool m of size c;
    Real coded crossover function is applied on m;
    Real coded mutation function is applied on m by using G_{best};
  end
end
Combine current and previous mating pools to constitute a combined pool;
From the combined pool, the fittest chromosome (S) is selected as the optimal sets of threshold values for image segmentation;
Calculate the global threshold values by taking mean of b number of solutions within S;
Number of clusters (K)= length of S;
img is segmented into K number of clusters by using the optimal threshold values for each cluster from S;
Classify each class in the segmented image to a LULC class by using ground truth information;
End;

class. Here, O_E is a maximization function. T_E is the Tsallis entropy measure. It is an elicitation of Boltzman–Gibbs entropy measure. It is a measure to a non-extensive system. It is presented in Eq. 4.

$$T_E = \sum_{i=0}^{m} S_q^i(t) \tag{4}$$

Here, q is the Tsallis parameter. t represents the threshold value, and S^i represents the Tsallis entropy. It is represented in Eq. 5.

$$S_i = \frac{1 - \sum_{i=1}^{t} P_i{}^q}{q - 1} \tag{5}$$

It is a multi-fractal theory-based formula. Here, P_i represents the probability of the system and is in state i. Here, T_E is also a maximization function. R_E is the Rènyi's entropy measure. It is an elicitation of Shannon entropy. It is presented in Eq. 6.

$$R_E = \sum_{i=0}^{m} R_\alpha(A_i) \text{ Where, } \alpha > 0 \tag{6}$$

Here, α is a non-substantial control parameter. The value for α should be greater than 0. A_i is the elicitation of the Shanon entropy. It is calculated by using Eq. 7.

$$R_\alpha = \frac{1}{1 - \alpha} \ln \sum_{i=1}^{L} P_i^\alpha \text{ Where, } \alpha > 0 \tag{7}$$

Here, P_i represents the probability distribution of the ith class. L defines the total number of threshold. R_E is also a maximization function. Fitness value for a chromosome is calculated by taking the mean of O_E, T_E and R_E. After that, for the first iteration, fitness proportionate selection is done only to select c number of solutions from the population to create a mating pool of size c. It is evident that the convergence rate of this selection strategy is higher than others. The chance of selection of a chromosome is proportionate to its fitness value. In the rest of the iteration, a combined mating pool is generated before selection. This combined pool is generated by combining all solutions from the previous generation with the solutions from the current generation. This operation ensures elitism. Then, reproduction is done by using genetic operator. In biological evaluation, crossover doesn't produce child chromosomes that are drastically different from the parent one. In literature, existing crossover methods doesn't consider this behavior. We have designed our algorithm in such a way that it exhibits the realistic behavior. In this method, two chromosomes (P_1 and P_2) are randomly selected first. Then, a probabilistic value C_j is randomly chosen. If this value is less than the thresholded crossover probability (C_p), then crossover is applied. In a crossover operation, first, the common gene casements (gc) between P_1 and P_2 are extracted. If the length of gc is greater than 1, then the minimum value between total number of gene casements (m_{P_1} and $m P_2$) of chromosome P_1 and P_2 is selected. Then, first, m number of gene casements are swapped between P_1 and P_2. Else, one among two different operations are executed based on a probabilistic value (act). If this value is greater than 0.5, then 0.5 is subtracted from all genes that are presented in gene casements of P_1 and P_2. Otherwise, 0.5 is added. This operation ensures the the avoidance of premature convergence. This process is repeated for $n/2$ times. Then, the length of each solution in a chromosome is restricted. For example, the input image has three bands. So, the chromosome is comprised with three solutions. Each solution represents a probable threshold values for a band. Let, the solution with minimum number of gene is P_S. Since our algorithm has variable-length behavior with respect to chromosomes and the chromosomes are represented as vector. All solutions within a chromosome should have equal length. So, the length of each solution is restricted with the length of P_S. After that, mutation is applied. In reality, a chromosome is mutated by updating its gene to make it more fit. So, we can get efficient genes if we update it with respect to the globally fittest chromosome (G_{best}). $G - best$ represents fittest chromosome among the all previous generations. In our method, a gene is mutated based on a probabilistic value (m_j). If it is less than the thresholded mutation probability (m_p), then g_i is updated by using the Lèvy flight formula. Otherwise, it is updated by subtracting 0.1% value of each threshold values g_i in g. Lèvy flight is an elicitation of Brownian motion. It produces short amount of (C) but random jumps toward the globally fit solution. It is presented in Eq. 8.

$$C = 0.01 \times S \times (g_i - b_g) \text{ Where, } 0 \leq i \leq b \tag{8}$$

Here, S is the random step. It is generated by using the symmetric Lèvy flight distribution. g_i and b_g are current gene and global best gene, respectively. The overall process is executed for 200 iterations. After completing all the iterations, a combined mating pool is generated by combining the mating pool of current and last generation. The fittest solution (S) from this pool is selected as the optimal set of threshold values for segmentation. Then, the global threshold values are calculated by taking the mean of threshold values obtained for b number of bands. After that, the input image is segmented by using S. If the length of S is n, then we get $n + 1$ number of segments in the segmented image.

3 Results and Discussions

In this section, the effectiveness of the proposed method is analyzed and validated with respect to three different state-of-the-art methods. All the experiments on the proposed method are done on a system with Software: Python 3.7 and ArcMap-10.5, Operating System: Windows-10, CPU: Intel(R)-Core(TM)-i5, and RAM: 8 GB. Apart from that, a GPS handheld device is used to perform the ground truthing to asses the accuracy of the segmentation. All the images are represented in the form of map. These maps are created by using ArcMap-10.5 software.

3.1 Experimental Results

We have considered VHR and PAN-sharpened multi-spectral image of Barasat, India, taken from WorlView-2 satellite sensor. This dataset is taken on February 22, 2011 and has a spatial resolution 0.46 m. It has 3 bands. It is shown in Fig. 1.

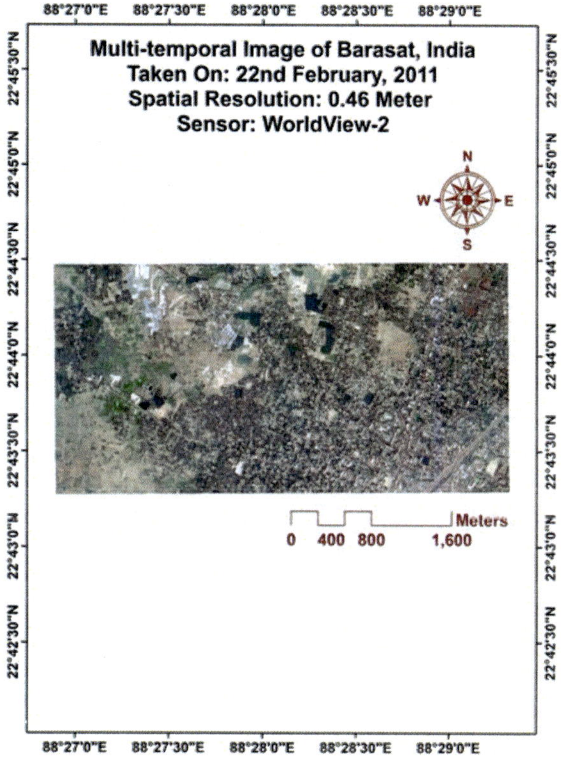

Fig. 1 Experimental VHR image of Barasat, West Bengal, India, with a spatial resolution 0.46 m, obtained from WorldView-2 satellite sensor, taken on December 14, 2020

The proposed method is executed on this dataset. We considered the size of the initial population as 100 and its executed for 200 iterations. The fittest solution from the final combined pool is selected. Then, global threshold values are calculated by considering the mean of all fittest chromosomes for all bands of the input image. Then, thresholding is performed to segment the input image using it. After that, it is classified into 4 different LULC classes (waterbody, built-up, vegetation, and bare soil) based on the ground truthing information. Then, methods of Eddine et al. [13], Xing et al. [14], and Mousavirad et al. [15] are executed on the input image, and the quality of segmentation is quantified for validation. The qualitative results obtained after classification are presented in Fig. 2.

We can observe that the quality of the results obtained by our proposed method is better than other methods. We have quantitatively validated this claim in Table 1. The internal performance of all segmentation methods is validated by using Dunn index [16], Davies Bouldin (DB) index [17], and Silhouette score [18]. We can observe that our proposed method outperforms other techniques in all cases. So, we can conclude that our proposed method can segment a satellite image with a complacent performance. Finally, the accuracy of the classification is quantified by using the Kappa Co-efficient (K) [19]. In this process, 100 random locations (latitude and longitude information) from the study area are selected. Then, we visited those

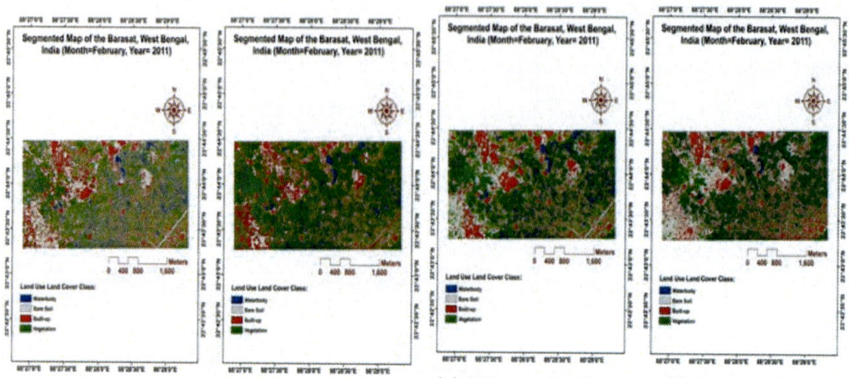

(a) Eddine et al. [13] (b) Xing et al. [14] (c) Mousavirad et al. [15] (d) Proposed method

Fig. 2 Segmented images generated by the: **a** Eddine et al. [13], **b** Xing et al. [14], **c** Mousavirad et al. [15], and **d** Proposed method

Table 1 Performance validation of the proposed method

Algorithm	Dunn score	DB score	Silhouette score	Kappa co-efficient
Eddine et al. [13]	1.0914324519	0.3865482180	0.4525618565	0.8616
Xing et al. [14]	0.9971457920	0.6519437862	0.4032873154	0.8002
Mousavirad et al. [15]	1.2722518245	0.1640250640	0.5502160820	0.9000
Proposed method	**1.4006205845**	**0.1109200425**	**0.5810586427**	**0.9275**

areas; check the real LULC class. Then, actual and predicted LULC classes are stored in an excel sheet. In the Table 1, superior performance of a method regarding a metric is presented using bold characters. Then, this data is used to calculate the value for K. We can observe that our proposed has the highest K score compared to other methods. So, we can also conclude that our proposed method is fair accurate.

4 Conclusion and Future Scope

We proposed a new technique for satellite VHR image thresholding by using our proposed improved GA. We have also discussed a way of defining realistic crossover and mutation operation by using the gene casements and Lèvy flight generation function, respectively. We have also shown that Tsallis, Otsu's, and Rènyi's entropic function collectively can produce better segmentation result. Our method can optimize the total number of output clusters automatically. Our proposed method is tested by using the dataset from WorldView-2 satellite sensor. It outperformed three different state-of-the-art methods in respect of four different evaluation parameters.

This method is based on ground truthing. So, a way of doing classification without ground truthing need to be find out in future.

References

1. Dey V, Zhang Y, Zhong M, A review on image segmentation techniques with remote sensing perspective. In: Proceedings of the international society for photogrammetry and remote sensing symposium (ISPRS10) XXXVIII (Part 7A)
2. Zhang D, Islam MM, Lu G (2012) A review on automatic image annotation techniques. Pattern Recogn 45(1):346–362. https://doi.org/10.1016/j.patcog.2011.05.013
3. Mitra P, Shankar BU, Pal SK (2004) Segmentation of multispectral remote sensing images using active support vector machines. Pattern Recogn Lett 25(9):1067–1074. https://doi.org/10.1016/j.patrec.2004.03.004
4. Satapathy S, Raja NSM, Rajinikanth V (2018) Multi-level image thresholding using otsu and chaotic bat algorithm. Neural Comput Appl 29:1285–1307. https://doi.org/10.1007/s00521-016-2645-5
5. Jain S, Laxmi V (2018) Color image segmentation techniques: a survey. In: Proceedings of the international conference on microelectronics, computing & communication systems. Springer Singapore, Singapore, pp 189–197
6. Yu H, He F, Pan Y (2018) A novel region-based active contour model via local patch similarity measure for image segmentation. Multimedia Tools Appl 77:24097–24119. https://doi.org/10.1007/s11042-018-5697-y
7. Guo Y, Şengür A, Akbulut Y, Shipley A (2018) An effective color image segmentation approach using neutrosophic adaptive mean shift clustering. Measurement 119:28–40. https://doi.org/10.1016/j.measurement.2018.01.025
8. Singh V, Misra A (2017) Detection of plant leaf diseases using image segmentation and soft computing techniques. Inf Process Agric 4(1):41–49. https://doi.org/10.1016/j.inpa.2016.10.005
9. Zhang HCM, Zhang L (2010) A neutrosophic approach to image segmentation based on watershed method. Signal Process 90(5):1510 – 1517, special Section on Statistical Signal and Array Processing. https://doi.org/10.1016/j.sigpro.2009.10.021
10. Liang Y, Zhang M, Browne WN (2014) Image segmentation: a survey of methods based on evolutionary computation. Simulated evolution and learning. Springer International Publishing, Cham, pp 847–859
11. Senthilkumaran N, Rajesh R (2009) Image segmentation–a survey of soft computing approaches. International conference on advances in recent technologies in communication and computing 2009:844–846. https://doi.org/10.1109/ARTCom.2009.219
12. Ruspini EH (1969) A new approach to clustering. Inf Control 15(1):22–32. https://doi.org/10.1016/S0019-9958(69)90591-9
13. Salah Eddine M, El Joumani S, Zennouhi R, Masmoudi L, Multi-objective optimization for worldview image segmentation funded on the entropies of tsallis and rényi, Multimedia Tools and Applications 79. https://doi.org/10.1007/s11042-020-09572-4.
14. Xing Z (2020) An improved emperor penguin optimization based multilevel thresholding for color image segmentation. Knowl-Based Syst 194:105570. https://doi.org/10.1016/j.knosys.2020.105570
15. Mousavirad SJ, Ebrahimpour-Komleh H (2020) Human mental search-based multilevel thresholding for image segmentation. Appl Soft Comput 97:105427. https://doi.org/10.1016/j.asoc.2019.04.002

16. Rivera-Borroto OM, Rabassa-Gutiérrez M, Grau-Ábalo R, Marrero-Ponce Y, Vega J (2012) Dunn's index for cluster tendency assessment of pharmacological data sets. Can J Physiol Pharmacol 90:425–33. https://doi.org/10.1139/y2012-002
17. Pakhira MK, Bandyopadhyay S, Maulik U (2004) Validity index for crisp and fuzzy clusters. Pattern Recogn 37(3):487–501. https://doi.org/10.1016/j.patcog.2003.06.005
18. LaHaye N, Ott J, Garay MJ, El-Askary HM, Linstead E (2019) Multi-modal object tracking and image fusion with unsupervised deep learning. IEEE J Sel Top Appl Earth Obs Remote Sens 12(8):3056–3066
19. Xu L, Jing W, Song H, Chen G (2019) High-resolution remote sensing image change detection combined with pixel-level and object-level. IEEE Access 7:78909–78918

Skeletonization and Its Application to Quantitative Structural Imaging

Punam K. Saha

Abstract Skeletonization provides a compact yet effective representation of an object using its medial loci defined by Blum's grassfire transform process. It has been popularly used in many low- as well as high-level imaging applications. Quantitative characterization of object structures is a popular application of skeletonization, where a skeleton serves two purposes—generations of a compact representation of an object and location of representative points for defining local structure morphology. This paper reviews the roles of skeletonization and digital topological analysis in characterizing individual trabecular bone plate-rod microstructure assessing overall bone quality and fracture risk at in vivo imaging.

Keywords Skeletonization · Digital topology · Trabecular bone · Microstructure · Plate-rod distribution · Osteoporosis

1 Introduction

Skeletonization provides a compact yet effective representation of important topologic and geometric features of an object by reducing its dimensionality to a "medial axis" or "skeleton" [1–3]. Skeletonization has been popularly used in several low- as well as high-level image processing and computer vision applications including object representation, retrieval, manipulation, matching, registration, tracking, recognition, compression, etc. The compact representation of a skeleton facilitates efficient assessment of local structure morphology, e.g., scale, orientation, topology, and geometry, which has been utilized for quantitative structural assessment in several image-related applications. This paper reviews the roles of skeletonization and digital topological analysis for characterization of individual trabecular plate-rod microstructure at in vivo imaging toward assessment of bone quality and fracture risk among patients with osteoporosis.

P. K. Saha (✉)
Departments of Electrical and Computer Engineering and Radiology, University of Iowa, Iowa City, IA 52242, USA
e-mail: pksaha@healthcare.uiowa.edu

S. Basu et al. (eds.), *Proceedings of International Conference on Frontiers in Computing and Systems*, Lecture Notes in Networks and Systems 404,
https://doi.org/10.1007/978-981-19-0105-8_23

Osteoporosis is an adult bone disease associated with an increased risk of fractures. The cost of osteoporotic fracture to the world healthcare system and its impact on an individual basis are both extremely high [4], and its incidence increases progressively with aging [5]. Clinically osteoporosis is linked to low bone mineral density (BMD), while BMD explains only 60–70% of the variability in bone strength [6]. The remainder is due to other factors, including cortical and trabecular bone (TB) structures [7]. Recent advances with medical imaging, such as magnetic resonance (MR) [6, 8], high-resolution peripheral quantitative computed tomography (HR-pQCT) [9], and multi-row detector computed tomography (MD-CT) [10], allow in vivo segmentation of individual TB structures for characterization of their microarchitectural integrity and assessment of bone strength and fracture risk.

There is histologic evidence confirming the relationship between the gradual conversion of trabecular plates to rods and increased fracture risk [11, 12]. Kleerekoper et al. [11] observed lower mean TB plate density among individuals with osteoporotic vertebral compression fractures compared with BMD-matched controls without fractures. Recker [12] reported reduced trabecular connectivity among patients with vertebral crush fractures as compared to healthy controls with matching TB volume. Digital topological analysis is applied to characterize individual trabecular plates and rods on skeletal representations of bone microstructure at in vivo image resolution.

2 Skeletonization

Most skeletonization algorithms are based on the basic principle of Blum's grassfire transform. Despite the commonality in the principle, many skeletonization algorithms are available literature, some of which are widely different in their basic computational approach. Several research groups [13–15] have adopted a continuous simulation of Blum's grassfire process, where the object boundary is approximated as a curve that goes through a curve evolution process, and the skeleton is formed at quench points where the evolution process is interrupted. A number of researchers [16, 17] have used geometric features, e.g., Voronoi diagram to directly identify the symmetry structures as the skeleton. Many researchers have embraced digital approaches of simulating the grassfire propagation as an iterative constrained erosion process [18–20]. Other researchers [21–25] have used digital distance transform [26] to locate centers of maximal balls (CMBs) [22] defining skeletal structures.

A DT-based 3-D skeletonization algorithm [24, 27, 28], applicable to both binary and fuzzy images, can be summarized using the following steps:

Primary skeletonization

- Locate centers of maximal balls or quench voxels.
- In a descending order of DT values, identify and mark quench voxels with a measure of high local shape significance.

- In an ascending order of DT values, sequentially delete unmarked voxels, which are simple points [29, 30], i.e., not necessary for topology preservation.

Final skeletonization

- Reduce two-voxel thick structures into single-voxel thin structures.
- Remove voxels with conflicting topologic and geometric properties [24].

Skeleton pruning

- Identify and remove noisy or spurious branches using a measure of global shape significance.

Primary skeletonization [20, 22, 29] produces a "thin set" [21], which is two-voxel thick medial representation of an object, where every voxel has at least one background neighbor except at very busy junctions [3]. Intermediate results of skeletonization are shown in Fig. 1.

The centers of maximal balls (CMBs) in an object are determined using the criterion originally proposed by Arcelli and Sanniti di Baja [31], which was extended to 5×5 weighted distances [32]. I was further generalized for fuzzy object [33]. The

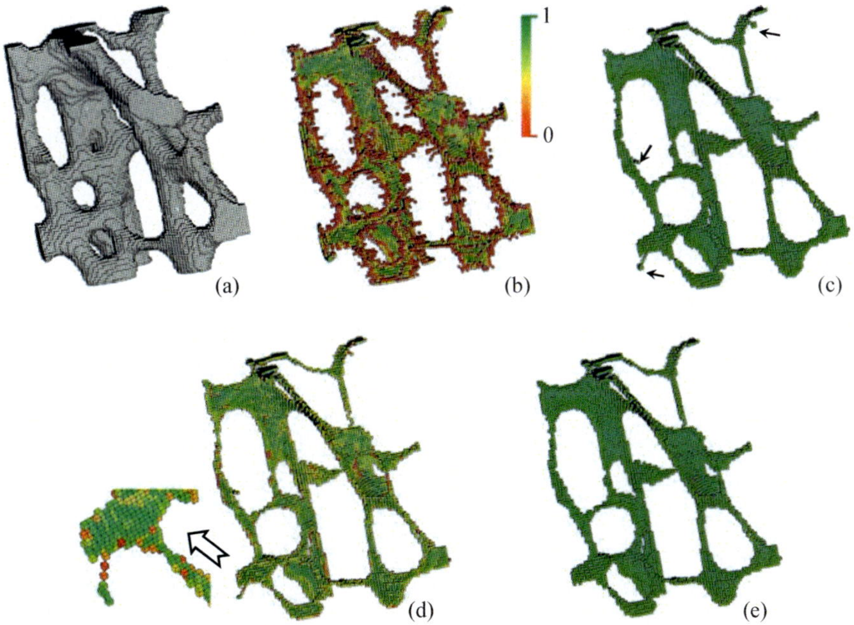

Fig. 1 Results of intermediate steps in skeletonization. **a** 3-D display of a trabecular bone region in a micro-CT image of a cadaveric ankle specimen. **b** Initial set of CMB voxels color-coded using their collision impact values; see the color-coding bar on the right. **c** The skeleton prior to pruning; noisy branches are indicated by arrows. It is encouraging to note that most of the noisy CMB voxels of (**b**) are eliminated using their local significance measures. **d** Collision impact value on the voxels of (**c**). The same color-coding of (**b**) is used. **e** Final skeleton after noisy branch pruning

most critical step in primary skeletonization is to identify locally significant skeletal voxels as it defines the overall morphology of the output skeleton. Each small protrusion or dent creates a false branch in a skeleton. It is a well-known challenge to avoid false branches while preserving significant ones. It is argued that, at this stage of the algorithm, only local significance features are sought to identify valid quench or skeletal voxels [3]. This challenge is further intensified for fuzzy objects, where each fluctuation in membership values creates an extra skeletal branch. Saha and his research group introduced a measure of *collision impact* to the membership of a skeletal voxel [24, 25, 34].

During *final skeletonization*, two-voxel thick structures are eroded under topology preservation and some additional geometric constraints [20, 22, 28] maintaining the overall shape of the skeleton. The aim of skeletal pruning step is to distinguish between significant and non-significant branches so that false branches can be removed. For this purpose, a collision impact-weighted length of each skeletal branch is used as a measure of its *global significance* to distinguish between false and valid skeletal branches.

Topology preservation in skeletonization is achieved by adding the constraints of simple points [29, 30, 35–39] to the deletion of non-significant skeletal voxels. Simple points or simple voxels are characterized using simple connectivity in $3 \times 3 \times 3$ neighborhood as follows.

Theorem 1 A voxel p is a (26, 6) simple point in a 3-D binary image if and only if it satisfies the following four conditions.

Condition 1.p has a background 6-neighbor.

Condition 2.p has an object 26-neighbor.

*Condition 3.*The set of object 26-neighbors of p is 26-connected.

*Condition 4.*The set of background 6-neighbors of p is 6-connected in the set of background 18-neighbors.

Bertrand and Malandain restated the same results, which was presented in a two-condition format [38].

3 Digital Topological Analysis

Digital topological analysis or DTA [40] accurately determines the topological class (e.g., surfaces, curves, junctions) of each individual voxel on the skeleton of 3-D digital object. This method has been popularly applied for quantifying quality of TB architectural makeup. Before applying DTA, a fuzzy segmented TB volume image is skeletonized using the algorithm described in Sect. 2. The principle of DTA is based on a simple observation on the skeletal representation, say S, of a 3-*D* object in the continuous space. It was observed that the topological class of a point $p \in S$ is uniquely defined by its local topological numbers, i.e., the numbers of object components, tunnels, and cavities in a sufficiently small neighborhood of p but

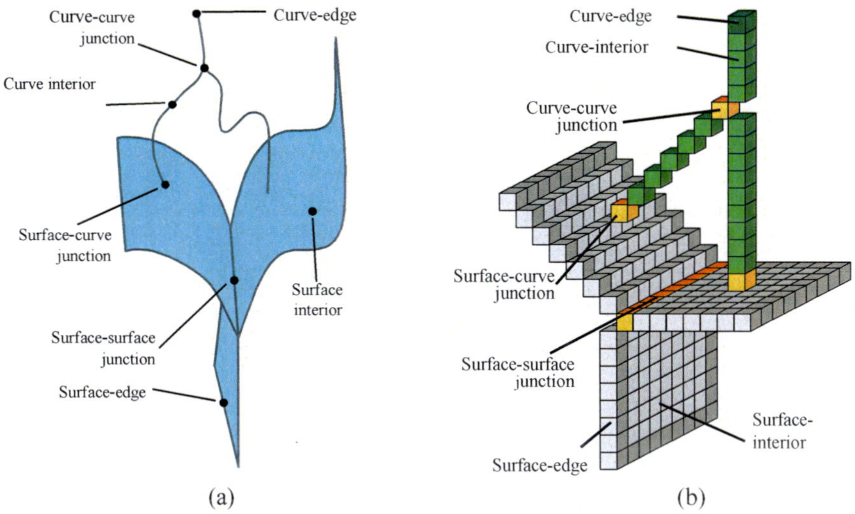

Fig. 2 Different topological classes in a surface-skeletal representation of an object. **a** An example in the continuous space. **b** An example in a 3-D digital grid

excluding p itself. Different topological classes on a skeletal structure are illustrated in Fig. 2. While the computations of the numbers of objects and cavities in the $3 \times 3 \times 3$ neighborhood are trivial, the number of tunnels is computed using the following theorem [36, 40, 41].

Theorem 2 If a voxel p has a background 6-neighbor, the number of tunnels in its excluded neighborhood is one less than the number of 6-components of background 18-neighbors of p that contain a 6-neighbor or zero otherwise.

In a digital case, the three local topological numbers fail to uniquely define the topological class of every skeletal voxel. To overcome this problem, a three-step approach is adopted to determine the unique topological class of every skeletal voxels. These three steps of the digital topological classification are—(1) determination of the local topological type, (2) the initial classification based on neighborhood configuration of topological types, and (3) the final classification after correction around junctions. The first step determines the initial topological type (see Table 1 in Saha et al. [40]) based on the three local topological numbers. This step fails to uniquely determine local topological class because, in a digital space, the local neighborhood configuration with one component, one tunnel, and no cavity may occur at a surface interior voxel as well as at a junction of two digital lines each parallel to a coordinate axis. These ambiguities are solved in the second step using a neighborhood configuration of topological types determined in the previous step. Essentially, these two steps together use the topological data over the $5 \times 5 \times 5$ neighborhood to determine the topological class of each voxel.

After the initial classification in the second, the process is almost complete except for possible corrections at some surface junctions. Specifically, during this step the surface–surface junction line is completed based on the topological principle of surface-edge lines around a junction. Also, the surface–curve junction voxel is missed using the previous two steps where a curve emanates from the edge of a surface. Such a junction voxel located when a component of curve voxels is adjacent to a surface voxel without a junction voxel gluing the two different topological classes. Separating the task of digital topological analysis in three steps makes it easier to understand and more efficient to implement.

Digital topological analysis has been widely applied for characterization of trabecular bone plate/rod micro-architecture from in vivo imaging [42–46]. Digital topological classes of individual bone voxels are used to compute several topological parameters characterizing the structural integrity of a trabecular bone network. The specific parameter, namely erosion index (EI), was found to be highly sensitive in most studies using DTA, and this DTA parameter is used in our experiments. The EI is defined as the ratio of all topological parameters expected to increase during the erosion process (specifically, curve, curve-edge, surface-edge, profile-edge, and curve–curve junction types) compared with those that are expected to decrease (surface and surface–surface junction types) [41].

Malandain et al. [47] have used topological analysis to segmental discrete skeletal surfaces. Serino et al. [48] have used distance labeled curve-skeletons to decompose 3-D objects into disjoint parts. More specifically, they used distance transform and curvature analysis to locate skeletal segments.

4 Volumetric Topological Analysis

Volumetric topological analysis or VTA [27] accurately measures local structure plate-width at individual voxels in a digitized object and characterizes the local structure type on the continuum between a perfect plate and a perfect rod; see Fig. 3. The original VTA algorithm [27] was developed for binary objects requiring thresholding on fuzzy representations of TB images, which is a sensitive and undesired step at in vivo resolution [49]. Recently, the algorithm has been modified to directly apply on fuzzy representation of an object eliminating the thresholding step—a major source of error, especially, at low spatial resolution common in in vivo imaging.

The VTA method applies on the surface-skeletal representation of an object and computes the geodesic manifold distance transform at curve-skeletal or axial locations from surface-skeletal edges to determine the local structure plate-width. This plate-width measures are then propagated from the curve-skeleton to the surface-skeleton and then from the surface-skeleton to the object volume. Specifically, the VTA algorithm is accomplished in five sequential steps [27]—(1) surface-skeletonization, (2) digital topological analysis, (3) geodesic manifold distance transform, (4) local structure plate-width computation at curve-skeletal locations, and (5) plate-width feature propagation from the curve-skeleton to the surface-skeleton, (6) plate-width feature propagation from the surface-skeleton to the entire object volume.

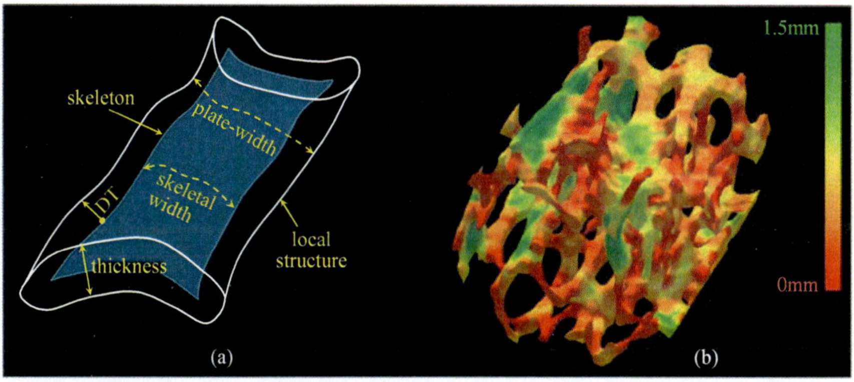

Fig. 3 Illustration of the plate-width measure. **a** Schematic description of different morphologic local measures on a platelike structure. The volumetric topological analysis or VTA algorithm computes the local plate-width measure of a structure. **b** A color-coded illustration of local plate-width computation for a CT image of trabecular bone network in a cadaveric ankle specimen

Basic principles of the VTA algorithm are described in the following. Let us consider that a 2-D manifold represents the surface-skeleton of a platelike object with varying widths along its axis. A surface-skeletal representation of an object may be obtained using the fuzzy skeletonization algorithm described in Sect. 2 or another surface-skeletonization algorithm [3]. After computing the surface-skeletonization, the DTA algorithm is applied to identify different topological entities including surface-edges. Geodesic distance transform (GDT) is computed at every surface-skeletal points. GDT value at a surface-skeletal point is equal to its geodesic distance from the nearest surface-edge point. The geodesic distance between two points on a surface-skeleton is the length of the shortest path between the two points such that the entire path lies on the skeletal surface. A front propagation approach similar to Dijkstra's algorithm [50] offers an efficient implementation of GDT computation. It was argued by Saha et al. [27] that the GDT front propagation should not cross a junction line. Allowing GDT fronts to propagate through a junction line leads to artefactual reduction in GDT values around the junction. A similar observation was noted by Arcelli et al. [28].

The formulation of the VTA algorithm is based on a simple observation that the GDT at a curve-skeletal or axial point provides the half-plate-width of the local skeleton. However, GDT fails to determine the local plate-width at non-skeletal points. A feature propagation step is applied, where a non-axial point inherits the local plate-width from its most representative axial point. Finally, another level of feature propagation is applied where the plate-width values are propagated from the surface-skeleton to the entire object volume. The feature propagation in Step 5 is accomplished using the principle established by Liu et al. [51] that is independent of scan or processing order. Lastly, it should be clarified that a VTA algorithm using GDT from surface-skeletal edges returns the value of the local skeletal width at each point. To compute local width of a volumetric structure, the algorithm is modified,

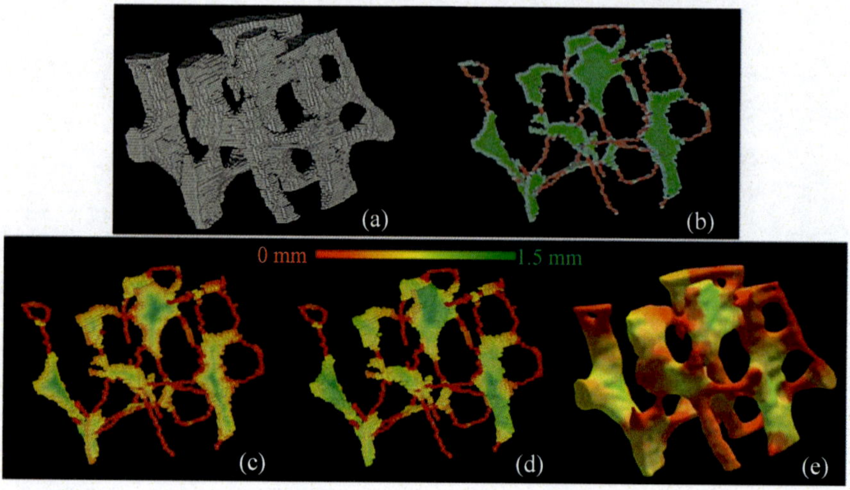

Fig. 4 Results of intermediate steps in VTA. **a** A trabecular bone region from a micro-CT image of a cadaveric ankle specimen. **b** DTA-computed topological classes including plates (green), rods (red), edges (light colors), and junctions (blue) on the skeleton of (**a**). **c–e** Color-coded displays of geodesic distance transform (**c**), plate-width outcome on the surface-skeleton after the first feature propagation step from curve-skeleton to surface-skeleton (**d**), and surface rendition of plate-width or VTA outcome over the entire TB volume after the second feature propagation step (**e**). In the panel (**e**), surface rendition is used instead of voxel display for better display

where the GDT value at a skeletal edge point is initialed with its distance transform or fuzzy distance transform value instead of zero; see Fig. 3a. Results of intermediate steps of VTA are presented in Fig. 4. VTA and individual trabecular plate-rod characterization have been successfully applied in a large number of human studies; for example, see [10, 42, 52–59].

5 Conclusion

Skeleton-based quantitative characterization of local structures has been used in a considerable number of applications seeking assessments of various morphologic aspects of an object. Along this research pathway, topologic and geometric approaches have been combined to quantitatively assess object morphometry and local structural properties in an object. In such methods, two aspects are critical—(1) determining the set of representative sample points and (2) accurate measurement of the target local morphologic feature.

References

1. Blum H (1967) A transformation for extracting new descriptors of shape. Models Percept Speech Vis Form 19:362–380
2. Siddiqi K, Pizer SM (2008) Medial representations: mathematics, algorithms and applications, vol 37. Springer, Berlin
3. Saha PK, Borgefors G, Sanniti di Baja G (2016) A survey on skeletonization algorithms and their applications. Pat Recognit Let 76:3–12
4. Melton LJ (1995) How many women have osteoporosis now. J Bone Miner Res 10:175–177
5. Riggs BL, Khosla S, Melton LJ 3rd (2012) Better tools for assessing osteoporosis. J Clin Invest 122:4323–4324
6. Wehrli FW, Saha PK, Gomberg BR, Song HK, Snyder PJ, Benito M, Wright A, Weening R (2002) Role of magnetic resonance for assessing structure and function of trabecular bone. Top Magn Reson Imaging 13:335–355
7. Seeman E, Delmas PD (2006) Bone quality–the material and structural basis of bone strength and fragility. N Engl J Med 354:2250–2261
8. Majumdar S (2002) Magnetic resonance imaging of trabecular bone structure. Top Magn Reson Imaging 13:323–334
9. Boutroy S, Bouxsein ML, Munoz F, Delmas PD (2005) In vivo assessment of trabecular bone microarchitecture by high-resolution peripheral quantitative computed tomography. J Clin Endocrinol Metab 90:6508–6515
10. Chen C, Zhang X, Guo J, Jin D, Letuchy EM, Burns TL, Levy SM, Hoffman EA, Saha PK (2018) Quantitative imaging of peripheral trabecular bone microarchitecture using MDCT. Med Phy 45:236–249
11. Kleerekoper M, Villanueva AR, Stanciu J, Rao DS, Parfitt AM (1985) The role of three-dimensional trabecular microstructure in the pathogenesis of vertebral compression fractures. Calcif Tissue Int 37:594–597
12. Recker RR (1993) Architecture and vertebral fracture. Calcif Tissue Int 53(Suppl 1):S139–S142
13. Kimia BB, Tannenbaum A, Zucker SW (1995) Shape, shocks, and deformations I: the components of two-dimensional shape and the reaction-diffusion space. Int J Comput Vis 15:189–224
14. Kimmel R, Shaked D, Kiryati N (1995) Skeletonization via distance maps and level sets. Comput Vis Image Process 62:382–391
15. Siddiqi K, Bouix S, Tannenbaum A, Zucker SW (2002) Hamilton-Jacobi skeletons. Int J Comput Vis 48:215–231
16. Brandt JW, Algazi VR (1992) Continuous skeleton computation by Voronoi diagram. CVGIP Image Underst 55:329–338
17. Ogniewicz R, Ilg M (1992) Voronoi skeletons: theory and applications. In: Proceedings of the computer vision and pattern recognition, pp 63–69
18. Tsao YF, Fu KS (1981) A parallel thinning algorithm for 3D pictures. Comput Graph Image Process 17:315–331
19. Palágyi K, Kuba A (1999) A parallel 3D 12-subiteration thinning algorithm. Graph Models Image Process 61:199–221
20. Saha PK, Chaudhuri BB, Majumder DD (1997) A new shape preserving parallel thinning algorithm for 3D digital images. Patt Recog 30:1939–1955
21. Arcelli C, Sanniti di Baja G (1985) A width-independent fast thinning algorithm. IEEE Trans Pattern Anal Mach Intell 7:463–474
22. Sanniti di Baja G (1994) Well-shaped, stable, and reversible skeletons from the (3,4)-distance transform. J Vis Commun Image Represent 5:107–115
23. Borgefors G, Nyström I, Sanniti di Baja G (1999) Computing skeletons in three dimensions. Pattern Recogn 32:1225–1236
24. Jin D, Saha PK (2013) A new fuzzy skeletonization algorithm and its applications to medical imaging. In: Proceeding of the 17th international conference image anal proc (ICIAP), Naples, Italy, LNCS 8156, 662–671

25. Saha PK, Jin D, Liu Y, Christensen GE, Chen C (2018) Fuzzy object skeletonization: theory, algorithms, and applications. IEEE Trans Vis Comput Graph 24:2298–2314
26. Borgefors G (1986) Distance transformations in digital images. Comput Vis Graph Image Process 34:344–371
27. Saha PK, Xu Y, Duan H, Heiner A, Liang G (2010) Volumetric topological analysis: a novel approach for trabecular bone classification on the continuum between plates and rods. IEEE Trans Med Imaging 29:1821–1838
28. Arcelli C, Sanniti di Baja G, Serino L (2011) Distance-driven skeletonization in voxel images. IEEE Trans Pattern Anal Mach Intell 33:709–720
29. Saha PK, Chaudhuri BB (1994) Detection of 3-D simple points for topology preserving transformations with application to thinning. IEEE Trans Pattern Anal Mach Intell 16:1028–1032
30. Saha PK, Chaudhuri BB, Chanda B, Majumder DD (1994) Topology preservation in 3D digital space. Pattern Recogn 27:295–300
31. Arcelli C, Sanniti di Baja G (1988) Finding local maxima in a pseudo- Euclidean distance transform. Comput Vis Graph Image Process 43:361–367
32. Borgefors G (1993) Centres of maximal discs in the 5-7-11 distance transform. In: Proceeding of the 8th Scandinavian conference on image analysis, Tromsø, Norway, pp 105–111
33. Saha PK, Wehrli FW, Gomberg BR (2002) Fuzzy distance transform: theory, algorithms, and applications. Comput Vis Image Underst 86:171–190
34. Jin D, Chen C, Saha PK (2015) Filtering non-significant quench points using collision impact in grassfire propagation. In: Murino V, Puppo E (eds) Proceedings of the 18th international conference imag anal proc (ICIAP), Genoa, Italy, vol 9279, pp 432–443
35. Morgenthaler DG (1981) Three-dimensional simple points: serial erosion, parallel thinning and skeletonization, Computer Vision Laboratory. University of Maryland, College Park, MD
36. Saha PK, Chanda B, Majumder DD (1991) Principles and algorithms for 2-D and 3-D shrinking. Indian Statistical Institute, Calcutta, India, TR/KBCS/2/91
37. Saha PK (1991) 2D thinning algorithms and 3D shrinking. INRIA, Sophia Antipolis Cedex, France
38. Bertrand G, Malandain G (1994) A new characterization of three-dimensional simple points. Pattern Recogn Lett 15:169–175
39. Kong TY (1997) Topology-preserving deletion of 1's from 2-, 3-and 4-dimensional binary images. In: Proceeding of the international workshop on discrete geometry for computer imagery. Montpellier, France, 1–18
40. Saha PK, Chaudhuri BB (1996) 3D digital topology under binary transformation with applications. Comput Vis Imag Und 63:418–429
41. Saha PK, Gomberg BR, Wehrli FW (2000) Three-dimensional digital topological characterization of cancellous bone architecture. Int J Imag Syst Tech 11:81–90
42. Dudley-Javoroski S, Saha PK, Liang G, Li C, Gao Z, Shields RK (2012) High dose compressive loads attenuate bone mineral loss in humans with spinal cord injury. Osteoporos Int 23:2335–2346
43. Lam SC, Wald MJ, Rajapakse CS, Liu Y, Saha PK, Wehrli FW (2011) Performance of the MRI-based virtual bone biopsy in the distal radius: serial reproducibility and reliability of structural and mechanical parameters in women representative of osteoporosis study populations. Bone 49:895–903
44. Chang G, Pakin SK, Schweitzer ME, Saha PK, Regatte RR (2008) Adaptations in trabecular bone microarchitecture in Olympic athletes determined by 7T MRI. J Magn Reson Imaging 27:1089–1095
45. Wehrli FW, Gomberg BR, Saha PK, Song HK, Hwang SN, Snyder PJ (2001) Digital topological analysis of in vivo magnetic resonance microimages of trabecular bone reveals structural implications of osteoporosis. J Bone Min Res 16:1520–1531
46. Liu XS, Sajda P, Saha PK, Wehrli FW, Bevill G, Keaveny TM, Guo XE (2008) Complete volumetric decomposition of individual trabecular plates and rods and its morphological correlations with anisotropic elastic moduli in human trabecular bone. J Bone Miner Res 23:223–235

47. Malandain G, Bertrand G, Ayache N (1993) Topological segmentation of discrete surfaces. Int J Comput Vis 10:183–197
48. Serino L, Arcelli C, Di Baja GS (2014) Decomposing 3d objects in simple parts characterized by rectilinear spines. Int J Pattern Recognit Artif Intell 28
49. Krug R, Burghardt AJ, Majumdar S, Link TM (2010) High-resolution imaging techniques for the assessment of osteoporosis. Radiol Clin North Am 48:601–621
50. Dijkstra E (1959) A note on two problems in connection with graphs. Numer Math 1:269–271
51. Liu Y, Jin D, Li C, Janz KF, Burns TL, Torner JC, Levy SM, Saha PK (2014) A robust algorithm for thickness computation at low resolution and its application to in vivo trabecular bone CT imaging. IEEE Trans Biomed Eng 61:2057–2069
52. Chang G, Xia D, Chen C, Madelin G, Abramson SB, Babb JS, Saha PK, Regatte RR (2015) 7T MRI detects deterioration in subchondral bone microarchitecture in subjects with mild knee osteoarthritis as compared with healthy controls. J Magn Reson Imaging 41:1311–1317
53. Chang G, Rajapakse CS, Chen C, Welbeck A, Egol K, Regatte RR, Saha PK, Honig S (2018) 3-T MR imaging of proximal femur microarchitecture in subjects with and without fragility fracture and nonosteoporotic proximal femur bone mineral density. Radiology 287:608–619
54. Chen C, Jin D, Liu Y, Wehrli FW, Chang G, Snyder PJ, Regatte RR, Saha PK (2016) Trabecular bone characterization on the continuum of plates and rods using in vivo MR imaging and volumetric topological analysis. Phys Med Biol 61:N478–N496
55. Dudley-Javoroski S, Petrie MA, McHenry CL, Amelon RE, Saha PK, Shields RK (2015) Bone architecture adaptations after spinal cord injury: impact of long-term vibration of a constrained lower limb. Osteoporos Int, pp 1–12
56. Saha PK, Liu Y, Chen C, Jin D, Letuchy EM, Xu Z, Amelon RE, Burns TL, Torner JC, Levy SM, Calarge CA (2015) Characterization of trabecular bone plate-rod microarchitecture using multirow detector CT and the tensor scale: algorithms, validation, and applications to pilot human studies. Med Phys 42:5410–5425
57. Zhang X, Comellas AP, Regan EA, Guha I, Shibli-Rahhal A, Rubin MR, DiCamillo PA, Letuchy EM, Barr RG, Hoffman EA (2021) Quantitative CT-based methods for bone microstructural measures and their relationships with vertebral fractures in a pilot study on smokers. JBMR Plus 5:10484
58. Holman M, Chang G, Ghatas M, Saha P, Zhang X, Khan M, Sima A, Adler R, Gorgey A (2021) Bone and non-contractile soft tissue changes following open kinetic chain resistance training and testosterone treatment in spinal cord injury: an exploratory study. Osteoporos Int, pp 1–12
59. Saha PK, Oweis RR, Zhang X, Letuchy E, Eichenberger-Gilmore JM, Burns TL, Warren JJ, Janz KF, Torner JC, Snetselaar LG (2021) Effects of fluoride intake on cortical and trabecular bone microstructure at early adulthood using multi-row detector computed tomography (MDCT). Bone 146:115882

Healthcare

Voting-Based Extreme Learning Machine Approach for the Analysis of Sensor Data in Healthcare Analytics

Tanuja Das, Ramesh Saha, and Vaskar Deka

Abstract There has been a huge surge in the production of sensor-based clinical devices for health monitoring systems over the last few years. This sudden spike has been due to many different factors such as development in sensor device technology and also the efforts to promote research that address the necessity for providing new ways for healthcare given the increasing challenges with an associated degree of an aging population. The processing and analysis of data is an important element of the research of such a system. The data generated from these healthcare devices are enormous and have the potential to ascertain well-being and to encourage effective management of health. In this work, a mechanism for the analysis of physiological sensor data from the healthcare devices, namely voting-based extreme learning machine, has been explored. The approach was also compared with the traditional extreme learning machine-based approach. Experimental results were very encouraging with respect to the performance accuracy as well as time taken by the voting-based extreme learning machine as compared to the traditional extreme learning machine to produce the output.

Keywords Machine learning · Sensor data · Vital signs data · Extreme learning machine · Voting-based extreme learning machine · Healthcare analytics

1 Introduction

With healthcare services increasing in clinical as well as non-clinical environments using vital signs provided by wearable sensor devices, the need for extracting and analyzing physiological measurements is rising dramatically. With the continuous evolution of the big data era, the healthcare systems have gained tremendous edge in the present times [1]. The ventures based on these healthcare systems gather massive

T. Das (✉) · R. Saha · V. Deka
Department of Information Technology, Gauhati University, Guwahati, Assam, India
e-mail: tanujadas55@gmail.com

S. Basu et al. (eds.), *Proceedings of International Conference on Frontiers in Computing and Systems*, Lecture Notes in Networks and Systems 404,
https://doi.org/10.1007/978-981-19-0105-8_24

volumes of data which are repositories of information concealed within them. These information, nowadays, are invaluable for efficient decision making.

Many sophisticated data mining methods are used to produce accurate information and to make very effective decisions on data. The fabrication of healthcare equipment has encountered a rapid shift in the present times, concentrating mainly on solving the primary goals of healthcare services, for instance, diabetes control or healthcare monitoring of the aged population. For achieving these objectives, many enthusiastic investigators put numerous attempts for developing the healthcare devices based on sensor technology [2]. From the ever-expanding frequency of novel proof-of-concept tests, the possible use of wearable technologies is clear.

Machine learning has changed the way data is treated by organizations and individuals to boost a system's performance. Machine learning for healthcare analytics projects is filled with new approaches and methodologies to build effective healthcare analytics solutions [3]. Because the literature in this field is so extensive, we have also narrowed this work's reach to include data from wearable sensors from the clinical environment that calculate health parameters such as vital signs for disease management and health monitoring systems for prevention.

The objective of this work to is use machine learning technique to predict the health status of an individual from physiological data available from wearable sensors. To achieve this, an approach, namely voting based on extreme learning machine, has been experimented to find out how accurate this model can detect the health status of an individual. Also, the approach has been compared with the traditional extreme learning machine technique.

The paper is arranged in the following manner. Section 2 provides a brief survey on some of the contemporary works done in this domain. Section 3 deals with the concepts used in our work. Section 4 gives the outline of the dataset used and the architecture used in the approach. Section 5 briefly discusses the outcomes, and finally, Sect. 6 draws the conclusion of our work.

2 Literature Review

There have been numerous studies on health monitoring systems using a variety of machine learning algorithms. Many such reviews specifically target or focus on providing the subject with a global overview. Some the recent works done are summarized in Table 1.

Table 1 Summary of related works

Author details	Year of publication	Vital signs data used	ML algorithms	Performance parameter	Results	Future scope
Thakur et al. [4]	2018	Heart rate, respiration rate, heart rate variability	Logistic regression, k-NN, AdaBoost, random forest, support vector machine	ROC, recall	90.16% ROC, 88.47% Recall	Can consider other important parameters like blood pressure, SpO2
Martinez et al. [5]	2018	Actigraphy signals	Bispectral entropy analysis	Bispectral entropy histograms	41.17% of the pairs correlated successfully	Number of samples can be increased to investigate for more associated diseases
Le et al. [6]	2019	Patient age, blood pressures, heart rate, tempera-ture, respiration rate, SpO2	Decision trees	AUROC	71.8% AUROC	The model can be tested on real-time data
Vistisen et al. [7]	2019	Heart rate, mean arterial pressure, respiratory rate, and peripheral oxygen saturation	Logistic regression, random forest	AUROC, recall	81% AUROC, 87% Recall	Results can be compared systemati-cally by allocating a label to the patient examined
Baker et al. [8]	2020	Heart rate, blood pressure, respiratory rate, blood oxygen levels, and temperature	Hybrid CNN-BiLSTM network	AUROC	88.4% AUROC	Clinical sessions can be tested to assure that data is stored consistently

3 Background Study

3.1 Extreme Learning Machine

In simple terms, extreme learning machine (ELM) [9, 10] is a feed forward neural network consisting only one layer. ELM has been used extensively for classification [11–13] as well as regression [14, 15] tasks by many researchers.

The weights that are to be input to the ELM are chosen in a random manner. This goes the same for the input of the hidden layers. With regards to the computation of the weights of the output layer, Moore-Prone generalization is utilized [9].

The representation of an ELM for a specific dataset with a count of N neurons present in the hidden layer and activation function G is given as:

$$f(x) = \sum_{i=1}^{N} \beta_i G(a_i, b_i, x_i) = \beta \cdot h(x) \tag{1}$$

The output weight of the ith hidden layer is denoted by β_i, whereas a_i denotes the input weight for the data x_i with basis b_i [9].

$$h(x) = G(a_i, b_i, x_i) \ldots G(a_N, b_N, x_N) \tag{2}$$

The computation of the output weights β_i is done by solving the least-squares solutions of linear equation as given below:

$$\min || \sum_{i=1}^{N} \beta_i \cdot h(x) - y_i || \tag{3}$$

The solution for Eq. (3) is given as:

$$HB = yY // \beta = H^{-1} Y \tag{4}$$

3.2 Voting-Based Extreme Learning Machine

The search for an optimal solution in the conventional ELM is inadequate. The reason being the random initialization of the input weight and the basis as well, both of which remains constant during the entire training [9]. So, the data located in the classification borderlines are tend to be classified erroneously.

To get around this obstacle, Cao et al. [16] first designed the voting-based ELM (V-ELM). In this approach, rather than making only one single training, multiple training is done. The final classification is based on the majority voting method. This concept helps to handle the incorrectly classified instances.

A number of distinct ELMs with the number of hidden nodes and the activation function being constant forms the backbone of the V-ELM. Identical training sets are used for each of the distinct ELM, but with autonomous learning parameters initialized in a random manner. Voting by the majority of the distinct ELMs leads to the final class label of a particular instance.

Assuming K distinct ELMs trained for the V-ELM, then for each test instance x^{test}, the number of results predicted will be K, each corresponding to a distinct ELM. The predictions computed for each class labels from each distinct ELM are then stored in a vector, $S_{K,x^{\text{test}}}$. The dimension of this vector is in accord with the number of class labels and is calculated by the equation below:

$$S_{K,x^{\text{test}}}(i) = S_{K,x^{\text{test}}}(i) + 1 \tag{5}$$

All the K predictions are stored in $S_{K,x^{\text{test}}}$ according to the class labels and then the final class of an instance x^{test} is decided by majority voting according to the equation below:

$$C^{\text{test}} \arg \max_{i \in [1 \ldots C]} S_{K,x^{\text{test}}}(i) \tag{6}$$

4 Materials and Methods

This section gives the summarized concept of the dataset used (given in Sect. 4.1) and the architecture of the model used (given in Sect. 4.2).

4.1 Dataset

The dataset considered here contains data stored from the tools used to observe a subject's vital signs [17]. The data employed for prognosis were reported from 32 cases. The data from the subjects included the vital signs data like heart rate, pulse, pulse oximeter, systolic blood pressure, respiration rate, and many more, containing a total of 47 vital signs [17]. The output as indicated in the dataset contains five patient alarms based on the patient's physiological status, which basically represents the risk level of a patient (0 being normal and 4 being high risk).

4.2 Model Architecture

The architecture of the V-ELM approach is depicted in Fig. 1. Assuming for K distinct ELMs with identical hidden neurons but unique initialization of hidden parameters,

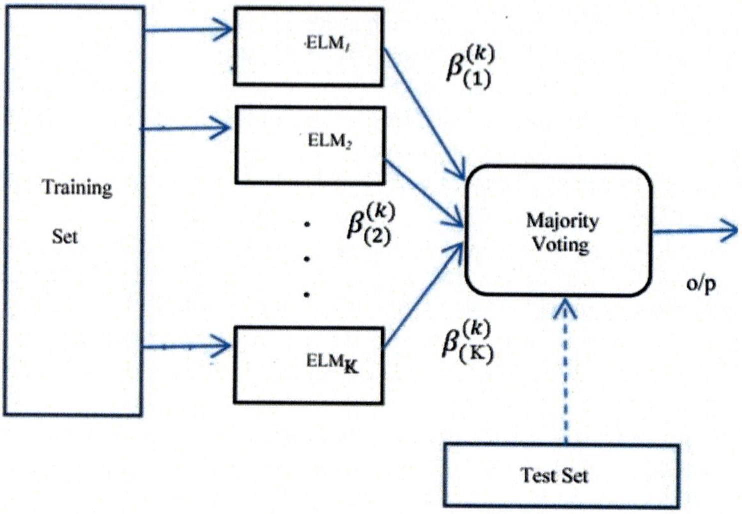

Fig. 1 Architecture of the voting-based extreme learning machine

Table 2 Results obtained by ELM and V-ELM approaches on University of Queensland database

Approach	Number of nodes	Number of ensembles	Training time (s)	Testing accuracy (%)	Deviation
ELM	1000	NA	137	85.61	0.93
V-ELM	1000	20	821	91.27	0.09

K number of outcomes are predicted. Then the final decision is made by the vote of the majority of the ELMs.

5 Results and Discussion

The efficiency of the V-ELM as well as ELM is determined by applying these approaches on the vital signs dataset [17]. The dataset consists of 201,818 instances with 161,454 training and 40,364 testing instances (80:20 ratio). The results have been averaged over 100 iterations of both the approaches with the shuffling of the data in a random manner.

All the implementations are done in the Python environment of Google colab [18]. As seen from Table 2, although the training time of ELM is quite fast, its efficiency in classification is quite poor with a large amount of deviation. In comparison the ELM, V-ELM is quite efficient for classification purposes with very little deviations from the results. The slowing down of V-ELM is due to the fact that an ensemble of distinct learners with parallel computation makes the process quite heavy.

6 Conclusion

There are new and interesting challenges for the machines to learn from the sensor data procured from the clinical environment. As such technologies are more quickly implemented in real-world settings, experimental testing will need to consider practical and long-term situations. The goal of this work was to check the health status of an individual from the physiological data. To achieve this, an approach namely, voting-based extreme learning machine has been implemented on the vital signs data of an individual. Results show that the system is quite robust in comparison to the conventional extreme learning machine approach in terms of performance accuracy and the time taken to produce the output.

Acknowledgements This research activity is a portion of the TEQIP Collaborative Research Scheme (CRS) project entitled, "Seamless Health Monitoring and Analysis of soldier using Machine Learning Approach" [CRS ID 1-5763896131]. The authors would like to thank NPIU, Government of India.

References

1. Raghupathi W, Raghupathi V (2014) Big data analytics in healthcare: promise and potential. Health Inf Sci Syst 2(1):1
2. Kim J, Campbell AS, de Ávila BE, Wang J (2019) Wearable biosensors for healthcare monitoring. Nat Biotechnol 37(4):389–406 Apr
3. Wiens J, Shenoy ES (2018) Machine learning for healthcare: on the verge of a major shift in healthcare epidemiology. Clin Infect Dis 66(1):149–53 Jan 1
4. Thakur SS, Abdul SS, Chiu HY, Roy RB, Huang PY, Malwade S, Nursetyo AA, Li YC (2018) Artificial-intelligence-based prediction of clinical events among hemodialysis patients using non-contact sensor data. Sensors 18(9):2833
5. Iglesias Martínez ME, García-Gomez JM, Sáez C, Fernández de Córdoba P, Alberto Conejero J (2018) Feature extraction and similarity of movement detection during sleep, based on higher order spectra and entropy of the actigraphy signal: Results of the Hispanic Community Health Study/Study of Latinos. Sensors 18(12):4310
6. Le S, Hoffman J, Barton C, Fitzgerald JC, Allen A, Pellegrini E, Calvert J, Das R (2019) Pediatric severe sepsis prediction using machine learning. Front Pediatr 7:413
7. Vistisen ST, Johnson AE, Scheeren TW (2019) Predicting vital sign deterioration with artificial intelligence or machine learning. J Clin Monit Comput 33(6):949–951
8. Baker S, Xiang W, Atkinson I (2020) Continuous and automatic mortality risk prediction using vital signs in the intensive care unit: a hybrid neural network approach. Sci Rep 10(1):1–2
9. Huang GB, Zhu QY, Siew CK (2004) Extreme learning machine: a new learning scheme of feedforward neural networks. In: IEEE international joint conference on neural networks (IEEE Cat. No. 04CH37541), 2, pp 985–990
10. Huang GB, Zhu QY, Siew CK (2006) Extreme learning machine: theory and applications. Neurocomputing 70(1–3):489–501
11. Zhang Y, Wang Y, Zhou G, Jin J, Wang B, Wang X, Cichocki A (2018) Multi-kernel extreme learning machine for EEG classification in brain-computer interfaces. Expert Syst Appl 96:302–10
12. Diker A, Avci D, Avci E, Gedikpinar M (2019) A new technique for ECG signal classification genetic algorithm wavelet Kernel extreme learning machine. Optik 180:46–55

13. Jin Z, Zhou G, Gao D, Zhang Y (2020) EEG classification using sparse Bayesian extreme learning machine for brain-computer interface. Neural Comput Appl 32(11):6601–9
14. Zhang J, Xiao W, Li Y, Zhang S (2018) Residual compensation extreme learning machine for regression. Neurocomputing 311:126–36
15. Armaghani DJ, Kumar D, Samui P, Hasanipanah M, Roy B (2020) A novel approach for forecasting of ground vibrations resulting from blasting: modified particle swarm optimization coupled extreme learning machine. Eng Comput 2:1–5
16. Cao J, Lin Z, Huang GB, Liu N (2012) Voting based extreme learning machine. Inf Sci 185(1):66–77
17. Liu D, Görges M, Jenkins SA (2012) University of Queensland vital signs dataset: development of an accessible repository of anesthesia patient monitoring data for research. Anesth Analg 114(3):584–9
18. Bisong E (2019) Google collaboratory. In: Building machine learning and deep learning models on google cloud platform. Apress, Berkeley, pp 59–64

U-Shaped Xception-Residual Network for Polyps Region Segmentation

Pallabi Sharma, Bunil Kumar Balabantary, and P. Rangababu

Abstract Segmenting the region of interest helps gastroenterologists for removing polyps during the surgery in the gastrointestinal tract. We propose a segmentation system to segment the area of the polyp from the informative frames. Informative frames are the frames that contain at least a polyp in colonoscopy still frames. Our proposed U-shaped convolution neural network model utilizes the concept of residual connection and Xception as a backbone structure. We consider colonoscopy still frames as input to the segmentation model and achieved a Dice and Jaccard score of 86.3 and 79, respectively. It outperformed the conventional U-net with a 3.7% performance gain with respect to Dice score. This proposed method can be used as a reliable alternative system to identify a region of polyps during colonoscopy analysis.

Keywords Colon cancer · Deep learning · Polyps segmentation · CNN · Colonoscopy

1 Introduction

Colon cancer is one of the leading cancer worldwide. Abnormalities in the colon begin as polyps. Initially, polyps start as benign, i.e., non-cancerous. But there is a high chance of becoming a polyp malignant, i.e., cancerous, if it is not treated in its

P. Sharma (✉) · B. K. Balabantary
Department of Computer Science and Engineering, National Institute of Technology Meghalaya, Shillong 793003, India
e-mail: pallabisharma@nitm.ac.in

B. K. Balabantary
e-mail: bunil@ntm.ac.in

P. Rangababu
Department of Electronics and Communication Engineering, National Institute of Technology Meghalaya, Shillong 793003, India
e-mail: p.rangababu@nitm.ac.in

S. Basu et al. (eds.), *Proceedings of International Conference on Frontiers in Computing and Systems*, Lecture Notes in Networks and Systems 404,
https://doi.org/10.1007/978-981-19-0105-8_25

early stage. So, it is possible to decrease the mortality rate due to colon cancer if it is identified early and follows the treatment process from its early stage onward. The colonoscopy test is the gold standard screening technique for identifying any abnormalities present in the large intestine—a camera attached to the colonoscope captures the inside view of the colon. Gastroenterologist stores the images sent by the colonoscope for analyzing the presence of abnormalities manually. In this analysis, doctors identify the region of polyps later on manually. If the polyp present in an image is benign, then they proceed for the further process of removing it through surgery. The manual identification of the polyp region from each and every image captured during colonoscopy is laborious, and human error cannot be denied due to the visual perception of the human eye. So, a computerized analyzing system is beneficial to reduce the workload of the gastroenterologist.

We have proposed a segmentation system to segment the region of polyps from the still images extracted from the colonoscopy video. The proposed segmentation model is U-shaped that utilizes the concept of Xception as a backbone structure along with residual connection. The motivation behind introducing this network is the superior performance of Xception [1] and ResNet-152 [2] than the other deep neural network architecture on ImageNet [3] classification challenge. The contribution of this work includes the following:

- A novel automatic polyps segmentation model from colonoscopy still images. Our model will segment the region of polyps from the informative frames.
- We have proposed a U-shaped segmentation model with Xception backbone that utilizes the concept of residualism.

2 Related Work

In recent years, due to the advancement of machine learning (ML), several works on literature introduce computer-aided design strategy in the analysis of disease from images. This advancement leads to the development of CAD application in colon cancer analysis as well. Literature related to colon cancer analysis from colonoscopy can be categorized based on three factors: (1) type of input consideration, i.e., video or still images, (2) type of analysis performed such as detection (localization and classification), or segmentation of the region of interest and (3) type of method adopted to analyze such as ML or deep learning (DL) or traditional image processing. In this work, our focus is on the segmentation of the region of interest, i.e., polyps from still images. In this direction, Graham et al. [4] introduced minimum information loss deep neural network (DNN) and achieved F1-score of 0.825 and an object-level Dice score of 0.875. Sornapudi et al. [5] used CNN-based transfer learning approach from ImageNet dataset and achieved 88.28% F1-score. Mahmud et al. [6] used U-net-based model, and their approach performs best on the VC-ColonDB database with a Dice score of 92.8%. Zhang et al. [7] and Kang et al. [8] used fully convolutional network (FCN) and achieved 0.701 Dice and 66.07% Jaccard score respectively.

Again, Brandao et al. [9] also used FCN and achieved 54.2% recall value.
From the literature, we observed that FCN and U-net [10]-based DNN methods are used in recent years to develop an efficient CAD system for polyps segmentation. But, a standard method to apply in colon cancer analysis is still not available. Thus, the development of an efficient polyp segmentation method is needed to contribute toward the advancement of colon cancer analysis.

3 Materials and Methods

Accurate representation of the characteristics of a target region is an important criterion to achieve the best result from a selected segmentation task. Conventional techniques for segmentation tasks are dependent on handcrafted features. But, features handcrafting is a difficult task in medical images due to lack of inter-group variability. The variable size and shape of polyps make it critical to find optimal features to represent each group of polyps. CNN, a subset of DNN, helps provide an end-to-end system from automatic feature extraction to segment the region of interest. Hence, a U-shaped CNN-based approach for polyps segmentation for colon cancer analysis is introduced in this work.

3.1 Dataset

In this work, a publicly available benchmark dataset "Kvasir-SEG" [11] is used that can be downloaded from the link https://datasets.simula.no/kvasir-seg/. The dataset contains 1000 polyp images and their corresponding ground truth.

3.2 Polyps Region Segmentation Model

The objective of this segmentation is to assign the same labels to each similar pixel to group the similar pixel together. In this work, the input images consist of two groups. The pixels of the polyps region are one group, and the background is the other group. Therefore, our problem statement is formulated as a binary problem, where the target region should be assigned a specific range of intensity values and the rest to the background.

The conventional U-net model contains three convolution blocks for scaling up the input feature maps and another three convolution blocks for scaling down feature map [10]. Each block consists of two 3×3 Conv layers followed by a max pooling layer.

The upscale module's output is connected to the previous downscale module's output, which promotes integration between low-level features and high-level features and accelerates the convergence of the model. Literature conveys that the VGG and ResNet outperform all other models [12], and this claim is supported by the performance of these models in ImageNet classification challenge [1]. Recent research shows an improvement of the basic ResNet model, i.e., Xception, where the convolution layer is replaced by depthwise separable convolution that yields a faster model with higher accuracy. In the depthwise separable convolution layer, the 3×3 convolution learns spatial interconnection, and the 1×1 convolution learns the interchannel correlation. Considering these advantages, in our proposed segmentation architecture, we used the Xception as a backbone that is optimized on the basis of ResNet. Similar to Xception, the proposed model excludes the nonlinear activation function.

Characteristics of Xception lead to efficiently extract pixels in the polyp's group even if it is occluded by noise. Bilinear upsampling is added in the upscale module for better extraction of spatial information.

In the proposed architecture, zero padding is utilized during convolution that keeps the dimension of input and output the same. In the decoder, the in place of max pool a 2×2 upsampling layer is used. This upsampling layer is followed by transpose convolution to maintain the feature map size of the encoding path concatenated to this layer. This is basically for the residual connection. The architecture for the network is shown in Fig. 1. The proposed architecture considers Adam as an optimizer, and for the accurate error propagation, sparse categorical cross entropy is used as a loss function.

4 Result and Discussion

This section reports the results of the proposed method. The models are implemented in Google Colab [13] that utilizes Keras deep learning framework and TensorFlow as back end.

4.1 Evaluation of Segmentation Performance

Dice and Jaccard coefficients are used to evaluate the segmentation performance. F1-score or Dice coefficient (DC) is the measure of spatial overlap between input image and the output image produced by the segmentation method. Jaccard coefficient (JC) is the ratio of area of overlap between the predicted segmentation and the ground truth to the area of union between the predicted segmentation and the ground truth. These can be calculated as follows:

$$\mathrm{DC} = \frac{2\,|\mathrm{TP}|}{2\,|\mathrm{TP}| + |\mathrm{FN}| + |\mathrm{FP}|} \tag{1}$$

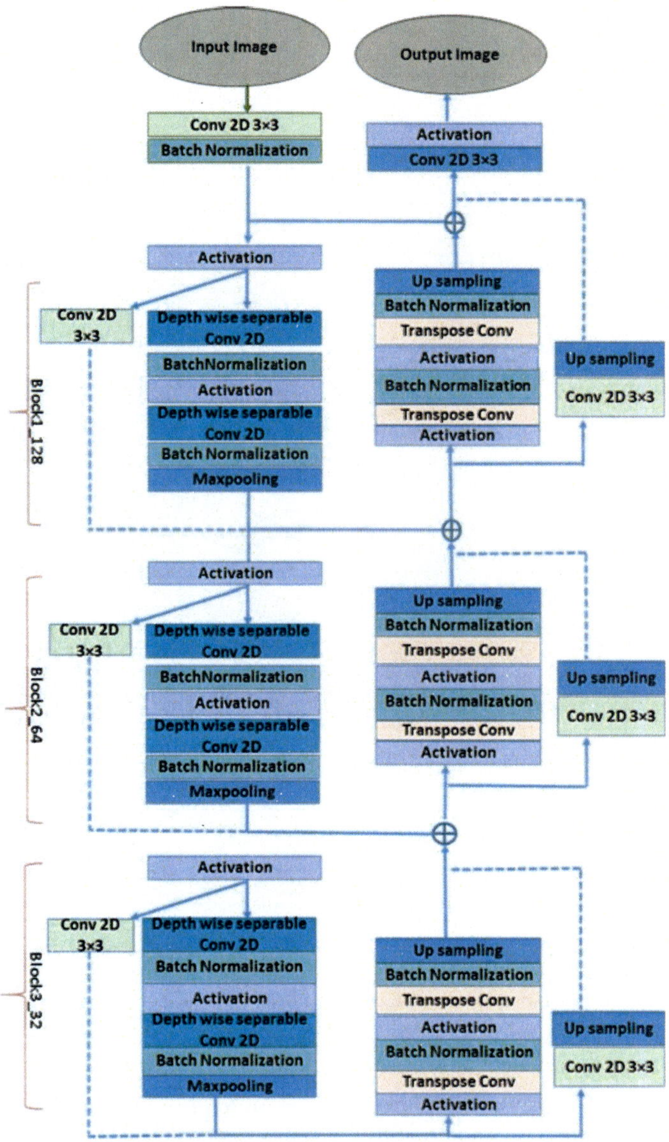

Fig. 1 Polyps segmentation model architecture

Table 1 Segmentation performance comparison with other methods

Methods	DC	JC
U-net	82.6	**80.6**
Proposed-net	**86.3**	79

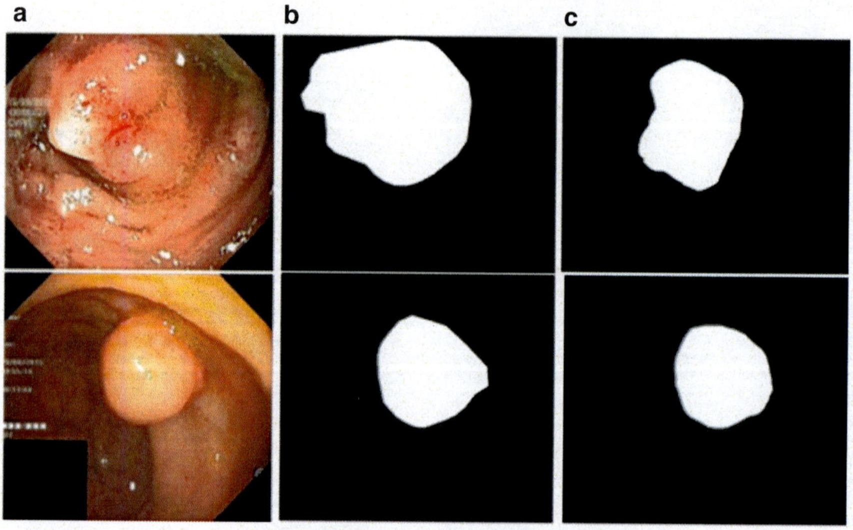

Fig. 2 Result of the proposed segmentation method. **a** Input image, **b** ground truth, **c** output image

$$JC = \frac{|TP|}{|TP| + |FN| + |FP|} \tag{2}$$

where TP = true positive, FP = false positive, TN = true negative and FN = false negative.

Table 1 includes the DC and JC value of proposed segmentation model along with the original U-net model. Proposed model achieved 86.3% DC 79% JC, whereas U-net achieved 82.6% DC and 80.6% JC. Hence, proposed method outperforms original U-net in terms of Dice score. Figure 2 shows the outcome of the proposed method. From the first column of Fig. 2, sample input image used as a test case for proposed segmentation model can be visualized. Second column represents the corresponding ground truth prepared by human expert, and in the third column, the respective output generated by the segmentation model is shown. A comparative analysis of the proposed model with similar prior work is introduced in Table 2. From the observation, it is seen that the proposed model shows superior performance to the other model based on the mentioned DC value in references [7, 16].

Table 2 Comparison of proposed method with similar work

Method	Model	DC
Proposed-net	CNN	**0.86**
Zha et al. [14]	ResUNet++	0.85
Zha et al. [15]	ColoSegNet	0.82
Akbari et al. [16]	CNN	0.81
Zhou et al. [17]	Customized U-net	0.80
Zhang et al. [7]	CNN	0.70
Ronneberger et al. [10]	U-net	0.70

5 Conclusion

In this paper, we introduced a new method for polyps segmentation from colonoscopy images. The proposed model automatically segments the region of polyps from the still images captured during the colonoscopy test. This is a U-shaped CNN model that contains Xception structure as the backbone. Along with Xception, the proposed model utilizes the concept of residual connection to avoid vanishing gradients. The achieved DC score of the proposed model is 86.3% that is 3.7% higher than the conventional U-net. Therefore, it is expected that the proposed model can reduce the workload of doctors by acting as an automatic tool in place of manual segmentation of polyps in colon cancer treatment.

References

1. Chollet F (2017) Xception: deep learning with depthwise separable convolutions. In: Proceedings of the IEEE conference on computer vision and pattern recognition, pp 1251–1258
2. Simonyan K, Zisserman A (2014) Very deep convolutional networks for large-scale image recognition. arXiv preprint arXiv: arXiv:1409.1556
3. Deng J, Dong W, Socher R, Li L-J, Li K, Fei-Fei L (2009) Imagenet: a large-scale hierarchical image database. In: 2009 IEEE conference on computer vision and pattern recognition. IEEE, pp 248–255
4. Graham S, Chen H, Gamper J, Dou Q, Heng P-A, Snead D, Tsang YW, Rajpoot N (2019) Mild-net: minimal information loss dilated network for gland instance segmentation in colon histology images. Med Image Anal 52:199–211
5. Sornapudi S, Meng F, Yi S (2019) Region-based automated localization of colonoscopy and wireless capsule endoscopy polyps. Appl Sci 9(12):2404
6. Mahmud T, Paul B, Fattah SA (2021) Polypsegnet: A modified encoder-decoder architecture for automated polyp segmentation from colonoscopy images. Comput Biol Med 128:104119
7. Zhang L, Dolwani S, Ye X (2017) Automated polyp segmentation in colonoscopy frames using fully convolutional neural network and Textons. In: Annual conference on medical image understanding and analysis. Springer, pp 707–717
8. Kang J, Gwak J (2019) Ensemble of instance segmentation models for polyp segmentation in colonoscopy images. IEEE Access 7:26440–26447

9. Brandao P, Mazomenos E, Ciuti G, Caliò R, Bianchi F, Menciassi A, Dario P, Koulaouzidis A, Arezzo A, Stoyanov D (2017) Fully convolutional neural networks for polyp segmentation in colonoscopy. In: Medical imaging 2017: computer-aided diagnosis, vol 10134. International Society for Optics and Photonics, p 101340F
10. Ronneberger O, Fischer P, Brox T (2015) U-net: convolutional networks for biomedical image segmentation. In: International conference on medical image computing and computer-assisted intervention. Springer, pp 234–241
11. Jha D, Smedsrud PH, Riegler MA, Halvorsen P, de Lange T, Johansen D, Johansen HD (202) Kvasir-seg: a segmented polyp dataset. In: International conference on multimedia modeling. Springer, pp 451–462
12. Kasugai K, Balabantaray BK, Sharma P, Bora K (2020) Two stage classification with CNN for colorectal cancer detection. Oncologie 22(3):129–145
13. Google, Google colab. https://colab.research.google.com
14. Jha D, Smedsrud PH, Johansen D, de Lange T, Johansen HD, Halvorsen P, Riegler MA (2021) A comprehensive study on colorectal polyp segmentation with resunet++, conditional random field and test-time augmentation. IEEE J Biomed Health Inform 25(6):2029–2040
15. Jha D, Ali S, Tomar NK, Johansen HD, Johansen D, Rittscher J, Riegler MA, Halvorsen P (2021) Real-time polyp detection, localization and segmentation in colonoscopy using deep learning. IEEE Access 9:40496–40510
16. Akbari M, Mohrekesh M, Nasr-Esfahani E, Soroushmehr SR, Karimi N, Samavi S, Najarian K (2018) Polyp segmentation in colonoscopy images using fully convolutional network. In: 2018 40th annual international conference of the IEEE engineering in medicine and biology society (EMBC). IEEE, pp 69–72
17. Zhou Z, Siddiquee MMR, Tajbakhsh N, Liang J (2019) Unet++: redesigning skip connections to exploit multiscale features in image segmentation. IEEE Trans Med Imaging 39(6):1856–1867

Motion Sensor-Based Android Game to Improve Fine Motor and Working Memory Skills of Children

Sudipta Saha, Saikat Basu, Koushik Majumder, and Debashish Chakravarty

Abstract Fine motor skills and working memory skills are determinants of scholastic performances. Educational achievement and high skill performances offer high salaries with comfortable lifestyles in a competitive economy. Presently in the competitive environment, children are often diagnosed with disorders like development coordination disorder (DCD), attention deficit hyperactivity disorder (ADHD), obesity, amblyopia, Tourette syndrome, etc. Children who suffer from the above disorders also lack fine motor and working memory skills. These skills can be improved by providing proper training with various tools and techniques. An Android game is developed in this literature to measure and improve children's working memory skills and fine motor skills at once. Android is the most widely used mobile operating system, and most handset manufacturers deliver Android operating systems with their smartphones. Due to the coverage of both skills and the use of Android as a gaming platform, this single game can be a cost-effective, time-efficient home-based training tool for improving both skills of children. This improvement may transfer a positive effect on children's scholastic performances.

Keywords Motion sensor · Accelerometer · DCD · Obesity · ADHD · Amblyopia · Tourette · Disorder · Fine motor · Working memory · Game

1 Introduction

Scholastic performance prepares the way to higher levels of education [9]. In the job market, academic performance is often used to determine a person's productiveness and salaries [9]. Poor fine motor skills are responsible for inferior scholastic achievements [17]. Fine motor tasks are movements of the body's parts like arm, leg, head,

S. Saha · S. Basu (✉) · K. Majumder
Maulana Abul Kalam Azad University of Technology, Kolkata, West Bengal, India
e-mail: saikat.basu@smst.iitkgp.ernet.in

D. Chakravarty
Indian Institute of Technology, Kharagpur, West Bengal, India
e-mail: dc@mining.iitkgp.ac.in

S. Basu et al. (eds.), *Proceedings of International Conference on Frontiers in Computing and Systems*, Lecture Notes in Networks and Systems 404,
https://doi.org/10.1007/978-981-19-0105-8_26

finger while keeping the body's centre of gravity unchanged [8]. Children have to perform several fine motor tasks daily in classrooms, such as copying from a book or board, taking notes, turning pages, answering questions, correcting their answers from hearing teachers, using a computer and doing craftworks[15]. The fine motor skills of children control their behaviour in school, affecting the learning quality and extent of them. A child with fine motor problems may suffer in any competition [15]. As a result, a child's confidence and motivation are negatively affected. The human's working memory (WM) is related to scholastic performances [11]. WM keeps necessary information at a readily accessible state and continuously processes new information [7]. WM has limited capacity, and the capacity varies from man to man [7]. The WM's capacity is measured by WM's span traditionally [24]. Many children with impairments of WM exhibit difficulty in reading and mathematics, fail to follow lengthy instructions and make more place-keeping errors [3]. Substantial research reported that children suffer from disorders like development coordination disorder (DCD), attention deficit hyperactivity disorder (ADHD), obesity, amblyopia, Tourette syndrome (TS), etc. Many of these studies indicated that children who suffer from the above disorders also possess marked impairment in fine motor skills and WM skills.

A child is said to be suffering from DCD if his performance is substantially poor than age-matched children in tasks that require motor coordination [16]. Type and severity of deficit vary among DCD children. Children with DCD are less confident, less active, less often utilized play equipment and have lower perceived self-worth and are more anxious [17]. Children with DCD are tested against the control group using five fine motor tasks and five gross motor tasks under McCarron Assessment of Neuromuscular Development (MAND). DCD-affected children performed significantly poorer in both fine and gross motor skill tasks [17]. Along with motor deficiency, children with DCD also suffer from verbal and visuospatial WM deficiency [4].

Attention deficit hyperactivity disorder (ADHD) is a mental disorder with inattentiveness, overactivity and uninhibition. It is the most frequent mental disorder in children. Three subtypes of ADHD are predominantly inattentive type (ADHD-PI), predominantly hyperactive/impulsive (ADHD-HI) and combined inattentive, hyperactive/ impulsive type (ADHD-C) [18]. Fifty per cent of ADHD children have motor deficiencies [18]. ADHD children are tested against the control group for comparing their fine motor ability using the Movement Assessment Battery for Children (MABC) and Purdue Pegboard Test (PPT) in [18]. ADHD-C and ADHD-PI groups performed remarkably more flawed than the control group in the manual dexterity task and ball skill task (two subcomponents of MABC) [18]. PPT's composite score was significantly more unsatisfactory for the ADHD-PI group than the control group [18]. ADHD children are often found with working memory deficits [20]. The dual request selective task (DRST) is a visual-spatial working memory task used to test ADHD children [20]. Children with ADHD gave significantly fewer correct responses and committed significantly more errors than controls [20].

Worldwide, nearly about one-tenth of school children and adolescents are obese or overweight [22]. BMI range 25.0–29.9 are classified as overweight, and BMI $\geq$30 is obese. Along with gross motor skills, obese children also experience fine motor impairment, although non-contributory mass movement is limited [10]. This fact is founded in [10], using the Bruininks-Oseretsky test (BOT) of motor proficiency. Worse Wide Range Assessment of Learning and Memory (WRAML) WM index of obese adolescents in [13] indicates lower WM capacity of obese children/adolescents.

Amblyopia is a vision impairment that affects about 3% of the population [27]. The practical definition of amblyopia is a two-line or more significant difference between the visual acuity of two eyes, where glasses or surgery cannot correct the visual acuity of patients immediately [27]. Children with amblyopia were assessed on eight fine motor tasks under visual motor control (VMC) and eight fine motor tasks under upper limb speed and dexterity (ULSD) of BOT [27]. This experiment revealed significantly lower fine motor performance of amblyopic children than same-aged control children. WM skills of amblyopic patients are not widely studied. Amblyopic patients exhibit a low processing speed when using the worst eye [19].

Tourette syndrome (TS) is a neurobehavioural disorder. Unintentional and repetitive motor and voice tics are primary troubles faced in Tourette syndrome [2]. Motor tic produces stereotyped movement, and vocal tics produce sound. TS is a childhood disease, starts at school age, reaches a peak during the preadolescent years and, in most cases, a significant decline of tics observed during the beginning of adulthood [6]. To test children's fine motor skills with a TS, [23] used the PPT on 50 children with TS. TS group scored significantly lower on dominant, non-dominant and bimanual conditions of the PPT. TS children were tested using the n-back task to determine their WM performance [12]. TS-moderate children made significantly more errors and took significantly more time to respond than TS-low and control children in both the 0-back and 1-back conditions [12].

The fine motor skills and working memory skills of children can be improved by physiotherapy/training. In [25], poor writers' handwriting quality and writing speed significantly improved after three months of training on fine motor tasks and gross motor tasks. This improvement also continued nine months after training ended. In [24], the authors concluded that WM capacity improvement through training is not task specific, but the training has a transfer effect in a wider variety of tasks. Previously, different studies were done on deficits of fine motor skills and WM skills of children who were suffering from different disorders. Some studies also designed solutions for their improvement from these two problems. These solutions are varied widely. Instead, a single solution may be effective for all of them since they share the same kind of skill deficiencies.

In this work, a game is developed using Android Studio. The name of the game is "Roll Ball into Fairy Holes". The game requires only an ordinary smartphone with Android OS. Android platform is chosen because it is the most popular platform and mobile phone is ubiquitous today. The main input of this game is the value of the smartphone inbuilt accelerometer sensor's readings. The game's purposes are testing and training fine motor skills and working memory skills for all kinds of disordered children suffering from the fine motor and working memory skill deficiency. Although several fine motor skills and working memory training materials already exist, a single game, which covers both, is rare. Most fine motor exercise training uses game consoles and stand-alone sensors, and these types of equipment may not be available in low/middle-income families. This reality mandates children to join a physiotherapy clinic. In COVID-19 pandemic situation, regularly visiting a physiotherapy clinic is too risky. Children's physical activities are nearly stopped in this situation as schools are closed for more than one year. Even in a normal situation, all the children families in countries like India are not economically capable of sending their child to physiotherapy clinic. It is expected that this single game will be a cost-effective, time-efficient home-based training tool for improving fine motor skills and working memory skills for children suffering from these two skill deficits.

2 Related Works

A few studies already tried to improve children's fine motor skills or working memory skills separately using computer games or mobile games. In [14], two interactive online typing computer games named "Dance Mat Typing" and "Typing Chef" were used to train the children with poor manual dexterity. The children were instructed to play any game for around 10 min, five times per week for four weeks. The intervention and control groups were tested baseline and after training with MABC-2 manual dexterity and the finger-tapping task. Compared to the control group, the trained group children considerably improved their manual dexterity scores in the MABC-2 but not in the finger-tapping task. A cognitive WM training (CWMT) was given to adolescents in [1] by providing them a CD. This CD contained 15 exercises on verbal and visuospatial WM with increasing difficulties. Participants were advised to perform a total of 90 trials of all exercises each day at home for 25 days. The researchers found that WM phonological loop and visuospatial sketchpad components improved more significantly using CWMT exercise than using medication. In [26], executive function training was given to obese children in the final phase of an obesity reduction programme. The training was embedded in a game named "Braingame Brian". Children were trained four times a week, over six weeks. The game had 25 sessions; each session was of about 40 min. Each session contained training on WM and on inhibition control (each 20 min). This executive function training improved the children's executive functioning skills, especially WM skills. Children who completed the training were in a better position to maintain their weight loss for eight weeks after training. Neuromotor task training (NTT) was given to children with DCD for a half-hour a week for nine weeks [16]. This training was task-based and used a cognitive neuroscience approach to motor control. While designing the tasks for training, the designer identified which cognitive and motor control processes were involved within each task. Before starting the training, physiotherapists discussed with the parent(s) to assess the problems suffered by the children in daily life. They also tried to determine whether cognitive or motor control processes are responsible for their motor deficit. Children's motor performances were assessed during two times—pre-intervention and post-intervention time—using Test of Gross Motor Development-Second Edition (TGMD2), MABC. The performance of trained children and the standard of the trained group's motor patterns improved significantly due to training.

A home-based binocular treatment was given to children with amblyopia for five weeks using a dichoptic game running on iPod [28]. Dichoptic training creates simultaneous and separate stimulation of both eyes. The children were instructed to play an altered version of the videogame Tetris for 1 h per day. In this game, a child had to manipulate the position and rotation of the falling red blocks for the tessellate with green base blocks. Amblyopic children needed to wear red and green filters to the amblyopic eye and other eyes as they played the game to allow the game's dichoptic presentation. High contrast red-coloured falling blocks were visible to the amblyopic eye. Low contrast fixed green base blocks were visible to the fellow eye. Children with better baseline visual acuity and better initial binocular function achieved better fine motor scores after treatment. Binocular function and visual acuity of amblyopic children are also improved after the treatment.

3 Game Details

3.1 Game Descriptions

The game "Roll Ball into Fairy Holes" has four levels. Each level's main tasks are dropping a red ball into some particular holes (good holes, increases score) while avoiding some other holes (evil holes, decreases score) by tilting the mobile device as many times as possible within 120 s. Good & evil holes depend upon the creatures/coin associated with each hole. Textual instructions and pictorial help for each level give indications to the player about good and evil holes. Figures 1 and 2 represent textual and pictorial instructions of level 4, respectively. At each level, before the game starts, stimuli are shown for 10 s and then disappear. In this game, stimuli are eight holes randomly spread on the mobile screen, and a creature/coin is randomly associated with each hole. The player has to memorize positions and associated creatures/coin for each of eight holes. On the game-playing screen, the positions of holes remain the same as stimuli, but associated creatures will disappear. The game's difficulty increases level by level by increasing the types of creatures associated with holes (maximum four types of creatures, present in highest level). When a player drops the red ball into a hole, the creature/coin associated with that hole will be displayed transiently, and the game will also play a short matching sound. The concept of old and new holes is present for the good holes only. It avoids the tendency of the repeated drops of the red ball in the same good holes. Dropping into a new good hole adds more scores than dropping into old good holes. A new good hole becomes an old good hole after first dropping in that hole. Level wise game details are shown in Table 1. The game-playing screen always displays the remaining available time, the number of available new good holes and the old good holes. Additionally, scores added/subtracted in each trial will be flashed after each

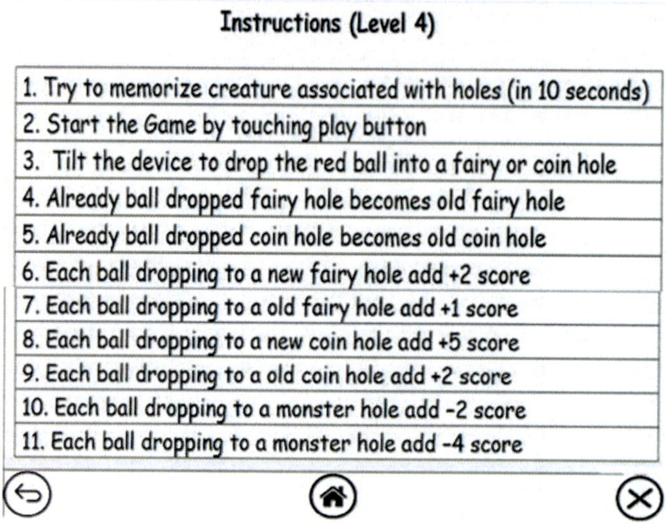

Fig. 1 Textual instruction of game level 4

Fig. 2 Pictorial instruction of game level 4

trial. Some fantasy elements are included in this game, such as fairy, monster and dragon. This is because several studies like [21] indicated that children prefer fantasy elements, and this game is targeted to children with different disorders. Figure 3 represents an arbitrary stimulus of level 4, and Fig. 4 is an instance of game screen of level 4. Table 2 compares some accelerometer ball games, which are available in Google Play Store with “Roll Ball into Fairy Holes”.

Table 1 Stimuli, tasks and scoring patterns of game "Roll Ball into fairy holes"

Level	Stimuli (Randomly distributed holes)	Tasks (Tilting the device to drop the red ball as many times as possible into holes according to preferences)	Scoring patterns (For ball dropping into)
Level 1	• Total holes: 8 • Fairy holes: 6 • Monster holes: 2	• First preference: new fairy holes • Second preference: old fairy holes • Avoid: monster holes	• New fairy holes: +2 • Old fairy holes: +1 • Monster holes: −2
Level 2	• Total holes: 8 • Fairy holes: 4 • Monster holes: 4	• First preference: new fairy holes • Second preference: old fairy holes • Avoid: monster holes	• New fairy holes: +2 • Old fairy holes: +1 • Monster holes: −2
Level 3	• Total holes: 8 • Fairy holes: 3 • Monster holes: 3 • Coin holes: 2	• First preference: new coin holes • Second preference: new fairy holes • Third preference: old fairy holes • Avoid: monster holes	• New coin holes: +5 • Old coin holes: +2 • New fairy holes: +2 • Old fairy holes: +1 • Monster holes: −2
Level 4	• Total holes: 8 • Fairy holes: 2 • Monster holes: 2 • Coin holes: 2 • Dragon holes: 2	• First preference: new coin holes • Second preference: new fairy holes • Third preference: old fairy holes • Avoid: monster holes and dragon holes	• New coin holes: +5 • Old coin holes: +2 • New fairy holes: +2 • Old fairy holes: +1 • Monster holes: −2 • Dragon holes: −4

Fig. 3 Arbitrary stimulus of game level 4

Fig. 4 Instance of the screen of game level 4

3.2 Technical Details

In this game, the position of the red ball changes according to the value of the smartphone's inbuilt accelerometer sensor along the X-axis and Y-axis. In this game, holes and red balls are both programmed as circles. The game "Roll Ball into Fairy Holes" broadly consists of three components—Create_Stimuli, Score_Assignment and Canvas_Draw. The function of the component Create_Stimuli is creating positions of eight holes utilizing the entire screen of the device. It also assigns creatures and coins with holes. The second component Score_Assignment has the primary responsibility of detecting the red ball dropping into holes and updating the score. When the distance between the centre of the red ball and a hole is less than 20 pixels, it is assumed that the red ball is dropped into that hole. The last one, the Canvas_draw component, is called by the system initially to draw a game view and, after that, many times whenever game views need to be redrawn.

3.3 Task Implemented in This Game

Fine Motor Task In this game, wrist movement occurs. This game implements fine motor tasks proposed by Cratty [8]. Wrists exercises involved in this game are extension, flexion, radial deviation and ulnar deviation. Bending palm towards the wrist is flexion, and bending the posterior side of the palm towards the backside is called extension [5]. Radial deviation and ulnar deviation are bending the wrist towards the thumb and little finger sides, respectively [5]. This game can be played using both hands or a single hand. Whenever a player tries to move the red ball towards down, wrist flexion occurs, whereas when he tries to move the red ball towards up, wrist extension occurs (plays with either single or both hands). Ulnar and radial deviations are not prominent when a player uses both hands, but these are prominent when he uses only a single hand while he is trying to move the red ball right

Table 2 Comparisons of some accelerometer ball games of Google Play Store with "Roll ball into fairy holes"

Game name	No. of levels	Main tasks (Tilting the device to control the movement of)	Obstacles	Motivational components	Memory component present?
Roll ball into fairy holes	4	The red ball so that it can be dropped into new good holes/old good holes while avoiding evil holes	Evil holes, which decrease scores	Fantasy creatures with corresponding sounds, variety of scores, no minimum scoring criteria to reach upper levels	Yes
Accelerometer ball pro	126	The red ball so that coin can be collected while avoiding deadly traps	Many obstacles present like oil, glue, saws, holes, ice walls, etc.	Medal awarded depending on level completion time	No
Accelerometer ball game	32	A number of balls so that it can be dropped into the particular green hole while avoiding other randomly distributed holes	All the holes except the particular green hole are obstacles	Developers claim that the game is suitable for stress treatment	No
Roll balls into a hole	7	Several white balls and a red ball so that these can be dropped into a hole in such a way that the red ball should be the last one to go in	On some level, mazes are present	Collection of gold coins	No
Roll balls in the hole	40	One or more ball(s) to reach the destination through a circular labyrinth	The movement of each ball creates movement of other balls problematic	Developers claim that the game brings back childhood memories	No

Fig. 5 Wrist flexion: movement of red ball downward, gameplay with right hand

or left. Figure 5 depicts wrist flexion at the time of moving the red ball downward, and Fig. 6 depicts ulnar deviation at the time of moving the red ball towards right.

Working Memory Task In the game, holes are not identical; some are good, while some are evil depending on associated creatures/coins. A player has to remember three kinds of information—creatures or coins associated with eight holes' position (shown in stimuli), which creatures/coin increase or decrease how much scores (from instruction), and already covered good holes (old good holes). Simultaneously, he/she has to tilt the device properly so that the red ball can fall into intended holes. Since both storage (storing task-related information) and processing (tilting the device to drop the ball) components are involved in this game, this game implements working memory tasks.

3.4 Usability as a Dichoptic Game

This game can be used as a dichoptic game for binocular treatment to amblyopic children. In this game, a player has to move a red ball in the green background to drop them into black holes. For using this game for binocular training purposes, children have to wear red filters to the amblyopic eye and green filters to the fellow eye while playing the game. The contrasts of red (ball) and green (background) colours may need to be readjusted to use it as a dichoptic game.

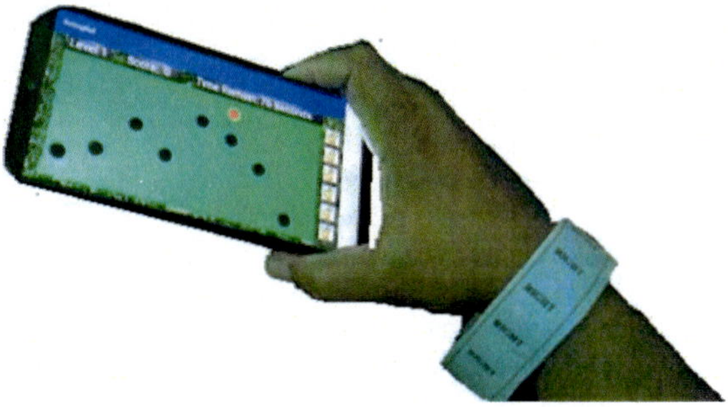

Fig. 6 Ulnar deviation: movement of red ball towards right, gameplay with right hand, backside of palm towards player

4 Conclusions and Future Scopes

Scholastic performances build the foundation of higher education. Highly educated people are more likely to be employed with higher salaries, and they have stable employment than those with lower education. The technologically demanding occupations need people with higher levels of education. Fine motor skills and working memory skills are two great predictors of children's scholastic performance and academic success. Several disorders, like DCD, ADHD, obesity, amblyopia, Tourette syndrome, etc., are observed among children. Unfortunately, these children frequently show fine motor and working memory skills deficiencies. Of course, these children need proper therapy to remove or minimize their disorderliness; these two specific skills can be improved using some home-based tools and techniques free of cost. In this context, an Android game is developed in the present work for measuring and improving fine motor skills and working memory skills of children at a time. No other special equipment except a smartphone is required to play this game. Therefore, it is expected that this game will be a low-cost, readily available household tool for children to improve fine motor and working memory performances. The game can be very time efficient since it covers exercises on two skills at the same time. The game's effectiveness depends on correctly measuring and improving the fine motor and working memory skills of children with different disorders. Soon, children with different disorders will be tested and trained with this game.

References

1. Ackermann S, Halfon O, Fornari E, Urben S, Bader M (2018) Cognitive working memory training (cwmt) in adolescents suffering from attention-deficit/hyperactivity disorder (adhd): a controlled trial taking into account concomitant medication effects. Psychiatry Res 269:79–85
2. Albin RL, Mink JW (2006) Recent advances in Tourette syndrome research. Trends Neurosci 29(3):175–182
3. Alloway TP (2006) How does working memory work in the classroom? Educ Res Rev 1(4):134–139

4. Alloway TP, Archibald L (2008) Working memory and learning in children with developmental coordination disorder and specific language impairment. J Learn Disabil 41(3):251–262
5. Batista TV, dos Santos Machado L, Valença AMG, de Moraes RM (2019) Farmyo: a serious game for hand and wrist rehabilitation using a low-cost electromyography device. Int J Serious Games 6(2):3–19
6. Bloch MH, Sukhodolsky DG, Leckman JF, Schultz RT (2006) Fine-motor skill deficits in childhood predict adulthood tic severity and global psychosocial functioning in tourette's syndrome. J Child Psychol Psychiatry 47(6):551–559
7. Conway A, Jarrold C, Miyake A (2008) Variation in working memory. Oxford University Press
8. Cratty BJ (1962) Comparison of learning a fine motor task with learning a similar gross motor task, using kinesthetic cues. Res Q Am Assoc Health Phys Educ Recreation 33(2):212–221
9. Dugan DJ, Jamison DT, Radner R (1976) Scholastic achievement: its determinants and effects in the education industry. In: Education as an industry. NBER, pp 53–94
10. Gentier I, D'Hondt E, Shultz S, Deforche B, Augustijn M, Hoorne S, Verlaecke K, De Bourdeaudhuij I, Lenoir M (2013) Fine and gross motor skills differ between healthy-weight and obese children. Res Dev Disabil 34(11):4043–4051
11. Hitch GJ, Towse JN, Hutton U (2001) What limits children's working memory span? Theoretical accounts and applications for scholastic development. J Exp Psychol Gen 130(2):184
12. Jeter CB, Patel SS, Morris JS, Chuang AZ, Butler IJ, Sereno AB (2015) Oculomotor executive function abnormalities with increased tic severity in Tourette syndrome. J Child Psychol Psychiatry 56(2):193–202
13. Maayan L, Hoogendoorn C, Sweat V, Convit A (2011) Disinhibited eating in obese adolescents is associated with orbitofrontal volume reductions and executive dysfunction. Obesity 19(7):1382–1387
14. McGlashan HL, Blanchard CC, Sycamore NJ, Lee R, French B, Holmes NP (2017) Improvement in children's fine motor skills following a computerized typing intervention. Hum Mov Sci 56:29–36
15. McHale K, Cermak SA (1992) Fine motor activities in elementary school: preliminary findings and provisional implications for children with fine motor problems. Am J Occup Ther 46(10):898–903
16. Niemeijer AS, Smits-Engelsman BC, Schoemaker MM (2007) Neuromotor task training for children with developmental coordination disorder: a controlled trial. Dev Med Child Neurol 49(6):406–411
17. Piek JP, Baynam GB, Barrett NC (2006) The relationship between fine and gross motor ability, self-perceptions and self-worth in children and adolescents. Hum Mov Sci 25(1):65–75
18. Pitcher TM, Piek JP, Hay DA (2003) Fine and gross motor ability in males with adhd. Dev Med Child Neurol 45(8):525–535
19. Popple AV, Levi DM (2008) The attentional blink in amblyopia. J Vis 8(13):12
20. Re A, De Franchis V, Cornoldi C (2010) Working memory control deficit in kindergarten adhd children. Child Neuropsychol 16(2):134–144
21. Rustina Y, Krianto T, Ayubi D et al (2018) Developing a health education game for preschoolers: What should we consider? Enfermeria Clin 28:1–4
22. Saraswathi Y, Najafi M, Gangadhar M, Malini SS (2011) Prevalence of childhood obesity in school children from rural and urban areas in Mysore, Karnataka, India. J Life Sci 3(1):51–55
23. Schultz RT, Carter AS, Gladstone M, Scahill L, Leckman JF, Peterson BS, Zhang H, Cohen DJ, Pauls D (1998) Visual-motor integration functioning in children with Tourette syndrome. Neuropsychology 12(1):134
24. Shipstead Z, Redick TS, Engle RW (2012) Is working memory training effective? Psychol Bull 138(4):628
25. Smits-Engelsman BC, Niemeijer AS, van Galen GP (2001) Fine motor deficiencies in children diagnosed as dcd based on poor grapho-motor ability. Hum Mov Sci 20(1–2):161–182
26. Verbeken S, Braet C, Goossens L, Van der Oord S (2013) Executive function training with game elements for obese children: a novel treatment to enhance self-regulatory abilities for weight-control. Behav Res Ther 51(6):290–299

27. Webber AL, Wood JM, Gole GA, Brown B (2008) The effect of amblyopia on fine motor skills in children. Invest Ophthalmol Vis Sci 49(2):594–603
28. Webber AL, Wood JM, Thompson B (2016) Fine motor skills of children with amblyopia improve following binocular treatment. Invest Ophthalmol Vis Sci 57(11):4713–4720

Specular Reflection Removal Techniques for Noisy and Non-ideal Iris Images: Two New Approaches

Md. Amir Sohail, Chinmoy Ghosh, Satyendranath Mandal, and Md. Maruf Mallick

Abstract Iris recognition is one of the most consistent and unique biometric-based recognition systems. Iris-based authentication has been a challenging problem especially for noisy and non-ideal iris images, as the images include realistic noise factors. Presence of noise like reflection of light on eye images, i.e. specular reflection, causes problem in iris recognition. It is one of the major pre-processing tasks in iris recognition system. In this paper, two new and efficient approaches have been proposed and compared their robustness with a few existing specular reflection removal techniques, considering the challenging situation of non-ideal iris images. The first approach uses inward interpolation, and the second approach uses bilinear interpolation with extension to remove reflection. These suggested approaches are faster in execution as well as robust enough to remove reflection without disturbing the texture pattern of original iris images.

Keywords Iris recognition · Biometrics · Reflection removal · Mean · Inward interpolation · Linear interpolation · Bilinear interpolation · Structural similarity

---

Md. Amir Sohail (✉)
Associate Data Scientist, Cognizant Technology Solutions India Pvt. Ltd., Pune, India
e-mail: mdamircoder@gmail.com

C. Ghosh
Department of Computer Science and Engineering, Jalpaiguri Government Engineering College, Jalpaiguri, India
e-mail: chinmoy.ghosh@cse.jgec.ac.in

S. Mandal
Department of Information Technology, Kalyani Government Engineering College, Nadia, India

Md. Maruf Mallick
Senior Associate, Cognizant Technology Solutions India Pvt. Ltd., Kolkata, India

S. Basu et al. (eds.), *Proceedings of International Conference on Frontiers in Computing and Systems*, Lecture Notes in Networks and Systems 404,
https://doi.org/10.1007/978-981-19-0105-8_27

1 Introduction

Biometric recognition systems were introduced for reliability in identification. There are so many biometrics [1] like signature, voice, keystroke (behavioural) fingerprint, palm print, face, iris, retina, DNA (physiological), etc. Among them, iris recognition system is the most unique and cost-effective biometric recognition systems. It involves mainly four steps: iris acquisition, localization and segmentation, feature extraction and identification or verification [2, 3]. Presence of specular reflection [4] decreases accuracy in segmentation as well as feature extraction [5] of iris images.

There are many reflection removal techniques for ideal iris database images. However, we cannot consider any ideal situation to remove reflection from non-ideal iris images, as non-ideal iris database considers realistic noise factors. Reflections in the image are always the brightest area, and some iris images may not have any reflection area at all.

Jarjes et al. [6] implemented a connectivity of four-connected pixels-based method to fill dark holes in the complement image. Jamaluddin et al. [7] used contrast enhancement, morphological closing and flood fill operation to remove pupil reflection and intensity adjustment or inward interpolation for removing iris reflection. Wang et al. [8] filled the reflection areas using the inpainting method based on Navier–Stokes [9] equations. Reddi et al. [10] removed specular reflection by using threshold-based reflection detection (pixel intensity more than 235) and then applying inpainting algorithm [11]. Sahmoud et al. [12] computed average intensity of effective iris region (segmented) for all three colour components of RGB images and then added a constant (adjusted once, for the whole iris database) to get the reflection intensity. For each colour space, all the pixels with intensity more than that reflection intensity are detected as reflection noise and then removed from the effective iris area. Raffei et al. [13] classified the reflection regions by using line intensity profile (LIP) method [14] along with support vector machine (SVM), because LIP method was not sufficient alone, and then used near intensity interpolation method [15] to fill the reflection spots with adjacent colours. Chun-Wei et al. [16] transformed the histogram function (with the help of mean and standard deviation of histogram) to detect the reflection noise and then excluded from the effective iris area during segmentation. He et al. [17] used top five% brightest pixels as reflection and then used bilinear interpolation. Aydi et al. [18] proposed morphological operation-based mask and used linear interpolation. Previously, we suggested reflection detection based on threshold and reflection removal with non-reflection neighbour mean [19].

2 Related Works

We have implemented few related works [17–19] to illustrate robustness of our proposed approach.

2.1 He et al.'s Bilinear Interpolation (He-Bilinear)

Authors [17] calculated a binary reflection mask $R_{(X,Y)}$ with top five% brightest pixels of the image $I_{(X,Y)}$ as reflection. Then four envelop points { $P^R_{(x_R,Y)}$, $P^L_{(x_L,Y)}$, $P^B_{(X,y_B)}$, $P^T_{(X,y_T)}$ } of a reflection pixel $P_{(X,Y)}$, are determined by below equations:

$$x_R = \min\{x : \sum_{i=0}^{L-1} R_{(x-i,Y)} = 0,\ R_{(x-L,Y)} = 1, x > X\} \tag{1}$$

$$x_L = \max\{x : \sum_{i=0}^{L-1} R_{(x+i,Y)} = 0,\ R_{(x+L,Y)} = 1, x < X\} \tag{2}$$

$$y_B = \min\{y : \sum_{i=0}^{L-1} R_{(X,y-i)} = 0,\ R_{(X,y-L)} = 1, y > Y\} \tag{3}$$

$$y_T = \max\{y : \sum_{i=0}^{L-1} R_{(X,y+i)} = 0,\ R_{(X,y+L)} = 1, y < Y\} \tag{4}$$

where L (set to 2) defines the required separation (one pixel away) between the reflection pixels and envelop points. It means envelop points are the second nearest non-reflection pixels along four directions (top, bottom, left and right) with respect to a reflection pixel. Those reflection areas are then filled with the help of bilinear interpolation following this Eq. (5):

$$I(P) = \frac{I(P^T)(y_B - Y) + I(P^B)(Y - y_T)}{2(y_B - y_T)} + \frac{I(P^L)(x_R - X) + I(P^R)(X - x_L)}{2(x_R - x_L)} \tag{5}$$

One interesting fact is that some iris images may not have any reflection area. Even if an image has reflection, there are some cases where this mask (He-mask) will recognize non-reflection areas (sclera or other brighter areas) as reflection noise. We calculated the reflection mask after converting the RGB input image into greyscale image and applied bilinear interpolation for each colour component of the RGB input image.

2.2 Aydi et al.'s Linear Interpolation with Extension (Aydi-LinearExt)

Aydi et al. [18] proposed a mask with the help of morphological top-hat filter, thresholding and dilation. For reflection pixel $P(x, y)$, four nearest non-reflection neighbours

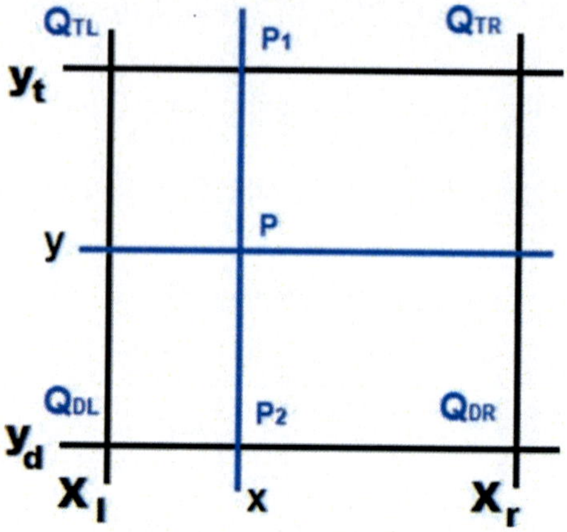

Fig. 1 Aydi et al.'s linear interpolation with extension

along top (left, right) and down (left, right) directions: $Q_{TL}(x_l, y_t)$, $Q_{TR}(x_r, y_t)$, $Q_{DL}(x_l, y_d)$, $Q_{DR}(x_r, y_d)$ are detected (shown in Fig. 1).

Authors used linear interpolation with extension to compute the function $f(P)$ for reflection pixel $P(x, y)$, by performing linear interpolation along x-axis direction first and then y-axis direction (or vice-versa), as shown in Eqs. (6–8). Their mask (Aydi-mask) can detect reflection in darker region (pupil or dark iris), but cannot properly detect reflection in bright regions (sclera or less dark iris). And we also need to use larger structuring element for morphological top-hat filtering to detect the complete reflection area.

$$f(P_1) = \frac{x - x_l}{x_r - x_l} f(Q_{DR}) + \frac{x_r - x}{x_r - x_l} f(Q_{DL}) \tag{6}$$

$$f(P_2) = \frac{x - x_l}{x_r - x_l} f(Q_{TR}) + \frac{x_r - x}{x_r - x_l} f(Q_{TL}) \tag{7}$$

$$f(P) = \frac{y_t - y}{y_t - y_b} f(P_1) + \frac{y - y_b}{y_t - y_b} f(P_2) \tag{8}$$

2.3 *Sohail's Direction-Wise Traverse Approach*

Reflection pixels located towards boundary are sometimes less bright than the reflection pixels at centre, for any specific reflection spot in an image. In our previous work [19] on specular reflection removal, we developed **EnlargeReflectionArea (ERA)** algorithm that uses (3 × 3) or (5 × 5) filters to enlarge the reflection areas by at most two or four pixels outward, to cover the less bright reflection areas as well. The ERA algorithm generates reflection enlarged image which acts like a mask to detect reflection spots. The reflection areas were detected by considering a reference pixel intensity threshold value of 230. In the reflection enlarged image, all the pixels with $intensity \geq 230$ were considered as reflection pixel. The reflection areas were then substituted with the mean of its closest non-reflection pixels, i.e. chosen direction pixels (CDP), using either one of the following two approaches. These reflection enlargement and removal approaches were applied on all colour components of RGB input image, and hence, they produce RGB output image.

2.3.1 4–Direction Traverse Approach (Sohail-ERA-4Dir)

This approach implies traverse in four directions (right, left, top, bottom) from any reflection pixel $P(x, y)$ and finds the CDP: $P_{\text{Right}}(x_{\text{Right}}, y)$, $P_{\text{left}}(x_{\text{Left}}, y)$, $P_{\text{Bottom}}(x, y_{\text{Bottom}})$ and $P_{\text{Top}}(x, y_{\text{Top}})$ [equation (9–12)].

$$x_{\text{Right}} = \min\{x_R : I(y, x_R) < 230, \forall (x+1) \leq x_R \leq \text{col}\} \tag{9}$$

$$x_{\text{Left}} = \max\{x_L : I(y, x_L) < 230, \forall 1 \leq x_L \leq (x-1)\} \tag{10}$$

$$y_{\text{Bottom}} = \min\{y_B : I(y_B, x) < 230, \forall (y+1) \leq y_B \leq \text{row}\} \tag{11}$$

$$y_{\text{Top}} = \max\{y_T : I(y_T, x) < 230, \forall 1 \leq y_T \leq (y-1)\} \tag{12}$$

where $I(y, x)$ represents reference pixel positioned at 'x'-th column and 'y'-th row of an image I with 'row' number of rows and 'col' number of columns, i.e. dimension (col, row). Then replace the reflection pixels by the mean of its four CDP [Eq. (13)]. Figure 2 explains selection of four CDP.

$$I(P) = \frac{I(P_{\text{Right}}) + I(P_{\text{Left}}) + I(P_{\text{Bottom}}) + I(P_{\text{Top}})}{4} \tag{13}$$

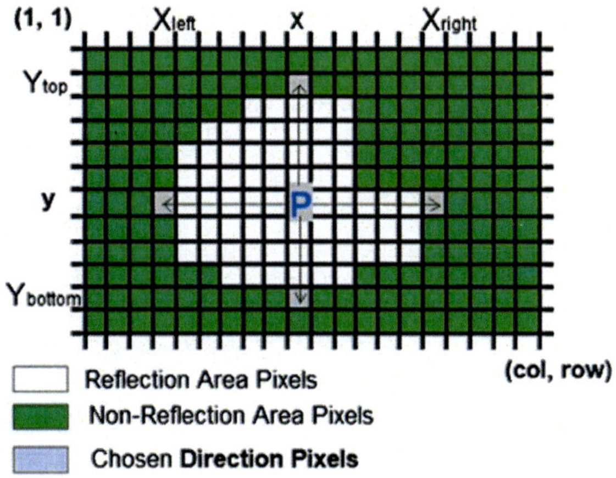

Fig. 2 4–direction traverse approach. *Source* Sohail et al. [19]

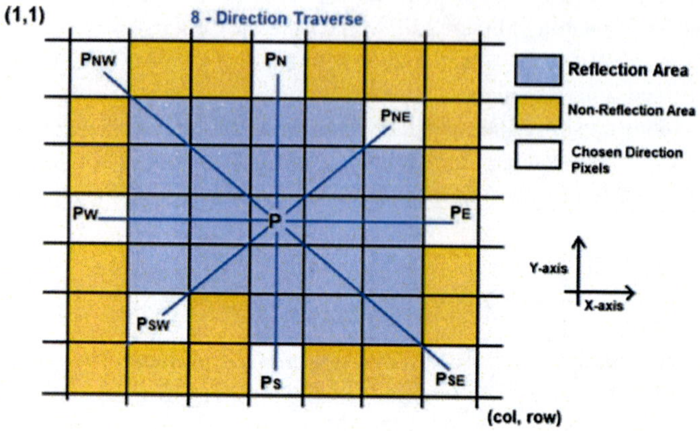

Fig. 3 8–direction traverse approach. *Source* Sohail et al. [19]

2.3.2 8–Direction Traverse Approach (Sohail-ERA-8Dir)

In a similar manner, this algorithm traverses in eight directions (south, north, east, west, south-east, south-west, north-east, north-west) from any reflection pixel P(x,y) and finds the closest non-reflection pixels (CDP): $P_S(x, y_S)$, $P_N(x, y_N)$, $P_E(x_E, y)$, $P_W(x_W, y)$, $P_{SE}(x_{SE}, y_{SE})$, $P_{SW}(x_{SW}, y_{SW})$, $P_{NE}(x_{NE}, y_{NE})$, $P_{NW}(x_{NW}, y_{NW})$. The non-diagonal pixels P_S, P_N, P_E and P_W are detected following Eq. 9–12. And the four other diagonal pixels P_{NE}, P_{NW}, P_{SE} and P_{SW} are determined based on the **Algorithm 2** and **Algorithm 3** mentioned in our previous work [19]. Figure 3 simplifies the selection process of eight CDP.

After finding the CDP, reflection pixel is replaced by mean of its eight CDP (Eq. 14). If there exists no such pixel in one or more directions, those directions are not considered to calculate the mean. It is also true for the previously discussed *Sohail-ERA-4Dir* approach.

$$I(P) = \frac{I(P_N) + I(P_S) + I(P_E) + I(P_W) + I(P_{NE}) + I(P_{NW}) + I(P_{SE}) + I(P_{SW})}{8} \tag{14}$$

3 Proposed Methodology

Every specular reflection removal technique basically consists of two phases: reflection detection and reflection removal. Reddi et al. [10] suggested a threshold intensity value 235 for reflection detection. As the reflection spots are the brightest pixels, generally their intensity varies between 220 and 255. After analysing different images of UBIRIS.v2 [20] database, we considered an intensity value of 230 as threshold and used previously mentioned *ERA* [19] algorithm with (5 × 5) filter for detecting the reflection noise. This algorithm searches

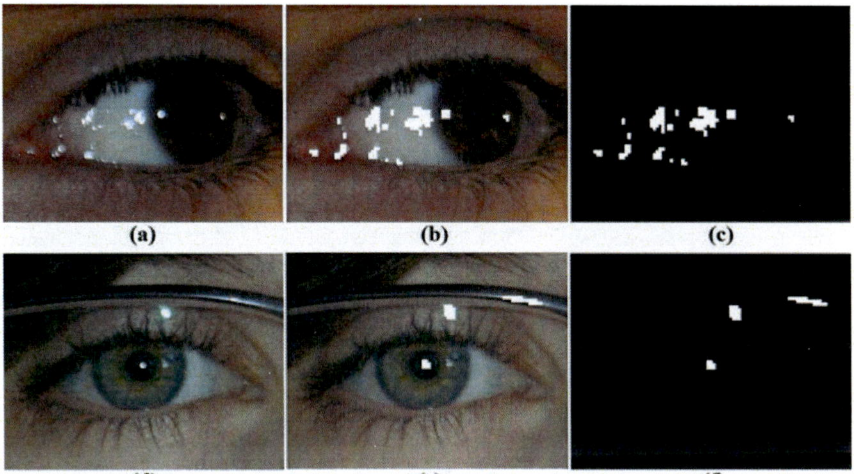

Fig. 4 **a**, **d** Input images; **b**, **e** reflection enlarged (5 × 5 filter) images; **c**, **f** ERA-mask (5 × 5 filter) images

across all three colour spaces of RGB input image for any reflection pixel among the pixels lying to that filter, and if found, all the pixels under the filter area are replaced with intensity value 255 (turned into reflection pixels). It gives a reflection enlarged image which is used as a reflection detection mask (*ERA-mask*), as shown in Fig. 4. In ERA-mask, all the pixels with intensity value ≥230 in any colour component of 'reflection enlarged' image are represented as white, and the rest of the pixels are turned into black. Then the reflection noise (white area of ERA-mask) is removed by our two proposed techniques. Non-reflection area (black area of ERA-mask) of the input image is not affected by our proposed methods.

3.1 Reflection Removal Using Inward Interpolation with ERA-mask (ERA-Inward)

In this approach, we performed inward interpolation on every reflection spot specified by ERA-mask, across all three colour spaces of RGB input image. At first, the ERA-mask is dilated by 1 pixel ($MASK_{\text{Dilated}}$) along north, south, east and west directions by using a lookup table corresponding to a function F applied to cross (five-point stencil) neighbourhood of every ERA-mask reflection (white) pixels [equation (15)].

$$F = \{1 : (M(P) + M(P_N) + M(P_S) + M(P_E) + M(P_W)) > 0\} \quad (15)$$

where M is a logical mask ($MASK_{ERA}$, i.e. ERA-mask). P is a pixel that belongs to reflection spots specified by that mask. P_N, P_S, P_E and P_W are four neighbours of P. F returns 1 if at least one pixel of the five-point stencil (P, P_N, P_S, P_E, P_W) neighbourhood is 1. Then a $MASK_{\text{Boundary}}$ is generated [Eq. (16)] containing only the outer boundary pixels, for every reflection spots specified in the ERA-mask image. More specifically, pixels specified (white) in $MASK_{\text{Boundary}}$ are the nearest non-reflection pixels of every reflection spot specified by the ERA-mask.

$$MASK_{\text{Boundary}} = MASK_{\text{Dilated}} \wedge \neg(MASK_{ERA}) \tag{16}$$

This approach smoothly interpolates the pixels of reflection region in inward direction, from the outer boundary pixels of that reflection region, without updating the outer boundary pixels. It computes *discrete Laplacian* over the reflection regions considering $MASK_{\text{Boundary}}$ as boundary condition. The reflection spot pixels are solved by the linear equation (17), with the help of its existing neighbours. All the interior pixels have four neighbours [P_N, P_S, P_E, P_W], border (non-corner) pixels have three neighbours [(P_S, P_E, P_W) or (P_N, P_E, P_W) or (P_N, P_S, P_W) or (P_N, P_S, P_E)], and corner pixels have two neighbours [(P_N, P_S) or (P_N, P_E) or (P_S, P_E) or (P_S, P_W)].

$$n \cdot X - \sum_{i=1}^{n} I(P_i) = 0 \tag{17}$$

where X is the unknown value of current specified pixel (reflection), coefficient n is the number of known neighbours of that pixel, and each valid P_i is an existing neighbour. The reflection pixels are interpolated without significantly affecting the characteristics of original image. The mask and output image samples of this approach are shown in Fig. 5.

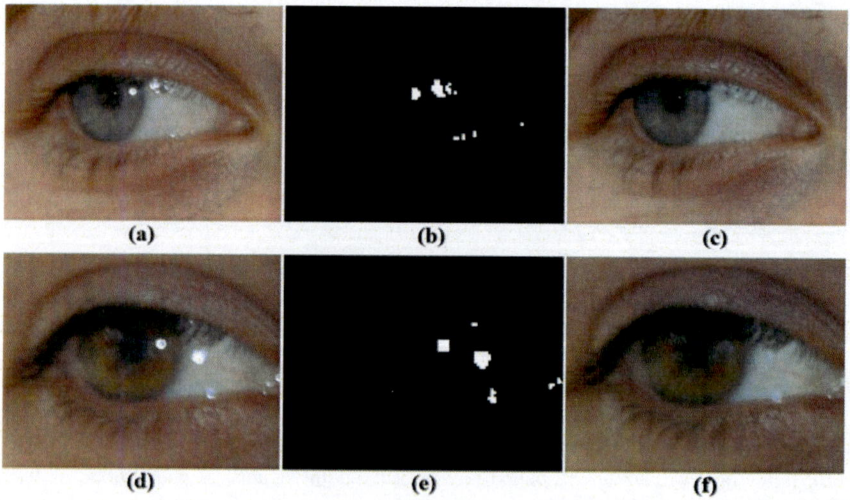

Fig. 5 ERA-inward method: **a**, **d** Input images; **b**, **e** ERA-mask (5 × 5 filter); **c**, **f** output images

3.2 8–Direction Bilinear Interpolation Using ERA-mask (ERA-BilinearExt)

It is an improvement of our previous work *Sohail-ERA-8Dir*. Here we used bilinear interpolation instead of mean to remove the reflection noise. We perform *bilinear interpolation with extension* and fill the reflection regions specified by ERA-mask.

In case of mean-based approach, all the neighbour pixels are given equal significance to determine the reference pixel. Contrarily for interpolation-based approach, the distance between neighbour pixel and reference pixel has an impact on determining the unknown value of reflection pixel. The eight CDP $P_S(x, y_S)$, $P_N(x, y_N)$, $P_E(x_E, y)$, $P_W(x_W, y)$, $P_{\mathrm{SE}}(x_{\mathrm{SE}}, y_{\mathrm{SE}})$, $P_{\mathrm{SW}}(x_{\mathrm{SW}}, y_{\mathrm{SW}})$, $P_{\mathrm{NE}}(x_{\mathrm{NE}}, y_{\mathrm{NE}})$ and $P_{\mathrm{NW}}(x_{\mathrm{NW}}, y_{\mathrm{NW}})$ are identified to interpolate the reflection pixel $P(x, y)$. The interpolation scheme is explained by following Eqs. (18–22). The mask and output image samples of this approach are shown in Fig. 6.

$$f_{N,S} = \frac{y - y_N}{y_S - y_N} \cdot I(P_S) + \frac{y_S - y}{y_S - y_N} \cdot I(P_N) \tag{18}$$

$$f_{E,W} = \frac{x_E - x}{x_E - x_W} \cdot I(P_W) + \frac{x - x_W}{x_E - x_W} \cdot I(P_E) \tag{19}$$

$$f_{\mathrm{NW,SE}} = \frac{\mathrm{Dist}(P_{\mathrm{NW}}, P) \cdot I(P_{\mathrm{SE}}) + \mathrm{Dist}(P, P_{\mathrm{SE}}) \cdot I(P_{\mathrm{NW}})}{\mathrm{Dist}(P_{\mathrm{NW}}, P_{\mathrm{SE}})} \tag{20}$$

$$f_{\mathrm{SW,NE}} = \frac{\mathrm{Dist}(P_{SW}, P) \cdot I(P_{NE}) + \mathrm{Dist}(P, P_{\mathrm{NE}}) \cdot I(P_{SW})}{\mathrm{Dist}(P_{SW}, P_{NE})} \tag{21}$$

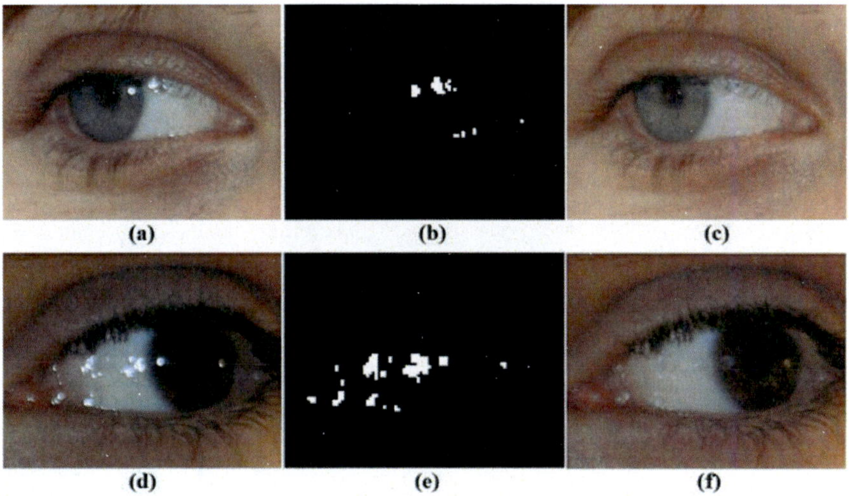

Fig. 6 ERA-BilinearExt method: **a**, **d** input images; **b**, **e** ERA-mask (5 × 5 filter); **c**, **f** output images

$$I(P) = \frac{f_{N,S} + f_{E,W} + f_{\text{NW,SE}} + f_{\text{SW,NE}}}{4} \tag{22}$$

where $\text{Dist}(P_i, P_j)$ measures the Cartesian coordinate distance between two pixels P_i and P_j. Rather than just considering x-axis and y-axis directions, we have also considered the diagonal directions for this bilinear interpolation scheme, to achieve more similarity with its neighbourhood.

4 Experiment Results

Experiments were performed on a system with Windows 10 OS (Core Intel i3, CPU @ 2 GHz, 8 GB RAM) and *MATLAB R2016a* software environment. We used UBIRIS.v2 [20] iris database images for our experiment. UBIRIS is a growing database. The images were captured in motion, at-a-distance, and on visible wavelength, thus considers more realistic noise factors. This database has a lot of variation in iris images, although we have chosen first thousand images based on the sequential order of image ID. Our proposed techniques are suitable for both greyscale and RGB format images and have shown improved results than other discussed approaches (Sect. 2).

ERA algorithm [19] can use either (3 × 3) or (5 × 5) filters. We have used (5 × 5) filter only, as it enlarges the reflection area more than the (3 × 3) filter. The (5 × 5) filter helps to provide a better ERA-mask and thus removes the reflection areas very well.

The outputs of all the six approaches have been displayed in Table 1. It clearly shows that ERA-Inward and ERA-BilinearExt methods have removed the pupil, iris, sclera and glass reflection noises very well, without altering the input image properties. Our approaches are both memory and time efficient.

'He-Bilinear' [17] always gives the worst results, and it always detects some non-reflection areas as reflection noise. 'Aydi-LinearExt' [18] has shown better result than [17], and it easily detects reflection spots surrounded by dark region when applying top-hat filtering with larger structuring element. But it cannot properly detect reflection spots embedded inside brighter areas. Our previous direction-wise traverse approach [19] is quite similar with respect to reflection detection, but our new approaches have shown improved results due to distance coefficients of interpolation-based reflection removal. The quality of output images has also been compared in terms of their similarity to input images, in Table 2. The similarity between input and output images has been measured with the help of structural similarity index measurement (SSIM) [21]. SSIM index is measured with three terms—luminance (l), contrast (c) and structure (s) along with their corresponding relative importance α, β, γ (> 0):

$$\text{SSIM}(I_x, I_y) = [l(I_x, I_y)]^{\alpha} \cdot [c(I_x, I_y)]^{\beta} \cdot [s(I_x, I_y)]^{\gamma} \tag{23}$$

Our purpose was to remove reflection spots without changing the characteristics of non-ideal iris images like Ubiris database. This Structural similarity comparison proves that our new approaches have performed better than our previous 'direction-wise traverse' approach [19] as well as 'He-Bilinear' [17] and 'Aydi-LinearExt' [18]. Between the two new approaches, 'ERA-Inward' shows better performance than 'ERA-BilinearExt', for most of the cases.

Table 1 Reflection removed image outputs comparison-related works versus proposed methods

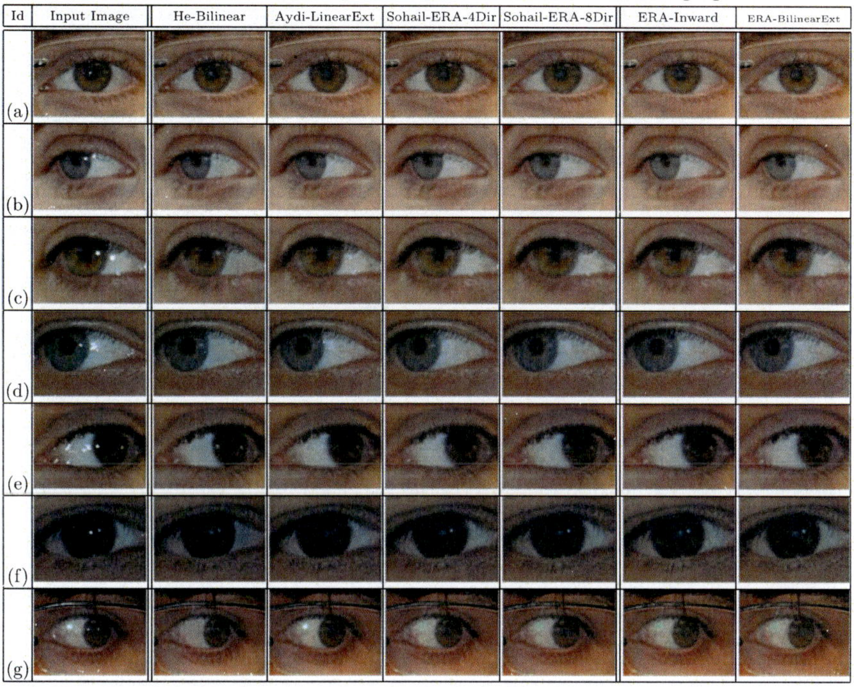

a Bright image, pupil reflection. **b** Bright image, off-angle (left), iris and sclera reflection. **c** Medium bright image, off-angle (left), partially blurred, iris and sclera reflection. **d** Medium bright image, off-angle (left), iris and sclera reflection. **e** Medium bright image,off-angle (right), iris and sclera reflection. **f** Poor contrast image, pupil and iris reflection. **g** Medium bright image, off-angle (right), iris, sclera and glass reflection

Table 2 Comparison of **structural similarity** (using SSIM index) to input image

Id	He-Bilinear	Aydi-LinearExt	Sohail-ERA-4Dir	Sohail-ERA-8Dir	ERA-inward	ERA-BilinearExt
a	0.98762733	0.98834462	0.99102518	0.99103507	0.99157529	0.99146441
b	0.98882453	0.99171079	0.99316342	0.99310177	0.99371242	0.99342029
c	0.97848009	0.99002977	0.99022583	0.99044474	0.99145340	0.99107121
d	0.98325969	0.98521130	0.98866971	0.98863516	0.98908003	0.98884698
e	0.97678696	0.97914198	0.97929765	0.97948692	0.98094977	0.98012577
f	0.98818954	0.99456806	0.99705331	0.99705351	0.99740573	0.99717793
g	0.97428097	0.98007689	0.98278922	0.98214907	0.98302111	0.98301230

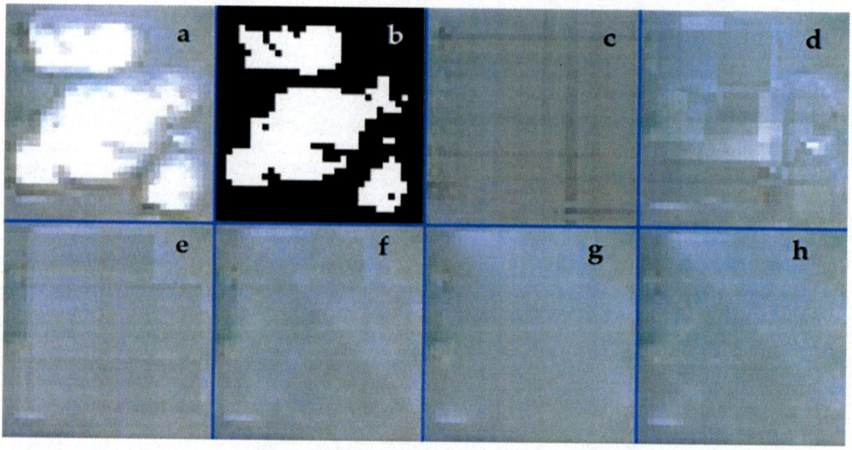

Fig. 7 Pixel-level analysis of reflection removal: **a** input image, **b** ERA-mask image, **c** He-Bilinear output image, **d** Aydi-LinearExt output image, **e** Sohail-ERA-4Dir output image, **f** Sohail-ERA-8Dir output image, **g** ERA-inward output image, **h** ERA-BilinearExt output image

In Fig. 7, we have also shown a pixel-level analysis of a reflection spot and reflection removal by different approaches. It helps us to understand how well (view perspective) our proposed methods have removed the reflection areas in comparison with the other described approaches.

5 Conclusion

We have developed two new strategies for specular reflection removal and compared their robustness and efficiency with four other existing techniques, two of them are our previous works. We have improved our performance with the proposed approaches, based on reflection removal (second phase). But reflection detection has always been a challenge. We applied a threshold pixel intensity of 230 on ERA-enlarged image for reflection detection. Though most of the reflection spots have pixel intensity$\geq$230, it may vary especially for poor contrast images. Cooperation of ERA algorithm has shown improved results for detection. In future, we will try to implement new strategies for reflection detection suitable for any non-ideal scenario. The proposed methods were applied on the UBIRIS.v2 database which is one of the best non-ideal data sets publicly available on the Internet. SSIM index comparison also justifies that the suggested methods have served the purpose of reflection removal retaining the characteristics of original image.

References

1. Li P, Zhang R (2010) The evolution of biometrics. In: Proceedings of the IEEE international conference on anti-counterfeiting security and identification in communication, China, pp 253–256
2. Hanho Sung J, Park LJ, Lee Y (2004) Iris recognition using collarette boundary localization. In: Proceedings of the 17th international conference on pattern recognition, vol 4, pp 857–860
3. Kumar A, Potnis A, Singh AP (2016) Iris recognition and feature extraction in iris recognition System by employing 2D DCT. Int Res J Eng Technol 3(12):503–510
4. Laughton MA, Warne DJ (2003) Electrical engineer's reference book, 16th edn. Elsevier Ltd., pp 21/22 (Chapter-Lighting, Subsection- Glare)
5. Shin KY et al (2012) New iris recognition method for noisy iris images. Pattern Recogn Lett 33(8):991–999. https://doi.org/10.1016/j.patrec.2011.08.016
6. Jarjes AA, Wang K, Mohammed GJ (2010) Gvf snake-based method for accurate pupil contour detection. Inf Technol J 9(8):1653–1658. https://doi.org/10.3923/itj.2010.1653.1658
7. Jamaludin S, Zainal N, Zaki WMDW, The removal of specular reflection in noisy iris image. J Telecommun Electron Comput Eng 8(4)
8. Wang N et al (2014) Toward accurate localization and high recognition performance for noisy iris images. Multimedia Tools Appl 71. https://doi.org/10.1007/s11042-012-1278-7
9. Bertalmio M, Bertozzi AL, Sapiro G (2001) Navier-stokes, fluid dynamics, and image and video inpainting. In: Proceedings- IEEE computer society conference on computer vision and pattern recognition (CVPR), USA, vol 1. https://doi.org/10.1109/CVPR.2001.990497
10. Reddi N, Rattani A, Derakshani R (2016) A robust scheme for iris segmentation in mobile environment. IEEE symposium on technologies for homeland security, USA, May, In, pp 1–6
11. Chui CK (2009) An MRA approach to surface completion and image inpainting. Appl Comput Harmon Anal 26(2):270–276. https://doi.org/10.1016/j.acha.2008.05.001
12. Sahmoud SA, Abuhaiba IS (2013) Efficient iris segmentation method in unconstrained environments. Pattern Recognit 46(12):3174–3185. https://doi.org/10.1016/j.patcog.2013.06.004
13. Raffei AFM et al (2013) Feature extraction for different distances of visible reflection iris using multiscale sparse representation of local Radon transform. Pattern Recogn 46(10):2622–2633. https://doi.org/10.1016/j.patcog.2013.03.009
14. Seong-Taek L, Tae-Ho Y, Kyeong-Seop K, Kee-Deog K, Wonse P (2010) Removal of specular reflections in tooth color image by perceptron neural nets. In: 2nd international conference on signal processing systems, Dalian, China, vol 1, pp V1-285-V1-289
15. Sankowski W, Grabowski K, Napieralska M, Zubert M, Napieralski A (2010) Reliable algorithm for iris segmentation in eye image. Image Vis Comput 28(2):231–237. https://doi.org/10.1016/j.imavis.2009.05.014
16. Tan C-W, Kumar A (2011) Automated segmentation of iris images using visible wavelength face images. CVPR 2011 Workshops, Colorado Springs, pp 9–14
17. He Z, Tan T, Sun Z, Qui X (2009) Toward accurate and fast iris segmentation for iris biometrics. IEEE Trans Pattern Anal Mach Intell 31(9). https://doi.org/10.1109/TPAMI.2008.183
18. Aydi W, Masmoudi N, Kamoun L (2011) New corneal reflection removal method used in iris recognition system. Int J Electron Commun Eng 5(5)
19. Sohail MdA, Ghosh C, Mandal S (2021) An efficient approach to remove specular reflection from non-ideal eye image. Proceedings of the international conference on computing and communication systems 2021:337–347. https://doi.org/10.1007/978-981-33-4084-8_32
20. Proença H et al (2010) The UBIRIS.v2 : a database of visible wavelength iris images captured on-the-move and at-a-distance. IEEE Trans Pattern Anal Mach Intell 32(8). https://doi.org/10.1109/TPAMI.2009.66
21. Zhou W et al (2004) Image qualifty assessment: from error visibility to structural similarity. IEEE Trans Image Process 13(4). https://doi.org/10.1109/TIP.2003.819861

Analyzing Behavior to Detect Cervical Cancer

Rup Kumar Deka

Abstract Cervical cancer is preventable if one can detect it at an early stage. Due to cervical cancers, the human body imparts various behaviors and seven different determinants of this behavior. Behaviors can be conscious and also subconscious. And also, all expressing behaviors might not be completely correlated. In this work, a strategy has been proposed to exclude non-correlated features to maintain or enhance the detection accuracy of cervical cancer at an early stage. Experimental results show a maximum of 94.4% accuracy, and thus, it proves the efficacy of the proposed work.

Keywords Cervical cancer · Correlation · Behavior analysis

1 Introduction

Cervical cancer is about 6–29% in India among all other cancers in women.[1] It is a concerning public health problem in the world. Fatality due to cervical cancer can be preventable if it is detected early. As per medical science, cervical cancer progresses slowly, and there are lots of identifiable cytological precursors. Cervical cancer prevention depends upon a pap-smear test, visual inspection of the cervix with acetic acid (VIA) test, and human papillomavirus (HPV) vaccination [1, 2]. But, lack of knowledge of the symptoms and various testing procedures, the low scope of screening in regions, embarrassment regarding the testing procedure can fuel more fatality due to cervical cancer [3, 4].

[1] https://www.ncbi.nlm.nih.gov/pmc/articles/PMC5234166/, Accessed on May 30, 2021.

R. K. Deka (✉)
Department of Computer Science and Engineering, Assam Don Bosco University, Azara Campus, GuwahatiAssam, Kamrup 781017, India
e-mail: rup.deka@dbuniversity.ac.in

S. Basu et al. (eds.), *Proceedings of International Conference on Frontiers in Computing and Systems*, Lecture Notes in Networks and Systems 404,
https://doi.org/10.1007/978-981-19-0105-8_28

1.1 Background Study

Machine learning algorithms are applied to detect various cancer diseases [5–8]. Singh and Goyal [9] have published their performance analysis of different machine learning algorithms to detect cervical cancer on feature-based datasets available in various repositories. William et al. [10] researched image-based automatics cervical cancer screening from pap-smear images using machine learning algorithms. Lu et al. [11] developed an ensemble approach to diagnose cervical cancer. In their work, they have strategized a voting technique, and also to improve the performance, they addressed a data correction mechanism. Ghoneim et al. [12] used a convolutional neural network and extreme learning methods to detect cervical cancer using image datasets. Khamparia et al. [13] described a deep learning system to detect and classify cervical cells using transfer learning. Mitra et al. [14] and Li et al. [15] described few image base techniques to deal with the cervical cancer detection processes.

1.2 Motivation

Human behaviors are correlated. So, having cervical cancer, a patient can impart various prominent, non-prominent behaviors along with non-related behaviors. It is necessary to establish the correlation among the behaviors to detect cervical cancer as early as possible.

1.3 Contribution

In this work, contributions are the following.

- Description of the importance of behavior analysis in the detection of cervical cancer.
- And, achievement of maximized accuracies of the cervical cancer detection process by excluding non-correlated features.

2 Behaviors Related to Cervical Cancer and Dataset Description

By analyzing human behavior, we can evaluate the risk of having cervical cancer. Behaviors are categorized into various variables or attributes to learn using a machine learning algorithm. As per sociology, the health belief model (HBM), protection motivation theory (PMT), theory of planned behavior (TPB), social cognitive theory (SCT), etc., are well-studied behavior models.

Table 1 Dataset description

Total number of proper feature category	Features			Class construction		
9	7 determinants	Perception (2 questions)	Translated into 19 questions and all numeric answers to these 19 questions are attributes	Cervical cancer persons (Class 1)	21 respondents	Total 72 samples
		Intention (2 questions)				
		Motivation (2 questions)				
		Subjective norm (2 questions)				
		Attitude (2 questions)		Non-cervical cancer persons (Class 0)	51 respondents	
		Social support (3 questions)				
		Empowerment (3 questions)				
	Behavior itself (3 questions)					

Perception due to threat of illness and counteract any health threat included in HBM. Lack of knowledge of cervical cancer, perceived seriousness, and susceptibility contribute to high-risk sexual behavior. As per the TPB model, intention can affect our behavior.

The intention is due to attitude, subjective norms, and behavioral control. SCT model is of goals, expectations, self-efficacy, and motivation. Machmud and Wijaya [16] described the dataset as having various behavioral components, as shown in Table 1 with other details of classes.

3 Problem Statement and Work Strategy

3.1 Problem Statement

To detect cervical cancer using machine learning algorithms based on human behaviors with better accuracy.

3.2 Work Strategy

Figure 1 shows the plot of class data. From this figure, we can visualize that more features are correlated, and a few of these are less correlated with all other features or

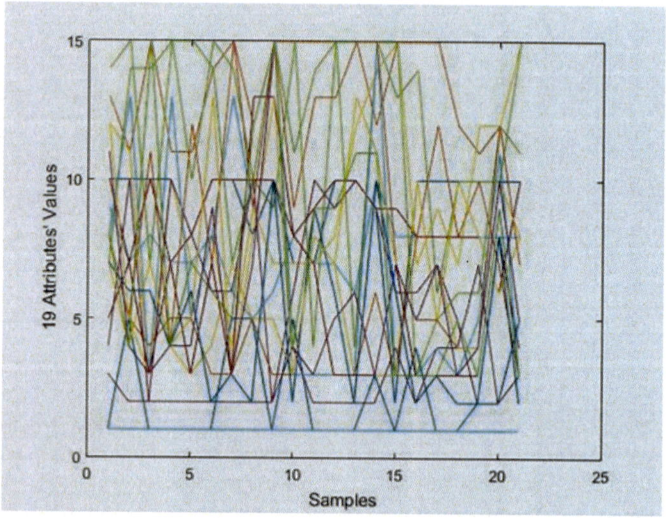

The Class of Cervical Cancer Persons

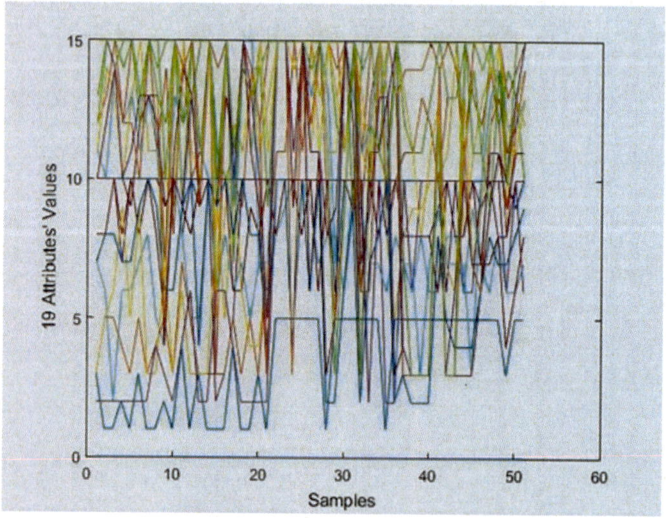

The Class of Non-Cervical Cancer Persons

Fig. 1 Plot of samples versus attributes of two classes

Table 2 Conceptual visualization of correlation links among features

S. no.	Features	Number of correlation link
1	w	2
2	x	1
3	y	1
4	z	0

attribute columns for both classes. So, it is necessary to find out how many features correlate with each other or not. There are 19 attributes in the dataset description. But in the figure, in the Y-axis, the values of those 19 attributes for different samples (X-axis) have been shown. And, the maximum value of any attribute variable in the dataset for sample data is 15. So, it is showing 15 on Y-axis. There are 19 lines (19 colors) in each subfigure.

For example, suppose there are four (4) features, w, x, y, and z. Table 2 shows the correlation links among these features or attributes. Supposedly, w has correlation links with x and y, and z has no correlation links with any three, then the correlation table will be as shown below in Table 2.

So, during the learning process, if a feature, z, is excluded, then there will be no effect on the accuracy value, and rather than that, the accuracy value might increase. So, all the features of the two classes are verified. Based on the number of correlation links among the attribute columns, having at least one correlation link with any other attribute column is included in the learning process to detect cervical cancer.

4 Experiment and Results

4.1 Experimental Set-up

The set-up of the whole experiment is on a workstation with a 2.10 GHz processor, 4 GB RAM, and a 64 bit Windows 10 operating system. We use MATLAB R2015a 64-bit edition for these experiments.

4.2 Results

The initial experiment calculates the correlation among the features, as shown in Table 3. Figure 2 depicts the proposed working methodology.

After getting the required correlated features, the dataset has preprocessed to eliminate the non-essential attribute columns and performed the learning and classification using various machine learning models. The results of the accuracies compared to all featured included accuracies are shown in Table 4. Table 5 lists the evaluated

Table 3 Correlation links among features

S. no.	Feature variable name	Correlation link	Inclusion/exclusion in learning process
1	behavior_sexualRisk	1	Inclusion
2	behavior_eating	1	Inclusion
3	behavior_personalHygine	0	Exclusion
4	intention_aggregation	0	Exclusion
5	intention_commitment	0	Exclusion
6	attitude_consistency	0	Exclusion
7	attitude_spontaneity	2	Inclusion
8	norm_significantPerson	0	Exclusion
9	norm_fulfillment	2	Inclusion
10	perception_vulnerability	1	Inclusion
11	perception_severity	1	Inclusion
12	motivation_strength	1	Inclusion
13	motivation_willingness	2	Inclusion
14	socialSupport_emotionality	0	Exclusion
15	socialSupport_appreciation	1	Inclusion
16	socialSupport_instrumental	2	Inclusion
17	empowerment_knowledge	2	Inclusion
18	empowerment_abilities	2	Inclusion
19	empowerment_desires	1	Inclusion

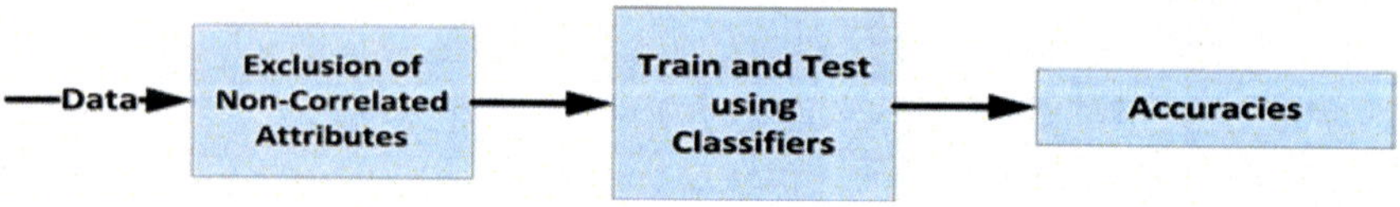

Fig. 2 Proposed framework

Table 4 Classification accuracies

S. no.	Machine learning models	Accuracy [all features include]	Accuracy [correlated features only]
1	Linear SVM	90.3	94.4
2	Fine KNN	86.1	87.5
3	Complex tree	76.4	83.3
4	Cubic SVM	90.3	91.7
5	Cubic KNN	84.7	86.1
6	Ensemble boosted trees	87.5	90.3

Table 5 Evaluated values of confusion matrix, precision, recall, *F*-measure score for all the six models

For correlated features included only

					TPR/FPR (%)	Accuracy (%)	Precision (%)	Recall (%)	F-measure (%)
Linear SVM	Class 0	True class	51	0	100/0	94.4	92.7	100	96.2
Fine KNN			47	4	92.2/7.8	87.5	90.3	92.2	91.2
Complex tree			47	4	92.2/7.8	83.3	85.4	92.2	88.7
Cubic SVM			50	1	98/2	91.7	90.9	98	94.3
Cubic KNN			51	0	100/0	86.1	83.6	100	91.1
Ensemble boosted trees			49	2	96.1/3.9	90.3	90.7	96.1	93.3
Linear SVM	Class 1		4	17	80.9/19.1				
Fine KNN			5	16	76.2/23.8				
Complex tree			8	13	61.9/38.1				
Cubic SVM			5	16	76.2/23.8				
Cubic KNN			10	11	52.4/47.6				
Ensemble boosted trees			5	16	76.2/23.8				
			Class 0	Class 1					
			Predicted class						

Table 6 Comparison with Akter et al. [7] and Machmud and Wijaya [16]

Author(s)	Aim of work	Technique used	Maximum accuracy
Akter et al. [7]	Cervical cancer detection based on behaviors	Decision tree, random forest, and XGBoost used	Accuracy is 93.3%
Machmud and Wijaya [16]	Cervical cancer detection based on behaviors	Naïve Bayes and logistic regression classifiers used	91.6% and 87.5%
This proposed work	Cervical cancer detection based on high correlated behavior	Linear SVM, Fine KNN, Complex tree, Cubic SVM, Cubic KNN, and ensemble boosted used	Maximum accuracy obtained 94.4%

values of confusion matrix, precision, recall, and the F-measure score for all the six classifying models.

4.3 Comparisons

Table 6 shows the comparison of this proposed work with Akter et al. [7] and Machmud and Wijaya [16] on behavior analysis to detect cervical cancer at an early stage.

5 Conclusion

Cervical cancer is a problematic health issue in women. The seriousness of this threat becomes risky when there is no adequate knowledge about the disease and low screenings in rural regions of India. But, all human beings impart various behavioral traits consciously or subconsciously. Many of those behaviors are correlated. Analyzing the correlated behaviors to detect cervical cancer as early as possible using machine learning models can reduce the risk of fatality.

References

1. Banura C, Mirembe FM, Katahoire AR, Namujju PB, Mbidde EK (2012) Universal routine HPV vaccination for young girls in Uganda: a review of opportunities and potential obstacles. Infect Agents Cancer 7(1):1–6
2. Balogun MR, Odukoya OO, Oyediran MA, Ujomu PI (2012) Cervical cancer awareness and preventive practices: a challenge for female urban slum dwellers in Lagos, Nigeria. Afr J Reprod Health 16(1)
3. Domingo EJ, Noviani R, Noor MRM, Ngelangel CA, Limpaphayom KK, Van Thuan T, Louie KS, Quinn MA (2008) Epidemiology and prevention of cervical cancer in Indonesia, Malaysia, the Philippines, Thailand, and Vietnam. Vaccine 26:M71–M79
4. Yanikkerem E, Goker A, Piro N, Dikayak S, Koyuncu FM (2013) Knowledge about cervical cancer, pap-test, and barriers towards cervical screening of women in Turkey. J Cancer Educ 28(2):375–383
5. Nithya B, Ilango V (2019) Evaluation of machine learning-based optimized feature selection approaches and classification methods for cervical cancer prediction. SN Appl Sci 1(6):1–16
6. Suguna C, Balamurugan SP (2020) An extensive review on machine learning and deep learning based cervical cancer diagnosis and classification models. J Comput Theor Nanosci 17(12):5438–5446
7. Akter L, Islam MM, Al-Rakhami MS, Haque MR (2021) Prediction of cervical cancer from behavior risk using machine learning techniques. SN Comput Sci 2(3):1–10
8. Adem K, Kiliçarslan S, Cömert O (2019) Classification and diagnosis of cervical cancer with stacked autoencoder and softmax classification. Expert Syst Appl 115:557–564
9. Singh SK, Goyal A (2020) Performance analysis of machine learning algorithms for cervical cancer detection. Int J Healthc Inf Syst Inform (IJHISI) 15(2):1–21
10. William W, Ware A, Basaza-Ejiri AH, Obungoloch J (2018) A review of image analysis and machine learning techniques for automated cervical cancer screening from pap-smear images. Comput Methods Programs Biomed 164:15–22
11. Lu J, Song E, Ghoneim A, Alrashoud M (2020) Machine learning for assisting cervical cancer diagnosis: an ensemble approach. Futur Gener Comput Syst 106:199–205
12. Ghoneim A, Muhammad G, Hossain MS (2020) Cervical cancer classification using convolutional neural networks and extreme learning machines. Futur Gener Comput Syst 102:643–649
13. Khamparia A, Gupta D, de Albuquerque VHC, Sangaiah AK, Jhaveri RH (2020) Internet of health things-driven deep learning system for detection and classification of cervical cells using transfer learning. J Supercomput 1–19
14. Mitra S, Das N, Dey S, Chakraborty S, Nasipuri M, Naskar MK (2021) Cytology image analysis techniques toward automation: systematically revisited. ACM Comput Surv (CSUR) 54(3):1–41
15. Li Y, Chen J, Xue P, Tang C, Chang J, Chu C, Ma K, Li Q, Zheng Y, Qiao Y (2020) Computer-aided cervical cancer diagnosis using time-lapsed colposcopic images. IEEE Trans Med Imaging 39(11):3403–3415
16. Machmud R, Wijaya A (2016) Behavior determinant based cervical cancer early detection with machine learning algorithms. Adv Sci Lett 22(10):3120–3123

Rapid Diagnosis of COVID-19 Using Radiographic Images

Debangshu Chakraborty and Indrajit Ghosh

Abstract A COVID-19 patient suffers from blockage of breathing and chest pain at a critical condition due to the formation of fibrosis in the lungs and needs emergency lifesaving treatment. Before starting an adequate treatment, a confirmed diagnosis of COVID-19 is a mandatory criterion. For a patient with critical respiratory syndrome, rapid and precise diagnosis is a prime challenge. Different manual methods of clinical diagnosis are in practice. However, these manual techniques suffer from serious drawbacks such as poor sensitivity, false negative results, and high turn-around time. The diagnosis based on the radiographic image (X-ray or computed tomography) of infected lungs is another clinical method for rapid diagnosis of COVID-19. However, it requires an expert radiologist for precise diagnosis. Instead of a prolonged clinical process, an alternative way of rapid diagnosis is the only way of some lifesaving. As an elegant solution, some radiographic image-based automated diagnostic systems have been suggested using deep learning techniques. However, they suffer from some unavoidable limitations concerned with deep learning. This paper suggests a user-friendly system for instant diagnosis of COVID-19 using radiographic images of infected lungs of a critical patient. The model is designed based on classical image processing techniques and machine learning techniques that have provided low complexity but a very high accuracy of 98.51%. In this pandemic situation, such a simple and instantaneous diagnostic system can become a silver lining to compensate for the scarcity of expert radiologists.

Keywords COVID-19 · Rapid diagnosis · Radiographic image-based system · Histogram

D. Chakraborty · I. Ghosh (✉)
Department of Computer Science, Ananda Chandra College, Jalpaiguri 735101, India
e-mail: ighosh2002@gmail.com

S. Basu et al. (eds.), *Proceedings of International Conference on Frontiers in Computing and Systems*, Lecture Notes in Networks and Systems 404,
https://doi.org/10.1007/978-981-19-0105-8_29

1 Introduction

The disease caused by the coronavirus (COVID-19) is a highly infectious disease that has spread over more than 200 countries. Because of the pandemic, more than 3.7 million people have lost their life worldwide. Global health is facing an unforeseen threat.

Once COVID-19 infects a person, several symptoms are observed. The most commonly found symptoms are fever, dry cough, tiredness, nasal congestion, conjunctivitis, aches and pains, headache, sore throat, diarrhea, loss of taste, loss of smell, discoloration of fingers and toes, skin rashes. All these symptoms are not fatal. However, in critical condition, the patient suffers from blockage of breathing and chest pain due to the formation of fibrosis in the lungs and needs emergency and adequate treatment as per COVID-19 protocol. For a patient with critical respiratory syndrome, rapid and precise diagnosis is a prime challenge.

Different methods of clinical diagnosis are in practice [1, 2]. The reverse transcriptase-polymerase chain reaction (RT-PCR) is a benchmark method for testing COVID-19. This manual technique suffers from serious disadvantages such as very poor sensitivity, false negative results, and high turn-around time [3]. Another variant of this technique is real-time reverse transcriptase-polymerase chain reaction (rRT-PCR). It is a popular laboratory method that uses an RT-PCR machine. The RT-PCR machine is very costly and takes nearly three hours for report generation. However, clinical laboratories take six to eight hours on average to generate a confirmed diagnosis report [4].

Clinical laboratory testing based on radiography is the third alternative method for rapid diagnosis of COVID-19. In this method, the radiologists study the chest X-ray or computed tomography (CT) image for diagnosis. The lungs of the critical patients are highly affected, and some abnormalities are observed in the lungs. These abnormalities are visually detectable in the chest radiographic images [5–7]. For example, the chest radiographic images of the lung of a healthy person and a person suffering from COVID-19 are presented in Figs. 1a and b, respectively. Similarly, the CT images of a healthy person and a COVID-19-infected person are depicted in Figs. 1c and d, respectively.

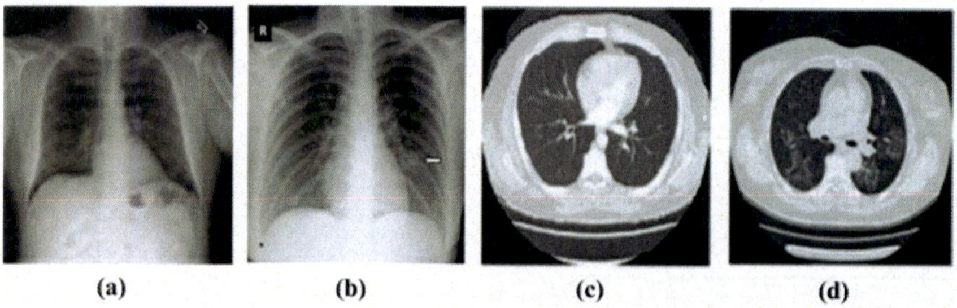

Fig. 1 **a** Chest X-ray of a healthy person. **b** Chest X-ray of a person with COVID-19. **c** CT image of a healthy person. **d** CT image of a person with COVID-19

Several remarkable image-based diagnosis systems have been proposed for COVID-19, including Linda Wang et al., Shuai Wang et al., Hussain et al., Xin Li et al., Ophir Gozes, Pandit et al., Ismael et al. [8–14]. Linda Wang et al. proposed a system that uses a tailor-made deep learning model called Covid-Net. Shuai Wang et al. designed a system that has used the convolutional neural network (CNN) with inception migration learning technique for diagnosis. Hussain et al. have proposed a novel model CoroDet based on CNN. All these COVID-19 diagnosis systems are designed using deep learning models. Like all other deep learning-based systems, these systems suffer from several drawbacks like lack of substantial training datasets, systems with very high configuration, more extensive training time, back box behavior.

This paper presents a user-friendly radiographic image-based rapid diagnosis system for COVID-19. The proposed work used the histogram-based features extracted from each image to train the MLP classifier. After training, the performance of the model is evaluated. When the best possible accuracy is achieved, the classifier is deployed in the real world, packaged in an application. This system has been designed with a completely different approach from the systems mentioned above. Instead of using deep learning models, this system uses classical image analysis techniques and machine learning techniques to achieve the automatic diagnosis and can easily bypass the limitations of the deep learning-based systems with better accuracy. Replacing the deep learning-based systems with systems designed using the machine learning techniques would lower the cost of diagnosis and be a graceful alternative to underdeveloped and developing countries [15].

2 Methodology

2.1 Datasets

The success of any intelligent image-based system depends on the availability of an extensive and robust dataset for its training and testing. A combination of five global benchmark datasets has been used for this system. The first dataset is taken from the COVID-19 Image Data Collection of Cornell University [16]. The second one is obtained from COVID-19 chest X-ray dataset initiative [17]. The third and the fourth datasets are collected from COVID-19 RADIOGRAPHY DATABASE [18] and Radiological Society of North America (RSNA) [19], respectively. The fifth dataset is obtained from Actualmed COVID-19 chest X-ray dataset initiative [20]. All these datasets contain the chest X-ray images of a normal person and persons infected with the COVID-19.

Finally, these five datasets are combined and randomly shuffled to make a single unbiased and robust master dataset. This combined master dataset consists of X-ray images of 1374 COVID-19 positive patients and 1341 normal persons, respectively.

The master dataset is split into two parts: training set and testing set. The training dataset consists of randomly selected 80% of the master dataset, and the rest 20% is used as a test dataset.

2.2 System Design

The proposed system is a radiographic image-based rapid diagnosis system for COVID-19. It consists of two modules. The first one is the machine vision module that constructs the histogram from the images and extracts the relevant features. The second one is the machine learning module that consists of intelligent classifier to diagnose COVID-19, considering the extracted features as inputs (see Fig. 2).

Histogram-based Feature Extraction. The histogram of an image is a function that maps intensity of each pixel to the number of times that occurs in the image [21]. A two-dimensional image is a collection of pixels. Each pixel has a specific intensity within a specific range of values. The histogram is a discrete function that is represented by $H(r_k) = n_k$, where r_k is the kth intensity of the image and n_k is the occurrence of r_k intensity in the image. In the histogram, the x-axis corresponds to different pixel intensities and the y-axis represents their occurrence. As an example,

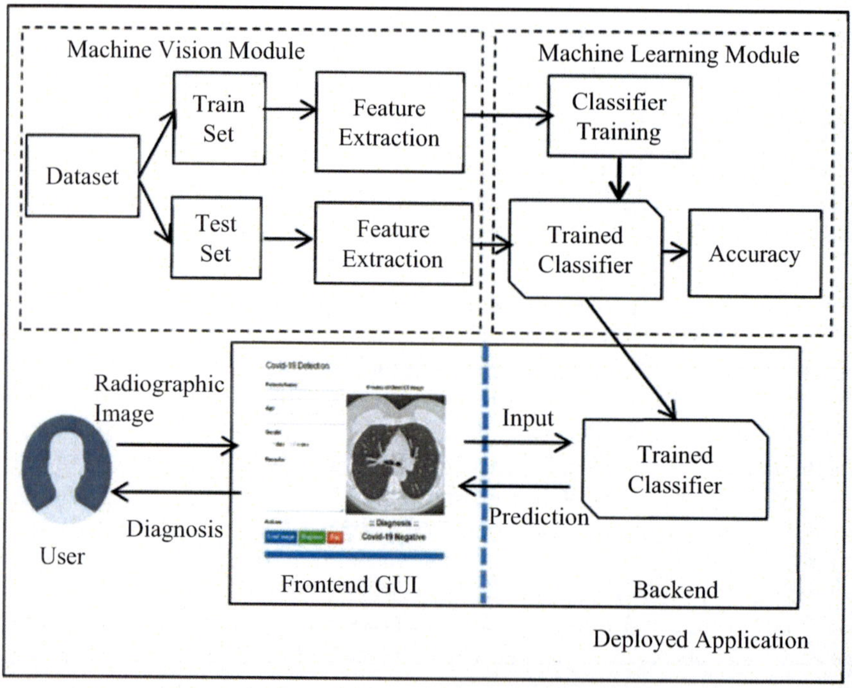

Fig. 2 System design

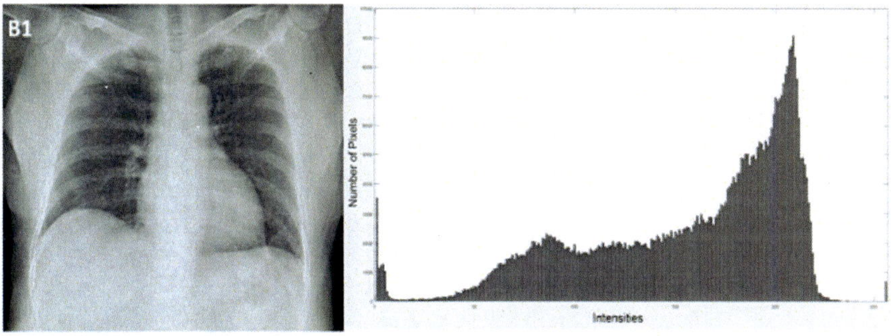

(a) (b)

Fig. 3 **a** Chest X-ray image, **b** respective histogram

a chest X-ray image and the corresponding histogram are presented in Fig. 3a and b, respectively.

Four statistical moments of the histogram of the X-ray image are considered as features. These four statistical moments are standard deviation, skewness, variance, and kurtosis, as these moments can represent the entire histogram. After the construction of the histograms, these four statistical moments are extracted that are used as features. These four moments are used because they can characterize a histogram, and when two histograms have to be compared, it makes more sense to compare their moments [22]. The short descriptions of the moments are presented below.

Standard deviation defines the nature of spread out of the histogram. It represents to which degree the histogram is spread out from the mean of the histogram. It is the square root of variance. Skewness is characterized by the degree to which the histogram differs from a normal distribution. It is the measure of symmetricity of the histogram. The variance represents how spread the histogram is. To calculate the variance, first, the mean of the histogram is calculated and then from this, the frequency (occurrence) of each pixel intensity is subtracted. These differences are squared, and their average is the variance of the histogram. Kurtosis of a histogram is the flatness of the top of the histogram compared to a histogram that has normal frequency distribution (normal distribution) [23].

All these features have distinct significance and represent a different aspect of the histogram. Combining these four values, we can get a good representative of the histogram, which in turn is representing the image. These features are computed from each image and are normalized between 0 and 1. The normalization ensures that small numbers do not override by large numbers, and the training process is not disrupted by premature saturation of hidden layer neurons [24, 25].

Designing the Classifier. The machine learning module consists of intelligent classifier to diagnose COVID-19, considering the extracted features as inputs. The classifier used here is a multilayer perceptron (MLP). The proposed MLP consists of an input layer, 98 hidden layers, and an output layer.

The input layer consists of four neurons for four input features. The output layer contains two neurons where each corresponding to one diagnosis of the healthy person and COVID-19 positive patient. To design the hidden layers, a trial and error method was adopted with the same dataset. Experimentally, it has been observed that the accuracy of the system was increased with the increasing number of hidden layers, and a maximum accuracy was achieved with 98 hidden layers having 92 neurons in each layer. Thus, the proposed system was designed with 98 hidden layers with 92 neurons in each layer. About the architecture, each neuron of one layer is connected to all neurons of the next layer, but a neuron has no connections with other neurons of the same layer. The back-propagation algorithm has been used as the learning algorithm.

In the training phase, the set of statistical features was extracted against each image of a class. Each image class is associated with a set of features. The values of the input features and output class pair of all images of the training set are fed to the classifier for training. The classifier learns from the set of features and its class. The task of assigning the appropriate weight is performed in the training phase of the classifier.

Implementation. The system consists of two components: a backend module and a frontend module. The backend module contains the trained MLP classifier. This backend is implemented in Python programming language using TensorFlow. A webpage is designed as the frontend that uses a user-friendly graphical user interface (GUI). The frontend GUI is developed using frontend languages (HTML, CSS, Bootstrap, JS). This GUI communicates with the backend by using a Python library Eel and allows the frontend to off-load the user from knowing the underlying details of the system. It enables even a layman user to perform rapid COVID-19 diagnosis. Figure 4 shows the graphical user interface of the system.

This is a user-friendly system. The chest X-ray image can be easily loaded by clicking on the "Load Image" button. The classifier will predict the output class of the loaded image when the "Diagnose" button is pressed. Based on the prediction of the classifier, a diagnosis will be displayed on the screen and can be printed by clicking on the "Print" button.

2.3 Performance Evaluation

Once the classifier was trained, the performance of the system was evaluated using the testing dataset. Several standard metrics are in use to measure the performance of a classifier [26–29]. For the present system, the performance of a classifier has been measured in terms of four well-accepted metrics: accuracy (A_c), Cohen's kappa (k), precision (Pr), and recall (R).

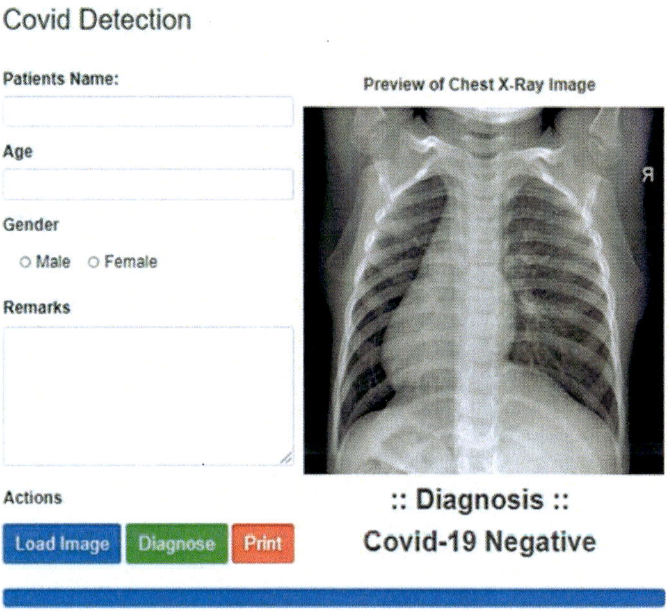

Fig. 4 Graphical user interface (GUI) of the system

The accuracy (A_c) can be mathematically defined as [27]:

$$A_c = \frac{1}{n}\sum_{i=1}^{n} I.(y_p = y_a) \tag{1}$$

where y_p is the predicted value, y_a is the actual value of an observation, and n is the number of samples. The function $I.(y_p = y_a)$ returns 1 if $(y_p = y_a)$ is true and zero otherwise.

Cohen's kappa (k) is a well-established metric for the performance measure of a classifier. For m classes, the value of Cohen's kappa (k) can be derived as [27]:

$$\text{Cohen's kappa } (k) = \frac{N.\sum_{i=1}^{m} C_{ii} - \sum_{i=1}^{m} C_{i_{\text{corr}}}.C_{i_{\text{pred}}}}{N^2 - \sum_{i=1}^{m} C_{i_{\text{corr}}}.C_{i_{\text{pred}}}} \tag{2}$$

where $\sum_{i=1}^{m} C_{ii}$ is the total number of instances correctly predicted, $C_{i_{\text{corr}}}$ is the number of instances correctly classified for class i, $C_{i_{\text{pred}}}$ is the total number of instances predicted as class i, and N is the total number of patterns [27].

Precision (Pr) is the ratio of true positive (Tp) to the sum of true positive (Tp) and false positive (Fp) [28]. A true positive outcome of a classifier indicates a correct prediction of the positive class, and a true negative outcome of a classifier indicates a correct prediction of the negative class. A wrong prediction of the positive class is known as a false positive, and a wrong prediction of the negative class is known as a false negative (Fn). Thus, the precision is defined as [28]:

$$Pr = \frac{T_p}{T_p + F_p}. \quad (3)$$

It indicates the classifier's capability of not labeling a negative sample positive. A higher precision indicates a better performance [28].

Recall (R) is a performance metric that correctly characterizes the classifier's ability to classify all the samples belonging to a positive class. The recall is defined as the ratio of true positive (Tp) to the sum of true positive and false negative (Fn). A higher value of recall characterizes a better classifier. Recall (R) is defined as [29]:

$$R = \frac{T_p}{T_p + F_n}. \quad (4)$$

3 Results and Discussion

To validate the efficiency of the proposed system, performance analysis has been carried out using test datasets. The performance of the system has been evaluated in terms of four metrics: accuracy (A_c), Cohen's kappa (k), precision (Pr), and recall (R) are tabulated in Table 1 along with the average values. The performance metrics of the

Table 1 Results of performances analysis of various COVID-19 diagnosis systems

Existing COVID-19 diagnosis systems proposed by	Accuracy (A_c)	Cohen's kappa (k)	Precision (Pr)	Recall (R)	Average
Wang et al. [8] Covid-Net	0.9330	–	–	–	0.9330
Wang et al. [9]	0.8250	–	–	–	0.8250
Hussain et al. [10] CoroDet	0.9910	–	0.9764	0.9530	0.9734
Li et al. [11]	0.9350	–	–	–	0.9350
Ophir Gozes [12]	–	–	–	–	–
Pandit et al. [13]	0.9600	–	–	–	0.9600
Ismael et al. [14]	0.9470	–	–	–	0.9470
Chakraborty & Ghosh	0.9851	0.9702	0.9962	0.9745	0.9815

"–" represents "not reported"

aforementioned existing deep learning-based diagnosis systems are also presented in Table 1 for comparative analysis.

It reveals from the above table that except "CoroDet" proposed by Hussain et al. [10], the performances of all other systems were evaluated only in terms of accuracy (A_c). The values of the other metrics were not considered.

For "CoroDet", the other two metrics, precision (Pr) and recall (R), were taken into account, along with the accuracy to evaluate the performance of the system. The values of these three metrics project an acceptable performance of the system. However, a well-recognized metric, Cohen's kappa (k), was not taken into account.

The performance of our proposed system has been evaluated in terms of four metrics: accuracy (A_c), Cohen's kappa (k), precision (Pr), and recall (R). The values of these four metrics project the robustness of the system. Only the accuracy value (A_c) for "CoroDet" is slightly higher than our system. Since the values of these metrics represent the efficiency of each system, the bigger the value, better the system. It reveals from the table that our system outperforms all the systems in terms of the average values of these metrics.

4 Conclusion

A COVID-19-infected patient in critical condition needs immediate lifesaving treatment. Before starting an adequate treatment, a confirmed diagnosis is a mandatory criterion. Instead of a prolonged clinical diagnostic process, an alternative way of rapid diagnosis is the only way to start immediate treatment. This paper suggests a model for rapid diagnosis of COVID-19 using radiographic images of infected lungs of critical patients within few minutes. Technicians can operate the system with minimal training that compensates for the scarcity of expert radiologists. In the future, a similar types of automatic medical diagnostic systems can be designed where a manual visual investigation is used.

Though, some rapid diagnosis systems have been suggested that use deep learning techniques for image classification. However, they suffer from some constraints associated with deep learning. This work proposes the simplest way of designing such a system based on the classical image processing techniques and machine learning techniques that provide very high accuracy with minimum instrumental and infrastructural supports. In this global crisis, such simple and rapid diagnostic systems can become a silver lining.

References

1. National Institute on Aging. https://www.nia.nih.gov/news/why-covid-19-testing-key-getting-back-normal. Last accessed 9 Oct 2020
2. Sharfstein J, Becker S, Mello M (2020) Diagnostic testing for the novel coronavirus. JAMA 323(15):1437–1438
3. Fang Y, Zhang H, Xie J, Lin M, Ying L, Pang P, Ji W (2020) Sensitivity of chest CT for COVID-19 Comparison to RT-PCR. Radiology 296(2):E115–E117
4. International Atomic Energy Agency. https://www.iaea.org/newscenter/news/how-is-the-covid-19-virus-detected-using-real-time-rt-pcr. Last accessed 9 Oct 2020
5. Ng M, Lee E, Yang J, Yang F, Li X, Wang H, Lui M, Lo C, Leung B, Khong P, others (2020) Imaging profile of the COVID-19 infection. Radiologic findings and literature review. Radiol Cardiothorac Imaging 2(1):e200034
6. Huang C, Wang Y, Li X, Ren L, Zhao J, Hu Y, Zhang L, Fan G, Xu J, Gu X, others (2020) Clinical features of patients infected with 2019 novel coronavirus in Wuhan, China. The Lancet 395(10223):497–506
7. Guan W, Ni Z, Hu Y, Liang W, Ou C, He J, Liu L, Shan H, Lei C, Hui D, others (2020) Clinical characteristics of coronavirus disease 2019 in China. N Engl J Med 382(18):1708–1720
8. Wang L, Lin Z, Wong A (2020) Covid-net: a tailored deep convolutional neural network design for detection of covid-19 cases from chest x-ray images. Sci Rep 10(1):1–12
9. Wang S, Kang B, Ma J, Zeng X, Xiao M, Guo J, Cai M, Yang J, Li Y, Meng X, others (2021) A deep learning algorithm using CT images to screen for Corona Virus Disease (COVID-19), pp 1–9
10. Hussain E, Hasan M, Rahman A, Lee I, Tamanna T, Parvez M (2021) CoroDet: a deep learning based classification for COVID-19 detection using chest X-ray images. Chaos Solitons Fractals 142:110495
11. Li X, Li C, Zhu D (2020) Covid-mobilexpert: on-device covid-19 screening using snapshots of chest x-ray. arXiv preprint arXiv:2004.03042
12. Gozes O, Frid-Adar M, Greenspan H, Browning P, Zhang H, Ji W, Bernheim A, Siegel E (2020) Rapid ai development cycle for the coronavirus (covid-19) pandemic: Initial results for automated detection & patient monitoring using deep learning ct image analysis. arXiv preprint arXiv:2003.05037
13. Pandit M, Banday S, Naaz R, Chishti M (2021) Automatic detection of COVID-19 from chest radiographs using deep learning. Radiography 27(2):483–489
14. Ismael A, Sengur A (2021) Deep learning approaches for COVID-19 detection based on chest X-ray images. Expert Syst Appl 164:114054
15. Towards Data Science. https://towardsdatascience.com/deep-learning-vs-classical-machine-learning-9a42c6d48aa. Last accessed 10 Oct 2020
16. Cohen JP, Morrison P, Dao L (2020) COVID-19 image data collection. arXiv:2003.11597
17. Chung A (2020) Figure 1 COVID-19 chest x-ray data initiative. https://github.com/agchung/Figure1-COVID-chestxray-dataset
18. Radiological Society of North America (2019) COVID-19 radiography database. https://www.kaggle.com/tawsifurrahman/covid19-radiography-database
19. Radiological Society of North America (2019) RSNA pneumonia detection challenge. https://www.kaggle.com/c/rsna-pneumonia-detection-challenge/data
20. Chung A (2020) Actualmed COVID-19 chest x-ray data initiative. https://github.com/agchung/Actualmed-COVID-chestxray-dataset
21. Da Silva E, Mendonca G (2004) The electrical engineering handbook. Academic Press, United States of America
22. Kapur J, Saxena H (1960) Mathematical statistics, 1st edn. S. Chand, and Company, India
23. Towards Data Science. https://towardsdatascience.com/intro-to-descriptive-statistics-252e9c464ac9. Last accessed 16 Oct 2020
24. Arif C, Mizoguchi M, Setiawan B, others (2012) Estimation of soil moisture in paddy field using artificial neural networks. Int J Adv Res Artif Intell 1(1):17–21

25. Francl L, Panigrahi S (1997) Artificial neural network models of wheat leaf wetness. Agric For Meteorol 88(1–4):57–65
26. Cohen J (1960) A coefficient of agreement for nominal scales. Educ Psychol Measur 20(1):37–46
27. Tallón-Ballesteros A, Riquelme J (2014) Data mining methods applied to a digital fo-rensics task for supervised machine learning. In: Computational intelligence in digital forensics: forensic investigation and applications, pp 413–428. Springer
28. Sci-kit Learn. https://scikit-learn.org/stable/modules/generated/sklearn.metrics.precision_score.html#sklearn.metrics.precision_score. Last accessed 12 June 2021
29. Sci-kit Learn. https://scikit-learn.org/stable/modules/generated/sklearn.metrics.recall_score.html#sklearn.metrics.recall_score. Last accessed 12 June 2021

BUS-Net: A Fusion-based Lesion Segmentation Model for Breast Ultrasound (BUS) Images

Kaushiki Roy, Debotosh Bhattacharjee, and Christian Kollmann

Abstract Breast cancer is the most common cancer(s) among women worldwide. The survival rate decreases if the cancer is not detected at an early stage. Breast ultrasound (BUS) is emerging as a popular modality for breast cancer detection owing to its several advantages over other modalities. We proposed a novel deep learning framework named BUS-Net for automated lesion segmentation in BUS images in this work. However, every deep learning framework has disadvantages of its own; however, the drawbacks associated with individual models can be overcome when combined. Our proposed BUS-Net is an ensemble of three popular deep learning frameworks, namely attention U-net, U-Net and SegNet. The final segmentation map generated by BUS-Net is a pixel-level fusion on the outputs of each of the individual frameworks. The potentiality of BUS-Net was tested on a publicly available dataset named BUSI dataset. This dataset consists of 647 tumor images collected from 600 different female patients. To prevent biased results, the training and test set were separate. BUS-Net framework achieved an accuracy—93.19%, precision—93.18%, recall—88.75%, dice—90.77%, and volume similarity—95.55% for lesion segmentation in the test set. The degree of correlation between the lesion region segmented by the medical experts and that segmented by BUS-Net was high ($R^2 = 0.9131$). Further, the performance of BUS-Net was also compared with the state-of-the-art techniques. This comparison showed that BUS-Net maintains a tradeoff between precision and recall, proving the robustness, efficiency, and reliability of the framework.

Keywords Breast cancer · Ultrasound · Attention U-Net · U-Net · SegNet · Fusion

K. Roy (✉) · D. Bhattacharjee
Department of Computer Science and Engineering, Jadavpur University, Kolkata 700032, India
e-mail: kroy.cse.rs@jadavpuruniversity.in

D. Bhattacharjee
e-mail: debotosh@ieee.org

C. Kollmann
Center for Medical Physics and Biomedical Engineering, Medical University Vienna, Waehringer Guertel 18-20, 1090 Vienna, Austria
e-mail: christian.kollmann@muv.ac.at

S. Basu et al. (eds.), *Proceedings of International Conference on Frontiers in Computing and Systems*, Lecture Notes in Networks and Systems 404,
https://doi.org/10.1007/978-981-19-0105-8_30

1 Introduction

Breast cancer is the most common cancer(s) in women worldwide. According to WHO [1], in 2020 alone, there were 2.3 million breast cancer cases. To reduce the mortality rate, it is highly required to detect cancer in its preliminary stages. Unlike the other imaging modalities available for breast cancer detection, breast ultrasound (BUS) is non-invasive, readily available, and inexpensive [2].

Qinghua et al. [3] have used a semantic classification-based approach for lesion segmentation in BUS images. In Ref. [4], authors have segmented BUS images using the active contour model (ACM). In Ref. [5], the authors improved the balloon force in the traditional DRLSE model for BUS segmentation. In another work [6], the authors proposed a hybrid approach for segmenting BUS lesion margins. However, BUS images are often corrupted with multiple artifacts and noise coupled with blurry boundaries and inhomogeneous intensity. These challenges make it difficult to extract promising discriminative features for BUS image segmentation using traditional methods.

Recently, deep learning has gained immense popularity in BUS image segmentation. This is mainly since deep learning techniques are devoid of complex feature extraction. In Ref. [7], authors have designed a sequential convolutional neural network (CNN)-based on pixel centric approach for pixel classification. In Ref. [8], authors have used an ensemble of two U-Net architectures for segmentation of lesion. In Ref. [9], authors have used a cascaded ensemble of various deep learning architectures for BUS lesion segmentation. In Ref. [10], a CNN network based on selective kernel U-Net was used for the segmentation task. In Ref. [11], the authors used dilated convolution in the deeper layer since dilated layers have a larger receptive field and can extract more spatial information. Each deep learning framework individually has one or more drawback(s). Thus, in this work, we propose a framework based on deep learning named BUS-Net, which is an ensemble of three deep learning frameworks, namely attention U-Net [12], U-Net [13], and SegNet [14] for lesion segmentation in BUS images. The BUS-Net segments the lesions in each input image by pixel-level fusion [15] on the segmentation map generated by each framework.

2 Proposed Methodology

This section describes the sequence of steps followed to design the proposed BUS-Net for lesion segmentation in BUS images in minute detail. The workflow of the proposed CNN model for segmentation is illustrated in Fig. 1.

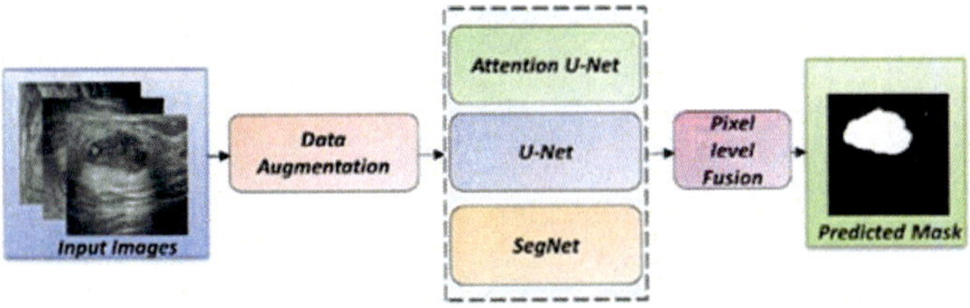

Fig. 1 Graphical illustration of the BUS lesion segmentation system

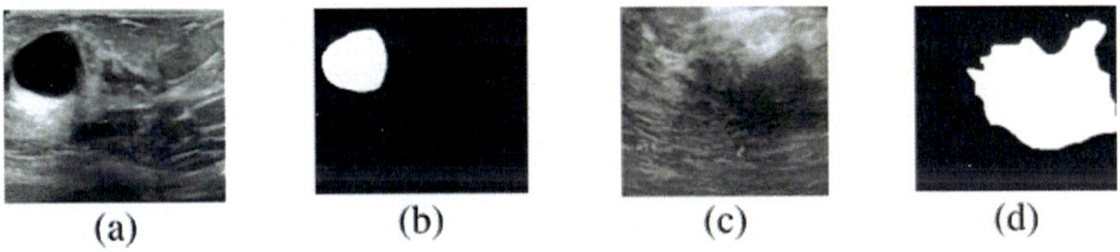

Fig. 2 Representative BUS lesion images used in the study: **a** input image (example 1), **b** lesion region, **c** input image (example 2), and **d** lesion region

2.1 Input Image

The input(s) to the BUS-Net are lesion images of average size 500×500, including benign and malignant lesions. Figure 2 has some of the representative images used in this work. The first example, $2a$, represents a benign lesion, whereas $2b$ represents a malignant lesion.

2.2 Data Augmentation

Augmentation of dataset is necessary to increase the number of input samples [16] and prevent overfitting. BUS images are invariant to rotation, so they were augmented with ${}^{\theta\pi}/_{4}$ rotations with θ varying between $\{0, 1, \ldots 8\}$, shown in Fig. 3.

2.3 Proposed BUS-Net Architecture

The BUS-Net proposed in this study uses an ensemble of attention U-Net, U-Net, and SegNet and predicts the unbiased lesion segmentation map by pixel-level fusion on the predicted output of the three different models.

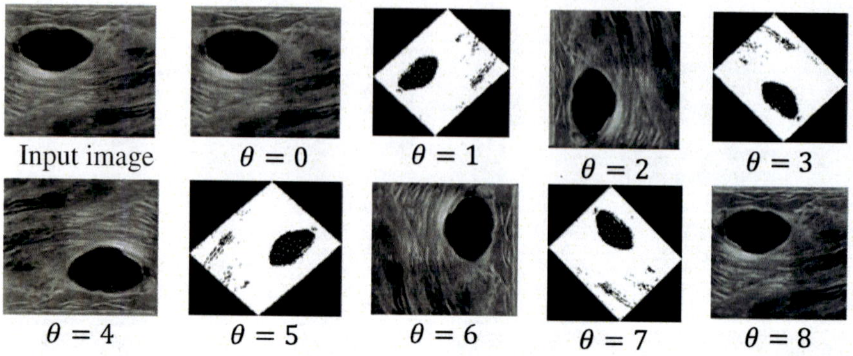

Fig. 3 Input image and its various rotation

Attention U-Net

Attention U-Net [12] prevents the extraction of redundant lower-level features between the encoder and the decoder. Attention U-Net uses attention gates in the skip connection layers before the concatenation to merge and focus only on the valuable and relevant activations, thereby suppressing any unnecessary low-level features. In Fig. 4, we present the architecture of attention U-Net.

U-Net Architecture

This framework [13, 17] combines the features extracted in the encoder layer with the decoder layer by means of skip connection. As a result, the low-level features

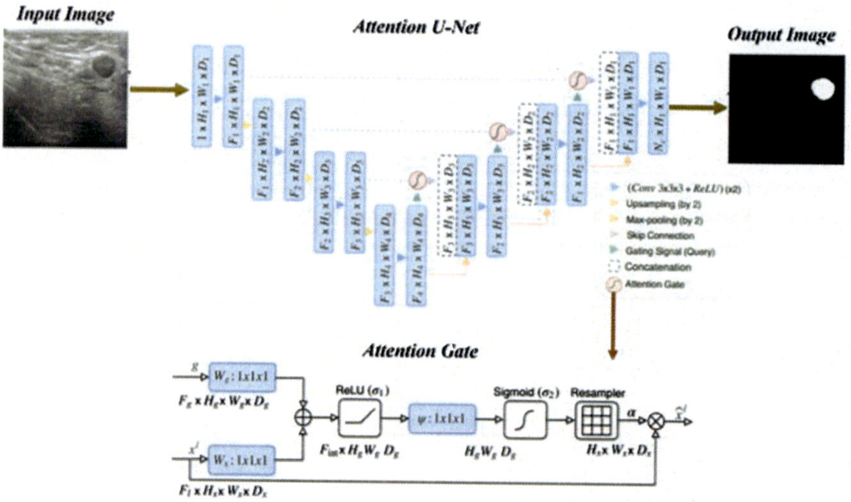

Fig. 4 Architecture of attention U-Net

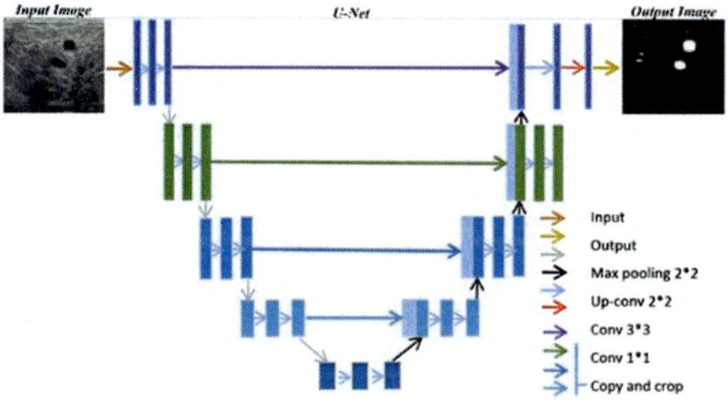

Fig. 5 Architecture of U-Net

that might carry useful information are retained and passed to the expansion/decoder layers. A semantic diagram of the U-Net architecture is presented in Fig. 5.

SegNet Architecture

SegNet [14] is another popular deep learning framework. Max pooling indices are transferred from the encoder to its corresponding level in the decoder and are used to perform non-linear upsampling of their input feature map. The main advantage of SegNet is its capability of boundary delineation, which is important for the accurate detection of lesion boundaries. The architecture of SegNet is presented in Fig. 6.

Pixel-Level Fusion

The final segmentation map by BUS-Net is generated by a pixel-level fusion of the segmentation map generated by each framework, as shown in Fig. 7.

In this study, Adam optimizer was used with a learning rate of 0.001. Each model was trained for 150 epochs. We have used Jaccard loss to train the BUS-Net.

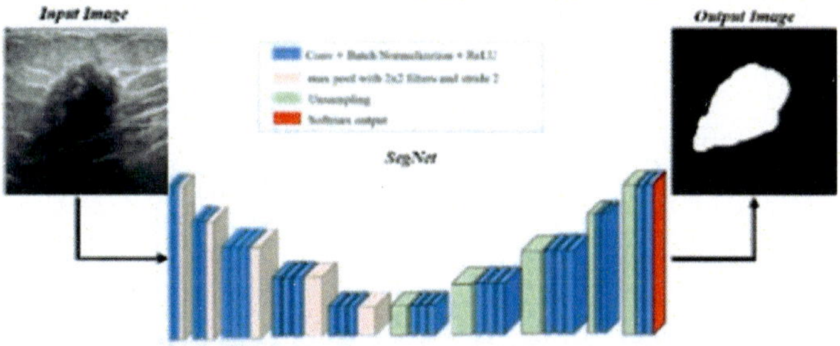

Fig. 6 Architecture of SegNet

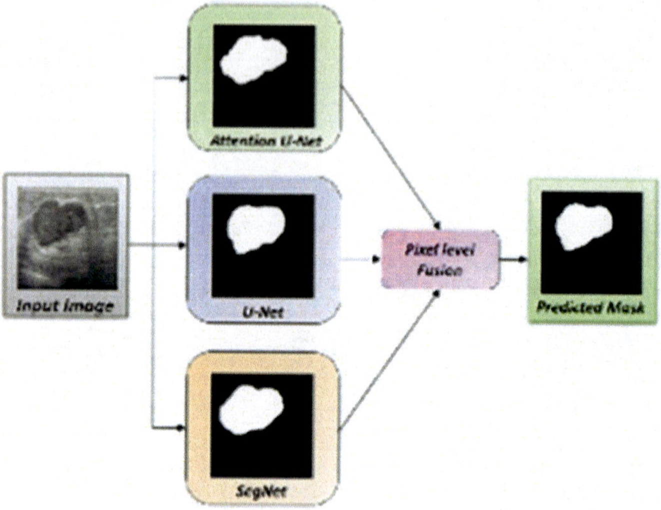

Fig. 7 Pixel-level fused segmentation map

3 Experimental Results

3.1 Dataset Description

The “BUSI” [18] dataset used for this research consists of 647 lesion images (437 benign images and 210 malignant lesion images) collected from 600 different female patients. Among them, 100 images from each class were separated for testing. Thus, our system was traid on 447 images and tested on 200 images.

3.2 Quantitative Evaluation of BUS-Net

Various discrepancy evaluation methods were used to evaluate the potentiality of BUS-Net for lesion segmentation on the test set. We constructed a linear regression plot in Fig. 8 between the area of lesion segmented by a medical expert against the area segmented by BUS-Net.

The high degree of correlation ($R^2 = 0.9131$) in the linear regression plot justifies the efficiency of our system. Further, in Table 1, we have compared of BUS-Net's performance with two state-of-the-art techniques using metrics like precision (Pre) [15], recall (Rec) [15], accuracy [15], specificity [15], Jaccard [15], dice coefficient [15], volume similarity (VS) [15], and area under the curve (AUC) [15].

It can be seen from Table 1 that BUS-Net maintains a tradeoff between Pre and Rec unlike [10], thereby justifying the consistency of BUS-Net.

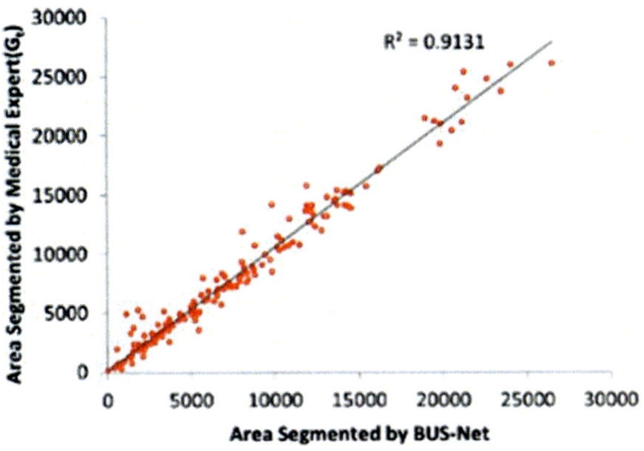

Fig. 8 Linear regression plot between total lesion area segmented by medical experts against the area segmented by BUS-Net

Table 1 Comparative analysis of BUS-Net with state-of-the-art

Method	Accuracy	Pre	Rec	Specificity	Jaccard	Dice	VS	AUC
M. Byra et.al. [10]	0.91	0.73	*0.97*	0.89	–	–	–	*0.95*
Y. Hu et.al. [11]	*0.95*	–	–	–	–	0.70	–	0.88
BUS-Net	0.93	*0.93*	0.89	*0.99*	*0.83*	*0.90*	*0.95*	0.93

Italic for best-performing methods

3.3 *Qualitative Evaluation of BUS-Net*

In Fig. 9, we presented qualitative analysis of BUS-Net for lesion segmentation by overlaying the lesion boundary segmented by experts and by BUS-Net on the original image. The first column in Fig. 9 represents the input image, the second represents the ground truth (segmented by medical experts), the third represents the segmentation map generated by BUS-Net, and the fourth column represents the overlay between ground truth and BUS-Net.

The red color in Fig. 9d represents the lesion contour in the ground truth, whereas the blue color represents the lesion contour segmented by BUS-Net. It is evident that the red and blue colors mostly coincide, implying the lesion boundary segmented by BUS-Net is like the lesion boundary in the ground truth.

4 Conclusion

In this study, we designed BUS-Net for the segmentation of BUS lesion images. This framework uses an ensemble of three deep learning architectures to segment the lesion region by pixel-level fusion on the output of the three individual networks.

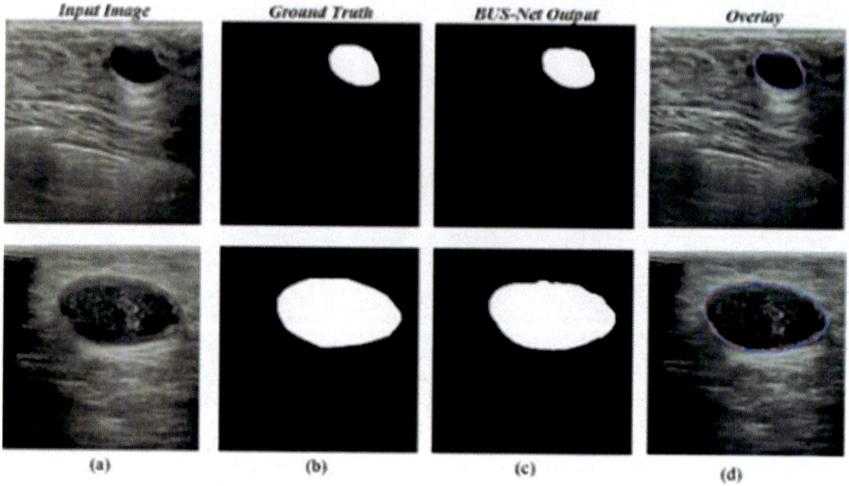

Fig. 9 Qualitative analysis of BUS-Net

Our future study will extract shape-based features from the segmented lesions to classify input BUS images as either malignant, benign, or normal.

Acknowledgements The first author is thankful to DST INSPIRE fellowship (IF170366). The authors are grateful to the DST, Government of India, and OeAD, Austria (INT/AUSTRIA/BMWF/P-25/2018) for providing support.

References

1. Breast cancer. https://www.who.int/news-room/fact-sheets/detail/breast-cancer. Accessed 30 June 2021
2. Breast Ultrasound. https://www.radiologyinfo.org/en/info/breastus. Accessed 30 June 2021
3. Huang Q, Huang Y, Luo Y, Yuan F, Li X (2020) Segmentation of breast ultrasound image with semantic classification of superpixels. Med Image Anal 61:101657. https://doi.org/10.1016/j.media.2020.101657
4. Fang L, Qiu T, Liu Y, Chen C (2018) Active contour model driven by global and local intensity information for ultrasound image segmentation. Comput Math with Appl 75(12):4286–4299. https://doi.org/10.1016/j.camwa.2018.03.029
5. Zhao W, Xu X, Liu P, Xu F, He L (2020) The improved level set evolution for ultrasound image segmentation in the high-intensity focused ultrasound ablation therapy. Optik (Stuttg) 202:163669. https://doi.org/10.1016/j.ijleo.2019.163669
6. Panigrahi L, Verma K, Singh BK (2019) Ultrasound image segmentation using a novel multi-scale Gaussian kernel fuzzy clustering and multi-scale vector field convolution. Expert Syst Appl 115:486–498. https://doi.org/10.1016/j.eswa.2018.08.013
7. Xu Y, Wang Y, Yuan J, Cheng Q, Wang X, Carson PL (2019) Medical breast ultrasound image segmentation by machine learning. Ultrasonics 91:1–9. https://doi.org/10.1016/j.ultras.2018.07.006

8. Amiri M, Brooks R, Behboodi B, Rivaz H (2020) Two-stage ultrasound image segmentation using U-Net and test time augmentation. Int J Comput Assist Radiol Surg 15(6):981–988. https://doi.org/10.1007/s11548-020-02158-3
9. Moon WK, Lee YW, Ke HH, Lee SH, Huang CS, Chang RF (2020) Computer-aided diagnosis of breast ultrasound images using ensemble learning from convolutional neural networks. Comput Methods Programs Biomed 190:105361. https://doi.org/10.1016/j.cmpb.2020.105361
10. Byra M et al (2020) Breast mass segmentation in ultrasound with selective kernel U-Net convolutional neural network. Biomed Signal Process Control 61:102027. https://doi.org/10.1016/j.bspc.2020.102027
11. Hu Y et al (2019) Automatic tumor segmentation in breast ultrasound images using a dilated fully convolutional network combined with an active contour model. Med Phys 46(1):215–228. https://doi.org/10.1002/mp.13268
12. Oktay O et al Attention U-Net: learning where to look for the pancreas
13. Ronneberger O, Fischer P, Brox T (2015) U-net: convolutional networks for biomedical image segmentation. In: Lecture Notes in Computer Science (including subseries Lecture Notes in Artificial Intelligence and Lecture Notes in Bioinformatics), vol 9351, pp 234–241.https://doi.org/10.1007/978-3-319-24574-4_28
14. Badrinarayanan V, Kendall A, Cipolla R (2017) SegNet: a deep convolutional encoder-decoder architecture for image segmentation. IEEE Trans Pattern Anal Mach Intell 39(12):2481–2495. https://doi.org/10.1109/TPAMI.2016.2644615
15. Banik D, Roy K, Bhattacharjee D, Nasipuri M, Krejcar O (2021) Polyp-Net: a multimodel fusion network for polyp segmentation. IEEE Trans Instrum Meas 70. https://doi.org/10.1109/TIM.2020.3015607
16. Roy K, Banik D, Bhattacharjee D, Nasipuri M (2019) Patch-based system for classification of breast histology images using deep learning. Comput Med Imaging Graph 71:90–103. https://doi.org/10.1016/j.compmedimag.2018.11.003
17. Ding Y, Chen F, Zhao Y, Wu Z, Zhang C, Wu D (2019) A stacked multi-connection simple reducing net for brain tumor segmentation. IEEE Access 7:104011–104024. https://doi.org/10.1109/ACCESS.2019.2926448
18. Al-Dhabyani W, Gomaa M, Khaled H, Fahmy A (2020) Dataset of breast ultrasound images. Data Br 28:104863. https://doi.org/10.1016/j.dib.2019.104863

Breast Cancer Detection from Histology Images Using Deep Feature Selection

Susovan Das, Akash Chatterjee, Samiran Dey, Shilpa Saha, and Samir Malakar

Abstract Screening of breast cancer from histology images is a popular research problem in medical imaging. Most of the methods in recent days used deep learning models for predicting the same. But, at times, such methods dealt with not only higher-dimensional features but also may suffer from containing irrelevant and sometimes redundant features. To overcome this shortcoming, in the present work, we employ a popularly used particle swarm optimization (PSO) algorithm to obtain the near-optimal feature set. To extract the features from images we first preprocess the images to obtain stain normalized images and then pass them through a pre-trained MobileNet model for extracting the features. We have evaluated our model on a recent dataset, published through ICIAR BACH 2018 grand challenge. The experimental results show an improvement of 6.25% recognition accuracy with around 54% reduced features. We have also compared our result with two state-of-the-art CNN models: InceptionResNet and DenseNet, and we found that the use of MobileNet is better. The capability of the present model is comparable with some state-of-the-art methods on the BACH dataset.

Keywords Breast cancer detection · Histology image · PSO · MobileNet · Feature selection

1 Introduction

One of the deadliest forms of cancer is breast cancer, which attacks mostly women. In 2020, the International Agency for Research on Cancer (IARC) estimated a total of 19,292,789 new cancer cases out of which 11.70% of the cases belong to breast cancer.[1] A death count of around 684,996 was also estimated from a total of 2,261,419

[1]https://gco.iarc.fr/today/data/factsheets/populations/900-world-fact-sheets.pdf.

S. Das · A. Chatterjee · S. Dey · S. Saha · S. Malakar (✉)
Department of Computer Science, Asutosh College, Kolkata, India
e-mail: malakarsamir@gmail.com

S. Basu et al. (eds.), *Proceedings of International Conference on Frontiers in Computing and Systems*, Lecture Notes in Networks and Systems 404,
https://doi.org/10.1007/978-981-19-0105-8_31

cases. For detecting breast cancer in pathology, a section of the biopsy material is cut and stained with hematoxylin and eosin (H&E). The hematoxylin solution helps to bind the DNA and highlight the nuclei, whereas eosin helps in binding the proteins and other structures. It is then analyzed manually by the learned pathologists by observing the highlighted regions in the stained tissues through a high magnifying microscope. One of the main information checked by the pathologists is the presence or absence of known features which helps in classifying breast tissues precisely into four broad categories: normal tissue, non-malignant (benign) tissue, invasive carcinoma, and in situ carcinoma.

However, analyzing tissue using a microscope is a tedious job. The complexity of histology images and a huge amount of data make the microscopic evaluation process costly and sometimes non-trivial. Expert diagnosticians must spend a substantial amount of time and energy to fulfill the piece of work. The efficiency and definiteness of evaluation also require the preceding knowledge of the diagnostician and the reliability of pathological reports. Here comes the need to develop an automatic diagnostic tool, though challenging, for accurate and speedy breast cancer detection in an economic way. In the past, many researchers provided some solutions using machine learning approaches to diagnose and predict breast cancer from histology images.

For example, Chennamsetty et al. [3] proposed an ensemble mechanization where they used pre-trained convolutional neural network (CNN) models to predict breast cancer from histology images. The mechanization is composed of two DenseNet-161 and ResNet-101 networks fine-tuned with images from varying preprocessing schemes. Brancati et al. [2] proposed another ensemble method where different versions of the ResNet model were used. In this work, the authors also used the concept of the fine-tuning scheme. The authors reduced the complexity of the problem by down-sampling the image size by a k factor and using only the central patch of size $m * m$ as input to the model. In another work, Wang et al. [15] preferred to use only VGG-16 to classify the images. Koné and Boulmane [8] proposed a hierarchical system of three ResNet50 models. This system is represented in a binary tree-like structure for the multi-class classification of breast cancer images. In another work, Rakhlin et al. [11] proposed an approach that utilizes several pre-trained CNN models as feature extractors and gradient boosted trees as classifiers for the said purpose.

All the abovementioned work fed the entire image to the system for the classification of microscopic histology images. Some authors have tried a patch-based approach to improve the breast cancer prediction accuracy from the histology images. For example, Roy et al. [12] proposed a method that used a patch-based classifier for the automatic classification of histology breast images. The patches of appropriate sizes carrying important distinguishing information were taken out from the original images. This patch-based classification method works in two different techniques: one patch in one decision (OPOD) and all patches in one decision (APOD). In another work, Sanyal et al. [13] present a novel hybrid ensemble mechanization consisting of multiple fine-tuned CNN models for supervised feature extraction and Eextreme gradient boosting trees (XGB) as a top-level classifier, for patch-wise classification of breast histology images.

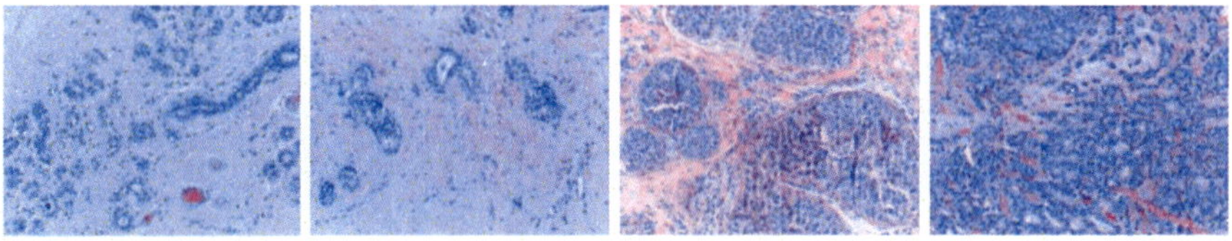

(a) Normal (b) Benign (c) In Situ (d) Invasive

Fig. 1 Hematoxylin and eosin stained breast histology microscopy images

From the above discussion, it is clear that automated histopathological image analysis and classification for breast cancer is a significant research problem in medical imaging where pre-trained CNN models are widely used to solve such problems. But, these CNN models might possess some redundant features that may lessen the model performance. Moreover, the redundant features increase the computational need. Therefore, in our present work, we select near-optimal deep features before final classification. We use the MobileNet model [5], pre-trained on ImageNet dataset, to extract deep features while popularly used binary particle swarm optimization (BPSO) technique is used to obtain the near-optimal features. The model is evaluated on a recent dataset, published through ICIAR BACH 2018 grand challenge [1].

2 Database Description

We have used the dataset that was made available as part of the ICIAR-2018 grand challenge [7] based on breast cancer histology image classification to evaluate the performance of the present method. The dataset was composed of hematoxylin and eosin (H&E) stained breast histology microscopy images that belong to one of the abovementioned four classes. It contains 400 images (100 images each class). Four sample images, one for each class, have been shown in Fig. 1.

3 Preprocessing of the Histology Images

Before sending the histology images to the feature extraction protocol, we have performed a set of preprocessing on the stained input images. The preprocessing pipeline is shown in Fig. 2.

At first, we normalized the amount of hematoxylin and eosin stained on the tissue as described in the work [10] to bring the microscopy images into a common space and thereby increasing quantitative analysis. Thereafter, the stain normalized images are reduced to the dimension 224×224 from their original dimension which is 2048×1536. The images are symmetrically padded through the width to generate an image of squared dimension using bilinear interpolation prior to resizing them.

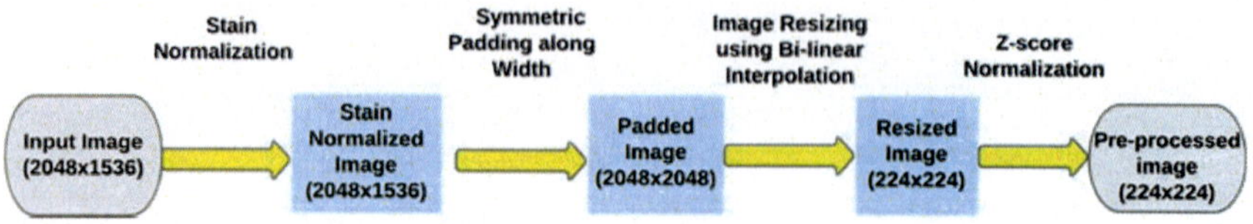

Fig. 2 Histology images' preprocessing pipeline

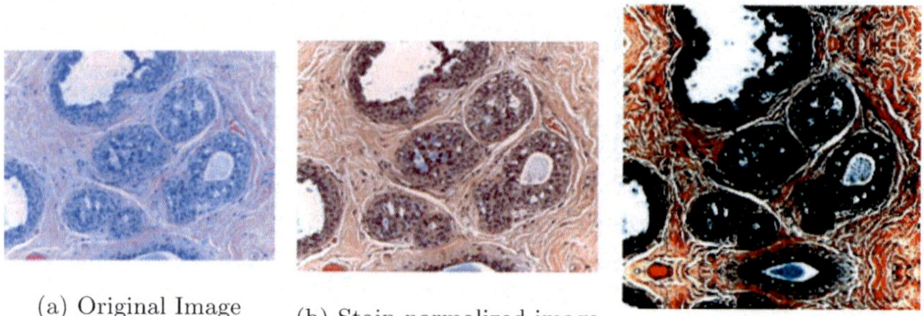

(a) Original Image (b) Stain normalized image (c) Z-score normalized image

Fig. 3 Transformation of the image in the preprocessing pipeline

At the end, the Z-score standardization method is employed to obtain zero mean and unit variance image using Eq. 1.

$$Z = \frac{x - \mu}{\sigma} \tag{1}$$

In Eq. 1, Z and x represent the resized image and input image, respectively, while μ and σ indicate mean and standard deviation of pixel intensities of all the images in a dataset, respectively. Figure 3 shows the transformation of an image throughout the preprocessing steps.

4 Proposed Work

In the present work, we have selected near-optimal deep features from histology images to classify them into one of the four breast cancer classes: normal, benign, invasive carcinoma, and in situ carcinoma. The overall model is shown in Fig. 4, and the sub-processes are described in the following subsections.

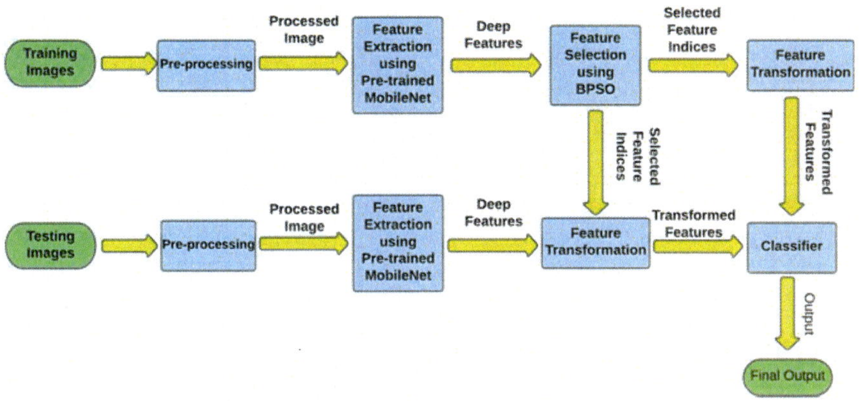

Fig. 4 Workflow of present work

4.1 Feature Extraction

After the preprocessing, the input images are resized and normalized. We extract features from the preprocessed images using the MobileNet model [6], pre-trained on the ImageNet dataset. MobileNet is a simple but efficient and not very computationally intensive convolutional neural network. In MobileNet, depth-wise separable convolution is used to minimize the model size and complexity, for that MobileNet has fewer parameters and fewer multiplications and additions. All these facts motivate us to use this CNN model for feature extraction purposes. In this CNN model, we remove fully connected layers up to the flatten layer for feature extraction. We get a feature vector of dimension 50,176.

4.2 Feature Selection

MobileNet extracts features that represent patterns in a better way, but it generates high-dimensional features, and the features may contain some redundant and/or irrelevant features. Hence, we use the binary version of particle swarm optimization (PSO) [9] algorithm to select a near-optimal feature set. It is a wrapper-based nature-inspired method that is very popular for the feature selection approach due to its efficiency and simplicity [14]. For the implementation of BPSO, we have used the Py_FS package [4], which is a Python package designed for feature selection purposes. It is to be noted that we have used the default parameters of PSO as set by the authors of the work [4]. During fitness value calculation, we have made use of a support vector machine (SVM) classifier.

4.3 Classification

In this process, we have classified the histology images into one of the four classes mentioned previously. For the classification of these images, we feed the optimized features, extracted from them, into an SVM classifier, chosen experimentally from a pool of three well-known and popularly used classifiers, viz SVM, random forest (RF), and XGB.

5 Experimental Results

As already mentioned, we have designed a model to predict breast cancer from histology images. In this model, first, we extract features using the MobileNet model, and then with the help of BPSO, we select the near-optimal features. For experimental needs, we use the BACH dataset. We consider 20% of each class of images as test samples randomly and the rest as train samples. During the experimentation, we first choose a suitable classifier for the same from a pool of three classifiers: SVM, RF, and XGB, and then the best classifier is carried forward to the next level. Figure 5 shows the recognition results before using BPSO-based feature selection. From the results, it is understandable that SVM brings off the best among the classifiers considered here, and therefore, we have used the SVM classifier for the feature selection purpose.

The classification accuracy after applying BPSO is shown in Fig. 6. To check the effectiveness of applying feature selection for breast cancer classification, we have also applied the same on features extracted using DenseNet201 and InceptionResnetV2 models, pre-trained on the ImageNet dataset. The results are shown in Fig. 6. From these results, it is clear that the use of feature selection for all the cases is beneficial as this helps to improve the accuracy 3–6% (see Fig. 6a) with less than 50% of the actual features (see Fig. 6b). We have also collated our results with some state-of-the-art methods, and our result is analogous with the state-of-the-art techniques (see Table 1).

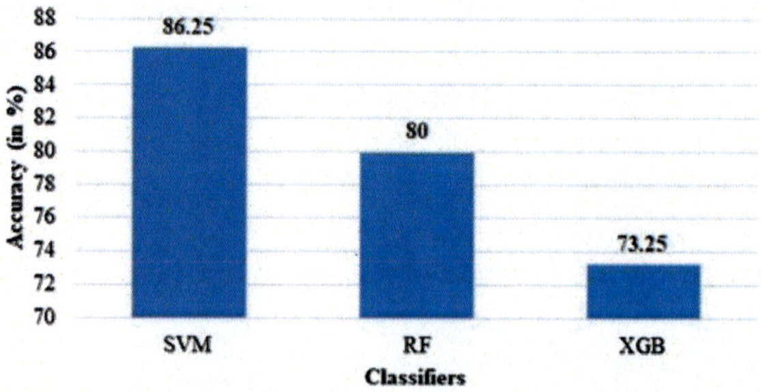

Fig. 5 Performance of different classifiers without using feature selection

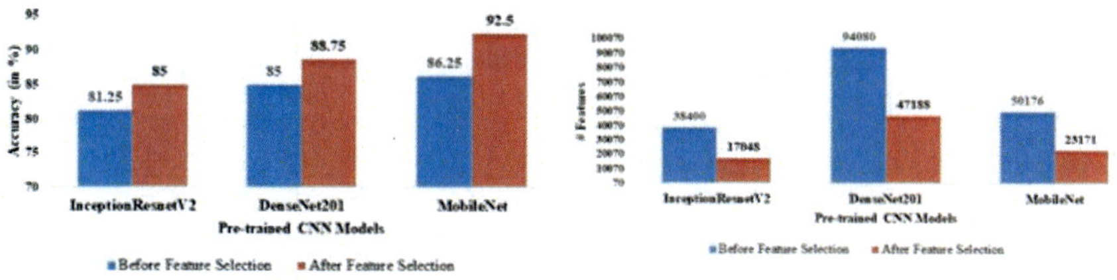

(a) Accuracy

(b) Feature Count

Fig. 6 Comparison of accuracy and feature count before and after using BPSO-based feature selection

Table 1 Comparison with state-of-the-art methods

Method	Technique used	Accuracy (in %)
Rakhlin et al. [11]	ResNet-400	84.20±4.2 using tenfold Cross validation
Roy et al. [12]	Patch-based classifier using CNN	77.40 using OPOD and 90.00 using APOD
Sanyal et al. [13]	Ensemble of deep features and XGB	86.50 at patch level and 95.00 at image level
Proposed method	MobileNet+PSO+SVM	92.25

6 Conclusion

In this work, BPSO-based deep feature selection method is proposed for prior detection of breast cancer tissue into one of the four categories: normal tissues, non-malignant (benign) tissues, invasive carcinomas, and in situ carcinomas from histology images. We have performed stain normalization before feeding the images to the feature extraction model. We have evaluated our model on the BACH dataset. BPSO-based feature selection helps the model to improve the accuracy by around 6% while using around 50% of the actual features. Also, the model performance is analogous to state-of-the-art methods. Although the present method works well, scopes are there for its betterment. In this work, we have only used BPSO as a feature selector. Hence, the use of some recent feature selection techniques might help in improving the performance in the future. Also, the leading methods on the BACH dataset used some ensemble mechanisms. Keeping this fact in mind, one can apply some ensemble techniques to improve performance. Lastly, we have evaluated our model only on the BACH dataset. Hence, the method may be applied to more datasets to establish its robustness.

References

1. Aresta G et al (2019) Bach: grand challenge on breast cancer histology images. Med Image Anal 56:122–139
2. Brancati N, Frucci M, Riccio D (2018) Multi-classification of breast cancer histology images by using a ne-tuning strategy. In: International conference image analysis and recognition. Springer, pp 771–778. https://doi.org/10.1007/978-3-319-93000-8_87
3. Chennamsetty SS, Safwan M, Alex V (2018) Classification of breast cancer histology image using ensemble of pre-trained neural networks. In: International conference image analysis and recognition. Springer, pp 804–811. https://doi.org/10.1007/978-3-319-93000-8_91
4. Guha R et al (2021) PyF S : A python package for feature selection using meta heuristic optimization algorithms. In: Accepted in 3rd international conference on computational intelligence in pattern recognition (CIPR-2021). Springer
5. Howard AG et al (2017) Mobilenets: efficient convolutional neural networks for mobile vision applications. In: arXiv preprint arXiv:1704.04861
6. Huang G et al (2017) Densely connected convolutional networks. In: Proceedings of the IEEE conference on computer vision and pattern recognition, pp 4700–4708. https://doi.org/10.1109/CVPR.2017.243
7. ICIAR 2018 grand challenge on breast cancer histology images. https://iciar2018-challenge.grand-challenge.org/
8. Koné I, Boulmane L (2018) Hierarchical resnext models for breast cancer histology image classification. In: International conference image analysis and recognition. Springer, pp 796–803. https://doi.org/10.1007/978-3-319-93000-8_90
9. Koohi I, Groza VZ (2014) Optimizing particle swarm optimization algorithm. In: 2014 IEEE 27th Canadian conference on electrical and computer engineering (CCECE), pp 1–5. https://doi.org/10.1109/CCECE. 2014.6901057
10. Macenko M (2009) A method for normalizing histology slides for quantitative analysis. In: IEEE international symposium on biomedical imaging: from nano to macro. IEEE, 1107–1110
11. Rakhlin A et al (2018) Deep convolutional neural networks for breast cancer histology image analysis. In: International conference image analysis and recognition. Springer, pp 737–744. https://doi.org/10.1007/978-3-319-93000-8_83
12. Roy K et al (2019) Patch-based system for classification of breast histology images using deep learning. In: Computerized medical imaging and graphics 71, pp 90–103. ISSN: 0895-6111. https://doi.org/10.1016/j.compmedimag.2018.11.003. https://www.sciencedirect.com/science/article/pii/S0895611118302039
13. Sanyal R, Kar D, Sarkar R (2021) Carcinoma type classification from high-resolution breast microscopy images using a hybrid ensemble of deep convolutional features and gradient boosting trees classifiers. In: IEEE/ACM transactions on computational biology and bioinformatics, pp 1–1. https://doi.org/10.1109/TCBB.2021.3071022
14. Sarkar S et al (2018) An advanced particle swarm optimization based feature selection method for tri-script handwritten digit recognition. In: International conference on computational intelligence, communications, and business analytics. Springer, pp 82–94
15. Wang Z et al (2018) Classification of breast cancer histopathological images using convolutional neural networks with hierarchical loss and global pooling. In: International conference image analysis and recognition. Springer, pp 745–753. https://doi.org/10.1007/978-3-319-93000-8_84

Using Cellular Automata to Compare SARS-CoV-2 Infectiousness in Different POIs and Under Different Conditions

Agnieszka Motyka, Aleksandra Bartnik, Aleksandra Żurko, Marta Giziewska, Paweł Gora, and Jacek Sroka

Abstract In this paper, we propose a stochastic model based on cellular automata and graphs to explore the spread of infectious viruses (like SARS-CoV-2) in closed rooms. We also present a simulator implementing this model which allows studying how different policies affect the spread of viruses. As we show, the simulator can be used to explore scenarios in various points of interest (POIs) like shops, public trams or fitness centres. It could be useful for policymakers to check (by changing the parameters of the simulations) the effectiveness of different regulations like limiting the maximum occupancy of POIs and mandating the usage of face masks to decrease the spread of aerosols. The simulator can also be used to compare the hazard level that different kinds of POIs pose. Also, the simulations can be visualised and showed to the public to increase support for the introduced measures and obedience to restrictions.

Keywords SARS-CoV-2 · Cellular automaton · Scientific simulations

---

A. Motyka · A. Bartnik · A. Żurko · M. Giziewska
Faculty of Physics, University of Warsaw, Warsaw, Poland
e-mail: a.motyka2@student.uw.edu.pl

A. Bartnik
e-mail: a.bartnik2@student.uw.edu.pl

A. Żurko
e-mail: a.zurko@student.uw.edu.pl

M. Giziewska
e-mail: m.giziewska@student.uw.edu.pl

P. Gora (✉) · J. Sroka
Faculty of Mathematics, Informatics and Mechanics, University of Warsaw, Warsaw, Poland
e-mail: p.gora@mimuw.edu.pl

J. Sroka
e-mail: j.sroka@mimuw.edu.pl

S. Basu et al. (eds.), *Proceedings of International Conference on Frontiers in Computing and Systems*, Lecture Notes in Networks and Systems 404,
https://doi.org/10.1007/978-981-19-0105-8_32

1 Introduction

The SARS-CoV-2 pandemic has brought many challenges for authorities who need to monitor the situation and react to it by introducing policies that limit the civil liberties when it is necessary to reduce the infections and avoid medical system collapse. There is a lot of ongoing research, e.g. [1–4], where models based on differential equations, cellular automata or their combinations are used to predict how the epidemic is developing and how it is likely to proceed if the current situation is maintained. Conclusions from such analyses are used by authorities to decide when to impose limitations and lockdowns. Yet, it is difficult to decide what kind of limitations are most efficient and at the same time the least burdensome to the population and economy. Furthermore, the motivation for the limitations needs to be understood by the society to be obeyed. It may seem at first that the experience of other countries can be reused. For example, an analysis of the human mobility patterns done in the USA has been published in [5]. It combines epidemiological model with a graph representing human mobility patterns. The graph is built based on individual cellphone locations from mobile operators. The study shows that there is a strong correlation between the reduction of infections and the reduction of maximum occupancy at points of interest (POIs) where people meet, like fitness centres, public transport trams and shops. It is also observed that the reduction of mobility has a weaker effect. As the results were published in Nature and they prioritised POIs types for risk to the population, they were widely discussed and used as a justification for introducing limitations for specific types of businesses.

At the same time, it is clear that the dynamics of epidemics differs much between countries which is mainly due to differences in lifestyles, age structure, general health and so on. For example, in some countries public transport is more widespread or open space offices can be less popular, and younger populations might have higher interaction rate of individuals. Thus, such results are not necessarily universal. Unfortunately, it is also not easy to repeat such experiments as the cellphone location data is not easily available for study due to privacy concerns.

In this paper, we take another approach. Instead of analysing data from the past and trying to find correlations, we propose a cellular automaton-based model and offer its simulator. The simulator can be used to compare the typical interaction patterns of people visiting different points of interest (POIs) like fitness centres, shops, public transport trams, classrooms, offices, etc. Many parameters can be tuned in the simulator like visitor frequency, infectious visitor percentage, infection probability, infectiousness of the virus and its rate of spread through the air. Simulations can be used to compare epidemiological significance of different types of POIs and the effectiveness of limitations for each type of POI. Simulations can also be played as built-in visualisations. This can be helpful in increasing public acceptance and obedience of restrictions.

The structure of the rest of the paper is as follows. In Sect. 2, we review other works that apply cellular automata to modelling epidemics. Then, in Sect. 3 we describe our model, and in Sect. 4, we present experiments and analyse their results. Section 5 concludes the paper and outlines further research plans.

2 Related Works

As the model presented in this paper and our simulator are based on a cellular automaton, we start from a quick overview of the available research works in which cellular automata were used to model the spread of viruses.

One of the first such approaches was presented in [6]. The authors proposed an epidemiological model based on probabilistic cellular automata. The transitions are driven by data available on chronology, symptoms, pathogenesis and transmissivity of the virus. By arguing that the lattice-based model captures the features of the dynamics along with the existing fluctuations, the authors performed computational analyses of the model to take into account the spatial dynamics of social distancing measures imposed on the people. Considering the probabilistic behavioural aspects associated with mitigation strategies, they studied the model considering factors like population density and testing efficiency. They focused on the variability of epidemic dynamics data for different countries and point out the reasons behind these contrasting observations.

In [7], the authors introduced two-dimensional cellular automata in which each cell stands for a square portion of the environment. They concluded that the results of laboratory simulations were in agreement with the real behaviour of epidemic spreading.

In [8], susceptible-exposed-infected-removed model is described in terms of probabilistic cellular automata and ordinary differential equations for the transmission of COVID-19. It is flexible enough for simulating different scenarios of social isolation according to the start day for the social isolation after the first death, the period for the social isolation campaign and the percentage of the population committed to the campaign. It also allows to estimate the impact of social isolation campaign adhesion. Results showed that efforts in the social isolation campaign must be concentrated both on the isolation percentage and campaign duration to delay the healthcare system failure.

Sree and Usha [9] introduced a classifier with nonlinear hybrid cellular automata, which is trained and tested on publicly available data sets to predict the effect of COVID-19 in terms of deaths, the number of people affected, the number of people that could be recovered, etc. The proposed classifier has reported an accuracy of 78.8%. This classifier can also predict the rate at which the SARS-CoV-2 virus spreads within and outside borders.

Sirakoulis et al. [10] present a cellular automaton model for the effects of population movement and vaccination on epidemic propagation. Each cellular automaton cell represents a part of the total population that may be found in one of three states: infected, immunised and susceptible. As parts of the population move randomly in the cellular automaton lattice, the disease spreads. Furthermore, the model is extended to include the effect of the vaccination of some parts of the population on epidemic propagation. The model establishes the acceleration of the epidemic propagation because of the increment, of the percentage of the moving population or of the maximum distance of population movement. On the contrary, the effect of population vaccination reduces the epidemic propagation.

The existing works on modelling pandemic and spread of viruses focus mostly on a long-term and global impact. In contrast, the presented research employs cellular automata to model real-time contagiousness in closed rooms.

3 Model of Human Movement, the Spread of Viruses and Infections

Due to space limit, we describe the model informally. Our model of contagion in closed spaces is based on cellular automata and directed graphs. The input to the model is a description of the studies POI with coordinates of every type of objects inside: excluded (people cannot enter them), active (people can stay in them), door (internal and external) and walls. An interior of a closed space POI is bounded by a polygon and represented by vertices which can be potential locations of people. Vertices do not need to cover the whole space as people may be excluded from some areas like the cash desk or shelves in a shop. Some vertices are connected using directed edges representing possible movement patterns (therefore, the vertices and edges represent a graph). Edges are weighted with probabilities of choosing the edge while being in a single location. The model's evolution is done in time steps. People enter and leave the room in a specific location and in each step either wait or traverse on edge. It can be assumed that one step of the simulation corresponds to 5 seconds in the real world.

Each person can be in one of the three possible states: S—susceptible, I—infectious, SI—initially susceptible but infected in the studied space. At the entrance, only states S and I are allowed. The rate of new visitors and percentage of infected visitors are configurable. It is also possible to set the maximum occupancy level. Infected people spread the virus in every vertex they visit. The amount of virus spread is specified as a parameter and is the same for all visitors.

The viruses can also spread in the air. For that we impose a grid covering the whole space. It divides the space into cells of the cellular automaton. With each cell, the current amount of the virus is associated. The spread of the virus in the air is computed interchangeably with people movement. The percentage of the virus transferred to neighbouring cells is also defined as a parameter. We assume that viruses spread in all directions with the same rate. If some direction is blocked, i.e. by a wall, it bounces off. There is also a parameter that defines percentage degradation of the airborne virus with time.

The amount of the virus in a given cell influences the probability of infection of susceptible people visiting that cell. The probability grows linearly with the amount of the virus until it reaches 1.

We have also allowed for the visitors to wear masks. It can be controlled by specifying the probability that a visitor wears a mask. We have assumed that the mask limits the aerosol emission by a fixed proportion (also configurable) and decreases in the same way the amount of viruses in a given cell that is taken into account when computing the probability of infection.

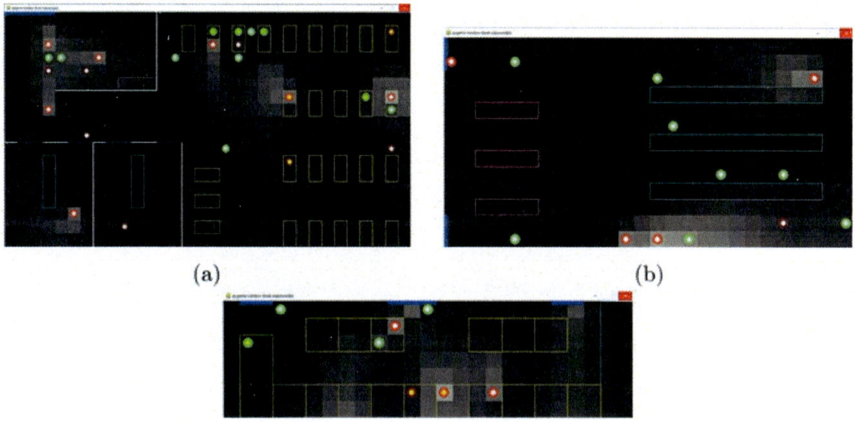

(a) (b) (c)

Fig. 1 Snapshots from simulations in **a** fitness centre, **b** shop, **c**, tram

The simulator comes with a built-in visualisation. Snapshots from simulations of public transport tram, a shop and fitness centre are presented in Fig. 1. The visitors are represented with dots and their state with dot colour (susceptible—green, infected—red). During simulation, colour of a green dot changes to claret if the person contract. The density of aerosols with the virus is depicted with background colour (the more the lighter the background).

4 Experiments

4.1 Set-up of Experiments

In this section, we describe experiments done with a dedicated tool that we developed (it is publicly available on GitHub [11], more information about the programme can be found on the project's website [12]). The goal of experiments was to compare different types of POI with respect to safety of visitors (the number of new infections). For every type of POI (fitness centre, shop and public transport tram), 3400 simulations were run. Every simulation lasted 5000 *steps*. Due to a stochastic character of the model, for each combination of parameters we ran 100 simulations; therefore, we tested 34 combinations of parameters. In each simulation, all parameters except one were set to their default values.

The following default values of parameters were used: *prob_in* (probability of a new visitor entering POI) set to 5.0, *max_humans* (max number of visitors who can stand in one position on a grid) set to 1.0, *mask* (probability that the new visitor is wearing a mask) set to 0.0, *status_I* (probability that the new visitor is infected) set to

5.0, *virus* (units of virus density left by an infected person in a single step) set to 1.0, *virus_death* (percentage of virus degradation in each step) set to 10.0, *virus_trans* (percentage of virus spreading to neighbouring cells) set to 20.0, *max_prob_I* (maximum probability of infection) set to 96.0, *factor_mask* (a number that divides probability of infection or amount of left virus, only if a person is wearing a mask) set to 5.0, *prob_I* (probability of infection) set to 0.1.

The experiments were carried out for the following variable parameter values (when one parameter had a value different than the default one, all other parameters were set to their default values): *prob_in*: 1.0, 5.0, 10.0, 20.0, 30.0, 40.0, 50.0, 60.0, 70.0; *mask*: 10.0, 40.0, 70.0, 90.0; *status_I*: 1.0, 5.0, 10.0, 15.0, 20.0, 30.0, 40.0, 50.0, *virus_death*: 1.0, 10.0, 20.0, 50.0, *virus_trans*: 1.0, 10.0, 20.0, 30.0, 50.0, *prob_I*: 0.05, 0.1, 0.5, 0.7.

In each simulation, the following data was collected: number of people who entered POI, number of infected people who entered POI, number of infected people recorded when programme stopped running (number of *steps* equalled 5000), number of people who left POI and the total number of people wearing a mask.

4.2 Experimental Results

Obtained results are presented on plots of dependence of the average number of the value of the modified parameter on newly infected people (SI) after leaving every POI. For each combination of parameters, the simulator computed the average and the standard deviation based on outputs of 100 simulations.

Figure 2a presents a plot of dependence on the probability of entrance. The average number of infected people grows with the increase of the modified parameter. The least growth of infection occurred in the shop.

Dependence on the probability of wearing a mask is presented in Fig. 2b. The value of this parameter does not really influence the output in the shop and the tram, but in the gym number of infected people falls with the increase of this parameter. The average number of infected people is similar in the shop and tram, but is greater in the gym.

The impact of the chances of entrance (in %) of an ill person on the number of newly infected people in POI is presented in Fig. 2c. In the shop and tram, a low increase of infections occurred, but in the gym, this increase was greater.

Figure 2d presents an impact of the per cent of viruses degrading in one step on the number of newly infected people. In all cases, the number of infections falls with the increase of this parameter, but in the gym, the values are greater.

A plot of dependence between the average number of infected people and the percentage of transmitted viruses to the surroundings is presented in Fig. 2e. The averages are similar for all values of the parameter. Results in the gym are greater than in the shop and the tram, where results are similar.

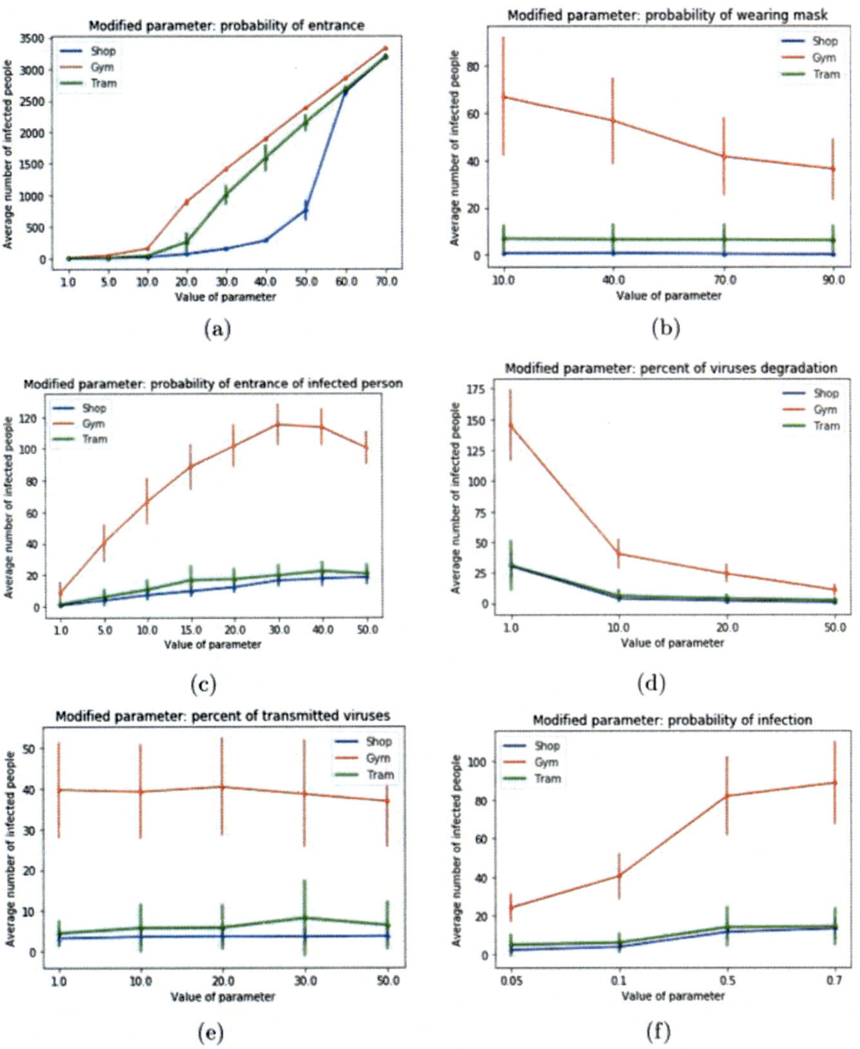

Fig. 2 Impact on the number of newly infected people of: **a** the entrance chances (in %), **b** the probability of wearing a mask, **c** the probability of entrance of infected person, **d** the per cent of virus degradation, **e** the per cent of transmitted viruses and **f** the probability of infection

Figure 2f presents the impact of the probability of infection on the average number of newly infected people during a stay in POI. In all cases, the growth occurred, but in the gym, this increase was greater than in other places.

4.3 Discussion of Results

As was expected, increasing percentage of virus degradation reduces the number of newly infected people. In addition, increasing probability of infection escalates amount of tainted. This interrelationships proof credibility of the performed simulations. Setting all the data outcome side by side marks major correlation between number of infected people and the probability of entrance as well as the probability of ill person entering the building. Prior relation might be rooting in transparent interconnection: if the entering group is sizeable, there will more likely be tainted person among them. In the fitness centre, probability of wearing mask has decreasing impact. Other parameters do not show any noticeable impact on contagiousness. Comparing examined places, there was striking contrast between the average number of people infected in the gym and averages in other localisations. It was speculated that outstanding time spent by people in the fitness centre (compared to other spots) was relevant to those characteristics.

5 Summary and Further Research

The presented model is based on cellular automata and graphs that define movement paths. Its input parameters describe spreading and infectiousness of the virus and attitudes of people, such as the probability of wearing a mask. The state of every person is stochastic and can change during the simulation. All simulations compute the number of newly infected people at the end. We observed that the largest impact on the number of infected people has the probability of entrance and the probability of initial infection. From all POIs, gym had the most infected people.

We plan to extend our simulator in several directions. Other types of POIs can be added like waiting rooms, queues, classrooms and churches. We also plan to include airflow coming from opened windows or ventilation and introduce different droplet sizes. The latter could allow us for more realistic modelling of effects of wearing a mask. Finally, we plan to introduce different aerosol emission rates associated with different activities like speaking or physical exercising.

References

1. Din A, Shah K, Seadawy A, Alrabaiah H, Baleanu D (2020) On a new conceptual mathematical model dealing the current novel Coronavirus-19 infectious disease. Results Phys 19:103510
2. Ghosh S, Bhattacharya S (2021) Computational model on COVID-19 pandemic using probabilistic cellular automata. SN Comput Sci 2(3):230
3. Giordano G, Blanchini F, Bruno R, Colaneri P, Di Filippo A, Di Matteo A, Colaneri M (2020) Modelling the Covid-19 epidemic and implementation of population-wide interventions in italy. Nat Med 26(6)

4. Schimit P (2021) A model based on cellular automata to estimate the social isolation impact on Covid-19 spreading in Brazil. Comput Methods Progr Biomed 200:105832
5. Chang S, Pierson E, Koh P, Gerardin J, Redbird B, Grusky D, Leskovec J (2021) Mobility network models of Covid-19 explain inequities and inform reopening. Nature 589(7840)
6. Ghosh S, Bhattacharya S (2021) Computational model on Covid-19 pandemic using probabilistic cellular automata. SN Comput Sci 2(3)
7. White SH, del Rey M, Sanchez GR (2009) Using cellular automata to simulate epidemic diseases. Appl Math Sci 3(20):959–968
8. Schimit PHT (2021) A model based on cellular automata to estimate the social isolation impact on Covid-19 spreading in Brazil. Comput Methods Progr Biomed 200
9. Sree PK, Usha SSSN (2019) A novel cellular automata classifier for Covid-19 trend prediction. J Health Sci 10:34–38
10. Sirakoulis GCh, Karafyllidis I, Thanailakis A (2000) A cellular automaton model for the effects of population movement and vaccination on epidemic propagation. Ecol Modell 133(3):209–223
11. Code repository of the developed simulation tool, https://github.com/aleksandrazurko/Model-epidemiologiczny/tree/main/Projekt
12. The official website of he project, https://sites.google.com/view/epidemicmodel/strona-g

Classification of Breast Tumor from Ultrasound Images Using No-Reference Image Quality Assessment

Ratnadeep Dey, Debotosh Bhattacharjee, Christian Kollmann, and Ondrej Krejcar

Abstract A computer-aided diagnosis (CAD) system can be helpful for the detection of malignant tumors in the breast. Ultrasound imaging is a type modality with low cost and lower health risk. In this paper, we have classified benign and malignant breast tumors from ultrasound images. We have used the image quality assessment approach for this purpose. No-reference image quality metrics have been used as features for the classification task. We have used a public database of ultrasound images of breast tumors containing 780 images. The classification of breast ultrasound images using image quality assessment is a very novel approach, producing significant results.

Keywords Image quality assessment · No-reference image quality metric · Breast ultrasound image · Breast cancer detection

1 Introduction

Breast cancer occurs by the unwanted growth of tissues in the breast. Mainly, the cells responsible for producing milk start growing abnormally very often, and this type of cancer is known as invasive ductal carcinoma. The other cells of the breast, called lobules, can also be affected. This type of breast cancer is known as invasive lobular carcinoma. As the causes of breast cancer, four reasons can be raised. The four

R. Dey (✉) · D. Bhattacharjee
Department of Computer Science and Engineering, Jadavpur University, Kolkata 700032, India
e-mail: rdeepdey.cse.rs@jadavpuruniversity.in

D. Bhattacharjee
e-mail: debotosh@ieee.org

D. Bhattacharjee · O. Krejcar
Center for Basic and Applied Science, Faculty of Informatics and Management, University of Hradec Kralove, Rokitanskeho 62, 500 03 Hradec Kralove, Czech Republic
e-mail: ondrej.krejcar@uhk.cz

C. Kollmann
Center for Medical Physics and Biomedical Engineering, Medical University Vienna, Waehringer Guertel 18-20, 1090 Vienna, Austria
e-mail: christian.kollmann@muv.ac.at

S. Basu et al. (eds.), *Proceedings of International Conference on Frontiers in Computing and Systems*, Lecture Notes in Networks and Systems 404,
https://doi.org/10.1007/978-981-19-0105-8_33

probable reasons are—genetic, hormonal, environmental, and lifestyle. However, there are some prominent cases of exceptions [1].

The unwanted growths in the breast tissues form tumors of different shapes and sizes. The tumors, in general, can be classified into two classes—benign and malignant. The benign tumors are not cancerous. The benign tumors do not have or have tiny live cells to divide and grow. Therefore, the growth of benign tumors is prolonged. The cells of benign tumors cannot invade, and this type of tumor can sufficiently respond to treatment.

On the other hand, malignant tumors are very fast growing, destroying healthy cells. The cells present in the malignant tumors invade through blood vessels. The malignant tumor contains cells of damaged DNAs that can divide abnormally. Therefore, cells are grown rapidly without the death of the old cell. The malignant tumor is cancerous and is a curse to human life. The identification of malignant tumor over benign tumor is an essential diagnosis step to diagnose breast cancer. The appropriate detection of malignant breast tumors can save the life of human beings.

Three image-based breast cancer screening tests have been used in general [2]. They are mammography, ultrasound imaging, and MRI. The MRI technique is not very appreciable due to its risk due to high radiation and colossal cost. Mammography is a more frequently used imaging technique for screening breast cancer. However, the case of dense breast mammograms does not provide great success. Ultrasound imaging can detect tumors in dense breasts, even for adolescent women. At the same time, an ultrasound imaging-based breast cancer screening system is less costly and has a shallow health risk. Therefore, ultrasound is one of the most frequently used breast cancer screening procedures nowadays. In Sect. 2, previous related research works have been discussed.

This paper proposed an image processing-based technique to detect malignant breast tumors over benign breast tumors using breast ultrasound imaging. We have used the image quality assessment technique for the classification task. No-reference image quality metrics have been used as features, and the two classifiers, kNN and SVM, have been used for classification. We have used a publicly available dataset containing 780 breast ultrasound images for evaluating our proposed system. Our proposed approach is very novel because breast tumor classification using image quality assessment is not available in the literature, and we have found promising results by the experiment of our proposed system on the public dataset.

The rest of the paper has been organized as follows. The following section contains literature survey, Sect. 3 contains methodology of the proposed system, Sect. 4 consists of experimental result, and Sect. 5 concludes the paper.

2 Literature Survey

Shi et al. [3] proposed a breast tumor classification system. In this system, the authors first segment the tumor region, which is the region of interest from the breast ultrasound images. Markov random field segmentation technique had been used to identify the suspicious region. Then, feature extraction had been done on the extracted regions. Three types of features, textural, fractal, and histogram-based had been extracted. Then, a stepwise regression method had been used to select the optimal subset of features. In this work, fuzzy SVM had been used for the classification task. The research work had been evaluated on a dataset of 87 images in which 36 images were malignant tumors and 51 images were benign. Silva et al. [4] extracted 22 features of five classes from the segmented area of breast tumor from an ultrasound image. The five classes were a morphological skeleton, normalized radial length, convex lesion polygon, circularity, and equivalent ellipse. Neural networks have been used for classification. The experiment was done on a dataset of 100 images, including 50 images each of malignant and benign. In work [5], multiple-ROI texture analysis of breast tumors was proposed. The work also proposed a probabilistic approach by fusing the tumor classification indicators obtained using multiple-ROI texture analysis and tumor classification indicators obtained using the multiple-ROI texture analysis of the entire tumor. In addition, morphological features were used for quantification of the shape and contour of breast tumors. This research work used breast ultrasound images of 54 benign and 46 malignant tumors. Becker et al., in their work [6], classified breast tumors using software named ViDi Suite v.2.0, ViDi system Inc. Villaz-Saint-Pierre, Switzerland. Yap et al. [7] proposed a lesion classification system using patch-based LeNet, UNet, and transfer learning with pre-trained fully connected AlexNet. They used two datasets containing 306 and 163 ultrasound breast images for the experiment. Daoud et al. [8] applied two approaches to classify breast tumors from ultrasound images. The first approach is to extract deep features, and the second approach is the use of transfer learning. AlexNet was used to extract deep features, and the authors reported that the in-depth feature-based approach performs better than the transfer learning-based approach does. In this work, two datasets having 136 and 74 images, respectively, had been used. In another work [9], deep features were combined with handcrafted features to classify breast tumors in ultrasound images. VGG-19 network had been used for the extraction of deep features. The features had been extracted at six different extraction levels. After that, the extracted features were processed using a feature selection algorithm and classified using SVM. Two types of handcrafted features—texture-based and morphological features—had been combined with the in-depth features. The system had been trained with one dataset of 360 breast ultrasound images and tested with another dataset of 163 images. Shia et al. [10] proposed a novel feature extractor pyramid histogram of oriented gradient and applied it to extract features from the breast ultrasound image.

3 Methodology

The methodology of the proposed system has been discussed in this section. We have classified the benign tumor and malignant tumor from the breast ultrasound image using image quality assessment. No-reference image quality metrics have been used as features for the classification task. Finally, two classifiers, kNN and SVM, have been used for the classification task. Figure 1 shows the complete workflow of the proposed methodology.

The breast ultrasound image is the input of the system. The complete image has been used for processing. No segmentation of the tumor area from the input image has not been done. The segmentation of the suspicious region raises the complexity of the system. Therefore, we avoid this segmentation process in preprocessing step and extract features from the complete input image. The first step of our proposed system is image quality assessment.

3.1 Image Quality Assessment

Image quality assessment is the research field of image processing where the quality of the image has been assessed. In general, image quality can be assessed in three ways—full reference, reduced reference, and no reference. In the case of full-reference image quality assessment, the reference image has been compared with the test image directly to assess the quality of the test image. Instead of the complete reference image, features of the reference image have been used for the quality assessment. The no-reference image quality assessment has no use of reference images. That means no-reference image quality is entirely independent of the reference image. In our case, a reference image for each breast ultrasound image is not available. Therefore, we have applied a no-reference image quality assessment for the proposed system.

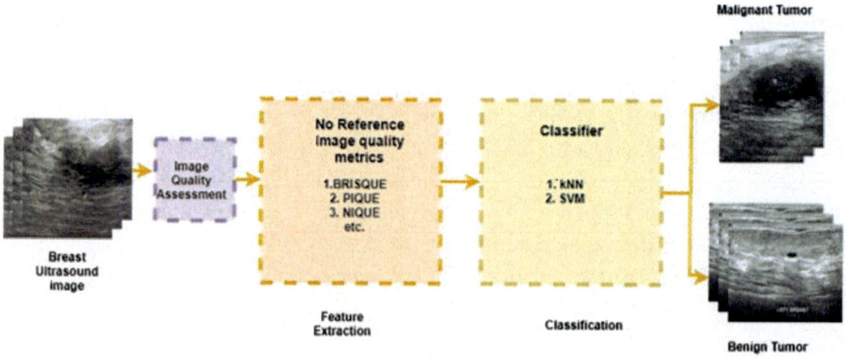

Fig. 1 Working methodology of the proposed system

3.2 *Feature Extraction*

We have used four no-reference image quality metrics as features. In this subsection, the metrics have been discussed briefly.

(i) Blind/Reference less Image Spatial Quality Evaluator (BRISQUE) [11]

BRISQUE is a no-reference image quality metrics based on natural scene statistics. The metric has been calculated in the spatial domain. There is no need for domain transformation. In addition, it is a distortion-independent image quality metric. These characteristics of the metric are suitable for this work.

(ii) Naturalness Image Quality Evaluator (NIQE) [12]

This is another no-reference image quality metric created on the natural scene statistics criteria. This metric is a no training-based image quality metric, which has been created without any supervised training process.

(iii) Perception-based Image Quality Evaluator (PIQUE) [13]

This is a no-reference image quality metric, which counts the human perception in the assessment of quality of a test image. According the methodology, the quality of local patches is calculated first. Then, the quality of the entire image is assessed.

(iv) Shannon Entropy [14]

Shannon entropy metric is on the concept of information theory. This metric measures the amount of information resides into an image. It follows a probabilistic approach for quantifying the amount of information in an image.

3.3 *Classifiers*

Our proposed system classifies between two classes—benign and malignant. We have used two classifiers for the classification task. We have used two classifiers for the comparison of results. We have used two standard classifiers—k-nearest neighbor [15] and support vector machine [16].

4 Experimental Result

4.1 *Dataset*

We have used a public dataset [17] containing 780 images, among which 487 images are benign, 210 images are malignant, and only 130 images are normal breast ultrasound images. The images were captured among women between the ages of 25 years to 75 years. The image resolution of the dataset is 500 $\times$ 500 pixels. We have used 170 images each from benign class and malignant class for training and 50 images

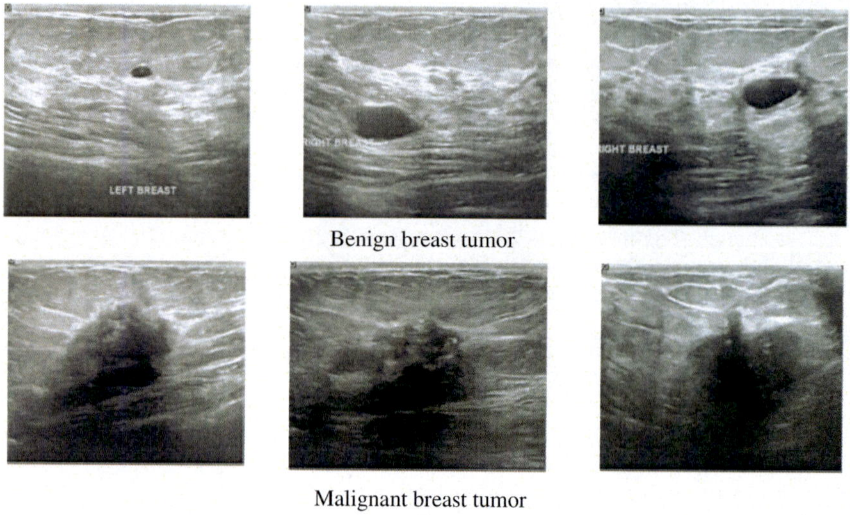

Fig. 2 Breast ultrasound images used in the experiment

from each class for testing—Fig. 2 displays images of benign breast tumors and malignant breast tumors.

4.2 *Evaluation Metric Used*

We have used different evaluation metrics for evaluating our proposed system. The evaluation metrics are as follows.

$$\text{Accuracy} = \frac{\text{TP} + \text{TN}}{\text{TP} + \text{FP} + \text{FN} + \text{TN}} \tag{1}$$

$$\text{Specificity} = \frac{\text{TN}}{\text{TN} + \text{FP}} \tag{2}$$

$$\text{Sensitivity} = \frac{\text{TP}}{\text{TP} + \text{FN}} \tag{3}$$

$$F\text{-measure} = \frac{2 \times \text{TP}}{(2 \times \text{TP}) + \text{FP} + \text{FN}} \tag{4}$$

In the above equations, the terminologies used are TP = True Positive, FP = False Positive, TN = True Negative, FN = False Negative.

Table 1 Experimental results and comparison with state-of-the-art methods

	Proposed method		Ensemble learning [18]	LBP based feature [19]
	kNN	SVM		
Accuracy	0.854	0.912	0.884	0.825
Specificity	0.885	0.924	0.903	0.81
Sensitivity	0.845	0.874	0.837	0.7
F-measure	0.778	0.811	0.805	0.8

4.3 Experimental Results

The experimental results have been shown in Table 1. Table 1 consists of values obtained for the evaluation metric from our experiment. We have used two classifiers, and results for each of the classifiers are shown. SVM performs better than the kNN. We have used a 0–1 scale for the evaluation metric values. We have compared our result with two state-of-the-art research work. In the paper [18], authors have used ensemble learning process, and LBP-based texture feature has been used in the paper [19].

5 Conclusion

In this paper, we have proposed a system for classifying breast tumors using ultrasound breast images. The proposed system uses an image quality assessment technique for the classification of breast tumors. No-reference image quality metrics have been used as features, and two classifiers, kNN and SVM, have been used for the classification task. This is a very novel approach for breast tumor classification from ultrasound images using the quality assessment approach. We have evaluated our proposed system with a public dataset containing 780 breast ultrasound images and experiment results showing that our proposed system performs significantly.

Acknowledgements The authors are grateful to the DST, Government of India, and OeAD, Austria (INT/AUSTRIA/BMWF/P-25/2018) for providing support. The work and the contribution were also supported by the SPEV project, University of Hradec Kralove, Faculty of Informatics and Management, Czech Republic (ID: 2102–2021), “Smart Solutions in Ubiquitous Computing Environments”. We are also grateful for the support of student Sebastien Mambou in consultations regarding application aspects.

References

1. Sung H, Ferlay J, Siegel RL, Laversanne M, Soerjomataram I, Jemal A, Bray F (2021) Global cancer statistics 2020: GLOBOCAN estimates of incidence and mortality worldwide for 36 cancers in 185 Countries. CA Cancer J Clin. https://doi.org/10.3322/caac.21660
2. Fiorica JV (2016) Breast cancer screening, mammography, and other modalities. Clin Obstet Gynecol 59(4):688–709. https://doi.org/10.1097/GRF.0000000000000246
3. Shi X, Cheng HD, Hu L, Ju W, Tian J (2010) Detection and classification of masses in breast ultrasound images. Digital Signal Process 20(3):824–836. https://doi.org/10.1016/j.dsp.2009.10.010
4. de Silva SD, Costa MGF, de Pereira WCA, Costa Filho CFF (2015) Breast tumor classification in ultrasound images using neural networks with improved generalization methods. In: 37th annual international conference of the IEEE engineering in medicine and biology society (EMBC). https://doi.org/10.1109/EMBC.2015.7319838
5. Daoud MI, Bdair TM, Al-Najar M, Alazrai R (2016) A fusion-based approach for breast ultrasound image classification using multiple-ROI texture and morphological analyses. In: Hindawi Publishing Corporation computational and mathematical methods in medicine. https://doi.org/10.1155/2016/6740956
6. Becker AS, Mueller M, Stoffel E, Marcon M, Ghafoor S, Boss A (2018) Classification of breast cancer in ultrasound imaging using a generic deep learning analysis software: apilot study. Br J Radiol 91:20170576
7. Yap MH, Pons G, Marti J, Ganau S, Sentis M, Zwiggelear R, Davison AK, Matri R (2018) Automated breast ultrasound lesions detection using convolutional neural networks. IEEE J Biomed Health Inf 22(4)
8. Daoud MI, Abdel-Rahman S, Alazrai R (2019) Breast ultrasound image classification using a pre-trained convolutional neural network. In: 15th international conference on signal-image technology & internet-based systems (SITIS)
9. Daoud MI, Abdel-Rahman S, Bdair TM, Al-Najar MS, Al-Hawari FH, Alazrai R (2020) Breast tumor classification in ultrasound images using combined deep and handcrafted features. Sensors 20:6838. https://doi.org/10.3390/s20236838
10. Shia WC, Lin LS, Chen DR (2021) Classification of malignant tumors in breast ultrasound using unsupervised machine learning approaches. Sci Rep 11:1418. https://doi.org/10.1038/s41598-021-81008-x
11. Mittal A, Moorthy AK, Bovik AC (2012) No-reference image quality assessment in the spatial domain. IEEE Trans Image Process
12. Mittal A, Soundararajan R, Bovik AC (2013) Making a completely blind image quality analyzer. IEEE Signal Process Lett 22(3):209–212
13. Venkatanath N, Praneeth D, Maruthi BC, Sumohana CS, Swarup MS Blind image quality evaluation using perception based features. In: Twenty First National Conference on Communications (NCC). https://doi.org/10.1109/NCC.2015.7084843
14. Vajapeyam S (2014) Understanding Shannon's Entropy metric for Information. Preprint arXiv:1405.2061
15. Cunningham P, Delany SJ (2007) k-Nearest neighbour classifiers. Technical Report UCD-CSI-2007-4 March 27
16. Hearst MA, Dumais ST, Osuna E, Platt J, Scholkopf B Support vector machines. IEEE Intell Syst Appl. https://doi.org/10.1109/5254.708428
17. Al-Dhabyani W, Gomaa M, Khaled H, Fahmy A (2020) Dataset of breast ultrasound images. Data Brief 28:104863. https://doi.org/10.1016/j.dib.2019.104863

18. Moon WK, Lee Y-W, Ke H-H, Lee SH, Huang C-S, Chang R-F (2020) Computer-aided diagnosis of breast ultrasound images using ensemble learning from convolutional neural networks. Comput Methods Programs Biomed 190:105361. ISSN 0169-2607.https://doi.org/10.1016/j.cmpb.2020.105361
19. Abdel-Nasser M, Melendez J, Moreno A, Omer OA, Puig D (2017) Breast tumor classification in ultrasound images using texture analysis and super-resolution methods. Eng Appl Artif Intell 59:84–92

Computer Networks, Communication and Security

A Switchable Bandpass Filter for Multiple Passband and Stopband Applications

Anjan Bandyopadhyay, Pankaj Sarkar, and Rowdra Ghatak

Abstract A four-state switchable microstrip filter is presented in this article. A pair of 0° fed uniform impedance resonators coupled with a pair of uniform impedance resonators are used to produce dual passband response at 2.4 GHz and 3.5 GHz, respectively. Four pin diodes are employed to switch the dual passband response to two single passband responses and one all stop response. Four transmission zeroes are observed at 2.1 GHz, 2.68 GHz, 3.26 GHz, 3.61 GHz, respectively, that help to get better selectivity and adjacent passbands isolation. In all switchable states, the filter shows an upper stopband response up to 6.0 GHz. The design layout is very compact in size of 0.21 × 0.25 λg at 2.4 GHz.

Keywords Uniform impedance resonator (UIR) · Pin diode · Bandpass filter (BPF)

1 Introduction

Reconfigurable filters are one of the essential components in modern wireless communication systems. Numerous researches are going on developing low cost, compact, and frequency agile bandpass filters. Frequency reconfiguration is carried out on the passive circuits using several types of active switches [1–8]. Microelectromechanical system (MEMS)--based switch is employed in Ref. [1] to achieve a switchable filter. A new kind of germanium telluride-based phase change switch was used in Ref. [2] to achieve X-band reconfigurable filter. A new class of tunable coplanar waveguide filter was reported in Ref. [3] using liquid metal as a tuning element. Independently switchable passband filters are discussed in Refs. [4–6] using solid-state pin diode switches. Tunable passband microstrip filters using solid-state varactor diode are discussed in Refs. [7, 8].

In this paper, pin diode switch-based four-state switchable microstrip filter is proposed. Primarily, a dual passband filter is designed using four uniform impedance

A. Bandyopadhyay (✉) · R. Ghatak
Department of ECE, NIT Durgapur, Durgapur 713209, India

P. Sarkar
Department of ECE, School of Technology, North-Eastern Hill University, Shillong 793022, India

S. Basu et al. (eds.), *Proceedings of International Conference on Frontiers in Computing and Systems*, Lecture Notes in Networks and Systems 404,
https://doi.org/10.1007/978-981-19-0105-8_34

resonators. Based on the four pin diodes on or off state, the dual-band response can switch to two different single-band responses and one all stop response. The proposed filter also offers an excellent passband selectivity and adjacent passband isolation with a wide upper stopband response up to 6.0 GHz.

2 Design of Dual-band BPF

Figure 1 shows the layout of the proposed reconfigurable filter. Four uniform impedance resonators are arranged on Teflon substrate of $\varepsilon_r = 2.2$, $\tan\delta = 0.001$, and height (h) of 0.8 mm to achieve a dual-band response. The simplified diagram is shown in Fig. 2. A pair of 0° fed UIR ($R1$ and $R2$) of impedance $Z1$, and an electrical length of $\theta 1$ is used to achieve one passband. At the same time, a pair of coupled UIR ($R3$ and $R4$) of impedance $Z2$ and electrical length of $\theta 2$ is used to achieve another passband. Under weak coupling, the resonance responses of individual resonators are shown in Fig. 3. As shown in Fig. 3, UIR $R1$ has a fundamental resonance at 2.38 GHz and a second harmonic at 4.73 GHz. On the other hand, the UIR $R2$ has its fundamental resonance at 3.48 GHz and spurious at 6.97 GHz. The optimized values of electrical lengths and impedance are $\theta 1 = 180°$, $\theta 2 = 122.4°$, $Z1 = 139.5\Omega$, $Z2 = 139.5\Omega$, respectively. Electrical lengths are calculated at 2.4 GHz. The simulated dual-band response of the proposed filter is shown in Fig. 4. The filter offers a 3 dB

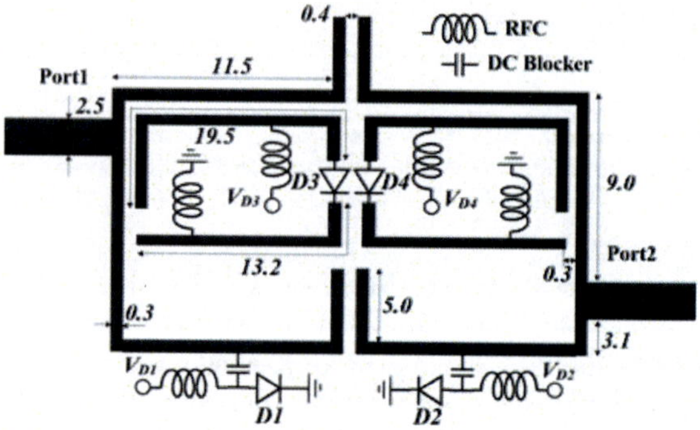

Fig. 1 Proposed layout of the reconfigurable filter. All dimensions are in *mm*

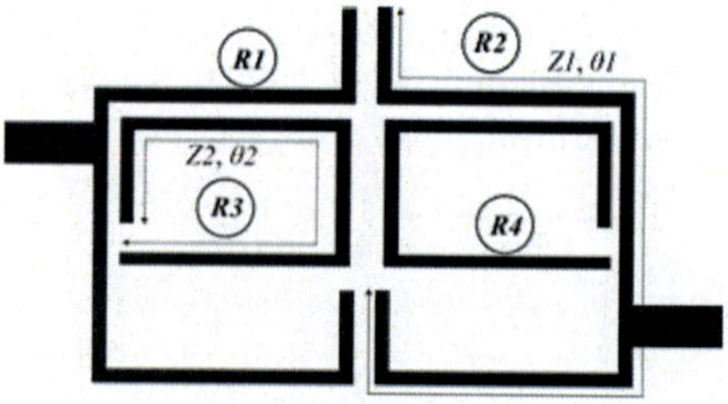

Fig. 2 Simplified layout of dual bandpass filter

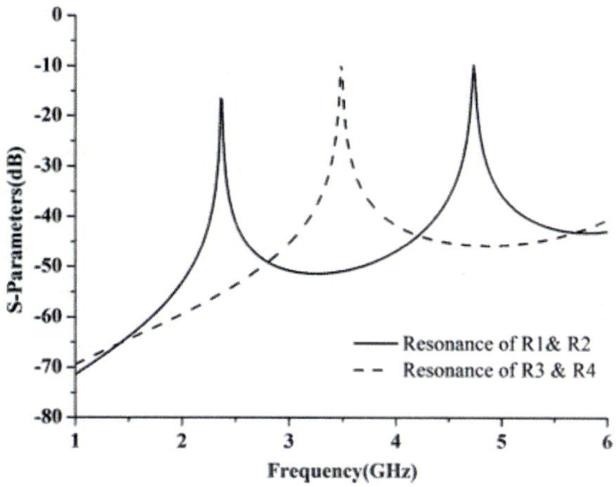

Fig. 3 Resonance modes of the resonators under weak coupling simulation

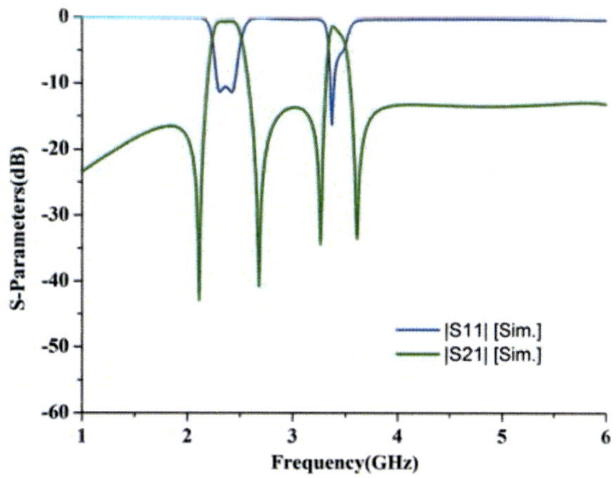

Fig. 4 Simulated dual-band response

fractional bandwidth of 12.2% and 5%, at 2.4 GHz and 3.5 GHz, respectively. Four transmission zeros at 2.1 GHz, 2.68 GHz, 3.26 GHz, and 3.61 GHz, respectively, help get excellent passband selectivity and adjacent passband isolation.

3 Reconfigurable BPF Design and Results

In order to achieve reconfigurability between the passbands, four pin diodes (*D*1–*D*4) are used as switching devices. The optimized design layout with all switching elements is shown in Fig. 1. In the simulation, the equivalent lumped model of pin diode SMP1340-079LF is used. A 2Ω resistance in series with the circuit is used to show pin diode in on state. In contrast, a 0.3pF capacitor in series with the circuit is used to make a pin diode in the off state. Figure 1 shows two pin diodes ($D1$ and $D2$) are mounted on the resonator $R1$ and resonator $R2$. Whereas, the other two pin diodes ($D3$ and $D4$) are mounted on the coupling arm of the resonator $R3$ and resonator $R4$.

To obtain dual-band response, diodes $D3$, $D4$ are kept on while diode $D1$ and $D2$ are kept off. The filter in dual-band response is shown in Fig. 4. The concept of switching of the first passband is shown in Fig. 5. As shown in Fig. 5, the pin diodes $D1$ and $D2$ are placed on the null voltage point of the second harmonic of UIR $R1$ and $R2$. While diodes $D1$ and $D2$ are on keeping $D3$ and $D4$ in on state, the fundamental frequency of resonators $R1$ and $R2$ is turned off. Eventually, 2.4 GHz band is turned off with an attenuation of more than 9.5 dB as shown in Fig. 6. The filter in single-band response at 3.5 GHz offers a 3-dB FBW of 5.8% with an upper rejection of 10 dB band up to 6 GHz. To switch off the 3.5 GHz band, diodes $D3$ and $D4$ are turned off. While pin diodes $D3$ and $D4$ are turned off, the resonators $R3$ and $R4$ are split into two sections, and eventually, the resonance frequency is shifted toward a higher frequency. As shown in Fig. 7, the proposed filter shows a single passband response at 2.4 GHz with a 3-dB FBW of 14.6%. The stopband response of more than 9.5 dB is visible up to 6 GHz. While diodes $D1$, $D2$ are kept on by

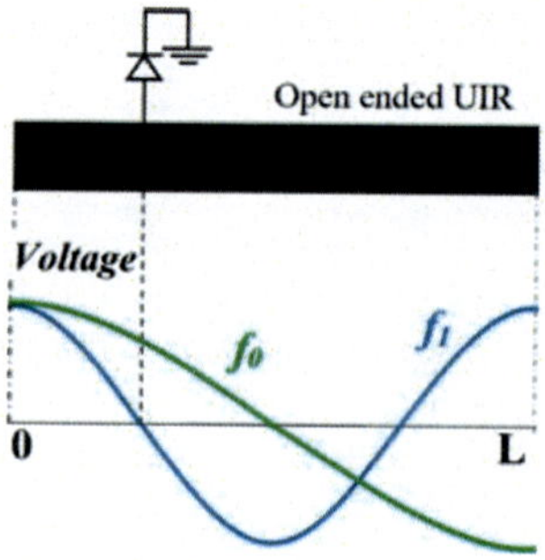

Fig. 5 Voltage distribution of the resonance modes on the UIR

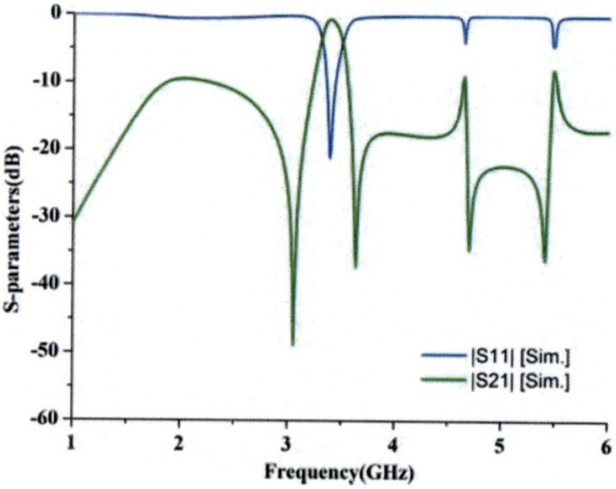

Fig. 6 Simulated response of the filter while all applied bias voltages (V_{D1}–V_{D4}) are on

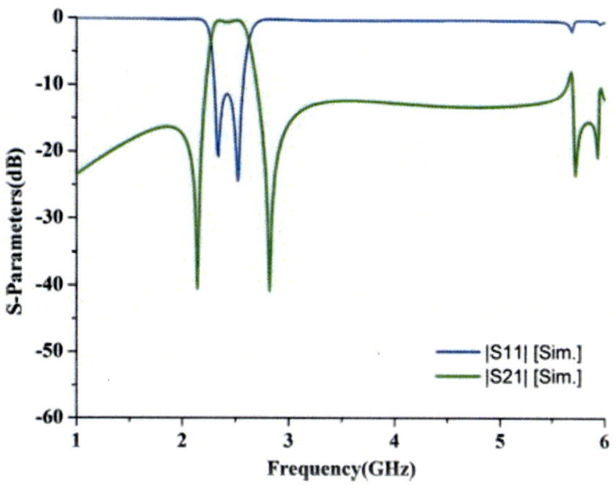

Fig. 7 Simulated response of the filter while all applied bias voltages (V_{D1}–V_{D4}) are off

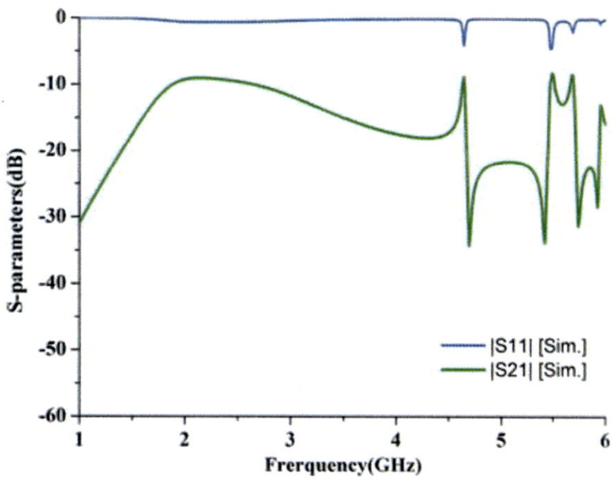

Fig. 8 Simulated all stop response while applied bias voltages V_{D1}, V_{D2} are on and V_{D3}, V_{D4} are off

keeping *D*3, *D*4 in the off state, the filter shows all stop response with an attenuation of 9 dB up to 6 GHz. The simulated all stop response is shown in Fig. 8.

4 Conclusion

The design of a compact, highly selective, four-state switchable low-cost microstrip filter has been discussed in this article. Switchability among the states has been achieved by switching on or off four pin diodes. Switchability along with a wide upper stopband characteristic makes the proposed filter suitable for wireless communications.

References

1. Chan KY, Ramer R, Mansour RR (2017) A switchable iris bandpass filter using RF MEMS switchable planar resonators. IEEE Microw Wirel Compon Lett 27(1):34–36
2. Wang M, Lin F, Rais-Zadeh M (2016) An X-band reconfigurable bandpass filter using phase change RF switches. In: 2016 IEEE 16th topical meeting on silicon monolithic integrated circuits in RF systems, Austin, TX, pp 38–41
3. Saghati AP, Batra JS, Kameoka J, Entesari K (2015) A miniaturized microfluidically reconfigurable coplanar wav guide bandpass filter with maximum power handling of 10 watts. IEEE Trans Microw Theory Tech 63(8):2515–2525
4. Kim CH, Chang K (2011) Independently controllable dual-band bandpass filters using asymmetric stepped-impedance resonators. IEEE Trans Microw Theory Tech 59(12):3037–3047
5. Chuang M-L, Wu M-T (2015) Switchable dual-band filter with common quarter-wavelength resonators. IEEE Trans Circuits Syst II Exp Briefs 62(4):347–351
6. Weng S-C, Hsu K-W, Tu W-H (2013) Independently switchable quad-band bandpass filter. IET Microw Antennas Propag 7(14):1113–1119
7. Chen Z-H, Chu Q-X (2016) Dual-band reconfigurable bandpass filter with independently controlled passbands and constant absolute bandwidths. IEEE Microw Wirel Compon Lett 26(2):92–94
8. Chen C-F (2013) A compact reconfigurable microstrip dual-band filter using varactor-tuned stub-loaded stepped-impedance resonators. IEEE Microw Wirel Compon Lett 23(1):16–18

Optimization of BBU-RRH Mapping for Load Balancing in 5G C-RAN Using Swarm Intelligence (SI) Algorithms

Voore Subba Rao and K. Srinivas

Abstract Cloud radio access network (C-RAN) is popularly known as centralized RAN which is an architecture for 5G network for process and manage cloud computing in a real-time environment. Cloud RAN (C-RAN) is popularly known as centralized RAN for providing flexibility for capital expenditure as well as operational expenditure. The benefits of C-RAN minimize the total cost ownership (TCO) and also improve the network performance. It provides benefit for low-latency network in 5G network as ultra-reliable low-latency communications (uRLLC). The 5G-CRAN enhances the benefits of not requirement of rebuild the transport networks again. C-RAN architecture is an essential and dynamically mapping of remote radio heads (RRHs) with baseband units (BBUs). Otherwise, it will cause call blocking and less quality of network connections. The proposed paper optimize to reduce the blocking of calls and also balance the load of BBUs by applying Swarm Intelligence(SI)algorithms. In proposed work, the simulation results are proved that nature-inspired computing algorithm will reduce the blocked calls and maximize the balance of processing load of BBUs.

Keywords C-RAN · BBUs · RRH · Swarm intelligence algorithms · Total cost ownership · Nature-inspired computing · Algorithms

1 Introduction

Due to increase of huge data traffic by mobile phones, real-time IoT applications and Industrial Internet of Things (IIoT) applications, the present 4G architecture is unable to store, access and process these data efficiently. The newly introduced 5G network supports the demands of client service requirements to handle such a huge data and applications efficiently. To manage such a huge data, 5G C-RAN introduces ultra-reliable and low-latency communications (uRRLCs) technology of

V. S. Rao (✉)
Department of Physics and Computer Science, Dayalbagh Educational Institute, Agra, India

K. Srinivas
Department of Electrical Engineering, Dayalbagh Educational Institute, Agra, India

S. Basu et al. (eds.), *Proceedings of International Conference on Frontiers in Computing and Systems*, Lecture Notes in Networks and Systems 404,
https://doi.org/10.1007/978-981-19-0105-8_35

5G to handle low-latency network. The 5G C-RAN architecture is reduce network energy consumption and maximize network energy efficiency [1–4].

In C-RAN, architecture is made of three components, i.e., base band unit (BBU), a remote radio unit (RRU), and a transport network called as Fronthaul [5]. BBU is a pool of centralized resources worked as a data center. RRU connect and communicate wireless devices. Fronthaul uses connectivity between BBU with RRU. The benefits of C-RAN include providing services such as resource pooling, reusable infrastructure, simple network operations, lower energy consumption, and lower capital expenditure (capex) as well as operational expenditure (opex).

In C-RAN architecture, remote radio head (RRH) is connected with base band unit (BBU) through Fronthaul. BBU connects and communicates with a host manager is a server to verify the load for every BBU. The responsible of host manager is to provide configure between BBU-RRH. Every BBU connects with number of sector, and every sector contains with various RRHs. These RRHs are be connected to single sector for a time duration. The network load of every BBU in BBU pool is depends on number of active users in BBU. The allotment of resources in C-RAN network will be managed by self-organizing network (SON) (Fig. 1).

Host Manager

- Host manager is a server to verify the load for every BBU. The responsible of host manager provides configure between BBU-RRH.
- RRHs are linked to a BBU pool through an interconnection protocol, such as the common public radio interface (CPRI).
- Every BBU connects with number of sector, and every sector contains with various RRHs. These RRHs are be connected to single sector for a time duration. The load of every BBU depends on number of active users in BBU.
- The mapping and allotment of resources in C-RAN network will be managed by self-organizing network (SON).

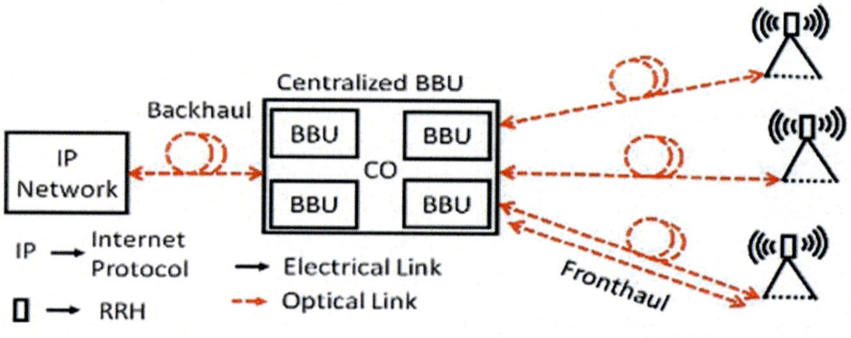

Fig. 1 C-RAN architecture

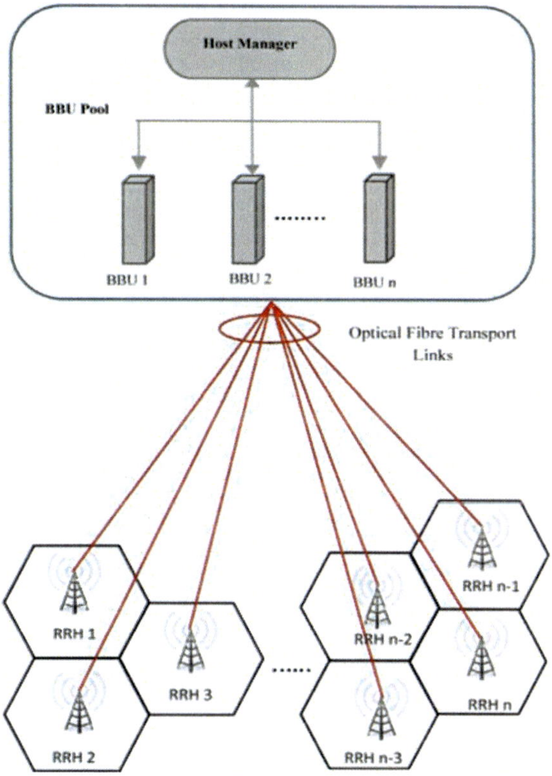

Fig. 2 C-RAN with Host manager

2 Literature Survey

In Ref. [6], authors proposed about traffic conditions of C-RAN dynamical communications between BBU and RRH. The objective of this paper is capacity routing of C-RAN for balancing of load in 5G network. Optimization is applied by genetic algorithm to optimize the balance of network traffic and minimize blocked calls in network and also improve the quality of service in C-RAN. The Quality of Server(QoS) is managed by Key Performance Indicator (KPI) is measure the efficiency of a network for serving the user. The KPI oppose the blocked calls.

In Ref. [3], authors proposed resource allocation method for 5G C-RAN. The authors address to balance network load, minimize network cost, and reach quality of services, proposed mapping between user equipment (UE) and remote radio head (RRH) and also mapping between remote radio head (RRH) and base band unit (BBU). This is NP hard problem. This should be optimized to decompose this problem into two resource allocation problems, i.e., UE-RRH association and RRH-BBU. Optimization of UE-RRH mapping (resource allocation) is optimized by artificial bee colony (ABC), and RRH-BBU mapping (clustering) is done optimized by ant colony optimization (ACO). The combination of ABC-ACO will be hybridized as bee-ant-C-RAN optimization method to minimize the resource wastage, improves spectral efficiency, and throughput.

In Ref. [7], authors proposed deep reinforcement learning (DRL) framework for C-RAN. The DRL agent is authorized on remote radio head (RRH) activates as per three phases like defined state, action, and reward function and take decision for transmit beamforming at active RRHs in every period. The proposed framework achieves higher sum rate in time-varying network.

3 Base Band Unit (BBU)—Remote Radio Head (RRH) Mapping

(i) Base Band Unit (BBU)—Remote Radio Head (RRH) Mapping

To improve quality of service (QoS) the KPI efficiently manage these blocked calls. As per key performance indicator (KPI), the host manager mapping of BBU-RRH for balancing of resources for BBUs. 20 RRHs randomly distributed to BBU pool contain two BBUs with three sectors [6]. The imbalance distribution may cause blocked calls. To reduce number of blocks calls, it is better to balancing of distribution of users of every sectors of BBU. For balancing of network the option is distribute no. of users for all sector in Base Band Unit (BBU). The balancing of users remote radio head (RRH) load in a sector is indicated by following equation as follows:

$$\text{Users}_{\text{sector}} = \sum_{j=1}^{N} \text{Users\_ RRH}_j \text{BV}_s^{i+1}, \;\; S = 1, 2, 3, \ldots K \tag{1}$$

$\text{Users}_{\text{sector}}$ = users in sector.
N = total number of RRHs.
$\text{Total}_{\text{sectors}}$ = total sectors.
Users_RRH_j = number of users connect to RRH_j.
BV = binary variable has value 1 whether RRH_j allocate in sectors.

The $\text{Users}_{\text{sector}}$ will vary as per number of users in network. Sectors automatically balance with BBUs.

$\text{Users}_{\text{sectors}}$ for K sectors will acquire lowest possible value ($\text{KPI}_{\text{minimize}}$). For that it will minimize the blocked calls and maximize quality of service (QoS) by Eq. (2) as follows.

$$\begin{gathered}\text{Key\_ Performance\_ Indicator}_{\text{minimize}} \doteq \sum_{s=1}^{k} \left(\text{users}_{\text{sector}} - \text{hard}_{\text{capacity\_ hc}}\right), \\ \left\{0, \;\; \text{if}\left(\text{users}_{\text{sector}} - \text{hard}_{\text{capacity\_ hc}}\right) < 0\right\} \text{ or } \\ \left\{\left(\text{users}_{\text{sector}} - \text{hard}_{\text{capacity\_ hc}}\right) \;\; \text{if}\left(\text{users}_{\text{sector}} - \text{hard}_{\text{capacity\_ hc}}\right) > 0\right\}.\end{gathered} \tag{2}$$

$\text{sector}_j^{i+1} = \{\text{sector}_1^{i+1}, \text{sector}_2^{i+1}, \text{sector}_3^{i+1} \ldots \text{sector}_N^{i+1}\}$ denoted as sectors of BBUs. sector_j^{i+1} and RRH allocated to these sectors.

Key Performance Indicator(KPI) is alternate to measure the quality of service (QoS) parameter. KPI for reducing the no. of blocked calls and maximizing the QoS.

4 Proposed Particle Swarm Optimization (PSO) Algorithm

To reduce blocked calls, the proposed paper implemented with particle swarm optimization(PSO) algorithm. The PSO algorithm balancing the processing load between the BBUs.

The proposed BBU-RRH mapping is implemented by swarm intelligence algorithm, i.e., particle swarm optimization (PSO). Swarm intelligence (SI) algorithms are one of the flavor of nature-inspired computing algorithms. The NIC algorithms are inspired from nature. These are model of solving computational problems of cloud computing with optimum results [8, 9]. SI algorithms are having features of self-organization, self-motivation, and collective behavior [10].

PSO is a meta-heuristic technique to solve optimization problem. Especially, the problem the linear, non-linear or mixed integer, or even our problem is a block optimization problem. The solution in optimization known as a particle or bird in swarm intelligence. Each particle or bird has position and velocity associated. In real-life particles keep change their positions by adjusting their velocity. There are many reasons particles change their positions. Do this either to seek food or avoid predators or identify environmental parameters.

Each particle keeps track its best position to identify it. So, all particles communicate their own best location and from this best location. The individual particles modified flying experience of that particle. Velocity is modified flying experience of that particular particle position and velocity of particles. The first step is to initialize the particle i is the best position and this particle has visited within the search space.

The PSO algorithms use swarm of initial particles. Every particle is similar to candidate solution. These solutions randomly have various speed and position. The swarm communicates to possible better positions among themselves and updating its own position as well as speed called as local best (P_{best}) and which swam best among all particles for position and speed called as global (G_{best}). The factors P_{best} and G_{best} are changed in every iteration [11]. The proposed work using particle swarm optimization (PSO) algorithm to optimization of BBU-RRH mapping with load balancing in 5G C-RAN [12, 13].

These following are the two equations that generation of new solution.

- Particle velocity (v) is determined as

  $Vi = wv + c_1 r_1 (P_{\text{best}}\text{-}X_i) + c_2 r_2 (G_{\text{best}}\text{-}X_i)$

  In above formula vi = velocity of ith particle

  ω = inertia of particles

$C1, c2$ = acceleration coefficients

$R1, r2$ = random numbers $\varepsilon[0, 1]$ of size $[1 \times D]$

P_{best}, i = personnel best of ith particle

G_{best}, i = global best of ith particle

Xi = position of ith particle

- Position of particle modified as
  $X_i = X_i + vi$

```
for i = 1 to |Δ| do
    Generate Particle i with random position and velocity
end
Initialise all pbest_i and gbest
I = 0
while I < I_max do
    for i = 1 to |Δ| do
        if f(x_i) > pbest_i then
            p_i = x_i
            pbest_i = f(x_i)
        end
        if pbest_i > gbest then
            gbest = pbest_i
            s = p_i
        end
    end
```

Algorithm: Particle Swarm Optimization (PSO).

The allover best particles substitute by vector is $\{\text{sector}_1^{i+1}, \text{sector}_2^{i+1}, \text{sector}_3^{i+1} \ldots \text{sector}_3^{i+1}\}$ in which every particle combination of RRHs allocate in sector of BBUs (Fig. 3 and Table 1).

In proposed work, 19 number of RRHs randomly distribute in geographical area manages by two number of BBUs with three number of sectors, hardware count (HC) = 20 for every BBU. In this approach, without considering of distribution in RRHS, consider only divide users between BBU sectors.

5 Results and Discussion

In Fig. 4a In existing work [6] produce sectors no call blocking but dis-advantage is imbalance of users. In proposed research work, sectors are uniformly allocated with no call blocking. In Fig. 4b shows existing work [6] no balancing network load of BBUs. But in proposed research work proper manage the balancing of network load in every BBU, this is evidence balancing load of all BBUs in a BBU pool of C-RAN.

The following Table 2 showing the proposed method proved an efficient in providing balancing of equal sectors.

In Fig. 5a and b, it shows the distribution of BBU sector. In Fig. 5a, it shows imbalance due to randomly distribution of sector. In proposed method, in Fig. 5b,

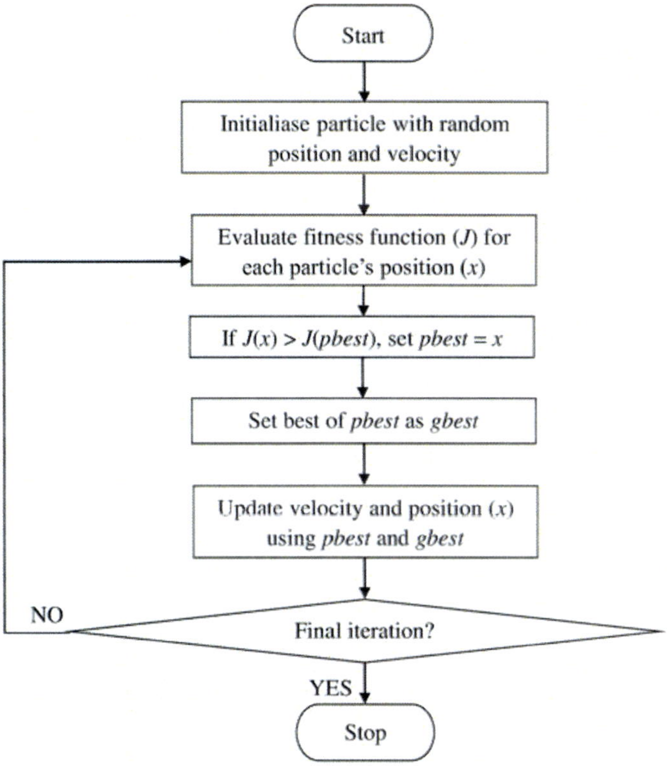

Fig. 3 Flowchart for PSO

Table 1 Parameter values

Parameter	Value
Local acceleration factor	1.8
Global acceleration factor	1.8
Population size	220
Limit number of interactions	100

results show the minimization of blocked calls and equally distributed load in BBU sectors.

The distribution of the BBU sectors can be observed in Fig. 6a and b. 20 RRHs governed by two BBUs in the BBU pool, each BBU deals with three sectors. In Fig. 6a, the sectors are distributed randomly, which can cause a possible imbalance. After the execution of the PSO algorithm, its final allocation can be noted in Fig. 6b. As well as being able to mitigate the problem of blocked calls, it can have a more equal load distribution in the BBU sectors.

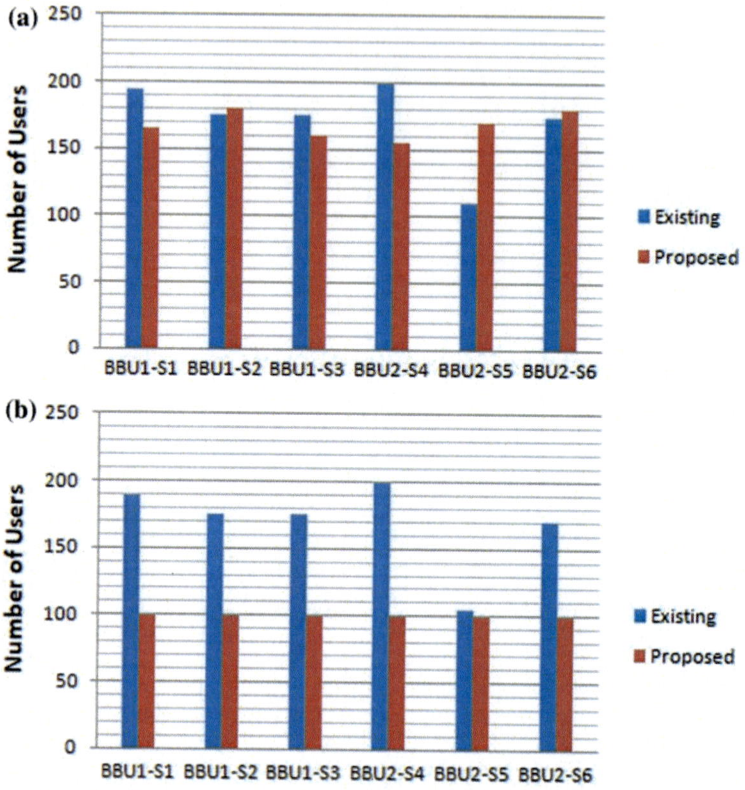

Fig. 4 **a** Number of users by sector. **b** Equal number of users balancing load of BBU

Table 2 Average number of users

Measurements (amt users)	S1	S2	S3	S4	S5	S6
Existing	195	175	175	199	110	174
Proposal	165	180	160	155	170	180

6 Conclusion

This paper is focusing on importance of nature-inspired computing algorithms to optimize cloud radio access network (C-RAN); mapping of resource allocation (RA) for BBU-RRH is a challenging task. The objective of this proposed paper is to reduce blocking calls, enhance quality of service (QoS), maximize energy efficiency and minimize energy consumption. In cloud radio access network (C-RAN), mapping of resource allocation (RA) for BBU-RRH is a challenging task. The proposed paper reduces no call blocking. For the gain of quality of service (QoS), the proposed

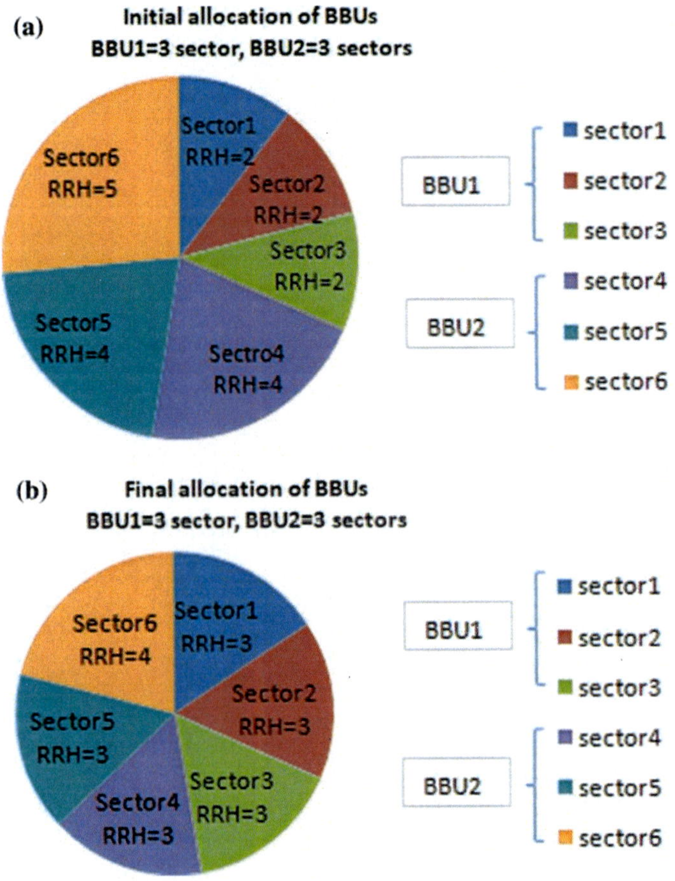

Fig. 5 **a** Starting allotment of BBU. **b** Final allocation of BBU

particle swarm optimization (PSO) is achieved equally which balance the sectors in BBU-RRH. The results showed better performance than that of existing research work [14]. In future work, we want to optimize load balancing of sectors for RRH-BBU by applying recent hybrid nature-inspired computing algorithms to gain better optimum results.

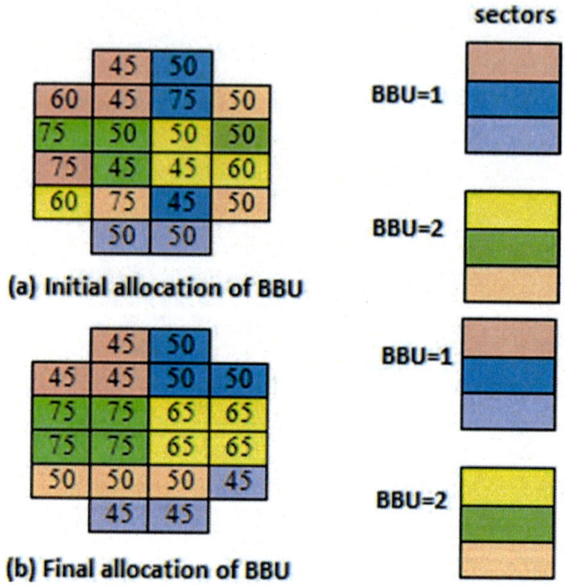

Fig. 6 RRHs governed by two BBUs in the BBU pool, each BBU deals with three sectors

References

1. C. M. R. Institute, C-RAN: the road towards green RAN, White Paper, version 3
2. Chih-Lin I, Huang J, Duan R, Cui C, Jiang JX, Li L (2014) Recent progress on C-RAN centralization and cloudification. IEEE Access 2:1030–1039
3. Ari AAA et al (2019) Resource allocation scheme for 5G C-RAN: a Swarm Intelligence based approach. Comput Netw 165:106957
4. Yuemeng T (2020) Dynamical resource allocation using modified artificial bee colony algorithm in 5G C-RAN. Waseda University, Diss
5. Ranaweera C et al (2017) 5G C-RAN architecture: A comparison of multiple optical fronthaul networks. In: 2017 International conference on optical network design and modeling (ONDM), IEEE
6. Khan M, Alhumaima RS, Al-Raweshidy HS (2015) Quality of service aware dynamic BBU-RRH mapping in cloud radio access network. In: 2015 international conference on emerging technologies (ICET), IEEE
7. Zhong C-H, Guo K, Zhao M (2021) Online sparse beamforming in C-RAN: A deep reinforcement learning approach. In: 2021 IEEE wireless communications and networking conference (WCNC), IEEE
8. Okwu MO, Tartibu LK (2021) Future of nature inspired algorithm, swarm and computational intelligence. In: Metaheuristic optimization: nature-inspired algorithms swarm and computational intelligence, theory and applications. Springer, Cham, pp 147–151
9. Abraham A, Das S, Roy S (2008) Swarm intelligence algorithms for data clustering. In: Soft computing for knowledge discovery and data mining. Springer, Boston, MA, pp 279–313
10. Waleed S et al (2021) Resource allocation of 5G network by exploiting particle swarm optimization. Iran J Comput Sci 1–9
11. Khan M, Sabir FA, Al-Raweshidy HS (2017) Load balancing by dynamic BBU-RRH mapping in a self-optimised cloud radio access network. In: 2017 24th international conference on telecommunications (ICT), IEEE

12. Khan M, Alhumaima RS, Al-Raweshidy HS (2015) Reducing energy consumption by dynamic resource allocation in C-RAN. In: 2015 European conference on networks and communications (EuCNC), IEEE
13. da Paixão EAR et al (2018) Optimized load balancing by dynamic BBU-RRH mapping in C-RAN architecture. In: 2018 third international conference on fog and mobile edge computing (FMEC), IEEE
14. Nayak J et al. (2018) Nature inspired optimizations in cloud computing: applications and challenges. In: Cloud computing for optimization: foundations, applications, and challenges. Springer, Cham, pp 1–26

Using Game Theory to Defend Elastic and Inelastic Services Against DDoS Attacks

Bhupender Kumar and Bubu Bhuyan

Abstract The QoS of elastic and inelastic service is badly affected when the network is under a DDoS attack, and minimum bandwidth requirements of these flows are not met. The current work uses game theory to address this issue of optimum allocation of bandwidth to these flows while simultaneously thwarting the DDoS attack. It quantifies the efficiency and effective data rates achieved by these flows as payoff. Similarly, it quantifies the attack payoff using the bandwidth occupancy by attack flows and the relative dropping of the elastic and inelastic flows. Further, the attacker's effort to compromise and capture a normal machine and turning it into a bot is quantified as the attack cost incurred. The scenario is modelled as a two players: nonzero sum and infinitely repeated game. The proposed mechanism enforces a data rate threshold on the flows using average available bandwidth during an interval of interest to lessen the network congestion at the bottleneck link in a dynamic manner. The corresponding payoffs are computed. Based on these payoffs, the Nash equilibrium at a particular instance of time is analysed using simulations. Subsequently, the Nash strategies are obtained which favour to enforce the optimum value of data rate threshold to maximize the combined payoff or QoS of elastic and inelastic services while making the attack costlier.

Keywords Elastic and inelastic service · DDoS attack · Game theory · Bandwidth optimization · Network security · Nash equilibrium

B. Kumar (✉) · B. Bhuyan
Department of Information Technology, North Eastern Hill University, Umshing Mawkynroh, Shillong, Meghalaya 793002, India
e-mail: bhupinder.nayak@gmail.com

B. Bhuyan
e-mail: bbhuyan@nehu.ac.in

S. Basu et al. (eds.), *Proceedings of International Conference on Frontiers in Computing and Systems*, Lecture Notes in Networks and Systems 404,
https://doi.org/10.1007/978-981-19-0105-8_36

1 Introduction

Recent digital evolution and proliferation of internet have added new dimensions to the popularity of real-time network services. The pandemic too has forced us to shift over to interactive and live communication technologies which are based on delay-sensitive Internet traffic and which are expected to reach their maximum in near future as compared to the delay-tolerant Internet traffic. Delay-tolerant Internet traffic offers services which are called as elastic services, for example, telnet, http, ftp and SMTP. Delay-sensitive Internet traffic offers inelastic services like VoIP, VoD, IPTV, videos streaming, online gaming, emergency response systems and multiparty conferencing. Availability of both of these services is severely threatened by DDoS attack. For example, DDoS on VoIP can be launched in a very similar way as on other data network such as by flooding the network bandwidth by higher payloads. It can result in inability of a legitimate user to place or receive a call [1]. The elastic services typically use TCP protocol and can tolerate the propagation delay and packet losses in case of a DDoS. They can also lower their sending rates per flow in case of network congestion. Inelastic services use RTP/UDP protocol suite. They do not tolerate any propagation delay or packet losses. Moreover, they do not decrease their sending rate per flow even if there is any network congestion. It further degrades the fair share of network bandwidth for elastic services resulting into a bandwidth starvation for them [2]. So the edge router has to decide how to allocate the fair share of the bandwidth among these services while simultaneously preventing a DDoS attack on the network. Allocating a fair share of bandwidth per flow to all services equally can prevent the elastic services from bandwidth starvation, but then the QoS of inelastic service is badly affected. One of the solutions to this problem can be increasing the bandwidth capacity or speed of the bottleneck link. However, due to dynamic nature of network traffic, it does not seem cost effective. To achieve an optimal network utility maximization (NUM), the maximum end-to-end latency of the inelastic service must be minimized while maximizing the minimum bandwidth availability to elastic service. It makes the network services available to all normal users while preventing a distributed denial of service (DDoS) attack. Abbas et al. [3] discussed the network utility maximization (NUM) issue by allocating bandwidth optimally to elastic and inelastic flows in a distributive manner. Higher dimension NUM function was decomposed into smaller subproblems. A stochastic method (surrogate sub-gradient) is used to solve the dual problem. To solve nonlinear and non-convex subproblems, a hybrid particle swarm optimization (PSO) and sequential quadratic programming (SQP) methods were used. Further, Chen et al. [4] dealt with maximizing QoS of a video conferencing among P2P nodes. QoS is modelled as a utility function which is to be maximized subjected to a node's uplink bandwidth. Similarly, Li et al. [5] modelled QoS of inelastic applications as a NUM problem in P2P network and used PSO algorithm to solve it. It was stated that the traditional gradient-based approaches may not resolve the problem of optimization due to non-convex nature of utility function in case of inelastic services. Further, Li et al. [6] considered bandwidth allocation problem for coexistent elastic and inelastic services to maximize QoS as utility in

P2P network. They formulated the non-convex problem into smaller convex problems using iterative approximation of the data rates of inelastic services. Since the problem is modified into a series of smaller subset of convex optimization problems, they used gradient-based bandwidth allocation approach to reach the suboptimal solution. A few approaches [7–9], which used game theory in this direction, were specific to TCP or TCP friendly flows only, leaving behind the issue of coexisting TCP and RTP/UDP flows. In practice, for coexisting RTP/UDP flows, fair allocation of bandwidth is a reasonable choice only if the network is not under DDoS. But if the network is under DDoS, restricting the RTP/UDP flows data rate may result into failure of real-time service. Secondly, these approaches were based on zero sum game. In reality, the payoffs may not add up to zero. It is clear that the recent approaches dealt with bandwidth allocation strategies to coexisting elastic and inelastic flows only in an attack-free network set-up. Severe QoS problem can occur if the network bandwidth is shared by coexisting, competitive attack flows. Centring on the limitations brought out above, we investigate the issue of optimal bandwidth allocation to achieve an optimal NUM; we call it network payoff maximization when the DDoS flows coexist together with elastic and inelastic flows. To the best of our knowledge, it is the first work which uses game theory for optimization of network payoff for elastic and inelastic flows while the network is under a DDoS attack. The main contribution of the paper can be summarized as follows:

1. We propose a game theoretical defence mechanism, GT4DDoS-Def, for optimal bandwidth allocation to coexisting elastic and inelastic flows when the network is under an ongoing DDoS attack. It is modelled as a nonzero sum game which is based on quantification of effective service data rates of the flows, efficiency obtained and cost incurred per bot.
2. Further, the mechanism enforces a data rate threshold based on the average available bandwidth to defend the network against DDoS attack while maximizing the combined payoff of elastic and inelastic services.

The rest of the paper is organized as follows. Section 2 presents the network model, Sect. 3 presents the corresponding game model, Sect. 4 presents the simulation and experimental results obtained, and Sect. 5 concludes the current paper while rendering some ideas to be taken up in future work.

2 The Network Model

The network model is represented by the tuple $T = (n, m, o, r_i, r_j, r_k, l, C, \tau, t)$, where n, m and o are the sets of elastic, inelastic and attack nodes, respectively. Their respective data rates are represented by r_i, r_j and r_k such that $i = (1, 2, .., n)$, $j = (1, 2, ..., m)$, $k = (1, 2, ..., o)$ and $\forall i, j, k \in N$. l is the bottleneck link having a constrained bandwidth capacity C kbps, τ is the flow rate threshold defined by our proposed GT4DDoS-Def, and t is the time interval of interest. The network traffic arriving at an edge router is a mixture of elastic, inelastic and attack flows originating from n, m and o nodes. It is to be noted that we do not differentiate between the flows originating from elastic or inelastic services. Both types of flow share the same link. However, they can be identified by the edge router based on the destination port number or socket number requested by a flow. Further, we assume arrival of a single flow from a single node to make the analysis simpler. The proposed defence mechanism GT4DDoS-Def is placed at the edge router of a LAN, inside which the destination host is located. GT4DDoS-Def can be soft coded in the edge router or it can be made as a stand-alone machine communicating with the edge router for the inputs and outputs. Initially, our mechanism assigns a fair share of bandwidth to all flows whether they are elastic, inelastic or attacking in nature. However, an elastic or inelastic flow may have less data to transmit than the allotted fair share resulting into a portion of network bandwidth un-utilized. The aggregate of such un-utilized bandwidth cause the link un-utilization during the time interval of interest t. So in the next step, the mechanism computes the average available bandwidth $B_{\text{avl}}^{\text{avg}}$ [11, 12] and diagnoses the bottleneck link's condition after time t. If the network is found to be congested, the GT4DDoS-Def enforces a τ on data rates of the flows. Based on the τ, the edge router makes the following decisions.

1. if $r_{(i,j,k)} \leq \tau$ Allow the flow to pass
2. if $r_{(i,j,k)} > \tau$ Do not allow the flow to pass. Drop it instantly.

3 The Game Model

The game is defined as $G \supseteq \{P, A, U\}$ consisting a player set P, action set A and set of payoffs U. P has two elements: GT4DDoS-Def and an attacker. For GT4DDoS-Def, A has a vector of different values of τ, and for an attacker, it has the vectors containing different values of r_k and o. U contains payoff $U_{\bar{r}}$ for GT4DDoS-Def and U_a for an attacker when they choose their actions. It is a nonzero sum infinitely repeated game which maximizes every player's payoffs.

The payoff is computed at every t interval. (1) If the payoff always increases and reaches to a global maxima, provided that n, m, o and r_i, r_k, r_j also reaches to ∞, the network is not under DoS, and fair share of the network bandwidth can be allocated in initial step. (2) If the payoff reaches to a global maxima at some finite n, m, o and r_i, r_k, r_j, the network at any time can go under DDoS. In the second case, a τ is enforced on data rates. τ is computed using Eq. 1.

Using $B_{\text{avl}}^{\text{avg}}$ to compute τ makes the attack costly. Secondly, it addresses the issue of fair share bandwidth allocation as is done in TDMA. A flow may have less data to send. Thus, it may finish data sending in less time than the time slot allotted. It makes the effective serving rate of a flow highly dynamic. So to optimally and effectively allot the bandwidth, the link's congestion state is analysed. The result obtained in the form of $B_{\text{avl}}^{\text{avg}}$ is thus used to determine τ. Conversely, same amount of data per flow may require more time slots if the total number of concurrent flows increases. It makes the bandwidth allocation more dynamic with two changing variables such as data rates and number of flows. The bandwidth has to be divided equally among all concurrent flows which affect their payoffs. Hence, τ is dynamically adjusted after every t period of time to make the attack costly.

$$\tau = \frac{C}{N} \times \left[1 - \frac{1}{B_{\text{avl}}^{\text{avg}}}\right] \tag{1}$$

3.1 Combined Payoff Function for Elastic and Inelastic Flows

To construct the payoff function of the players, one needs to quantify the cost and benefits of the player using real network parameters. For GT4DDoS-Def, we have considered the benefits solely in terms of the combined bandwidth occupied by the elastic and inelastic flows that means the effective data rates achieved by them. Consequently, we first quantify the efficiency achieved of the bottleneck link under consideration. A link's efficiency is the ratio of the actual average data that reach to a destination host to the maximum amount of data that can reach to a destination host, i.e. the full capacity of the link [10]. We denote the efficiency by E. It is given by Eq. 2, where $\bar{f}, \bar{r}$ are the number of flows that successfully reach to destination host with an average effective data rate $\bar{r}$, respectively. The $\bar{r}$ is dependent upon E of a bottleneck link. Let us suppose the fair share of bandwidth allotted to a flow is B_i fraction out of C of the link. Thus, we get Eq. 3, where r is the data rate that a flow can achieve when full fraction of C is allotted to it that is $B_i = 1$.

$$E = \frac{\bar{f}\bar{r}}{C} \tag{2}$$

$$\bar{r} = E \times B_i \times r \tag{3}$$

Now we put the value of E as given in Eq. 2 into Eq. 3 and construct the payoff function of a flow.

$$U_{\bar{r}} = \frac{\bar{f}\bar{r}}{C} \times B_i \times r \tag{4}$$

The payoff function is although same for elastic and inelastic flows but enforcing τ may yield different payoff to them. That means the effective data rates achieved may be different even when the same fraction of B_i is allotted to them. The shapes of payoff function of the elastic and inelastic services are S and flipped Z shape, respectively. It is because elastic flows are delay-tolerant flows. On the other hand, inelastic flows are delay-sensitive flows which are abysmally hit if less bandwidth is allotted to them. Their payoff suddenly increases when the required amount of bandwidth allocation is made to these flows. Following two generalized statements can be given in respect of the payoff function, where payoff function is denoted by U, number of flow by f and their respective data rates by r.

1. If, for a given value of ϵ, where $\epsilon > 0$ and the payoff function is convex (but never concave) in the neighbourhood of $[\epsilon, 0)$, then there exists some $(\bar{f}, \bar{r})$, such that

$$U_{(\bar{f},\bar{r})} > U_{(f,r)}; \quad \forall\, \bar{f} > f, \bar{r} > r \tag{5}$$

   In this case, the network is running smoothly and not under DDoS till the time $f > \bar{f}, r > \bar{r}$
2. Again if, for a given value of ϵ, where $\epsilon > 0$ and the payoff function is strictly concave and monotonically increasing everywhere in the neighbourhood of $[\epsilon, 0)$, then the network is under congestion and may suffer a sever DDoS any time.

Consequently, we can construe that the success probability of DDoS in a network depends on the shape of the curve obtained by the payoff function. In our case, n is number of elastic flows, m is number of inelastic flows and their respective data rates are r_i, r_j. Then the link efficiencies E achieved by them are given by

1. **No attack, No threshold, All flows are passed**

$$E^n_{r_i} = \frac{\sum_{i=1}^{n} r_i}{C} \tag{6}$$

$$E^m_{r_j} = \frac{\sum_{j=1}^{m} r_j}{C} \tag{7}$$

2. **Under attack, Threshold enforced, Some flows may drop**

$$E^n_{r_i} = \frac{\sum_{i=1}^{n} r_i(Pr(r_i \leq \tau))}{C} \tag{8}$$

$$E^m_{r_j} = \frac{\sum_{j=1}^{m} r_j(Pr(r_j \leq \tau))}{C} \tag{9}$$

   where Pr is the probability of the data rate of a flow to be less than or equal to τ. Consequently, the combined payoff function which is to be optimized is defined using Eqs. 4, 8 and 9.

$$U^{n,m}_{r_i,r_j} = U^n_{r_i} + U^m_{r_j} \tag{10}$$

$$U^{n,m}_{r_i,r_j} = E_{n,r_i} \cdot B_i \cdot r_i + E_{m,r_j} \cdot B_j \cdot r_j \tag{11}$$

$$U^{n,m}_{r_i,r_j} = \frac{\sum_{i=1}^{n} r_i(Pr(r_i \leq \tau)) \cdot B_i \cdot r_i}{C} + \frac{\sum_{j=1}^{m} r_j(Pr(r_j \leq \tau)) \cdot B_j \cdot r_j}{C} \tag{12}$$

3.2 Payoff Function of the Attacker

It is pretty clear from the above discussion that increasing the number of concurrent flows and increasing the data rates per flow beyond a certain limit do not increase the payoff in terms of E or $\bar{r}$ if τ is imposed on data rates. The attacker has to incur some cost ω to compromise and maintain a bot. Since, we have assumed single flow per source machine, we interchangeably use the terms flow and bot. Furthermore, a rational attacker is always interested in maximizing the attack payoff with minimum cost; it does trade-off analysis and decides whether to increase the flows (bots) o or increase the attack rates r_k per flow. In the first case, detection chances are less, but it is a costly affair. In the later case, detection and drop probabilities of flows are high, but it is less costly. Thus, the attacker has to decide optimal values of o and r_k when τ is enforced by GT4DDoS-Def. We quantify two main goals of the attacker to define its benefits. First is causing an absolute congestion on bottleneck link by consuming maximum portion of the bandwidth. Similar to elastic and inelastic flow, it is constructed using Eqs. 2 and 3. Second is causing a relative dropping of elastic and inelastic flows. It is constructed using the combined efficiency achieved by the elastic and inelastic flows. If their combined efficiency is 1, the relative impact caused on them is 0 and vice versa. Using these quantifications, the payoff function of the attacker is defined as below.

$$U^{o}_{r_k} = \alpha\left(\frac{\sum_{k=1}^{o} r_k(Pr(r_k \leq \tau))}{C}\right) + \beta\left(1 - (E^{n}_{r_i} + E^{m}_{r_j})\right) - (\omega \cdot o) \qquad (13)$$

where α and β are used as scaling parameters. The above payoff function can further be simplified by Eqs. 8 and 9. We are leaving the Nash equilibrium analysis due to space constrains but showing it in simulation and experimental analysis carried out in section follows.

4 Simulation and Experimental Results

We have conducted a set of simulations using MATLAB to verify and show how the network behaves under some given inputs when the proposed GT4DDoS-Def is invoked. Initially, we have taken n, m, o equals to 10 each with varying data rates and a fixed 10% of $B^{\text{avg}}_{\text{avl}}$. The data rate of the flows is considered as continuous random variable and modelled using normal distribution. The mean data rates are provided by the users, respectively, for all flows. We assume standard deviation $\rho = 5$ each for elastic and attack flows and $\rho = 1$ for inelastic flow to model the delay sensitivity. The CDF is used to compute various probabilities of the data rates to be less than or equal to τ. Further, we computed the maximum payoff that can be achieved by the elastic and inelastic flows separately. Thereafter, the combined payoff is also computed to show the optimum τ that can be set. The other initial user inputs are

$C = 1000$ kbps, $\alpha = \beta = 1$, $\omega = 1$ and the time interval of interest t is 100 s after which $B_{\text{avl}}^{\text{avg}}$ is rechecked and the defence game continues. We discuss four cases based on these initial inputs by a user.

1. Equal number of concurrent flows with equal mean data rates: The graph in Fig. 1a shows the resultant payoff achieved by each flow when payoff functions discussed in Sect. 3 are applied. The input values of number of flows are $n = m = o = 10$ with their respective mean data rates $r_i = r_j = r_k = 20$ kbps. The maximum payoff achieved by inelastic flows is 112.2 when the value of τ is 24 kbps while the maximum payoff that the elastic flows can achieve is 133.2 when the τ is set equals to 35 kbps. However, their combined maximum payoff is 245.5 when τ is 38 kbps. Further increasing the value of τ is no more beneficial. If that is done, the attacker can increase its mean attack data rate per flow to occupy maximum bandwidth at least for the time interval t as discussed next. However, with the given o and r_k, the attacker in this case can also maximize its attack payoff which is equal to 145.5 when τ is 38. But what we should be more interested in is the optimum value of τ, when the $B_{\text{avl}}^{\text{avg}}$ is only 10% during previous interval of interest. It is so because the network must not go under DDoS attack in case the attacker floods it with unsolicited attack data rates. So using Eq. 1, GT4DDoS-Def enforces τ equals to 30 kbps which yields an optimized combined payoff equals to 242.5. It is evident from the figure that delay-sensitive application or inelastic flows are all passed at this value of τ. However, doing so, some elastic flows can drop or wait until network conditions are improved in terms of $B_{\text{avl}}^{\text{avg}}$.

2. Equal number of concurrent flows with equal r_i and r_j but higher mean attack data rate, $r_k = 30$ kbps: The graph in Fig. 1b shows the resultant payoff of attack flow when its mean attack data rate r_k is increased to 30 kbps. Payoff and other inputs in respect of elastic and inelastic flows are same as in case 1. It is evident that when the r_k is increased to 30 kbps, the attack payoff reaches to maximum equals to 312.1 when the value of τ is set to 41 kbps. It exceeds the combined payoff obtained by the elastic and inelastic flows even at the lower value of τ that is 33 kbps. This result clearly demonstrates the effectiveness of the τ as suggested by GT4DDoS-Def and favours it. Attack payoff remains lesser than the combined payoff of elastic and inelastic flows when the τ is set at 30 kbps as computed by the GT4DDoS-Def.

3. Equal number of concurrent flows with equal r_i and r_j but highest mean attack data rate, $r_k = 60$ kbps: We further analyse one more case where the mean attack data rate is increased further to 60 kbps. The graph in Fig. 2a shows that at $r_k = 60$, the maximum achievable payoff by the attacker is 55.06 when the τ is set to 18 kbps. Question arises: How the less value of τ gives maximum payoff to attacker? The answer lies in the dropping of elastic flows at this τ, which is considered as one of the attacker's motive in designing the payoff function in respect of the attacker. The effective sending rates or efficiency achieved by these flows is badly affected at this τ. However, it is not the attacker, whose efforts have fetched it that payoff. It is on the part of the defender who sets a very low τ. The attack payoff otherwise is either negative or remains less than the combined payoff of elastic and inelastic

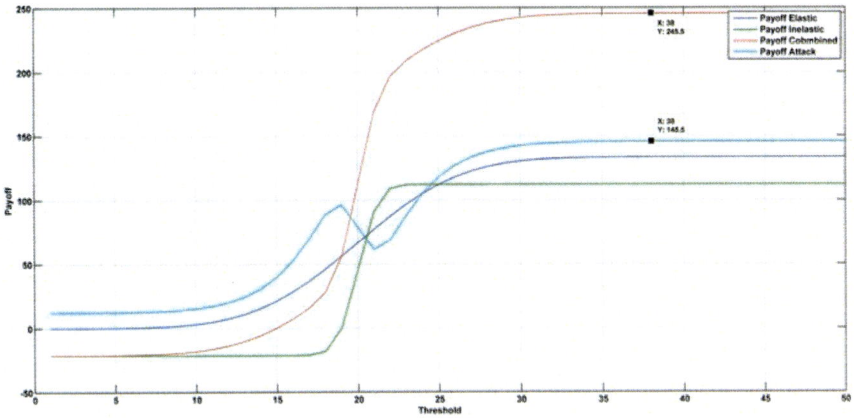

(a) Case 1: (n,m,o=10), $(r_i, r_j, r_k = 20)$

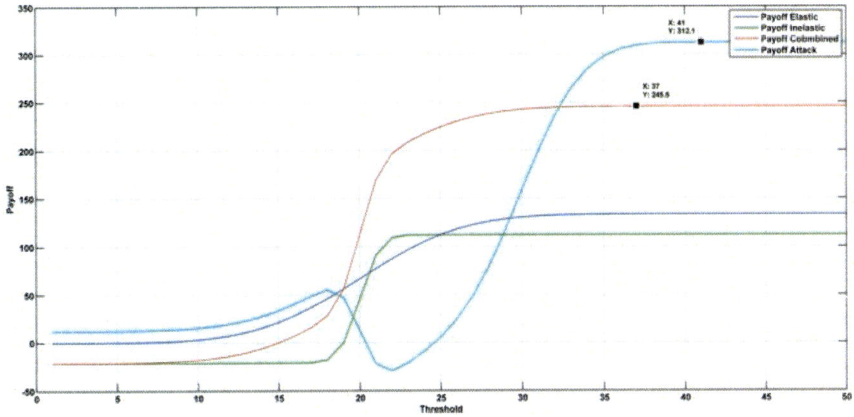

(b) Case 2: (n,m,o=10),r_k = 30 while $(r_i, r_j = 20)$

Fig. 1 Payoff variations of the flows at different data rates

flows as discussed in previous two cases. This result also favours the effectiveness of the GT4DDoS-Def.

4. Searching optimum r_k and o when $r_i = r_j = 20$ kbps, $n = m = 10$ and optimum value of τ is set at 30 kbps: Now we analyse further the maximum payoff that can be achieved by an attacker while optimizing its number of flows and the corresponding mean attack data rates. Since the GT4DDoS-Def has set τ equal to 30 kbps for an interval of 100 s, the attacker can achieve the maximum payoff 662.9 by increasing the number of flows to 50 with a mean attack data rate equals to 26.5 kbps as shown in Fig. 2b. Deviating to any other combination of o and r_k yields either negative of less payoff. This indicates that the attacker is forced to keep its sending rates below than the threshold if it wants to increase its payoff. However, it can increase the total number of flows. As discussed already, the later is a costly affair, and hence, a rational attacker refrains from choosing o beyond an optimum value. Thus, the attack is proactively prevented.

Nash Equilibrium Strategies: Now at a given instance of time and network conditions, GT4DDoS-Def and the attacker choose the best action profiles. Deviating from these action profiles gives them the negative or less payoff. In the four cases discussed above, selecting $\tau = 30$ kbps, $o = 50$, $r_k = 26.5$ are in subgame Nash equilibrium and only for the given interval of interest. Since the game runs for infinite duration, a learning mechanism or use of hash table which can store best action profiles with highest payoff can be implemented together with GT4DDoS-Def to achieve an overall Nash equilibrium of the game.

The results obtained in the cases discussed above show that higher attack data rates generated per flow can guarantee a maximum attack payoff if the value of τ is also large enough. Similarly, the maximum payoff can be obtained by the elastic and inelastic flows at the large value of τ. It is because setting large value of τ makes more bandwidth available per flow resulting in more effective sending rates and more efficiency achieved by the flows by capturing more bandwidth. However, it is not feasible keeping the link's capacity constraints in mind. Note that since the network is highly dynamic where old flows disconnect and new flows join over a time period, which results into fluctuations in fair share bandwidth allocation. It further imposes a deviation in $B_{\text{avl}}^{\text{avg}}$ and hence τ. Thus, the game is repeated over infinite number of iterations.

5 Conclusion and Future Work

Coexistent DDoS attack flows in a network environment severely affect the QoS and efficiency of elastic and inelastic flows. Flows of elastic services can tolerate transmission delay, but inelastic services get hit abysmally if their minimum bandwidth requirement is not met. Various network utility maximization schemes are proposed to solve the issue of bandwidth allotment in an attack-free network environment, but none addressed it when the network is under DDoS attack. To address it, the current work used game theory to propose a defence mechanism which enforces a data rate threshold on the flows and optimize the bandwidth allocation while defending against a DDoS. The cost and benefits of the attacker and defender are quantified to design payoff functions. Different payoffs are calculated. Through simulation, optimal threshold and Nash strategies are derived. Further, we intend to extend the scheme for infinite iterations of the game to arrive at a Pareto optimal solution and, implementing a heuristic or belief learning process also needs further study.

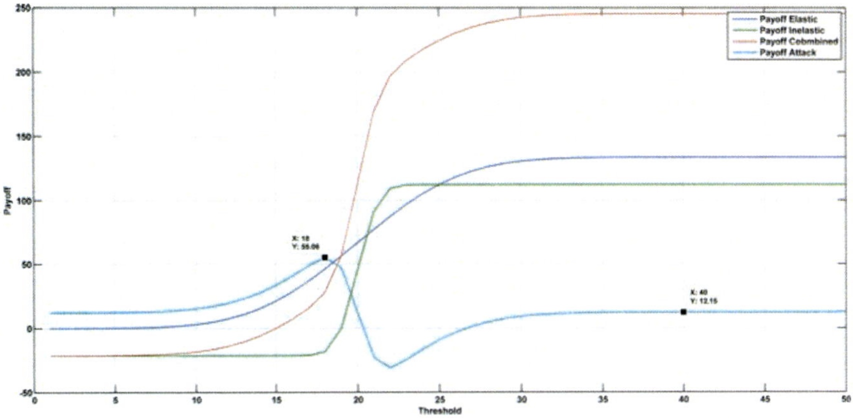

(a) Case 3: (n,m,o=10), $r_k = 60$ while $(r_i, r_j = 20)$

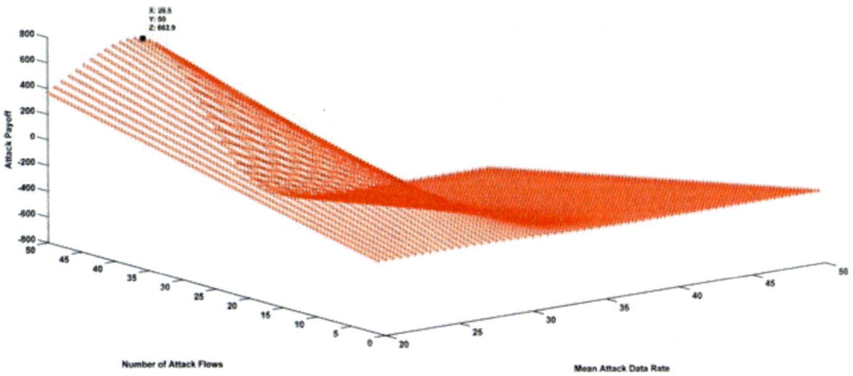

(b) Case 4: Attack payoff optimization at $\tau = 30$ while o and r_k varies and (n,m=10), $(r_i, r_j = 20)$

Fig. 2 Payoff variations of flows at different data rates and attack payoff optimization

References

1. Butcher D, LiX, Guo J (2007) Security challenge and defense in VoIP infrastructures. IEEE Trans Syst Man Cybern Part C (Appl Rev) 37(6):1152–1162
2. Chodorek A, Chodorek RR, Krempa A (2008) An analysis of elastic and inelastic traffic in shared link. In: Conference on human system interactions. IEEE, pp 873–878
3. Abbas G, Nagar AK, Tawfik H et al (2009) Quality of service issues and nonconvex Network utility maximization for inelastic services in the Internet. In: IEEE international symposium on modeling, analysis & simulation of computer and telecommunication systems, pp 1–11
4. Chen M, Ponec M, Sengupta S et al (2012) Utility maximization in peer-to-peer systems with applications to video conferencing. IEEE/ACM Trans Netw 20(6):1681–1694

5. Li S, Jiao L, Zhang Y et al (2014) A scheme of resource allocation for heterogeneous services in peer-to-peer networks using particle swarm optimization. IAENG Int J Comput Sci 44(4):482–488
6. Li S, Yuan K, Zhang Y et al (2019) A novel bandwidth allocation scheme for elastic and inelastic services in peer-to-peer networks. IAENG Int J Comput Sci 46(2):163–9
7. Kumar B, Bhuyan B (2019) Using game theory to model DoS attack and defence. Sadhana 44(12):245
8. Kumar B, Bhuyan B (2021) Game theoretical defense mechanism against bandwidth based DDoS attacks. In: Maji AK, Saha G, Das S et al (eds) Proceedings of the international conference on computing and communication systems. LNNS, vol 170. Springer
9. Bedi HS, Roy S, Shiva S (2011) Game theory-based defense mechanisms against DDoS attacks on TCP/TCP-friendly flows. In: 2011 IEEE symposium on computational intelligence in cyber security. IEEE, pp 129–136
10. Jiang Z, Ge Y, Li Y (2005) Max-utility wireless resource management for best-effort traffic. IEEE Trans Wirel Commun 4(1):100–111
11. Prasad R, Constantinos D, Margaret M et al (2003) Bandwidth estimation: metrics, measurement techniques, and tools. IEEE Netw 17(6):27–35
12. Antoniades D, Manos A, Papadogiannakis A et al (2006) Available bandwidth measurement as simple as running wget. In: Proceedings of 7th international conference on passive and active measurements (PAM), Adelaide, Australia, 30–31 Mar 2006, pp 61–70

Outage Probability of Multihop Communication System with MRC at Relays and Destination Over Correlated Nakagami-*m* Fading Channels

Rajkishur Mudoi and Hubha Saikia

Abstract This paper analyzes the outage probability of a multihop transmission link with the maximal ratio combining (MRC) diversity technique associated at the intermediate relays and the destination. The exponential and equal correlation models are considered among the input receiving antennas of MRC diversity reception at the relays and end user. The source node is assumed to have a single antenna for transmitting the signal to the node at destination. The source node transmits the signal through decode and forward type of relay nodes to the destination node. The exact-form explanations have been acquired for the outage probability of the system considering the communication through Nakagami-*m* wireless fading channels with two correlation models. The response of correlation at the outage probability presentation is analyzed. It is noticed that outage performance degrades with an increase in correlation for both cases. The consequence of outage probability for separate stages of fading parameter, diversity order, and the different number of hops has been observed.

Keywords Correlation · Maximal ratio combining (MRC) · Multihop communications · Outage probability · Nakagami-*m* fading

1 Introduction

In multihop communication, a lengthy transmission link is divided into many trustworthy short communication links, maintained by the relays. The multihop broadcasting is suitable to extend the coverage area without the requirement for enormous power at the transmitter, and it improves energy efficiency of the system [1]. In case of decode and forward (DF) relays, the received signals are decoded by the relay node and thereafter transmits the decoded signals. The multihop transmission system has earned adequate recognition at the literature in the recent time period.

R. Mudoi (✉) · H. Saikia
Department of Electronics and Communication Engineering, North-Eastern Hill University, Shillong, Meghalaya 793022, India

S. Basu et al. (eds.), *Proceedings of International Conference on Frontiers in Computing and Systems*, Lecture Notes in Networks and Systems 404,
https://doi.org/10.1007/978-981-19-0105-8_37

The article in [2] addressed a mathematical expression for the average bit error rate (ABER) of multihop DF routes for wireless networks under Nakagami-m fading channels considering different modulation techniques. The ASEP of multihop communication systems with regenerative relays for M-ary QAM was investigated in [3] over AWGN channels as well as Nakagami-m fading on all hops. In [4], ABER of M-QAM expression over the $\alpha - \mu$ fading channels depending on additive white-generalized Gaussian noise (AWGGN) was derived. In [5], the ABEP expression for BPSK and DPSK schemes in multihop amplify-and-forward (AF) system over i.n.i.d $\kappa - \mu$shadowed channels was derived. The outage probability and ASER of multihop cooperative type relay-based systems communicating under generalized-K fading channels were conferred in [6].

All these analyses consider independent fading paths for every relay and destination. However, in actual life, this is challenging to undergo an independent fading path owing to various practical limitations [7]. This paper seeks for providing the outage probability analysis of a multihop communication system with spatial correlation for an arbitrary value of m over Nakagami-m fading channels. Nakagami-m fading channel can be observed in urban environment and moreover perfectly fits many experimental statistics based on different physical distribution [7]. In this paper, we have considered two important practical correlation models, for instance exponential correlation as well as equal correlation. The equal correlation model is considered by a closely placed set of antennas, or for an antenna array placed on an equilateral triangle [8], and exponential correlation is realized, while antennas are located in a linear array [9].

The paper also introduces the channels and system model in Sect. 2, and the outage probabilities have been analyzed for both equal and exponential correlation in Sect. 3. In Sect. 4, the analytical outcomes and analysis are conferred, followed by a conclusion in Sect. 5.

2 Channels and System Model

A multihop communication link is considered with the H number of hops. The source node S communicates messages to the user or destination node D over $H-1$ regenerative, decode, and forward intermediate relays. The source node is equipped with a single antenna to transfer information via a relay network connected on a straight line to the destination. The relays are connected in series in order that one node may accept signals only from their preceding adjoining node. The intermediate relays are located at an equal distance, and therefore, all the H-hops are at the same distance. Each relay and the destination are equipped with Rh receiving antennas for MRC. Each relay node and the destination node are served by a single transmit antenna of the preceding node. The SNR of the hth hop can be indicated by γ_h where $h = 1, 2, ..., H$. The channel of the hth hop experiences flat Nakagami-m fading channels. All the relays and the destination node of the multihop system perform

MRC to raise the status of the received signal. In this MRC diversity technique, received signals from the whole diversity antennas are proportionally weighted, co-phased, as well as added together to magnify the SNR at the output of combiner. The instantaneous SNR γ_h at the output of MRC receiver is obtained as [10]:

$$\gamma_h = \sum_{r=1}^{R} \gamma_r. \tag{1}$$

where R is the total number of input diversity branches in an MRC receiver.

3 Outage Probability Analysis

The outage effect takes place when the instantaneous SNR of at least a hop falls below a determined threshold SNR γ_{th}. The outage probability maintains a necessary performance standard of service quality for wireless cellular radio communication systems [11]. Outage probability can be written as:

$$\begin{aligned} P_{\text{out}} &= 1 - \Pr(\gamma_1 > \gamma_{th}, \gamma_2 > \gamma_{th}, \ldots, \gamma_H > \gamma_{th}) \\ &= 1 - \prod_{h=1}^{H} \left\{1 - F_{\gamma_h}(\gamma_{th})\right\}. \end{aligned} \tag{2}$$

where $F_{\gamma_h}(.)$ is the CDF of the SNR for hth hop.

3a. Exponential Correlation: The receiving antennas of all the relays and the destination node are located on a linear array; therefore, the receiving antennas are influenced by exponential correlation. For exponential correlation in MRC receiver, the PDF of SNR for the hth hop can be given as [7]:

$$f_{\gamma_h}(\gamma_h) = \frac{\gamma_h^{\frac{mR^2}{a}-1} \exp\left(\frac{-mR\gamma_h}{a\overline{\gamma}}\right)}{\Gamma\left(\frac{mR^2}{a}\right)\left(\frac{a\overline{\gamma}}{mR}\right)^{\frac{mR^2}{a}}}. \tag{3}$$

where $\Gamma(.)$ is the gamma function, $\overline{\gamma}$ is the average SNR in each input branch of MRC receiver, $a = R + \frac{2\rho}{1-\rho}\left(R - \frac{1-\rho^R}{1-\rho}\right)$, and ρ is the correlation coefficient for the input diversity antennas of an MRC receiver. Applying [12, (3.381.1)], the CDF of RV γ_h can be given as

$$F_{\gamma_h}(\gamma_h) = \frac{1}{\Gamma\left(\frac{mR^2}{a}\right)} g\left(\frac{mR^2}{a}, \frac{mR\gamma_h}{a\overline{\gamma}}\right). \tag{4}$$

where the lower incomplete gamma function is symbolized as $g(.,.)$. Defining $\overline{\gamma}_N = \frac{\overline{\gamma}}{\gamma_{th}}$ as the normalized average SNR, expression of CDF can be given as:

$$F_{\gamma_h}(\gamma_{th}) = \frac{1}{\Gamma\left(\frac{mR^2}{a}\right)} g\left(\frac{mR^2}{a}, \frac{mR}{a\overline{\gamma}_N}\right). \tag{5}$$

Substituting the expression in (5) into (2), the outage probability can be expressed.

3b. Equal Correlation: Equal correlation is noticed in the diversity reception node by a closely placed set of antennas or by the arrangement of three antennas positioned on an equal-sided triangle. Considering MRC receiver and equal correlation among all the input branches, the PDF for SNR of the hth hop is written as [7]:

$$f_{\gamma_h}(\gamma_h) = \frac{\left(\frac{\gamma_h m}{\overline{\gamma}}\right)_{t,k}^{Rm-1} \exp\left(\frac{-m\gamma_h}{\overline{\gamma}(1-\rho)}\right)}{\left(\frac{\overline{\gamma}}{m}\right)(1-\rho)^{m(L-1)}(1-\rho+R\rho)^m\Gamma(Rm)} \times {}_1F_1\left(m, Rm, \frac{Rm\rho\gamma_h}{\overline{\gamma}(1-\rho)(1-\rho+R\rho)}\right). \tag{6}$$

where the confluent hypergeometric function is symbolized by ${}_1F_1(\alpha; \beta; z)$. Writing the confluent hypergeometric function in terms of infinite series representation as well as applying [12, (3.381.1)], the CDF of RV γ_h can be given as:

$$F_{\gamma_h}(\gamma_h) = \sum_{w=0}^{\infty} \frac{R^w \rho^w \Gamma(m+w)(1-\rho)}{\Gamma m \Gamma(Rm+w)(1-\rho+R\rho)^{m+w} w!} \times g\left(Rm+w, \frac{m}{\overline{\gamma}(1-\rho)}\gamma_h\right). \tag{7}$$

In terms of normalized average branch SNR $\overline{\gamma}_N$, the formulation for CDF is rewritten as:

$$F_{\gamma_h}(\gamma_{th}) = \sum_{w=0}^{\infty} \frac{R^w \rho^w \Gamma(m+w)(1-\rho)}{\Gamma m \Gamma(Rm+w)(1-\rho+R\rho)^{m+w} \mathrm{w}!} \times g\left(Rm+w, \frac{m}{\overline{\gamma}_N(1-\rho)}\right). \tag{8}$$

Substituting the expression in (8) into (2), the outage probability can be derived.

4 Analytical Consequences and Analysis

Arithmetical analysis of outage probability expressions derived in the earlier section has been carried out for exponential and equal correlation of receiving antennas. The results are framed for various numbers of R the diversity order, varying numbers of hops H, fading parameter m, and taking into account various levels of correlation coefficient ρ. Figure 1 shows, P_{out} versus $\overline{\gamma}_N$ plot of five-hop ($H = 5$) links communicating over Nakagami-m fading channels including antenna exponential correlation for different values of ρ, different fading parameters m and $R = 2$. The response of received antennas correlation on the outage can be revealed by analyzing outage probability data for $\rho = 0.5$ and $\rho = 0.9$ with reference to the data for $\rho = 0$ (uncorrelated position). Certainly, for the increment in ρ, the destination node or receiver experiences higher outage, for a definite value of γ_{th}. The outage performance develops upon advancement in fading parameter m, which signifies a better channel. The behavior of ρ is similar for different m. It can also be observed concerning the figure that larger fading parameter with large correlation performs better than low fading parameter with no correlation.

In Fig. 2, P_{out} versus $\overline{\gamma}_N$ has been characterized for the various numbers of hops (H) with exponential correlation for $\rho = 0.7$ and $m = 2$. As expected, outage performance deteriorates with an increase in H. The best performance can be obtained using one-hop link only. Again, it can be observed from Fig. 2 that outage performance

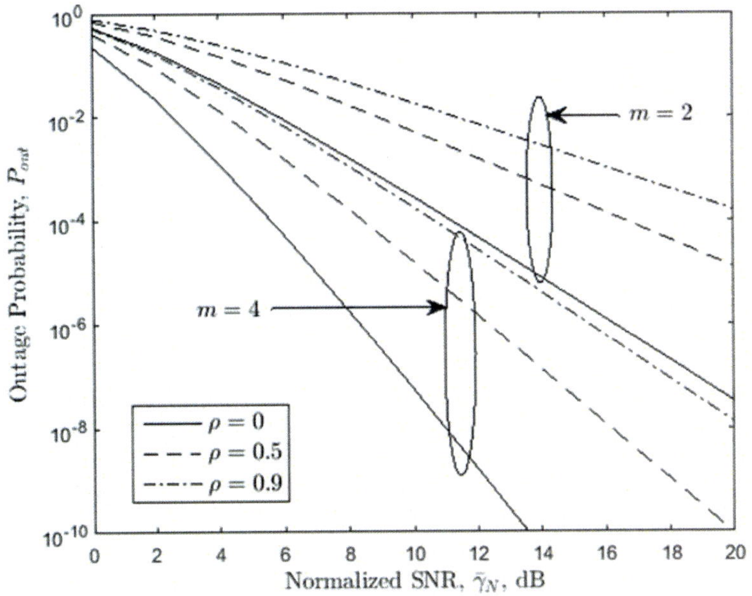

Fig. 1 Outage probability of five-hop ($H = 5$) links communication over Nakagami-m fading channels considering exponential correlation and $R = 2$

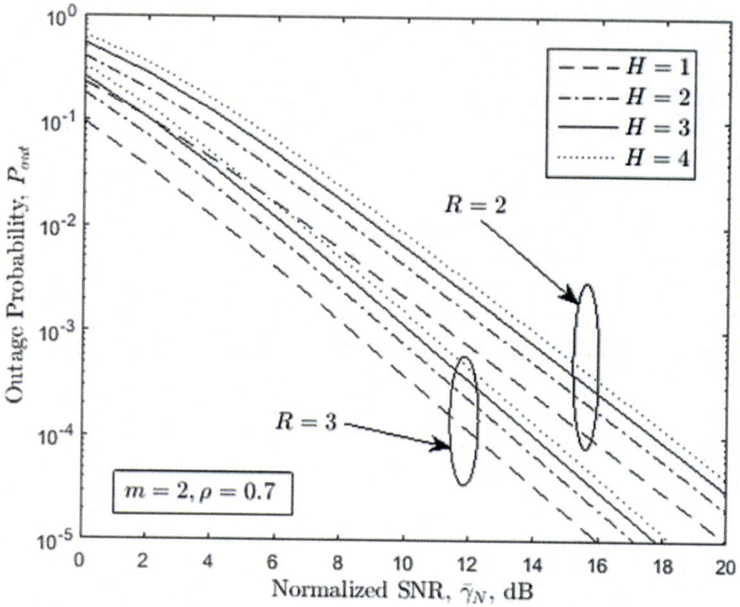

Fig. 2 Outage probability of one-hop, two-hop, three-hop, four-hop links communication over Nakagami-m fading channels considering exponential correlation, $\rho = 0.7$ and $m = 2$

develops with the rise in the diversity order or the number of receiving antennas R of a node.

In Fig. 3, P_{out} versus $\overline{\gamma}_N$ has been plotted for five-hop ($H = 5$) links communicating under Nakagami-m fading channels including equal correlation for different values of ρ, and $R = 2$. Outage performance upgrades with the increase in the fading parameter m. It can also be observed from the figure that higher fading parameter ($m = 4$) upon large correlation functions low as related to small fading parameter ($m = 2$) with no correlation in low SNR region.

In Fig. 4, P_{out} versus $\overline{\gamma}_N$ has been plotted for the different numbers of hops (H) with equal correlation for $\rho = 0.7$ and $m = 2$. As predicted, the outage performance deteriorates with increment in H, whereas the outage performance upgrades upon the increment in the diversity order R.

From Figs. 1–4, it is apparent that with $\rho = 0$, outage probability is almost equal for both the exponential and equal correlation models due to sufficient spacing among the receiving antennas of a node. The results show that an increment in the quantity of diversity branches R results in better outage performance. The outage performance degrades with a hike in the number of hops (H).

A good fading channel is always acceptable for a superior system. Enlarging the number of input antennas on a node and avoiding the antennas correlation, i.e., by putting suitable distance between any two antennas, the system functioning may be

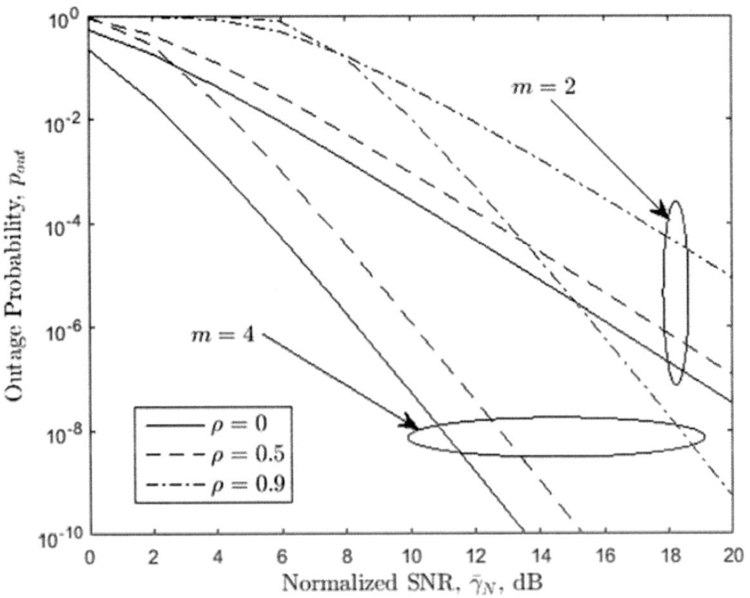

Fig. 3 Outage probability of five-hop ($H = 5$) links under Nakagami-m fading channels with equal correlation and $R = 2$

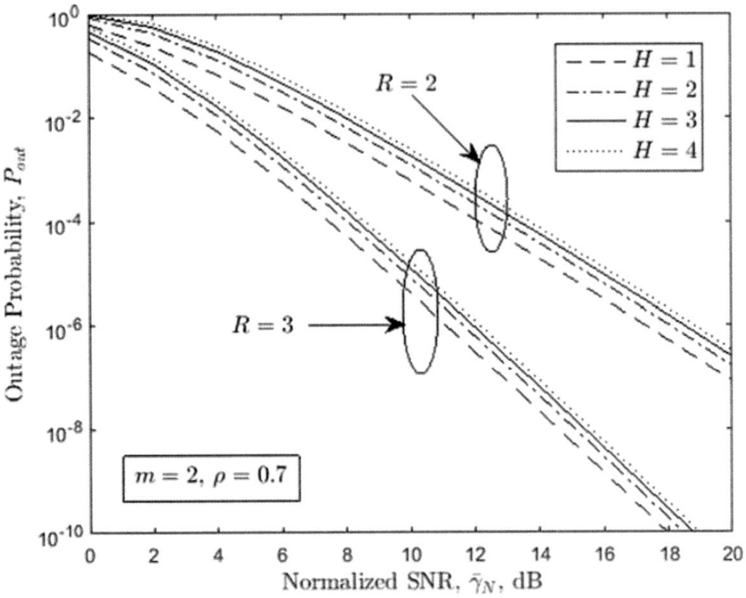

Fig. 4 Outage probability of one-hop, two-hop, three-hop, four-hop links communication under Nakagami-m fading channels considering equal correlation, $\rho = 0.7$ and $m = 2$

developed. Although high weightage on the hops (H) causes frequent outage of the system, the transmitting power can be minimized, and coverage area may also be expanded.

5 Conclusions

This paper analyzed outage probability for multihop links communicating over Nakagami-m fading channels considering exponential and equal correlation amidst the received antennas of MRC relays and the destination. Expressions are obtained in terms of incomplete gamma function. Numerically evaluated data have been arrayed to illustrate the effect of several parameters on the system outage performance.

References

1. Cao J, Yang LL, Zhong Z (2012) Performance analysis of multihop wireless links over Generalized-*K* fading channels. IEEE Trans Veh Technol 61(4):1590–1598
2. Morgado E, Mora-Jimenez I, Vinagre JJ, Ramos J, Caamano AJ (2010) End-to-end average BER in multihop wireless networks over fading channels. IEEE Trans Wirel Commun 9(8):2478–2487
3. Muller A, Speidel J (2008) Symbol error probability of M-QAM in multihop communication systems with regenerative relays. In: VTC Spring 2008—IEEE Vehicular Technology Conference, pp 1004–1008
4. Badarneh OS, Almehmadi FS (2016) Performance of multihop wireless networks in α-μ fading channels perturbed by an additive generalized Gaussian noise. IEEE Commun Lett 20(5):986–989
5. Shaik RH, Naidu KR (2019) Performance analysis of multi-hop cooperative system under κ-μ shadowed fading channels. In: International conference on communication and signal processing, pp 587–591
6. Lateef HY, Ghogho M, McLernon D (2011) On the performance analysis of multi-hop cooperative relay networks over Generalized-*K* fading channels. IEEE Commun Lett 15(9):968–970
7. Aalo VA (1995) Performance of maximal-ratio diversity systems in a correlated Nakagami-fading environment. IEEE Trans Commun 43(8):2360–2369
8. Alexandropoulos GC, Sagias NC, Lazarakis FI, Berberidis K (2009) New results for the multivariate Nakagami-*m* fading model with arbitrary correlation matrix and applications. IEEE Trans Wirel Commun 8(1):245–255
9. Karagiannidis GK, Zogas DA, Kotsopoulos SA (2003) On the multivariate Nakagami-*m* distribution with exponential correlation. IEEE Trans Commun 51(8):1240–1244
10. Simon MK, Alouini M-S (2005) Digital communications over fading channels, 2nd edn. Wiley-Interscience. John Wiley & Sons, Inc
11. Zhang QT (1996) Outage probability in cellular mobile radio due to Nakagami signal and interferers with arbitrary parameters. IEEE Trans Veh Technol 45(2):364–372
12. Gradshteyn IS, Ryzhik IM (2000) Table of integrals, series, products, 6th edn. Academic Press Inc., New York

Outage Probability Analysis of Dual-Hop Transmission Links with Decode-and-Forward Relaying over Fisher–Snedecor *F* Fading Channels

Hubha Saikia and Rajkishur Mudoi

Abstract Cooperative communication has become more important because it extends the coverage areas of wireless communication, thus decreasing the transmission power of the base station. This paper analyzes the outage probability of decode-and-forward type relaying cooperative protocol over the Fisher–Snedecor *F* fading channels. A dual-hop transmission link is considered, where there are three nodes, i.e., a source, an intermediate relay, and a destination. The source–relay and relay–destination channels experience Fisher–Snedecor *F* fading. The outage probability performance in relation to fading parameter and shaping parameter is analyzed. Furthermore, the effects of Rayleigh fading and one-sided Gaussian fading on the outage performance of the communication system are investigated.

Keywords Cooperative communication · Decode-and-forward · Dual hop · Fisher–Snedecor *F* fading · Outage probability

1 Introduction

Cooperative communication is beneficial for extending the coverage area of wireless communication [1]. In Ref. [2], the outage probability and ABER are derived for a two-hop wireless system applying a nonregenerative relay node over an $\alpha - \mu$ fading channels. In Ref. [3], the dual-hop sending links employing variable gain amplify-and-forward (AF) type relays are analyzed, which works over the existence of both multipath fading as well as shadowing of Fisher–Snedecor *F* fading distribution. Thus, the Fisher–Snedecor *F* fading distribution is suitable to outline the mixed responses of multipath fading as well as shadowing for recent wireless communication schemes. In Ref. [4], it is made clear that Fisher–Snedecor *F* distribution may also be employed to design device-to-device transmission links at 5.8 GHz for both indoor as well as outdoor surroundings. Further, the Fisher–Snedecor *F* distribution could be formulated to several standard fading distributions for instance

H. Saikia · R. Mudoi (✉)
Department of Electronics and Communication Engineering, North-Eastern Hill University, Shillong, Meghalaya 793022, India

S. Basu et al. (eds.), *Proceedings of International Conference on Frontiers in Computing and Systems*, Lecture Notes in Networks and Systems 404,
https://doi.org/10.1007/978-981-19-0105-8_38

Nakagami-m, one-sided Gaussian, and Rayleigh distribution [4–6]. In Ref. [7], multi-hop and cooperative relay-based systems are applied to enhance the coverage and the performance gain of wireless transmission systems and cognitive radio networks over Rayleigh fading channels. The authors analyze outage probability as well as ASER of an opportunistic relaying scheme (ORS) in Ref. [8], considering outdated channel state information (CSI) in source-relay as well as relay-destination links for transmission of data. In Ref. [9], the physical layer security is examined under Fisher–Snedecor F composite fading channels. A NOMA-based relay-assisted transmission employing MRC diversity is studied in Ref. [10].

The other sections in this paper are positioned as given: The system model is conferred in Sect. 2. In Sect. 3, the outage probability of the system is evaluated. The numerical developments with discussion are analyzed in Sect. 4. Finally, the opinions extracted from the paper are added in Sect. 5.

2 System Model

A wireless system is examined that contains a single source (S), an intermediate relay (R), and a destination (D). The message is transmitted from the source to the end user by the relay, which acts as decode-and-forward (DF) type. The communication happens at two-time slots. The source node sends the message signal, and the intermediate relay receives the message, in the first time slot. The message is decoded by the relay in the second slot and forward after decoding the message to the destination node. No straight link occurs joining the source with the destination node because of intense shadowing. The instantaneous SNRs of S–R as well as R–D hops are given by [11]

$$\gamma_{S,R} = \frac{P_S}{\eta_0}\left|h_{S,R}\right|^2 \tag{1}$$

and

$$\gamma_{R,D} = \frac{P_R}{\eta_0}\left|h_{R,D}\right|^2 \tag{2}$$

where $h_{S,R}$ and $h_{R,D}$ are the channel coefficients between S to R and from R to D, respectively. P_S as well as P_R are the source power and the relay power, respectively. η_0 is the power provided by AWGN to relay as well as destination node. Both the S–R as well as R–D link experience Fisher–Snedecor F channel. The PDF of SNR for Fisher–Snedecor F channel is stated as [4]

$$f_{\gamma_k}(\gamma_k) = \frac{m^m (p\overline{\gamma}_k)^p \gamma_k^{m-1}}{B(m, p)(m\gamma_k + p\overline{\gamma}_k)^{m+p}} \tag{3}$$

In the expression (3), $k = 1$ is for the channel joining the source and relay, $k = 2$ for the channel joining the relay and destination. $\overline{\gamma}_1 = E[\gamma_1]$ and $\overline{\gamma}_2 = E[\gamma_2]$ are the average SNRs, where statistical expectation is symbolized by $E[.]$; $B(.,.)$ symbolizes the beta function [12]; the fading severity parameter and shape parameter are m and p, respectively.

Some exceptional cases of the Fisher–Snedecor F model are $p \to \infty$ for Nakagami-m fading distribution; $p \to \infty$, $m = 1$ are Rayleigh fading distribution, and $p \to \infty$, $m = 0.5$ are one-sided Gaussian distribution.

3 Outage Probability

In a dual-hop communication link, outage probability is the primary performance measure for examining the system's operation. The communication is suitable through dual-hop routing, when the channel joining directly the source and destination becomes weak. The outage probability in DF cooperative system is the probability that the SNR of one-hop falls below a specific value γ_{th}. The outage probability is given as,

$$P_{\text{out}} = \text{Prob}\{\min(\gamma_1, \gamma_2) < \gamma_{th}\} \tag{4}$$

where $\text{Prob}(\gamma_1 > \gamma_{th})$ can be defined as

$$\text{Prob}(\gamma_1 > \gamma_{th}) = \int_{\gamma_{th}}^{\infty} f_{\gamma_1}(\gamma_1) \mathrm{d}\gamma_1 \tag{5}$$

With the help of (3), it can be rewritten as,

$$\text{Prob}(\gamma_1 > \gamma_{th}) = \frac{m^m (p\overline{\gamma}_1)^p}{B(m, p)} \int_{\gamma_{th}}^{\infty} \frac{\gamma_1^{m-1}}{(m\gamma_1 + p\overline{\gamma}_1)^{m+p}} \mathrm{d}\gamma_1 \tag{6}$$

Solving the integral using [13, (3.194.2)], the expression in (6) can be obtained as

$$\text{Prob}(\gamma_1 > \gamma_{th}) = \frac{p^{2p+m-1} (\overline{\gamma}_1)^{2p+m}}{B(m, p)(m\gamma_{th})^p} {}_2F_1\left(m + p, p; p + 1; -\frac{p\overline{\gamma}_1}{m\gamma_{th}}\right) \tag{7}$$

where ${}_2F_1(.,.;.;.)$ is the hypergeometric function. Similarly, $\text{Prob}(\gamma_2 > \gamma_{th})$ can be expressed as,

$$\text{Prob}(\gamma_2 > \gamma_{th}) = \int_{\gamma_{th}}^{\infty} f_{\gamma_2}(\gamma_2)\mathrm{d}\gamma_2 \tag{8}$$

Substituting (3) into (8) and simplifying with the help of [13, (3.194.2)], it can be written as

$$\text{Prob}(\gamma_2 > \gamma_{th}) = \frac{p^{2p+m-1}(\overline{\gamma}_2)^{2p+m}}{B(m,p)(m\gamma_{th})^p} {}_2F_1\left(m+p, p; p+1; -\frac{p\overline{\gamma}_2}{m\gamma_{th}}\right) \tag{9}$$

Putting the values of (7) and (9) in (4), the outage probability can be obtained as

$$P_{\text{out}} = 1 - \left[\frac{p^{2(2p+m-1)}(\overline{\gamma}_1\overline{\gamma}_2)^{2p+m}}{\{B(m,p)\}^2(m\gamma_{th})^{2p}} \times {}_2F_1\left(m+p, p; p+1; -\frac{p\overline{\gamma}_1}{m\gamma_{th}}\right) {}_2F_1\left(m+p, p; p+1; -\frac{p\overline{\gamma}_2}{m\gamma_{th}}\right)\right] \tag{10}$$

4 Results and Discussion

The analytical presentation derived in the foregoing section has been figured out. The results are depicted for various data of fading severity and shadowing parameter. In Fig. 1, the outage probability versus average SNR for every hop is computed considering different values of m and p with $\gamma_{th} = 10$ dB. Outage probability develops into superior with a progress in fading parameter m and a fall in shadowing parameter p. In Fig. 2, the value of fading severity is varied, whereas the shadowing parameter is

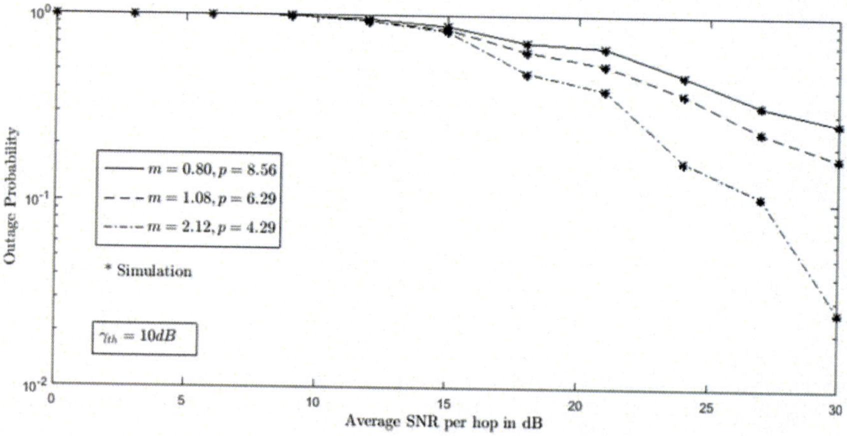

Fig. 1 Outage probability versus average SNR per hop for various values of m and p with $\overline{\gamma}_1 = \overline{\gamma}_2$

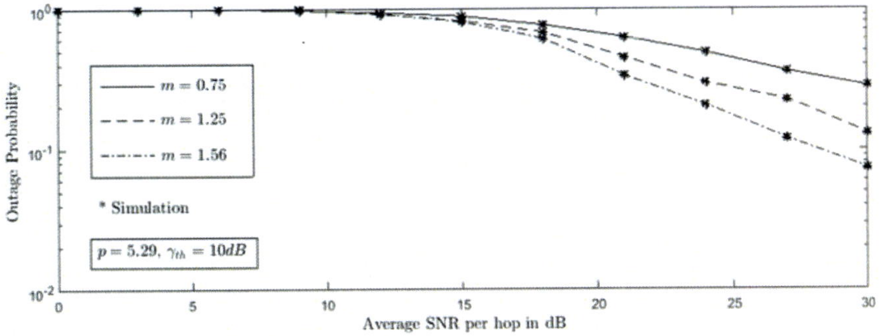

Fig. 2 Outage probability versus average SNR per hop for different values of m, keeping p constant with $\overline{\gamma}_1 = \overline{\gamma}_2$

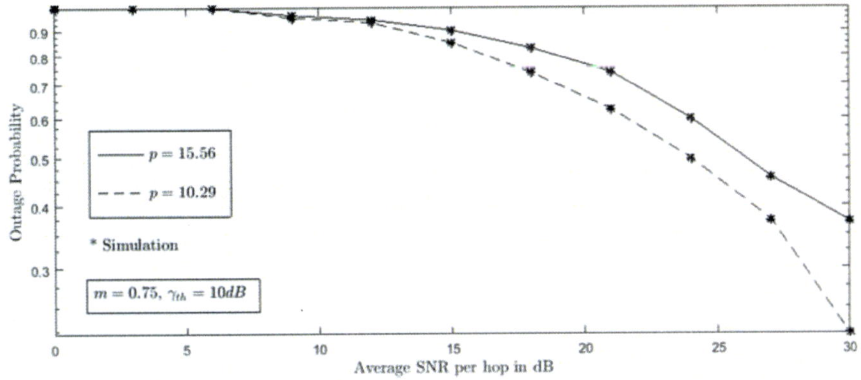

Fig. 3 Outage probability versus average SNR per hop for various values of p, keeping m constant with $\overline{\gamma}_1 = \overline{\gamma}_2$

kept constant at $p = 5.29$. In Fig. 3, the value of fading severity is kept constant at $m = 0.75$, and the shadowing parameter is varied. The average threshold SNR per hop in all two cases are kept equal at 10 dB. In both the slopes, higher the value of m and lower the value of p better the appearance of outage probability. In the Fig. 4, outage probability for the special cases of Fisher–Snedecor F fading channel, i.e., for Rayleigh fading distribution, and one-sided Gaussian distribution, is shown with $\gamma_{th} = 10$ dB. It is seen that Rayleigh gives better outage performance as compared to the one-sided Gaussian fading channel.

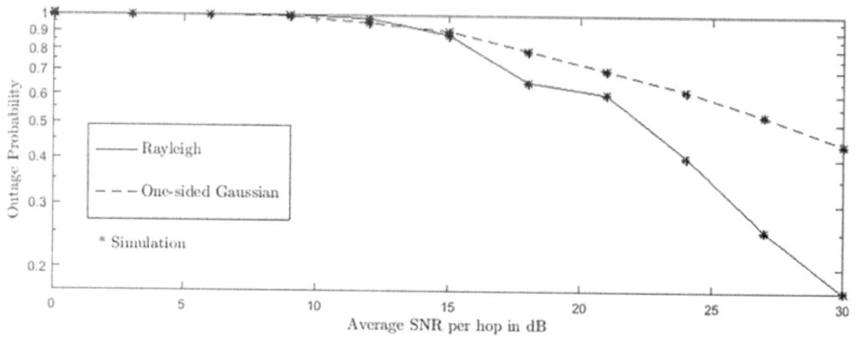

Fig. 4 Special case of Fisher–Snedecor F fading for Rayleigh and one-sided Gaussian fading

5 Conclusions

This paper examines the performance of outage probability for the Fisher–Snedecor F fading channel using dual-hop decode-and-forward type relaying communication. The expression is derived in terms of hypergeometric function and is evaluated by mathematical software. Estimated outcomes of the expression and computer-simulated outcomes are arranged in the figures. These results are in conformity with one another. Furthermore, it is recognized from the outage figures that the performance gets improvement with the increment in fading parameter and with the decrease in shadowing parameter.

References

1. Liu KJR, Kwasinski A, Su W, Sadek AK (2008) Co-operative communications and networking, First. Cambridge University Press, New York
2. Magableh AM, Aldalgamouni T, Jafreh NM (2014) Performance of dual-hop wireless communication systems over the fading channels. Int J Electron 101(6):808–819
3. Zhang P, Zhang J, Peppas KP, Ng DWK, Ai B (2020) Dual hop relaying communications over Fisher-Snedecor -fading channels. IEEE Trans Commun 68(5)
4. Yoo SK, Cotton SL, Sofotasios PC, Matthaiou M, Valkama M, Karagiannidis GK (2017) The Fisher-Snedecor distribution: a simple and accurate composite fading model. IEEE Commun Lett 21(7):1661–1664
5. Kong L, Kaddoum G (2018) On physical layer security over the Fisher-Snedecor wiretap fading channels. IEEE Access 6:39466–39472
6. Badarneh OS, da Costa DB, Sofotasios PC, Muhaidat S, Cotton SL (2018) On the sum of fisher-snedecor variates and its application to maximal-ratio-combining. IEEE Wirel Commun Lett 7(6):966–969
7. Boddapati HK, Bhatnagar MR, Prakriya S (2018) Performance of cooperative multi-hop cognitive radio networks with selective decode-and-forward relays. IET Commun 12(20):2538–2545
8. Kim S, Park S, Hong D (2013) Performance analysis of opportunistic relaying scheme with outdated channel information. IEEE Trans Wirel Commun 12(2):538–549
9. Badarneh OS, Sofotasios PC, Muhaidat S, Cotton SL, Rabie KM, Aldhahir N (2020) Achievable physical-layer security over composite fading channels. IEEE Access 8:195772–195787

10. Lv G, Li X, Xue P, Jin Y (2021) Outage analysis and optimisation of NOMA-based amplify-and-forward relay systems. IET Commun 15:410–420
11. Thakur A, Singh A (2018) Performance evaluation of dual hop decode and forward relaying protocol under generalized gamma fading channels. In: 9th ICCCNT, IISc Bengaluru
12. Petkovic MI, Ivanis PN, Diordievic GT (2018) Outage probability analysis of mixed RF-FSO system influenced by Fisher-Snedecor fading and gamma-gamma atmospheric turbulence. In: 26th telecommunications forum (TELFOR)
13. Gradesteyn IS, Ryzhik IM (2000) Table of integrals, series and products, 6th edn. Academic Press Inc., New York

Simultaneous Wireless Information and Power Transfer for Selection Combining Receiver Over Nakagami-M Fading Channels

Nandita Deka and Rupaban Subadar

Abstract In this paper, we derive the outage probability (OP) and average bit error rate (ABER) expression in closed form, considering arbitrary number of branches for selection combining receiver over Nakagami-m channels. Here, we assume wireless information and power transfer (WIPT) system having power splitter (PS) at the receiver side. The power splitter separates the received signal into information transmission and energy harvesting (EH) receiver. The derived expressions of our system consider arbitrary channel parameters and diversity branches. Monte Carlo simulation for the OP and ABER curves with SC receiver is matched with the closed forms. The analytical results correlate with the simulation for different variables of the diversity order, fading parameters, and power splitter factor. This type of system will be beneficial for reliable data transmission in an energy-limited scenario.

Keywords Nakagami-m fading · Selection combining · Power splitter · WIPT · Outage probability · ABER

1 Introduction

Radio signals propagate through wireless channels becomes sophisticated phenomenon caused by numerous effects including multipath fading due to reflections, refractions, and scattering. The main objective of the wireless receiver is to reduce the impact of fading as much as possible. In wireless channels, the accurate modeling helps the design engineer to reduce this effect. Various models are offered to describe the statistical behavior of multipath fading [1]. Practically, Nakagami-m fading is a basic distribution model which describes Rayleigh, Rician fading model. However, the Nakagami distribution gives more flexibility and accurate experimental data for many physical propagation channels compared to other distributions [2].

N. Deka (✉) · R. Subadar
Department of ECE, NEHU, Shillong, India

S. Basu et al. (eds.), *Proceedings of International Conference on Frontiers in Computing and Systems*, Lecture Notes in Networks and Systems 404,
https://doi.org/10.1007/978-981-19-0105-8_39

Diversity combining technique is an effective way to conflict multipath fading due to the improvement of signal-to-noise ratio (SNR) which is related to the selection branches at the receiver [3]. Selection combining (SC) selects the signal from the diversity branches having high instantaneous SNR and is considered for its simpler implementation and is frequently used in practice [4].

Nowadays, EH becomes a prominent technology to extend the lifetime battery of sensor networks [5]. In WIPT system, the radiated RF signals carry both information and energy signal simultaneously [6, 7]. For this reason, the concept of WIPT is preferred and rapidly becomes an important research area in both industry and academic fields. The WIPT technique can also simplify the complication of the wireless networks and reduce the cost of recharging and replacement of batteries. However, the energy harvesting with WIPT system has been presented with two advanced protocols, such as power splitting and time switching (TS) in Ref. [8]. In most of the works, the analysis of wireless powered communication (WPC) systems using TS technique has been focused in Refs. [9, 10]. However, to our knowledge, WIPT systems over Nakagami-m fading channels with SC diversity techniques using power splitters are not available in other literature. This motivates us to carry out the current research work.

In the context, we have derived the novel closed-form expression for OP and ABER over Nakagami-m fading channels for $L - SC$ receiver using a power splitter at the receiver side. Further, the impact of the system and fading parameters on the performance of the system have been verified through simulation results. The remaining parts of the paper are presented as described: Sect. 2 presents the proposed model for analysis purposes. Section 3 shows the output SNR of the proposed receiver based on both CDF and PDF approaches. Performance of the proposed system has been derived in Sect. 4. In Sect. 5, the analytical results of the OP and ABER are matched with Monte Carlo simulation for different variables of the fading parameter, the number of L-branches, and the power splitter ratio β. Our work is concluded in Sect. 6.

2 System Model

We consider a slow and frequency nonselective characteristics having Nakagami-m fading distribution. Figure 1 shows the proposed model that has been considered for evaluation of the system performance.

As shown in Fig. 1, the model contains of one transmitter and the receiver set up with L antennas, respectively. Without loss of generality, assuming that there is no correlation among these antennas. The receiver uses power splitter protocol to collect the signal and breaks into two parts: one part for energy harvester with power splitter factor β and the other part for information receiver with $(1 - \beta)$.

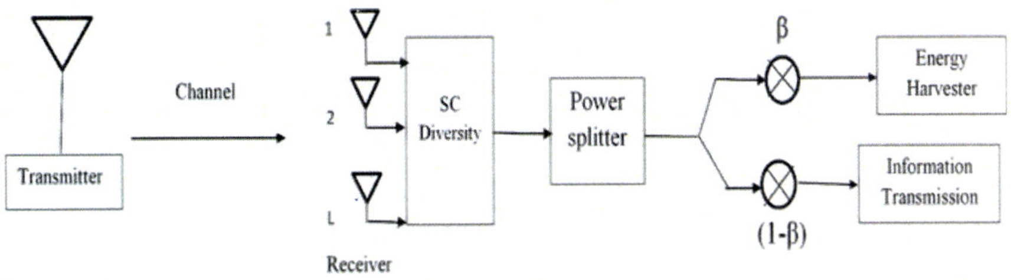

Fig. 1 System model

Here, we assume that the signal envelope a_l is the Nakagami-m fading distribution whose probability density function (PDF) is given as [11]

$$p_a(a_l) = \frac{2m^m a_l^{2m-1}}{\Psi_l^m \Gamma(m)} \exp(-\frac{ma_l^2}{\Psi_l}), \qquad a_l \geq 0 \tag{1}$$

where $\Psi_l = E[a_l^2]$ is the mean signal power, m ∈ (1/2,∞).

The PDF of SNR in the fading channel [4] is

$$f_\gamma(\gamma) = \left(\frac{m}{\bar{\gamma}}\right)^m \frac{\gamma^{m-1}}{\Gamma(m)} e^{-\frac{m\gamma}{\bar{\gamma}}}, \qquad \gamma \geq 0, m \geq \frac{1}{2} \tag{2}$$

The selection combiner selects the best signal with maximum SNR [1] which can be mathematically expressed as

$$\gamma_{SC} = 1 \leq \overset{\max}{l} \leq L\{\gamma_l\} \tag{3}$$

For Nakagami-m fading channel the PDF given in Eq. (2), the corresponding CDF can be obtained, by using [12, (3.381.1)],

$$F_\gamma(\gamma_0) = \frac{(m-1)!}{\Gamma(m)} \left(1 - e^{-\frac{m}{\bar{\gamma}_N}} \sum_{c=0}^{m-1} \frac{\left(\frac{m}{\bar{\gamma}_N}\right)^c}{c!}\right) \tag{4}$$

where g (.,.) is the incomplete gamma function.

3 The CDF and PDF of SC Receiver

This receiver selects diversity branches having high SNR. Thus, the CDF of output SNR can be obtained as

$$F_{\gamma_{SC}}(\gamma) = \prod_{l=1}^{L} F_{\gamma}(\gamma_0) \tag{5}$$

In the above equation, by inserting Eq. (4), and applying binomial theorem, we get

$$F_{\gamma_{SC}}(\gamma) = \sum_{b=0}^{L} \sum_{\substack{c_n=0 \\ n=1,2,\ldots b}}^{m-1} (-1)^b \binom{L}{b} \frac{\left(\frac{m}{\overline{\gamma}}\right)^{\sum_{n=1}^{b} c_n} \prod_{i=1}^{L} (m_i - 1)!}{\{\Gamma(m)\}^L \prod_{n=1}^{b} c_n!} \mathrm{e}^{-\frac{mb\gamma_{SC}}{\overline{\gamma}}} \gamma_{SC}^{\sum_{n=1}^{b} c_n} \tag{6}$$

The output SNR PDF can be obtained by differentiating Eq. (6) with γ. After simplification, the obtained expression gives

$$f_{\gamma_{SC}}(\gamma) = \sum_{b=0}^{L} \sum_{\substack{c_n=0 \\ n=1,2,\ldots b}}^{m-1} (-1)^b \binom{L}{b} \frac{\left(\frac{m}{\overline{\gamma}}\right)^{\sum_{n=1}^{b} c_n} \prod_{i=1}^{L} (m_i - 1)!}{\{\Gamma(m)\}^L \prod_{n=1}^{b} c_n!} \times \mathrm{e}^{-\frac{mb\gamma_{SC}}{\overline{\gamma}}} \left[\sum_{n=1}^{b} c_n \gamma_{SC}^{\sum_{n=1}^{b} c_n - 1} - \frac{mb}{\overline{\gamma}} \gamma_{SC}^{\sum_{n=1}^{b} c_n} \right] \tag{7}$$

Further, the received signal is inserted into a PS, where the signals are split into the information and energy harvester separately. Applying RV transformation, the expression for the PDF can be given as

$$f_{SC}(s) = \sum_{b=0}^{L} \sum_{\substack{c_n=0 \\ n=1,2,\ldots b}}^{m-1} (-1)^b \binom{L}{b} \frac{\left(\frac{m}{\overline{\gamma}}\right)^{\sum_{n=1}^{b} c_n} \prod_{i=1}^{L} (m_i - 1)!}{\{\Gamma(m)\}^L \prod_{n=1}^{b} c_n!} \times \mathrm{e}^{-\frac{mbs}{\overline{\gamma}\beta}} \left[\sum_{n=1}^{b} c_n \frac{s^{\sum_{n=1}^{b} c_n - 1}}{\beta^{\sum_{n=1}^{b} c_n - 1}} - \frac{mb}{\overline{\gamma}\beta^{\sum_{n=1}^{b} c_n}} s^{\sum_{n=1}^{b} c_n} \right] \tag{8}$$

4 Performance Evaluation

4.1 Outage Probability (OP)

Outage probability is defined as the probability with instantaneous probability of error exceeding a specific value of the output SNR γ that falls below γ_{th} [1]. The expression can be obtained by substituting $\gamma_{SC} = \gamma_{th}$ from the CDF of output SNR. The outage probability can be obtained as

$$P_{\text{out}} = \prod_{i=1}^{L} \frac{1}{\Gamma(m_i)} g\left(m_i, \frac{m_i}{\beta \overline{\gamma}_N}\right) \tag{9}$$

4.2 Aber

The ABER can be obtained using CDF of the output SNR taken from [13] that is given as

$$\overline{P_e}(s) = -\int_0^{\infty} p_e'(s) F_{sc}(s) \mathrm{d}s \tag{10}$$

where $p_e'(s)$ is the conditional error probability given by [13]

$$p_e'(s) = -\frac{\xi^{\eta} s^{\eta-1} \mathrm{e}^{-\xi s}}{2\Gamma(\eta)} \tag{11}$$

where ξ and η are constant and depend on modulation scheme. For BPSK modulation, the value is given as: $(\xi,\eta) = (1, 0.5)$.

Putting the value of $F_{SC}(s)$ and $p_e'(s)$ from Eqs. (6) and (11) and using [12, (3.462.1)], the results of ABER can be expressed as

$$\overline{P}_e = \sum_{b=0}^{L} \sum_{\substack{c_n=0 \\ n=1,2,\ldots b}}^{m-1} (-1)^b \binom{L}{b} \left(\frac{m}{\overline{\gamma}\beta}\right)^{\sum_{n=1}^{b} c_n} \times \frac{\delta^{\eta} \Gamma\left(\sum_{n=1}^{b} c_n + \eta\right) \prod_{i=1}^{L} (m_i - 1)!}{2\Gamma(\eta)\{\Gamma(m)\}^L \prod_{n=1}^{b} c_n! \left(\frac{mb}{\overline{\gamma}\beta} + \delta\right)^{\sum_{n=1}^{b} c_n + \eta}} \tag{12}$$

5 Numerical Results and Discussion

In the present section, we present the numerical and simulations results of the OP and the ABER. From the obtained results, we discuss the effect of different system parameters on the performance of the system. Figure 2 shows the impacts of diversity order L with assumption that shape parameters $m = 1$ and power splitter factor $\beta = 0.5$ on P_{out}. To consider the effect of L in P_{out}, the outage probability reduces with increasing value of the diversity order. It can be recognized that the outage probability is a decreasing function of L. The outage probability versus average SNR for $L = 2$ have been plotted in Fig. 3, with variable of m and β. It can be observed from the figure that the OP performs better with the increased value of m and β assuming $L = 2$. Also, with increase in m and β the receiver suffers more outage. Figure 4 shows the impact of the shape parameter and power splitter factor on ABER of the proposed system using BPSK with $L = 2$ SC receiver over Nakagami-m fading . From the observation, the ABER gives good performance as the value of

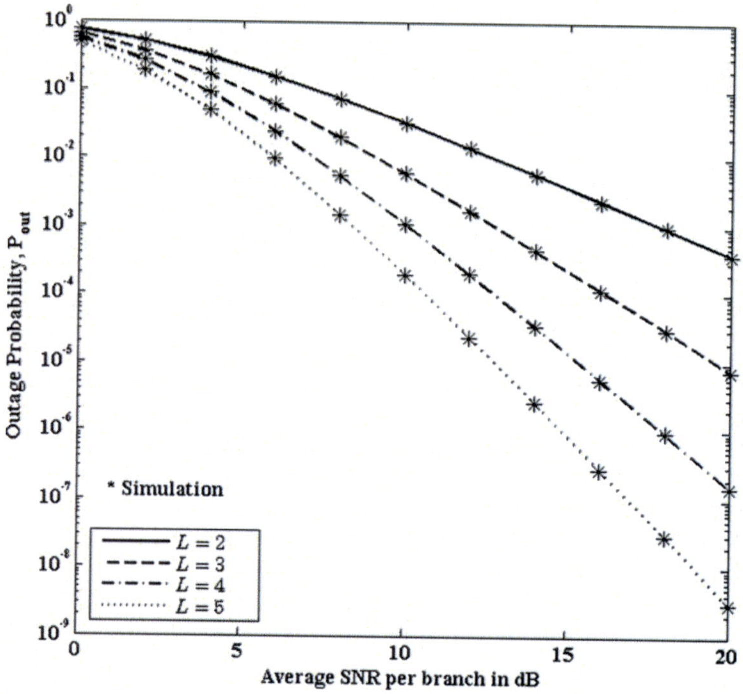

Fig. 2 Outage probability with different variables of L and $m = 1$, $\beta = 0.5$

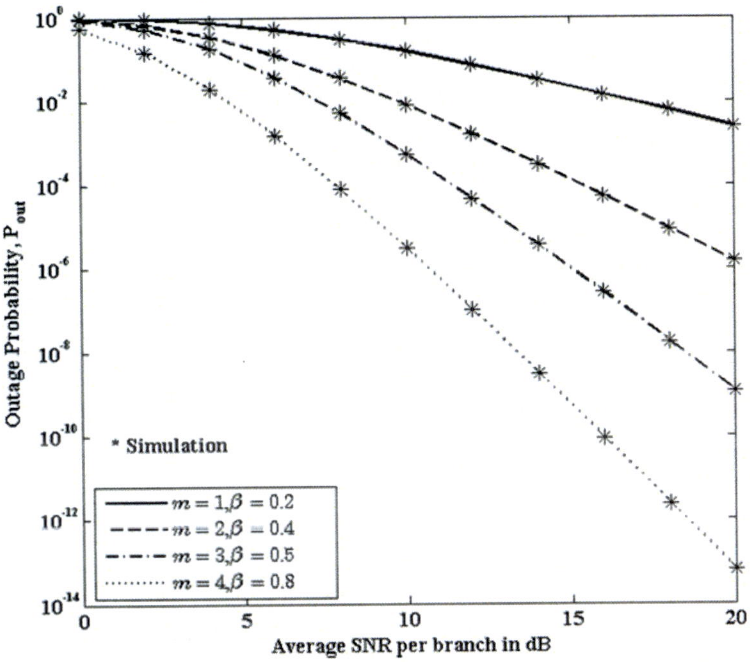

Fig. 3 Outage probability with $L = 2$ and values of m and β changes

m and β increases. Increases in m and β means decrease in shadowing effect and the performance improves. Therefore, the ABER increases when variable of m and β increases. Figure 5 shows the impact of the shape parameter and power splitter factor on ABER with $L = 4$, in similar environment. The observations are similar from both Figs. 4 and 5 except the ABER with $L = 4$ performs better than the ABER with $L = 2$. This means as the diversity order increases, the receiver performance improves. Further, with increasing diversity order the complication of the system will also increase. Thus, for sensible application there is always an exchange between the system performance and the order of diversity.

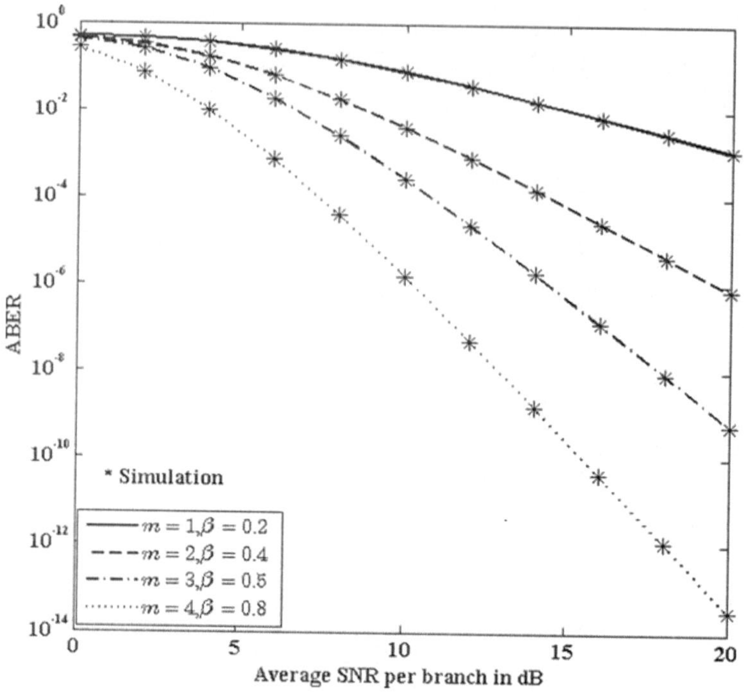

Fig. 4 ABER with $L = 2$ with variables of m and β for BPSK

6 Conclusions

In this work, WIPT system for $L - SC$ receiver over Nakagami-m fading channels has been obtained assuming that the receiver uses power splitter scheme for information transmission and energy harvesting parts. The analytical expressions of outage probability and ABER have been presented. The performance of the proposed system in the influence of channel's fading parameter, power splitter factor, and the order of diversity is evaluated. The proposed analytical expressions are validated and compared with Monte Carlo simulations.

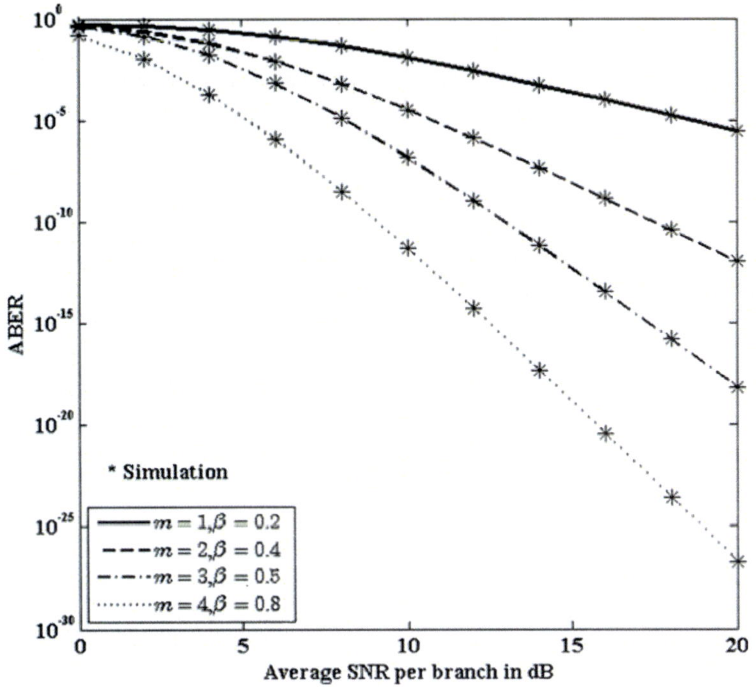

Fig. 5 ABER of the proposed scheme with $L = 4$ for variables of m, β

References

1. Simon MK, Alouini MS (2005) Digital communication over fading channels, Second edn.https://doi.org/10.1002/0471715220
2. Al-Hussaini EK, Al-Bassiouni AA (1985) Performance of MRC diversity systems for the detection of signals with nakagami fading. IEEE Trans Commun 33:1315–1319. https://doi.org/10.1109/TCOM.1985.1096243
3. Kong N, Milstein LB (1999) Average SNR of a generalized diversity selection combining scheme. IEEE Commun Lett 3:57–79. https://doi.org/10.1109/4234.752901
4. Tiwari D, Soni S, Chauhan PS (2017) A new closed-form expressions of channel capacity with MRC, EGC and SC over lognormal fading channel. Wirel Pers Commun 97:4183–4197. https://doi.org/10.1007/s11277-017-4719-9
5. Mishra D, De S, Jana S, Basagni S, Chowdhury K, Heinzelman W (2015) Smart RF energy harvesting communications: challenges and opportunities. IEEE Commun Mag. https://doi.org/10.1109/MCOM.2015.7081078
6. Zhang R, Ho CK (2013) MIMO broadcasting for simultaneous wireless information and power transfer. IEEE Trans Wirel Commun 12:1989–2001. https://doi.org/10.1109/TWC.2013.031813.120224
7. Zhou X, Zhang R, Ho CK (2013) Wireless information and power transfer: architecture design and rate-energy tradeoff. IEEE Trans Commun 61:4754–4767. https://doi.org/10.1109/TCOMM.2013.13.120855
8. Nasir AA, Zhou X, Durrani S, Kennedy RA (2013) Relaying protocols for wireless energy harvesting and information processing. IEEE Trans Wirel Commun. https://doi.org/10.1109/TWC.2013.062413.122042

9. Van PT, Le HHN, Le MDN, Ha DB (2016) Performance analysis in wireless power transfer system over Nakagami fading channels. In: International conference on electronics, information, and communications, ICEIC 2016. https://doi.org/10.1109/ELINFOCOM.2016.7562971
10. Zhong C, Chen X, Zhang Z, Karagiannidis GK (2015) Wireless-powered communications: performance analysis and optimization. IEEE Trans Commun 63:5178–5190. https://doi.org/10.1109/TCOMM.2015.2488640
11. Aalo VA (1995) Performance of maximal-ratio diversity systems in a correlated nakagami-fading environment. IEEE Trans Commun 43:2360–2369. https://doi.org/10.1109/26.403769
12. Table of Integrals, Series, and Products (2000). https://doi.org/10.1016/b978-0-12-294757-5.x5000-4
13. Peña-Martín JP, Romero-Jerez JM, Téllez-Labao C (2013) Performance of TAS/MRC wireless systems under hoyt fading channels. IEEE Trans Wirel Commun 12. https://doi.org/10.1109/TWC.2013.062713121487

A High-Selective Dual-Band Reconfigurable Filtering Antenna for WiMax and WLAN Application

Sangeeta Das and Pankaj Sarkar

Abstract The paper presents a compact dual-band reconfigurable filtering antenna. First and foremost, a novel reconfigurable filter with two switchable pass bands is designed. The proposed filter is accomplished by using two open-ended half-wavelength stepped-impedance resonators (SIRs) at the top and two open-ended half-wavelength uniform impedance resonators (UIRs) at the bottom to operate at 3.2 GHz and 5.5 GHz used for WiMax/WLAN applications, respectively. The PIN diodes are connected to the resonators and biased properly to switch the operation from dual-pass band to two single-pass bands and all stop operation. Second, a broadband antenna is proposed using a rectangular and a trapezoidal radiating patch to radiate from 2.5 GHz to 6 GHz. Finally, both the filter and the antenna structure is cascaded together to implement the proposed dual-band reconfigurable filtering antenna. The filtering antenna is simulated, and a well agreement is observed with the predicted response.

Keywords Reconfigurable filtering antenna · Stepped-impedance resonator · Uniform impedance resonator · PIN diode · Dual-band reconfigurable filter

1 Introduction

There is an enormous demand to realize antennas that are awful to tune their operational mechanism according to the dynamic communication requirements. Moreover, using multiple antennas to cover each of the different wireless services that are scattered over wide frequency bands increases the system cost and the space requirements. Furthermore, a significant amount of isolation among the antennas is very difficult to implement. Reconfigurable antennas are promising candidates for future RF front-end solution as it minimizes the number of antennas required in a particular system [1–3]. Reconfigurable antennas have been extensively studied

S. Das (✉) · P. Sarkar
Electronics and Communication Engineering Department, School of Technology, North-Eastern Hill University, Shillong 793022, India

S. Basu et al. (eds.), *Proceedings of International Conference on Frontiers in Computing and Systems*, Lecture Notes in Networks and Systems 404,
https://doi.org/10.1007/978-981-19-0105-8_40

throughout the last two decades. Electrically reconfigurable antennas use radio-frequency microelectromechanical system (RF MEMS), PIN diodes, or varactors used as switching elements to initiate reconfigurable operation in the antenna functionality. The on/off mechanism of such switching elements requires biasing lines which produces unwanted resonances in the operating band. Moreover, the radiation pattern also changes due to presence of biasing network [4]. Modern communication systems require reconfigurable microwave components in order to satisfy the demand of multiband systems. In among, band pass filter (BPF) is one of the most influential devices that can transmit desired signal and eliminate unwanted interference. Hence, the BPF carrying various reconfigurabilities have been widely investigated in recent years [5–7]. A triband high-selective BPF is presented in Ref. [6] by assembling two λg/2 open-ended and four short-ended λg/4 uniform impedance resonators. However, in Ref. [7], the terminated cross-shaped resonators are used to implement bandwidth reconfiguration. Varactor diode in conjunction with PIN diode is used in Ref. [8] to achieve tunability and switchability simultaneously. A new concept named filtering antenna has been explored recently which can realize the radiation functionality and the filtering mechanism simultaneously [9–15]. The integrated topology helps to realize more compact filter and antenna structure as a result the performance of the whole system improves significantly. Cascaded topology is a suitable choice of reconfigurable filtering antenna design where BPF plays a crucial role on system performance. In Ref. [9], a dipole radiator is reformed as a stepped-impedance dipole (SID), and it is further placed in proximity to a parasitic SIR. This harmonic band is then fully suppressed by an integrated low-pass filter (LPF). Likewise, a reconfigurable two-pole filtering slot antenna (FSA) with the inset-coupling structure is proposed in Ref. [10] for wideband communication systems, and a GaAs FET switch is used in Ref. [11] for the realization of filtering antenna. Single-PIN diode can also be used to switch the state between two bands [12]. In Ref. [14], a reconfigurable filtering antenna is presented using three-*p-i-n* diodes to switch the operation between wideband states to two narrowband states.

Based on the many research above, a new compact dual-band reconfigurable filtering antenna is presented. The novelty of the proposed filtering antenna is that it can initiate high selectivity switchable operation in single compact structure using PIN diodes for multiband operations. Depending on the bias voltage given to the PIN diode, the proposed filtering antenna can be easily switched from dual-band to single-band configuration. The structure is implemented on FR4 substrate of height 1.0 mm. For the purpose of EM simulation, CST Microwave Studio is used.

2 Filter design

The basic configuration and resonance properties of a half-wavelength open-ended SIR are calculated from [15]. The proposed filter is designed using two $\frac{\lambda_g}{2}$ open–ended SIRs and two $\frac{\lambda_g}{2}$ open–ended UIRs. For the top SIRs, the Resonator1 and the

Resonator2 are uniform. Both the resonators are realized to operate at fundamental frequency and first spurious at 3.2 GHz and 7.5 GHz, respectively. The impedance ratio of the SIR is calculated as 0.62 to achieve the desired resonance characteristic. The bottom UIRs, Resonator3 and Resonator4, are also uniform and constructed to operate at fundamental resonance of 5.5 GHz. Both the top and the bottom resonators are designed such a way that the spurious frequencies cannot be coupled to the output line. In Fig. 1, the coupling scheme is depicted for better understanding the working principle of the proposed filter. It can be observed that the first pass band is generated by fundamental resonating frequency of Resonator1 and Resonator2. The second pass band is created with the help of fundamental resonating frequency of Resonator3 and Resonator4. The first spurious frequency of all the SIRs and UIRs is kept at higher side to achieve wide upper stop band characteristics. The proposed filter is realized on a FR4 substrate of height 1 mm (relative dielectric constant $\varepsilon_r = 4.4$ and loss tangent $\tan\delta = 0.015$), and Ansoft designer is used for EM simulation purpose.

The layout of the proposed dual-band reconfigurable filter is revealed in Fig. 2. So, as to switch the operation from dual band to two single bands or vice versa, PIN diodes (SMP1340-079LF) are connected to resonators as displayed in Fig. 2. Total

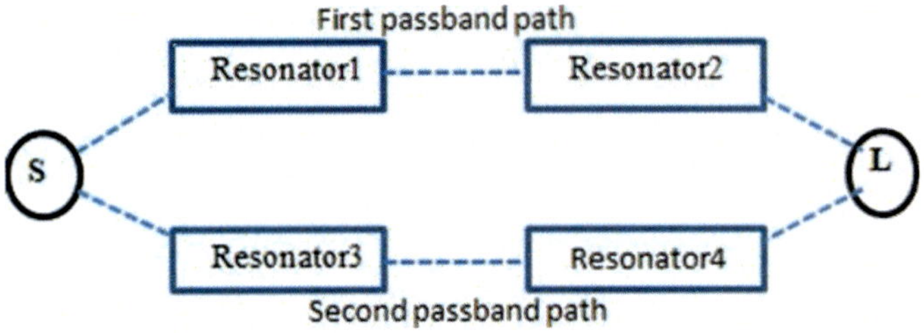

Fig. 1 Coupling scheme for the proposed filter

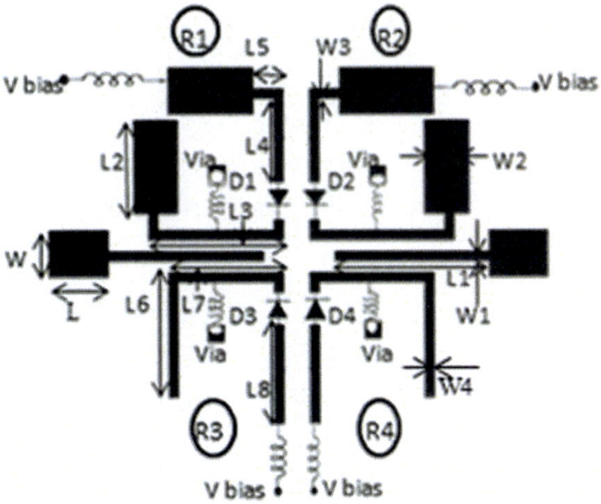

Fig. 2 Layout of the dual-band reconfigurable band pass filter with W = 1.9, L = 2.5, W1 = 0.3, W2 = 1.8, W3 = 0.3, W4 = 0.3, L1 = 7, L2 = 3.7, L3 = 6, L4 = 4.8, L5 = 0.7, L6 = 5.5, L6 = 5, L7 = 5.5 (*all dimensions are in mm.*)

four numbers of PIN diodes are used to accomplish the switching operation. The bias voltage is applied to the PIN diode through 2.4 nH inductor. The diodes are forward biased by applying a forward bias voltage of 5 V. Another terminal of the PIN diode is connected to ground with the help of inductor. Inductors have been used to allow DC steady current to flow through the circuit and to block AC given to the input port. For simulation purpose, the PIN diode is modeled with a series resistor of 2 Ω in the on state and a series capacitance of 0.3 pF in the off state.

3 Simulated Results for the Proposed Filter

Reconfigurable operation is carried out by changing biasing voltage across the PIN diode. The proposed filter preserves four different output characteristics. The bias voltage V_{dc} of 5 V is applied to make the PIN diode on. The bias voltage must be detached to turn the PIN diode off. For the first configuration, when all diodes are turned off, a dual-band characteristic can be accomplished from the proposed filter. The frequency response of the filter is shown in Fig. 3a when diodes D1, D2, D3, and D4 are turned off. The pass bands are centered at 3.2 GHz and 5.5 GHz with 10 dB fractional bandwidth (FBW) of 2.5% and 5.5%, respectively. The selectivity is quite acceptable due to the presence of transmission zeroes near to the pass bands. For the first pass band, the skirt factor (ratio of 3 dB bandwidth to 20 dB bandwidth) of 0.44 is obtained, whereas for the second pass band, the skirt factor is 0.5. The insertion loss for both the bands is below 0.5 dB which determines very less loss transmission. The switching of the filter state from dual band to single band centered at 3.2 GHz is carried out by switching diodes *D1* and *D2* on, whereas the diodes D3 and D4 are kept off. The frequency response of the filter is shown in Fig. 3b for this diode configuration. It is seen that the pass band is centered at 3.2 GHz with a 10 dB fractional bandwidth of 5.4%. The second pass band centered at 5.5 GHz is switched off with an attenuation level of 35 dB. Therefore, the proposed filter behaves as a single-pass band filter centered at 3.7 GHz. Similarly, when D1 and D2 are off and D3 and D4 are on, the proposed filter acts like a single-pass band filter with resonating frequency centered at 5.5 GHz with 10 a dB fractional bandwidth of 5.5%. The characteristic is illustrated in Fig. 3c. As can be seen, the first pass band is destructed with an attenuation level of 40 dB. For the last configuration, when all diodes are kept off, both pass band is collapsed with an excellent attenuation level. The frequency response is plotted in Fig. 3d. The filter, therefore, can be employed as an all stop filter for this configuration.

4 Broadband Antenna Design

Figure 4 illustrates the proposed broadband antenna covering the frequency range of 2.5–6 GHz. The structure of this antenna composes a rectangular patch and an

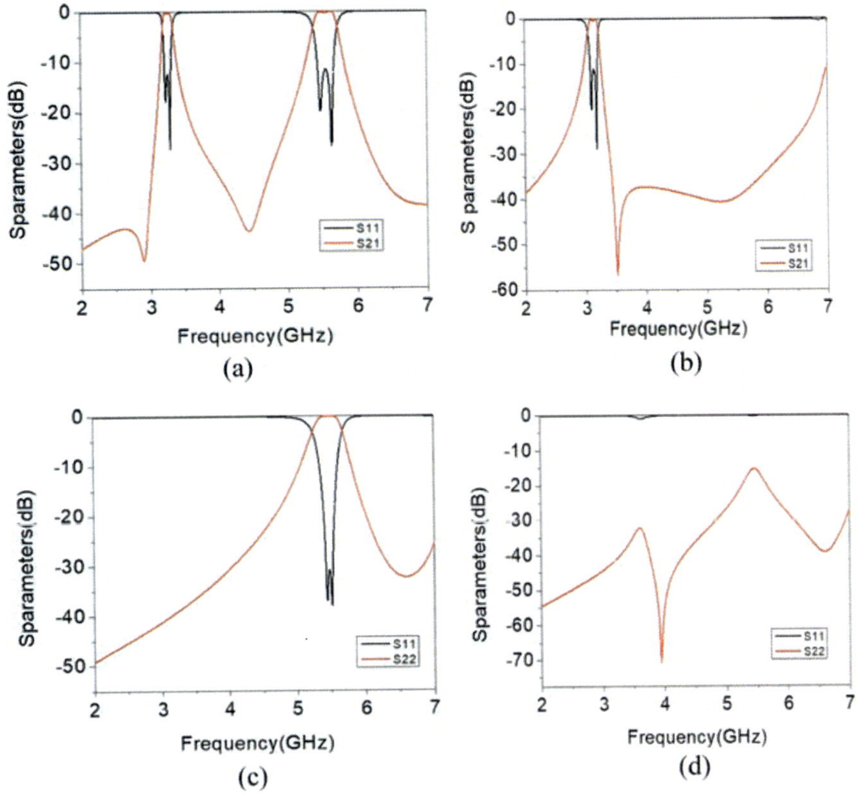

Fig. 3 EM-simulated results of the suggested filter **a** when D1, D2, D3 and D4 are off **b** when D1, D2 are on and D3, D4 are off **c** when D3, D4 are on and D1, D2 is off **d** all diodes are off

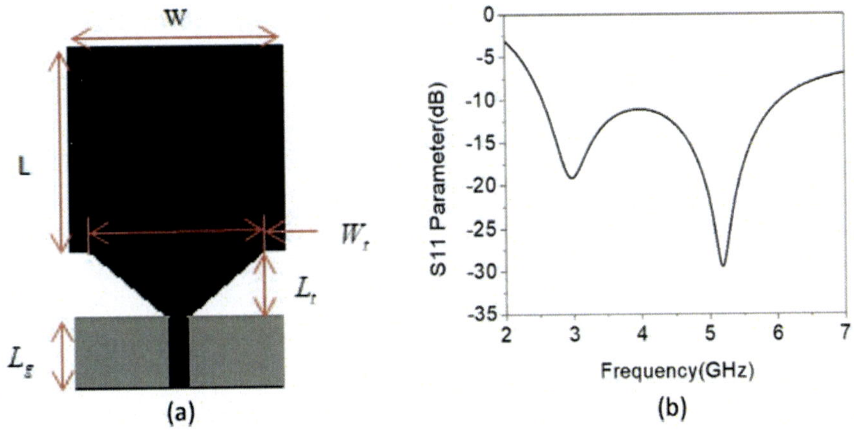

Fig. 4 **a** Proposed antenna structure with $L = 22$, $W = 24$, $L_t = 7.5$, $W_t = 20$, $L_g = 7.5$ (*all dimensions are in mm*). **b** EM-simulated response of the proposed antenna

isosceles trapezoidal patch. The proposed antenna holds two resonant paths, viz., the rectangular patch and the trapezoid section. First, a rectangular patch is designed with a length L that permits the antenna radiating at lower frequency 2.5 GHz and a width W that regulates the radiation pattern, the bandwidth and the input impedance. Next, a trapezoid section is inserted between the feed and the rectangular patch. The length L_t of the isosceles trapezoid section is parametrically studied to allow the antenna working at high frequency around 6 GHz. The resonance of rectangular patch merges with the resonance of trapezoid section, thereby creating a broadband from 2.5 to 6 GHz. Thus, the proposed antenna can be utilized as a broadband antenna. The width W_t plays identical role as the width of the rectangular patch antenna. The proposed broadband antenna is designed on the same FR4 substrate of height 1.0 mm. The radiating patch is excited through 50Ω microstrip transmission line. The ground plane has been tapered to obtain proper impedance matching. Finally, the antenna structure is EM simulated, and the S11 parameter is illustrated in Fig. 5a. It is observed that the proposed antenna can be utilized as a broadband antenna with a 10 dB fractional bandwidth of 90.34%. The size of the proposed antenna is very compact, i.e., $(40 \times 26)\ mm^2$.

Figure 5a and b presents radiation pattern of the two resonant modes at 3.2 GHz and 5.5 GHz, respectively, which reflect omnidirectional radiation pattern of the proposed broadband antenna. Figure 5c shows the gain of the proposed broadband antenna for different frequencies. It is observed that simulated gain of 2.2 dBi, 2.8 dBi, 3.0 dBi, 4.2 dBi, 4.95 dBi, and 4.50 dBi is achieved at 2.5 GHz, 3.2 GHz, 3.5 GHz, 4.5 GHz, 5.5 GHz, and 6.5 GHz, respectively. The variation of gain is from 2.2 dBi to 4.95 dBi. As can be seen in Fig. 6a, the antenna structure is cascaded

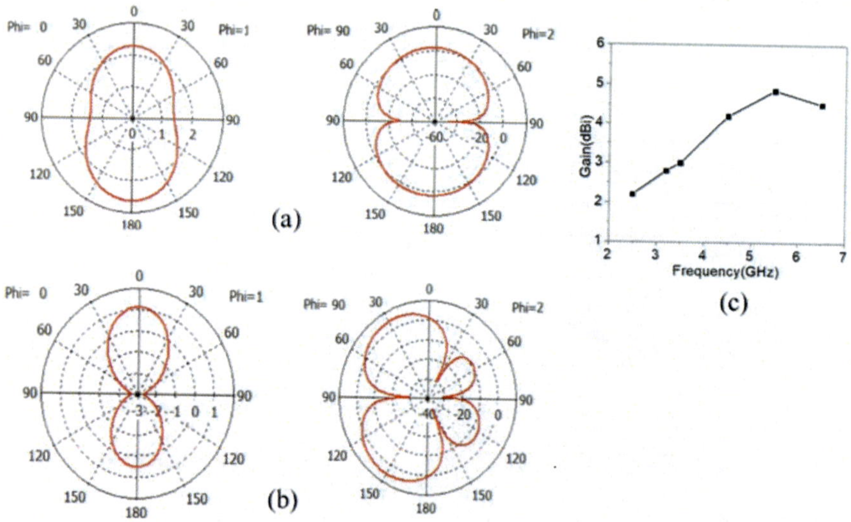

Fig. 5 Radiation pattern of the proposed broadband antenna for phi = 0° and phi = 90° radiating at **a** 3.2 GHz and **b** 5.5 GHz **c** Gain of the proposed broadband antenna for different frequencies

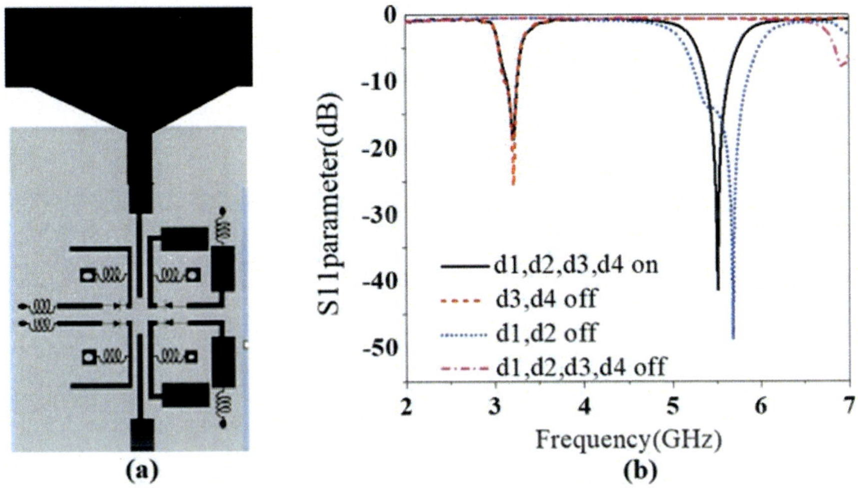

Fig. 6 **a** Layout of the proposed filtering antenna, **b** S11 parameters of the proposed filtering antenna for four switching configurations of the diodes D1, D2, D3, and D4

to the dual-band reconfigurable filter in the end to develop the proposed dual-band reconfigurable filtering antenna. The filter structure is inserted near to the antenna for minimal space requirement. To obtain switching operation, the two PIN diodes are biased externally through a RFC of inductance 2.4 nH. Another terminal for both the diode is connected to the metallic via with the help of inductors.

5 Results and Discussions

For the purpose of EM simulation of the combined structure of the reconfigurable filter and the antenna, CST microwave studio tool is used. The S11 parameter is displayed in Fig. 6b. It can be stated that the proposed composite structure also maintains identical switching characteristics as discussed above. The return loss characteristics is extremely fine for all diode configurations. The antenna can be reconfigured to the dual band, two different single radiating bands and no radiation states by properly biasing PIN diodes. Figure 7 represents the radiation pattern for dual-band configurations at the two center frequencies 3.2 GHz and 5.5 GHz for the two principle planes of $\phi = 0°$ and $\phi = 90°$. It portrays the dumbbell-shaped radiation pattern that makes the proposed antenna to obtain greater omnidirectional radiation characteristics and can be used for WiMax and WLAN multiband operation as well.

The simulated gain at 3.2 GHz is 4.12 dBi and at 5.5 GHz is 5.00 dBi for dual pass band configuration. For single-pass band configurations, the radiation patterns are plotted in Fig. 8. It can be observed that gains of 4.0 dBi and 4.6 dBi are obtained for the radiating bands centered at 3.2 GHz and 5.5 GHz, respectively. The variation

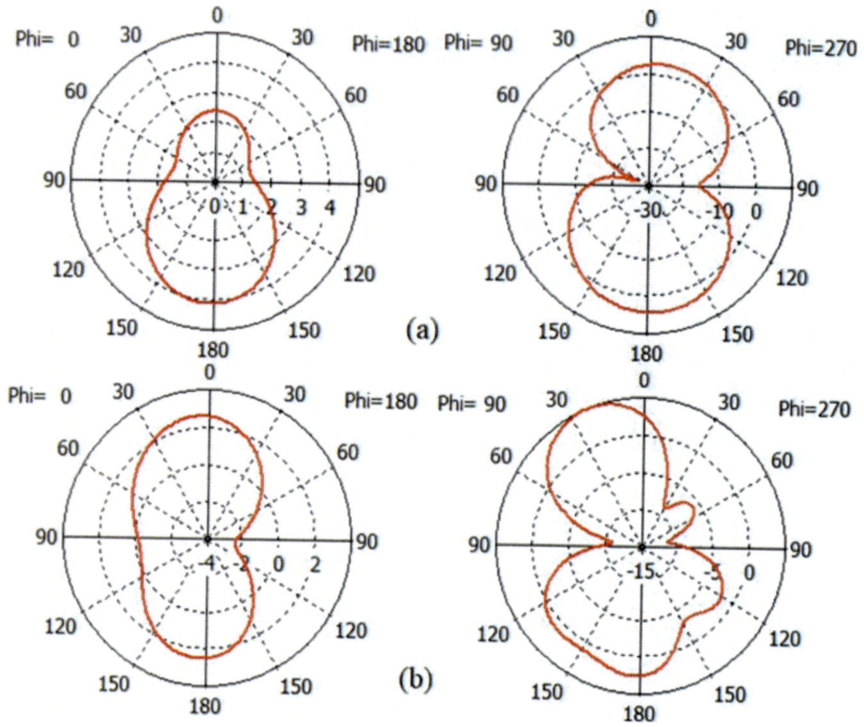

Fig. 7 Radiation pattern of the proposed filtering antenna for phi = 0° and phi = 90° for dual pass band configuration **a** for first pass band **b** for second pass band

of the gain over radiating bands is found very less, i.e., 0.12 dBi at 3.2 GHz and 0.4 dBi at 5.5 GHz.

Figure 9 shows the gain of the proposed dual-band reconfigurable filtering antenna for different frequencies. For dual-band configuration, simulated gains are 5.21 dBi, 4.12 dBi, 3.5 dBi, 6.00 dBi, 5.00 dBi, and 4.31 dBi at 2.5 GHz, 3.2 GHz, 3.5 GHz, 4.5 GHz, 5.5 GHz, and 6.5 GHz, respectively. The variation of gain is from 3.5 dBi to 6.0 dBi.

6 Conclusions

In this work, a compact dual-band reconfigurable filtering antenna is realized using four PIN diodes. A rectangular patch and an isosceles trapezoidal patch are attached to each other and inserted to the feeding line to design the proposed antenna where the dimension of each patches furnishes the flexibility on the selection of radiating band. The structure has an omnidirectional radiation pattern in both H-plane and E-plane. The proposed filtering antenna is compact (60 × 26) mm^2 which makes the

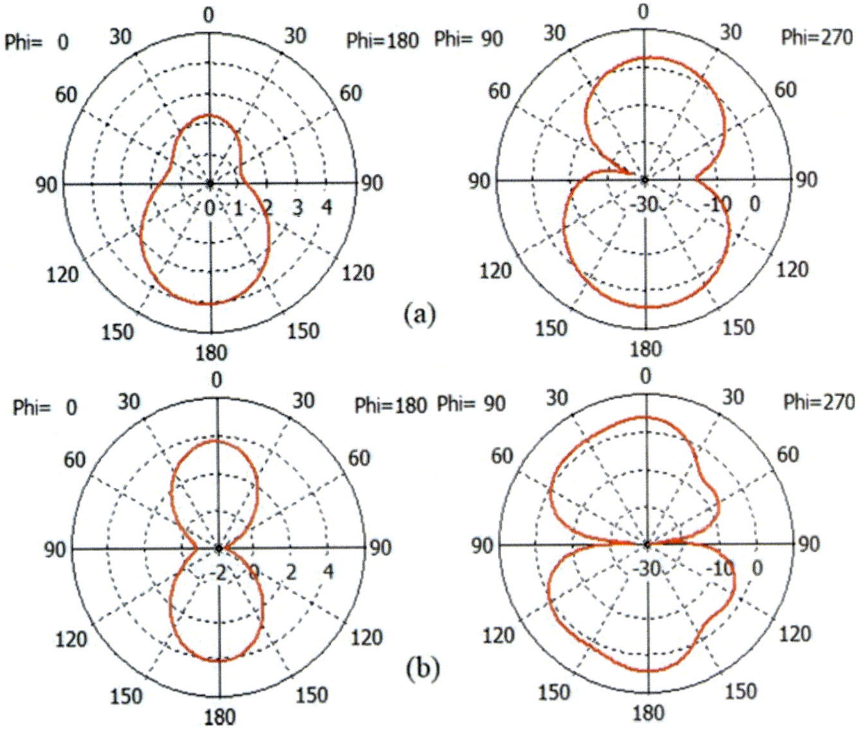

Fig. 8 Radiation pattern of the proposed filtering antenna for phi = 0° and phi = 90° at **a** 3.2 GHz **b** 5.5 GHz for two single-pass band configurations

Fig. 9 Gain of the proposed dual-band filtering antenna for different frequencies

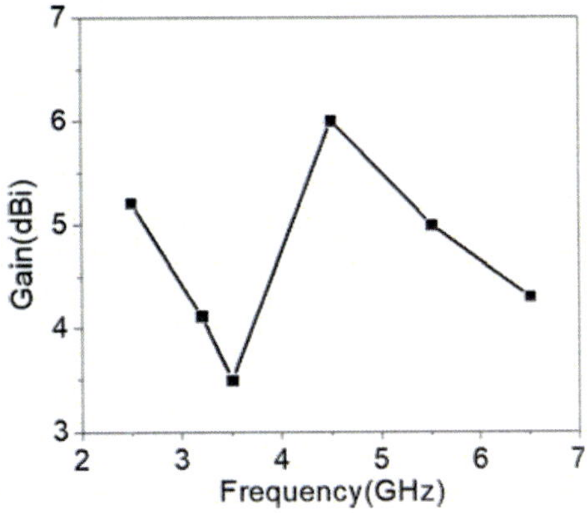

structure very convenient for wireless communication appliances. It offers different multiband performances such as dual band, single band, and all stop operation. For all diode configurations, the structure facilitates excellent pass band and stop band characteristics as well. The non-radiating bands originate numerous transmission zeroes that enhance the selectivity of the filtering antenna structure. The proposed structure has less insertion loss, fine return loss, and wide upper stop band characteristics. Hence, a comfortably reconfigurable filtering antenna can be achieved to meet the modern multiband communication systems.

References

1. Hong JSG, Lancaster MJ (2004) Microstrip Filters for RF/microwave, 2nd edn. John Wiley & Sons
2. Balanis CA (2016) Antenna theory: analysis and design, 4th edn. John Wiley & Sons
3. Tawk Y, Costantine J, Christodoulou C (2012) A varactor-based reconfigurable filtenna. IEEE Antennas Wirel Propag Lett 11:716–719
4. Costantine J, Tawk Y, Barbin SE, Christodoulou CG (2015) Reconfigurable antennas: design and applications. Proc IEEE 103(3):424–437
5. Horestani AK, Shaterian Z, Naqui J, Martín F, Fumeaux C (2016) Reconfigurable and tunable S-shaped split-ring resonators and application in band-notched UWB antennas. IEEE Trans Antennas Propag 64(9):3766–3776
6. Bandyopadhyay A, Sarkar P, Mondal T, Ghatak R (2020) A high selective tri-band bandpass filter with switchable passband states. Int J Microw Wirel Technol 12(2):103–108
7. Bi X-K, Cheng T, Cheong P, Ho S-K, Tam K-W (2019) Design of dual-band bandpass filters with fixed and reconfigurable bandwidths based on terminated cross-shaped resonators. IEEE Trans Circ Syst II Exp Briefs 66(3):317–321
8. Zhang Y-J, Cai J, Chen J-X (2019) Design of novel reconfigurable filter with simultaneously tunable and switchable passband. IEEE Access
9. Sun G-H, Wong S-W, Zhu L, Chu Q-X (2016) A compact printed filtering antenna with good suppression of upper harmonic band. IEEE Antennas Wirel Propag Lett 15:1349–1352
10. Fakharian MM, Rezaei P, Orouji AA, Soltanpur M (2016) A wideband and reconfigurable filtering slot antenna. IEEE Antennas Wirel Propag Lett 15:1610–1613
11. Yang XL, Lin JC, Chen G, Kong FL (2015) Frequency reconfigurable antenna for wireless communications using GaAs FET switch. IEEE Antennas Wirel Propag Lett 14:807–810
12. Wen Z, Tang MC, Ziolkowski RW (2019) Band and frequency-reconfigurable circularly polarized filtenna for cognitive radio applications. IET Microw Antennas Propag 13(7):1003–1008
13. Qin PY, Guo YJ, Wei F (2014) Frequency agile monopole antenna using a reconfigurable bandpass filter. In: 2014 IEEE antennas and propagation society international symposium (APSURSI). IEEE, pp 1250–1251
14. Deng J, Hou S, Zhao L, Guo L (2018) A reconfigurable filtering antenna with integrated bandpass filters for UWB/WLAN applications. IEEE Trans Antennas Propag 66(1)
15. Das S, Sarkar P () A compact dual-band reconfigurable filtering antenna using PIN diode. In: Proceedings of the international conference on computing and communication systems, vol 170

Synthesis and Characterization of Uniform Size Gold Nanoparticles for Colorimetric Detection of Pregnancy from Urine

Shyamal Mandal and Juwesh Binong

Abstract A gold nanoparticles (AuNPs)-based device for pregnancy detection form urine sample is described. Human chorionic gonadotropin (hCG) is released by placenta during pregnancy. The gold nanoparticle-based device detects the hormone by changing of its color. The rapid color changes detected by naked eye, and it also confirmed the pregnancy. However, in this method, reaction rate is very fast rather than any conventional method. In this work, we are introducing the application of a new device that mainly contains immobilized gold nanoparticles conjugated with primary antibody and the antigen to measure concentration of hCG from urine sample. The level of hCG for normal women 5mlU/ml, but during pregnancy, it increased up to 25–50 mlU/ml. The line color intensity of hCG at 10 pg/ml tested with device was almost same to the intensity that we measured at 30 pg/ml with the normal device. The same device can be used in future with proper marker for early detection of cancer.

Keywords Early detection · Pregnancy · Sensor device · Gold nanoparticles · hCG

1 Introduction

Initial symptoms of pregnancy can not be detected without any confirmation test. It may be somewhat similar to the symptoms of an approaching menstrual cycle. Sometimes, they may be think about it may be onset of a cold or flu or in the case of women in their age above 40 s, hormonal changes that precede menopause. Gold nanoparticles-based biosensor is a non-invasive device that plays a major role to detect early stage of pregnancy. Nowadays, lots of company try to make pregnancy

S. Mandal (✉)
Department of Biomedical Engineering, North Eastern Hill University, Shillong, Meghalaya 732202, India
e-mail: shyamal.mandal.iit@gmail.com

J. Binong
Department of Electronics and Communication Engineering, North Eastern Hill University, Shillong, Meghalaya 732202, India

S. Basu et al. (eds.), *Proceedings of International Conference on Frontiers in Computing and Systems*, Lecture Notes in Networks and Systems 404,
https://doi.org/10.1007/978-981-19-0105-8_41

kits, and some of these are very costly. Design low-cost pregnancy kits are challenging task for the scientist. Gold nanoparticles-based pregnancy device may be the best solution for the problems. The gold nanoparticles (AuNPs) are the promising material for designed colorimetric biosensor due to SPR band [1]. The polyionic drug, protamine, has been detected with optical sensor, and it is also reversible by used of heparin [2]. Gold nanoparticles shape and size also depend on electro catalytic activity [3]. Gold nanoparticles are vastly used for electrochemical sensing high conductivity [4]. Gold nanoparticle itself is also highly sensitive, and when merged with carbon nanotube (CNT), sensitivity also increased [5, 6]. AuNPs are also used as a carrier of drug during drug delivery, and it is also used for disease diagnosis by using its SPR band [7]. It is also immobilized when attached with thiol [8]. Toxic elements free gold nanoparticle can be synthesized from fruit extract [9]. Turkevich research shows that formation of gold nanoparticles via chemical root method. He used tri-sodium citrate as a reducing agent as well as capping agent [10]. The gold micro-gap array is prepared with a printed circuit board (PCB) chip for detection of myocardial infection [11]. Bhalinge et al. discussed about different types of biosensor and their application and functionality [12]. Biosensor is also used for detection of cell structure in certain electric field [13]. Human chorionic gonadotropin (hCG) is a hormone that gets to shine as it is normally produced in women's body during pregnancy. Thus, hormone amps up the production of estrogen and progesterone. Early on in pregnancy, hCG levels are low, but during pregnancy, it soon raise and double every two days. And it goes in peak value between 7 and 12 weeks. Antigen-packed gold nanoparticles have used as marker top detection presence of that hormone in urine [14]. Village children's mouth and feet are affected by different virus due to walking with barefoot. Gold nanoparticles-based ultrasonic biosensor was designed for detect and diagnose that virus [15]. AuNPs have special property to sense the chemical by quenching SPR band. Colorimetric detected of biological same have done with help of gold nanoparticles [16]. It is very interesting to note that Au nanoparticles are useful alternative method for pregnancy test from urine sample.

2 Materials and Methods

2.1 Preparation of Gold Nanoparticles

Gold nanoparticles were synthesized by reducing chloroauric acid using various stabilizers. All the reagents were used as obtained without further purification unless stated otherwise. De-ionized Milli Q water (DI) with a resistivity of 18.2 MΏ-cm was used throughout the experiment. ($HAuCl_4.3H_2O$) and tri-sodium citrate ($Na_3C_6H_5O_7$) were bought from SRL Chemicals Mumbai, India. Synthesis process has been carried out according to the procedures described by Turkevich et al. [10]. 2 ml of chloroauric acid ($HAuCl_4$) acid was added into boiling water with vigorous mechanical stirring for 20 min. 10 ml tri-sodium citrate was added and boiled for an

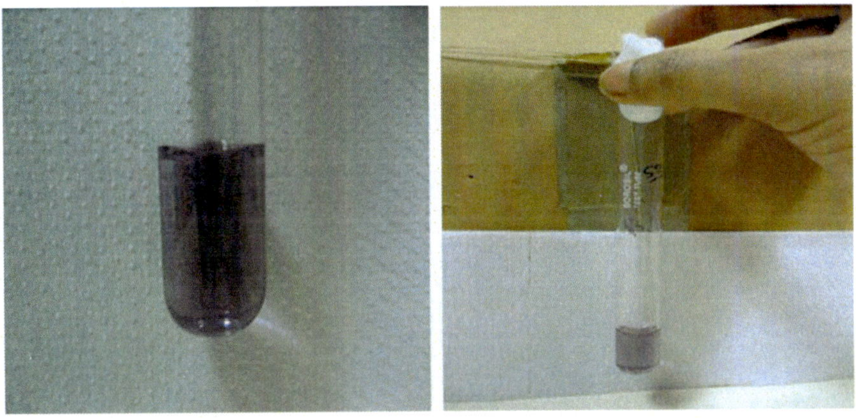

Fig. 1 Picture of gold nanoparticles solution

additional 20 min. After that, the solution temperature was slowly cooled down to room temperature. Gold atoms released from the chloroauric acid during the reduction process, and it is aggregated to form nanoparticles (NPs).This step continued until all the tetra-chloroauric acid was reduced. Then, it was stored in a dark place at 4 °C. Figure 1 shows that uniformly disperse gold nanoparticle in water solvent. The end product of the reaction are gold nanoparticles (AuNPs) sodium ketogluterate, sodium chlorite, chlorine, hydrogen, and carbon dioxide.

$$2HAuCl_4 + 3Na_3C_6H_5O_7 = 2Au + 3Na_2C_5H_6O_5(\text{Sodium ketoglutarate}) + 3NaCl + 5Cl^- + 5H^+ + 3CO_2$$

2.2 *Preparation of Sensor Device*

Monoclonal anti-human α-subunit (MabHαS) and monoclonal anti-human chorionic gonadotropin (MabhCG) were procured from Merck (USA). Potassium dihydrogen phosphate (KH_2PO_4) and hydro fluoric acid were procured from Merck (USA). MabhCG solution of 40 μg/ml have prepared by diluting with 4 mM KH_2PO_4 solution (pH 7.02) and added Millipore water (18.2 MΩ-cm) to make final volume of 200 μl. The gold nanoparticle solution (2.0 ml) was added into the 200 μl solution of MabhCG and immediately mix the two solutions with help of magnetic stirrer. Two vertical lines were made in the glass slide; the line dimensions are $L \times B \times W$ (20 mm × 3 mm × 0.3 mm), and the lines are marked as 1 (one) and 2 (two) one for test sample and two for blank. The mixer solution's droplets pour into the line of glass slide and kept for 20 min in normal temperature for immobilization of mixer solution. The slides are kept in clean chamber for overnight for drying, and

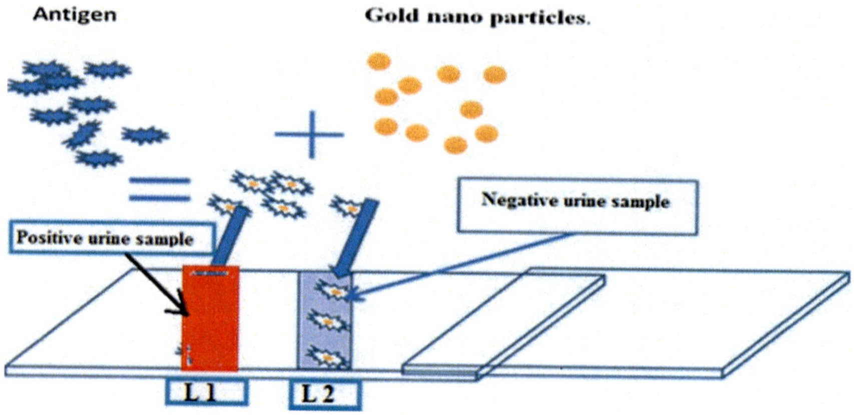

Fig. 2 Device for pregnancy detection

next morning, it is observed that solution was dried in the glass slide. After 24 h, the device are ready for pregnancy test (Fig. 2).

3 Results and Discussion

A total 3 mL of gold nanoparticles solution was pour into the quartz cell from absorption analysis. The absorbance spectra were recorded at time interval of 1 min. Figure 3 shows the optical absorbance spectra obtain for $HAuCl_4$ and colloidal Au nanoparticles. The absorbance band at 298 $_{nm}$ for residue $HAuCl_4$ [14] and 527.2 nm

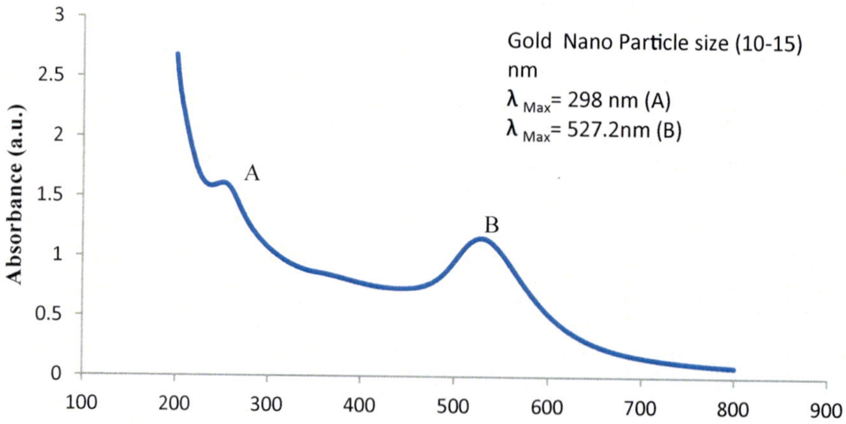

Fig. 3 UV—Visible spectra of the as synthesized gold nanoparticles, A = absorbance peak of $HAuCl_4$, B = absorbance peak of AuNPs

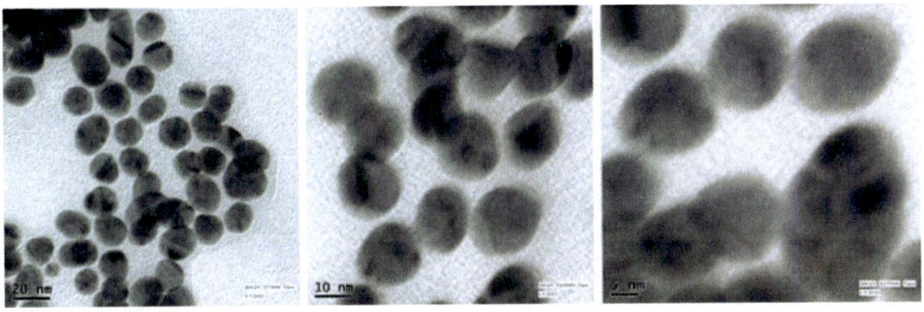

Fig. 4 TEM micrograph of the as synthesized gold nanoparticle

for size of the Au nanoparticle. The samples absorbance graph were recorded on a UV–vis spectrometer (Shimadzu UV-1601).

The TEM micrographs (JEOL, JEM 2010) of the gold nanoparticles were performed, and the size distribution around 10–15 nm were seen in the images Fig. 4. Gold nanoparticles are suspended in slightly acidic solvent pH 6.30. The exact shape and morphology can be clearly obtained from these photographs. The spread gold nanoparticles are almost round shape.

3.1 Evaluation of the Device

The positive sample prepared by mixing human chorionic gonadotropin hormone and negative urine samples were prepared without mixing of hCG hormone. Figure 4 shows that hCG-rich urine solution put into the line one (L1) and negative urine solution put line two (L2). After drying the solution, hCG-rich urine line become red (positive) and without hCG mixed line color not change (negative). The colloidal AuNPs show an intense SPR in the visible region of the electromagnetic spectrum. The red colour intencity depend's upon the quantity of gold nanoparticles binds with MabhCG molecules on the slide.

4 Conclusion

Gold nanoparticles were synthesized in different pH condition. Variation concentration of gold chloroauric acid and tri-sodium citrate not only parameters of size and shape of the gold nanoparticles. Temperature and pH also played major role to control size and shape of the Au nanoparticles. The application can help for the non-invasive diagnosis of pregnancy and can be an alternative method for a urine pregnancy test . Advantage of that device is reusable by washing with buffer solution.

The sensor device could enhance the sensitivity of immunochromatographic assay for detection of hCG model case. Apart from glass slide Au can immobilize in the cellulose filterpaper and this technique may useful for making paper sensor.

References

1. Aldewachi H, Chalati T, Woodroofe MN, Bricklebank N, Sharrack B, Gardiner P (2018) Gold nanoparticle-based colorimetric biosensors. Nanoscale 10:18–33
2. Jena BK, Raj CR (2008) Optical sensing of biomedically importance polyionic drug during using nano size gold particles. Biosensor Bioelectronics 23:1285–1290
3. Das AK, Raj CR (2013) Shape and surface structure dependent electrocatalytic activity of Au Nano particles. Electrochim Acta 107:592–598
4. Alagiri M, Rameshkumar P, Pandikumar A (2017) Gold nanorod-based electrochemical sensing of small biomolecules: a review. Microchim Acta 184:3069–3092
5. Katz E, Willner I (2004) Integrated nanoparticles-biomolecules hybrid systems synthesis property and applications. Angew Chem Int Ed 43:6042–6108
6. Sinha N, John T, Yeow W (2005) Carbon nano tube for biomedical application. IEEE Trans Nanobio Sci 4:180–194
7. Kuldeep M, Sahil N, Mahero AS, Ananya S, Pawan M, Shounak R, Amit J, Remu S, Pranjal C (2019) Gold nano particles surface engineering strategies and their application in biomedicine and diagnostics, Springer review article, vol 57, pp 1–19
8. Mahsa MR, Mehedi SK, Hossein RM, Amin P (2017) Effect of surface modification with various thiol compound on colloidal stability of gold nanoparticles, Appl Organometalic Chem 1–11
9. Khandanlou R, Murthy V, Wang H (2020) Gold nanoparticle-assisted enhancement in bioactive properties of Australian native plant extracts, Tasmannia lanceolata and Backhousia citriodora. Mater Sci Eng 112:37–46
10. Turkevich J, Stevenson PC, Hillier J (1951) A study of nucleation and growth processes in the synthesis of colloidal gold. Discuss Farady Soc 11:55–75
11. Lee T, Lee Y, Park YS, Hong K, Kim Y, Park C, Chung HY, Lee ML, Min J (2019) Fabrication of electrochemical biosensor composed of multi-functional DNA structure/Au nanospike on micro-gap/PCB system for detecting troponin I in human serum. Colloids Surf B 175:343–350
12. Bhalinge P, Kumar S, Jadhav A, Suman S, Gujjar P, Perla N, Biosensors: nanotools of detection—A review, Int J Healthcare Biomed Res 04:26–39
13. Keese CR, Giaever I (1994) A biosensor that monitors cell morphology with electric field. IEEE engineering in medicine and biology, pp 435–445
14. Chia CC, Chie PC, Chung HL, Chen YC, Chii WL (2014) Colorimetric detection of human chorionic gonadotropin using catalytic gold nanoparticles and a peptide aptamer. ChemComm 1443–1448
15. Mervat EH, Michele DC, Hussein AH, Taher AS, Aman HED, Mohamed ME, Gullia P, Davia C (2018) Development of gold nanoparticles biosensor for ultrasensitive diagnosis of foot and mouth virus. J Nano Biotech 1–12
16. Chia CC, Chia PC, Tzu HU, Ching H, Chi WL, Chen YC (2019) Gold nanoparticle based colorimetric strategies for chemical and biological sensing application. Nanomaterials 9:1–24

A Data Hiding Technique Based on QR Code Decomposition in Transform Domain

Sakhi Bandyopadhyay, Subhadip Mukherjee, Biswapati Jana, and Partha Chowdhuri

Abstract Designed and created by Toyota subsidiary Denso Wave, quick response (QR) code has now been commonly used in many suitable applications to deal with real-time data because it can be read quickly by smartphones or tablets. The covering of information is indeed the technique of keeping secret knowledge inside the source media. Two qualities would have a strong approach to data hiding: high consistency of the stego image as well as high potentiality for concealing. In this paper, a redundant discrete wavelet transform (RDWT) domain-based information hiding method using QR code decomposition is proposed. An appropriate cover is torn down by RDWT mechanism, and the private image is concealed within the sub-bands of the factorized QR code. For the purpose of demonstrating the effectiveness of the proposed approach with respect to imperceptibility, capability and robustness to different signal processing assaults, multiple tests and case analysis are conducted.

Keywords Steganography · RDWT · PSNR · Information Security

1 Introduction

In order to secure the hidden message from an unwanted user, information hiding [1] is used to conceal and secure the communication by cleverly avoiding the human visual system. The problem of Internet security is growing due to the tremendous growth of the Internet and its utilization. In certain cases, it needs more encryption to convert the data from the sender to the receiver. Many information security techniques such as steganography, watermarking and cryptography have been frequently used to encrypt the data. Steganography plays a significant role in information security within these three techniques. Steganography is the strategy for concealing data in source

S. Bandyopadhyay (✉) · B. Jana · P. Chowdhuri
Department of Computer Science, Vidyasagar University, Midnapore 721102, India
e-mail: banerjeesakhi18@gmail.com

S. Mukherjee
Department of Computer Science, Kharagpur College, Kharagpur 721305, India

S. Basu et al. (eds.), *Proceedings of International Conference on Frontiers in Computing and Systems*, Lecture Notes in Networks and Systems 404,
https://doi.org/10.1007/978-981-19-0105-8_42

images that provides forward and backward functionality. Mainly in the frequency domain and spatial domain, the steganography schemes conduct the data hiding operations. For the techniques within the spatial domain, without any alteration, the information is concealed directly inside the pixels of the source image. For the LSB of least significant bit matching and LSB substitution, these two methods are widely used in this domain. Moreover, few changes are made in the system of LSB replacement based on the process of pixel adjustment with optimality. In the frequency domain, using frequency-oriented mechanisms, the private data is concealed in the transformation coefficients. It is important to achieve a reasonable balance among visual quality, robustness and embedding capacity.

The remaining portion of this article is organized as follows: In Sect. 2, we have explained the related works of data hiding in transform domain. The QR code decomposition and redundant discrete wavelet transform are described in Sect. 3. In Sect. 4, our proposed methodology is illustrated. The results and comparisons are displayed in Sect. 5. And lastly, in Sect. 6, the overall conclusion is given.

1.1 Motivations

We have observed that, till now, most of the existing QR code-based data hiding schemes are presented with pure SVD technique. For achieving better visual quality, Subhedar et al. [2] suggested a steganographic approach by employing SVD and DWT. Subhedar et al. concealed the private message within the DWT sub-bands' singular values. Though this approach performs well in terms of capacity and imperceptibility, it has a very serious design flaw or issue of false positive. It can be proved that the SVD-based data hiding methods cannot gain good results for the three special attacks discussed by Guo et al. [3]. Therefore, we have motivated to propose this work to increase the hiding capacity and improve the stego image quality.

2 Related Works

In transform domain, the private information is concealed within the transform coefficients with the help of a frequency-centric technique such as discrete cosine transform (DCT), Z-transform, Fresnelet transform and DWT or discrete wavelet transform. For hiding information with reversibility in DCT coefficients, Lin [4] has used a histogram shifting tool for increasing the concealing capacity. Mukherjee et al.[5] also suggested hiding data on the basis of differential expansion, which increases the ability of embedding. In order to increase protection, Banerjee and Jana [6] have used the Reed–Solomon technique in various color images. By using the slantlet transform (SLT) technique, Thabit [7] has suggested a method of lossless data hiding. Here, the information embedding is achieved by changing the difference in high-frequency sub-bands among the SLT coefficients' mean values and gained 49,152 bits of payload. For regular test images, the payload of 49,152 bits is substantially poor and requires enhancement. Xiao et al. [8] suggest a high-capacity method of data concealing based on compressive sensing and framelet transformation. The compressive sensing was applied to the singular values of the private data and then incorporated with the cover sub-bands of low frequency obtained using framelet transformation. The stego image was created in the original image of 512×512 size after hiding a private image of 256×256. It has been found that 2 dB of PSNR increased as the sampling rate ranges from 1 to 0.6, but the normalized correlation decreases to 0.9927 from 1.0. In addition, Xiao et al. did not elaborate on the efficiency of this approach in terms of various statistical attack, steganalysis and robustness. To overcome these problems, we have suggested a new information hiding methodology which is mainly based on QR decomposition.

3 QR Decomposition and Redundant Discrete Wavelet Transform

The mechanism of decomposing an square matrix A into a multiplication of an upper triangular R and an orthogonal matrix Q, i.e., $A = QR$, is known as QR decomposition, also referred to as QR factorization or QU factorization. The columns of matrix Q are the vectors of orthogonal units meaning $Q^T Q = Q Q^T = I$. This mechanism is also applied to address the problem of least squares in linear form because it is the foundation of the QR algorithm, an unique eigenvalue algorithm. If A is invertible, then if we require the diagonal elements of R to be positive, the factorization is unique. With the development of JPEG 2000, a number of image processing applications such as image compression, scanning and recognition have undoubtedly attracted decimated bi-orthogonal wavelet transformation. DWT gives a sparse representation of time–frequency and improved compaction of energy compared to DCT. The absence of shift variation and low directional selectivity, however, are major disadvantages of DWT. The consequence of essential subsampling or downsampling is shift variance. Minor differences in the input signals cause significant changes in the coefficients of wavelets, leading to differences in the energy distribution at different scales among the coefficients of DWT. In addition, it results in incorrect image restoration after wavelet coefficients have been updated. One of the alternatives to this downside is to eliminate the decimation stage, and RDWT is the resulting outcome.

4 Proposed Scheme

Data hiding on the basis of QR decomposition in RDWT domain is illustrated in this section. The structure for the hiding and retrieval of hidden images is shown in Fig. 1. The suggested scheme benefits from a suitable option of renowned image database for original image, low redundancy in RDWT and less complexity in QR factorization.

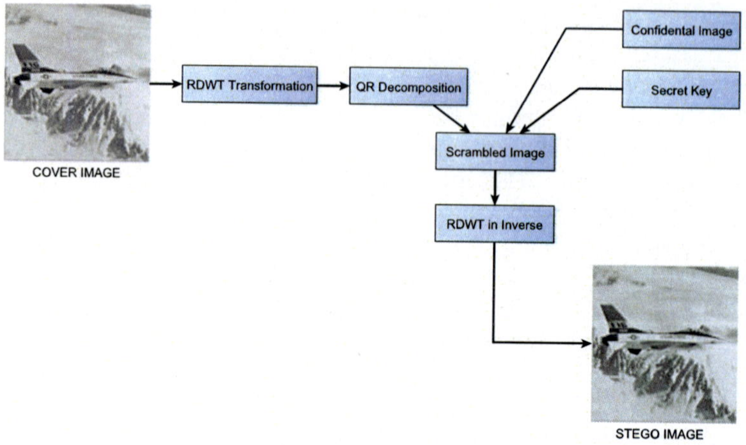

Fig. 1 Flowchart of the proposed scheme

4.1 Embedding Procedure

1. Input an original cover image of size 512 × 512.
2. The scrambling operation using Arnold transformation is performed. It generates the second level of security to proposed scheme.
3. Perform the RDWT transformation and shift invariant transform to decompose the original image. The decomposition generates four sub-bands of the original image of size 512 × 512 each.
4. Perform the *QR* factorization which has a less computational complexity of $O(n^2)$. After that, the sub-band *HH* is construed into the matrices *R* and *Q*.
5. With the scaling factor (S_f), the private data is concealed into the *R* matrix using Eq. (1), where *I* is the identity matrix. To form the stego image, obtain the updated coefficients of the sub-bands and then apply the inverse RDWT using Eq. (2).

$$R' \Longrightarrow (S_f \times I + R) \tag{1}$$

$$L_1^r L_2^{-1} \Longrightarrow Y \tag{2}$$

4.2 Extraction Procedure

1. Obtain modified sub-band by computing the level 1 of RDWT of the stego image.
2. Compute the *R* coefficients by applying the *QR* decomposition and extract secret using Eqs. (3) and (4), where Y' is extracted data.

3. Retrieve complete secret image by applying the secret key in Arnold transformation, where $r = \{LL, LH, HL, HH\}$.

$$Y' = \frac{1}{S_{\mathrm{f}}} \times (|R' - R|) \tag{3}$$

$$pq[Y^{\mathrm{r}}] \Longleftarrow [R \times Q] \tag{4}$$

5 Experimental Results

The suggested method is experimented with five test images (see Fig. 2), namely Peppers, Airplane, Barbara, Boat and Sailboat, from the USCID database [9] in MATLAB R2011a in a system with Intel i7 processor and Windows 10 operating system. The size of the confidential image is 256 × 256 which is a pattern image. To judge invisibility of stego image, the metric for judging the quality of the stego image, i.e., PSNR or peak signal-to-noise ratio, is taken in our experiments. PSNR is the most often utilized image quality statistic for evaluating stego picture quality. The PSNR for the original image $V_{a,b}$ of size $A \times B$ and stego image $V'_{a,b}$ is defined in Eq. (5).

$$\mathrm{PSNR} = 20\log_{10}\frac{255}{\sqrt{\frac{1}{AB}\sum_{a=1}^{A}\sum_{b=1}^{B}(V_{a,b} - V'_{a,b})^2}} \tag{5}$$

Normalized absolute error is an important metric for determining the imperceptibility of the suggested method, and it is defined in Eq. (6).

$$\mathrm{NAE} = \frac{\sum_{a=1}^{A}\sum_{b=1}^{B}(V_{a,b} - V'_{a,b})}{\sum_{a=1}^{A}\sum_{b=1}^{B}V_{a,b}} \tag{6}$$

(i) Peppers: NAE = 0.0014 PSNR = 46.19 dB
(ii) Airplane: NAE = 0.0016 PSNR = 46.15 dB
(iii) Barbara: NAE = 0.0013 PSNR = 46.14 dB
(v) Boat: NAE = 0.0017 PSNR = 46.20 dB
(v) Sailboat: NAE = 0.0019 PSNR = 46.16 dB

Fig. 2 NAE and PSNR for all test images

Table 1 Outcome comparison of the proposed method

Images	Schemes	Secret image size	PSNR
Peppers	Su et al. [10]	32 × 32	36.37
	Xiao et al. [8]	256 × 256	41.28
	Kanan et al. [11]	256 × 256	45.21
	Singh et al. [12]	128 × 128	45.05
	Huang et al. [13]	130 × 130	43.86
	Proposed	256 × 256	48.19
Airplane	Su et al. [10]	32 × 32	36.32
	Xiao et al. [8]	256 × 256	41.25
	Kanan et al. [11]	256 × 256	45.24
	Singh et al. [12]	128 × 128	45.02
	Huang et al. [13]	130 × 130	43.87
	Proposed	256 × 256	48.15
Barbara	Su et al. [10]	32 × 32	36.34
	Xiao et al. [8]	256 × 256	41.26
	Kanan et al. [11]	256 × 256	45.22
	Singh et al. [12]	128 × 128	45.04
	Huang et al. [13]	130 × 130	43.88
	Proposed	256 × 256	48.14
Boat	Su et al. [10]	32 × 32	36.35
	Xiao et al. [8]	256 × 256	41.21
	Kanan et al. [11]	256 × 256	45.20
	Singh et al. [12]	128 × 128	45.05
	Huang et al. [13]	130 × 130	43.85
	Proposed	256 × 256	48.20
Sailboat	Su et al. [10]	32 × 32	36.34
	Xiao et al. [8]	256 × 256	41.27
	Kanan et al. [11]	256 × 256	45.28
	Singh et al. [12]	128 × 128	45.03
	Huang et al. [13]	130 × 130	43.86
	Proposed	256 × 256	48.16

In Table 1, the comparison in terms of secret image size and PSNR of the proposed method with other existing methods [8, 10–13] is illustrated. From this table, it can be clearly stated that our suggested methodology has achieved average 48.17 dB of PSNR. In comparison with Su et al. [10], Singh et al. [12] and Huang et al. [13], the proposed scheme not only can hide higher sized secret images but also can produce higher-quality stego image with an average 32.55, 6.95 and 9.83% higher PSNR value, respectively. However, in comparison with Xiao et al. [8] and Kanan et al. [11], though the proposed scheme can hide same sized secret images, it can generate higher-quality stego image with an average 16.78 and 6.50% higher PSNR value,

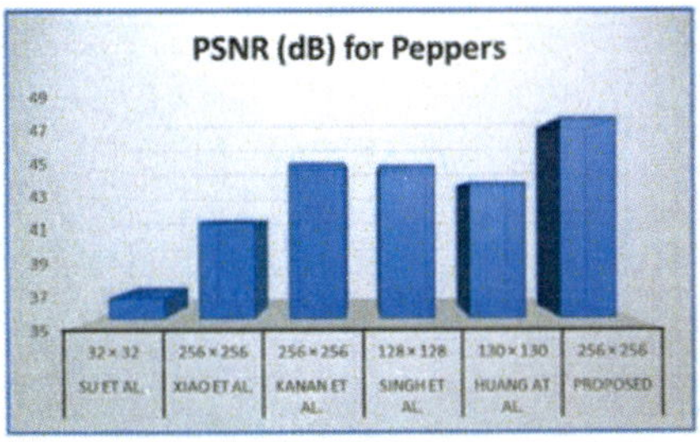

(a) PSNR comparison for the image Peppers.

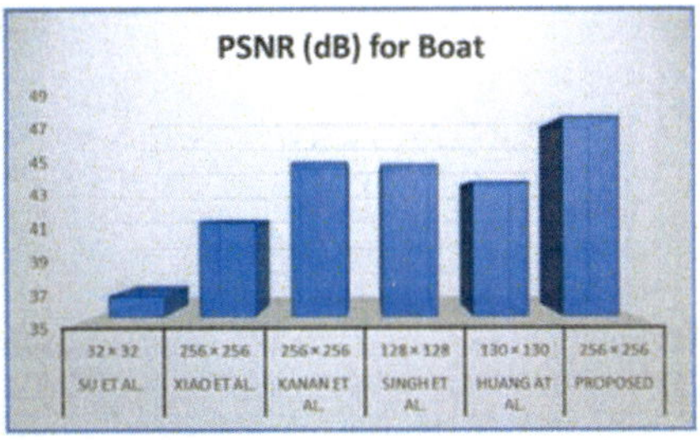

(b) PSNR comparison for the image Boat.

Fig. 3 PSNR comparison for the images peppers and boat

respectively. Hence, the proposed scheme can generate better quality stego image than other existing schemes. To establish the security strength of our suggested method, the NAE is determined for each test image. In Fig. 3, the values of PSNR of [8, 10–13] are compared with the proposed scheme for the images Peppers and Boat. From Fig. 2, we can clearly see that the outcome of NAE for each image tends to zero; therefore, it is proved that our suggested method is highly secure.

6 Conclusion

Data hiding within an image based on the decomposition of QR in RDWT domain is proposed in this article. Stego image restoration is satisfactory, and steganographic capability is enhanced due to the change in invariant design and improved redundancy in RDWT. Low computing complexity and no uncertainty in hidden information retrieval are the most important benefits of our method. Hidden image size 256 $\times$ 256 is inserted in the proposed algorithm, and PSNR is found to be high, while NAE values are found to appear to zero, meaning greater imperceptibility. For both stego attacks, error rates are very low and demonstrate that the proposed system has good robustness for a number of signal processing operations. Experimental findings indicate that the clean and stego picture cannot be separated, and the scheme is extremely resistant to steganalysis.

References

1. Jana B, Giri D, Mondal SK (2018) Dual image based reversible data hiding scheme using (7, 4) hamming code. Multimedia Tools Appl 77(1):763–785
2. Subhedar MS, Mankar VH (2014) High capacity image steganography based on discrete wavelet transform and singular value decomposition. In: Proceedings of the 2014 international conference on information and communication technology for competitive strategies, pp 1–7

3. Guo J-M, Prasetyo H (2014) Security analyses of the watermarking scheme based on redundant discrete wavelet transform and singular value decomposition. AEU-Int J Electron Commun 68(9):816–834
4. Hwang J, Kim J, Choi J (2006) A reversible watermarking based on histogram shifting. In: International workshop on digital watermarking, Springer, pp 348–361
5. Mukherjee S, Jana B (2019) A novel method for high capacity reversible data hiding scheme using difference expansion. Int J of Nat Comput Res (IJNCR) 8(4):13–27
6. Banerjee A, Jana B (2019) A robust reversible data hiding scheme for color image using reed-solomon code. Multimedia Tools Appl 78(17):24903–24922
7. Thabit R, Khoo BE (2015) A new robust lossless data hiding scheme and its application to color medical images. Digital Signal Process 38:77–94
8. Xiao M, He Z (2015) High capacity image steganography method based on framelet and compressive sensing. In: MIPPR 2015: multispectral image acquisition, Pprocessing, and analysis, vol 9811. International Society for Optics and Photonics, p 98110Y
9. USCID image database. http://sipi.usc.edu/database/
10. Su Q, Niu Y, Zou H, Zhao Y, Yao T (2014) A blind double color image watermarking algorithm based on qr decomposition. Multimedia Tools Appl 72(1):987–1009
11. Kanan HR, Nazeri B (2014) A novel image steganography scheme with high embedding capacity and tunable visual image quality based on a genetic algorithm. Expert Syst Appl 41(14):6123–6130
12. Singh N, Sharma D (2017) An efficient multiple data hiding technique for medical images using qr code authentication. Int J Sci Res Sci Eng Technol 3(1):135–139
13. Huang H-C, Chen Y-H, Chang F-C, Tseng C-T (2020) Multi-purpose watermarking with qr code applications. In: 2020 IEEE 2nd global conference on life sciences and technologies (LifeTech). IEEE, pp 42–45

ADT-SQLi : An Automated Detection of SQL Injection Vulnerability in Web Applications

Md. Maruf Hassan, Rafika Risha, and Ashrafia Esha

Abstract Web applications are constantly being developed to make life easier and more convenient for businesses and customers; it makes intruders involved in conducting malicious activities. Intruders use vulnerabilities to perform malicious attacks, and injection is the top-ranked vulnerability of web applications. SQL injection is a technique of code injection that places malicious code through web page input in SQL statements. Several numbers of case studies are found in previous research on vulnerability in the web application layer. Various models are introduced, built, and compared with many current SQL injection models and other vulnerabilities in the web application layer. However, there are few automation detections works on SQL injection that provide high precision and no finite state model-based works. This research aimed to propose a model and develop an automated SQL injection detection tool called ADT-SQLi based on the proposed model. In addition, this work was intended to simulate the proposed model with automata called a finite state machine. ADT-SQLi provides better efficacy on the identification of SQL injection and found ADT-SQLi as a finite model as it has exactly one of a finite number of states at any given time.

Keywords SQL injection · Automated detection tool · Cyber security · Finite state machine · Web application vulnerability

1 Introduction

The use of web applications has turned very prevalent in past years due to knowledge dissemination and management of trade operations. Through web applications, institutions including governments, banks, and business organizations are extending their usefulness to the common people. As web applications manage precise and confidential data of people and institutions, intruders are interested in generating attacks

Md. M. Hassan (✉) · R. Risha · A. Esha
Daffodil International University, Dhaka, Bangladesh
e-mail: maruf.swe@diu.edu.bd

S. Basu et al. (eds.), *Proceedings of International Conference on Frontiers in Computing and Systems*, Lecture Notes in Networks and Systems 404,
https://doi.org/10.1007/978-981-19-0105-8_43

on web applications which compels security as a fundamental concern. Nonetheless, security has not run with the accelerated progress of web applications.

Varieties of attacks materialize in order to exploit the vulnerabilities in web applications. The most prevalent vulnerabilities of web applications are injection, broken authentication and session management, cross-site scripting (XSS), broken access control, security misconfiguration, sensitive data exposure, etc. [1, 2]. When a user accepts malicious feedback to interfere with the application's routine operation, and due to inadequate server-side validation, a critical danger may be reported; this kind of vulnerability is known as injection vulnerability which permits an attacker to steal private data or inject malicious data in to the application by adjusting the security of the application. An innovative directory scoring architecture has the advantage of allowing both benign and malicious traffic to be used in model training while maintaining a satisfactory detection rate and an acceptable false positive rate [3]. A case study-based research presented an analysis on three major SQLi techniques in 309 web applications to determine the security condition for vulnerability exploitation [4]. Some studies proposed a model for detecting and protecting SQLIA on the web application layer based on SQL syntax and negative taint at the database layer [5–8]. An experiment described the SQL injection attacks with steps and developed a method to prevent SQL injection attacks [9]. A model is designed to detect SQLi attacks using feature selection [10]. A study designed a new black box testing analyzer based on calculating hash value with the scanning process [11]. A diverse static analysis tool is developed to prevent SQL injection vulnerabilities by differentiating between common and uncommon execution, analyzing and collecting insecure coding statements [12, 13]. A static analysis tool and a WASP tool are developed to detect vulnerabilities automatically by using machine learning with the combination of taint analysis and data mining [14, 15]. A dynamic analyzer, tester, and runtime mitigation techniques are proposed to detect and block the SQLi attacks [16, 17]. Two studies discussed a method for detecting SQL injection attacks that removes the SQL query parameter values [18, 19]. A novel adaptive model AMOD is presented to detect malicious queries in web attacks [20]. Two tools—NoTamper and WAPTEC—are presented to identify parameter tampering vulnerabilities [21, 22]. A black box approach is used in NoTamper, while a white box approach for WAPTEC to detect vulnerabilities.

It is observed in recent studies that several attempts to examine the problems have been triggered as well as various approaches such as MITRE, SANS, and OWASP have developed and overseen security awareness programs [23, 24]. Despite these attempts, the continual attack on web applications has been faced due to poor architecture and careless coding by the web application architects, designers, and developers. Therefore, there is a need to figure out an automated detection solution that can detect the SQL injection vulnerability and firewall bypass. Though the implemented models can detect SQL injection vulnerability, these models faced some unwanted anomalies in analyzing false negative/positive results to improve accuracy and firewall bypassing. Also, there is lacking simulation of sequential logic and functions of detection models.

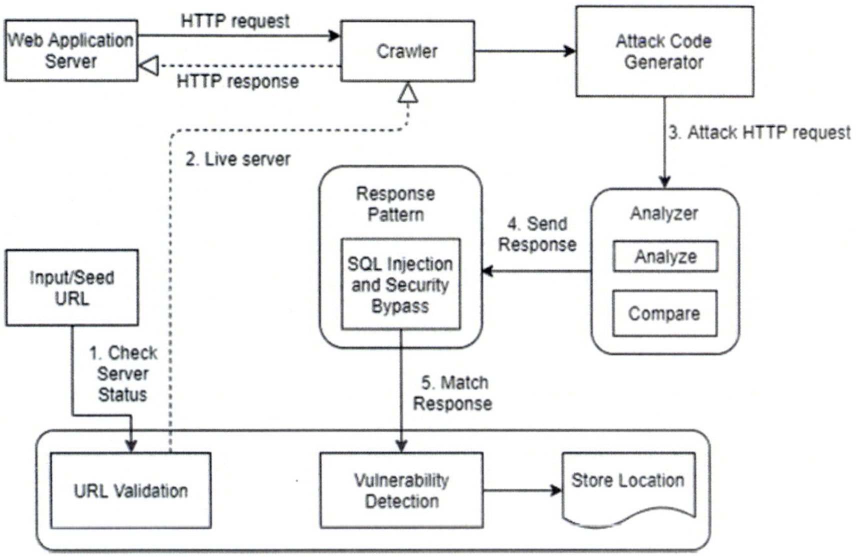

Fig. 1 System architecture of ADT-SQLi

An automated web application vulnerability detection model has been proposed to analyze SQL injection vulnerability and firewall bypass, considering the advantages and disadvantages of the given solutions. To the best of our knowledge, this model provides better accuracy where the evaluation of SQL injection vulnerability and firewall bypassing has been compared with the accuracy of manual penetration testing and also simulates the proposed model with automata called the finite state machine.

The remaining portions of this paper are structured as follows: Sect. 2 presents the methodology of ADT-SQLi with the proposed approach and implementation details. Section 3 serves the result and discussion part. The paper concludes in Sect. 4.

2 Methodology

The system architecture of SQL injection tools and the problems solved by the components are discussed here. Figure 1 demonstrates the architecture based on the five components, and the way they relate to each other. The problems in each of our components are discussed in more detail below.

Crawler. In this system, BeautifulSoup4 (BS4) extracts every possible URLs from the source code as crawler and stores those in a temp_variable. Then the system sends the source code into the detection mechanism to find out the parameterized URLs and define malicious payload.

HTML/JavaScript Analyzer. To determine the working process of a parameter, we observe parameters passed in a web application through HTTP request and response analysis. In this process, every request is waiting for a certain number of parameters in which it contains a name and value. Several pieces of research are available for the detection of SQLi vulnerabilities conducted through HTML/JavaScript analyzer [10]. This analyzer is used for designing web pages to achieve the restrictions in each form field. The HTML/JavaScript parser is designed for parsing a web page to make restrictions (if any) in each form field.

Identifying JavaScript validation code. The analyzer must allocate the code fragment, which is important for validating the parameters and appreciating the working process of these code fragments accumulating the variable temp_variable. This can be complex since the validation procedure can be performed in two directions: (1) when a form is submitted and (2) when user inputs or modifies data on the form using managers of events.

Analyzing validation code. After identifying the temp_variable, they must be analyzed. BeautifulSoup (BS4) addresses these challenges by implementing a mixed web scraper using the SQLi detection mechanism. This application provides scope for all monitoring paths in the verification code and reproduces verification of data used by users.

Resolving SQL Injection. BS4 extracts and stores all possible URLs from source code in a variable named temp_variable. The existence of parameterized URLs is identified from the stored source code. If any parameterized URL is found, then ADT-SQLi selects and splits the parameterized URL to generate possible vulnerable links. It then compares the created script to a predefined script and returns vulnerable links with potential SQL injection if the script matches.

Attack Code Generator. The SQLi detection tool identifies the parameterized URLs with the help of this detection mechanism procedure named "attack code generator." The attack code generator mechanism covers individual tasks by generating unique input data to solve these formulas. Afterward, it splits the parameterized URL and also regenerates the URL. Then it defines malicious payload and response patterns. To continue this process, it goes to the next step called "analyzer."

Analyzer. The analysis segment has two phases—analyzer and compare. The analyzer phase starts its process by generating formulas. The formula will be executed in compare phase to combine regenerated URL, payloads, and firewall checker through the user agent. The user agent collects responses for the remaining process.

Response Pattern. In this segment, predefined responses for SQL and firewall will be defined. The main problem is that when the server is processed as a package, the SQLi detection mechanism must have to decide whether the server will carry a specific hostile scenario. The detection mechanism will solve this complication by formulating a malicious script on how the responses on the host script server match the predefined response. After getting the result as SQLi vulnerability detected, it ends up the procedure with storing the location.

2.1 FSM Construction

ADT-SQLi's flow and mechanism are theoretically constructed as a finite state machine (FSM) and will be briefly described in this section which is demonstrated in Fig. 2. FSM is compatible with designing any model or system's constitution; for this reason, FSM is employed to map the flow and function of ADT-SQLi. An annotated FSM is modeled with six tuples $(Q, \Sigma, A, \Delta, q0, F)$ to check SQL injection vulnerability.

Q refers to a finite set of states. Q denotes a finite set of states. Activity or destination nodes of this model are identified as states. Σ denotes a finite set of inputs. The input symbols are making the transition between states to destination states. A is the set of annotations. Annotations are used to support the additional conditions to assist the transition from one state to another. In this model, annotation is represented with a tuple—P representing the set of HTTP parameters and their respective values those passing along the HTTP request. Δ is the transition function. The transition function indicates which state to go to for each symbol or input from each state. $q0$ defines the initial state. Here seed URL is the initial state from which the FSM begins. F indicates the final set of states. Final state(s) refers to the state(s) marking the end state(s) of a model.

In the annotated FSM, the activity nodes are represented as states. The parameters are marked on the edges of the FSM as annotations reflect the flow of data. The edges are classified as input symbols, HTTP parameters are user or system input or set of inputs, and post-parameters are marked as user discrete or dynamic input from the system or user. On a specific web page, the input symbols and the annotation functions determine the path to the following web pages.

2.2 Algorithm

This section provides the central algorithm of the ADT-SQLi model, which is illustrated in Algorithm 1.

URL Extraction. The extraction of the URL is done via multiple steps. First (1) checking the server status of the URL supplied and (2) technology incorporated on the web page. Second (1) URL analyzer extracts from the specified web page all potential links and (2) creating potential URLs using a predefined library.

2.3 Implementation

The HTML analyzer was developed based on the API that HTML parser provides [25]. The JavaScript study was performed via an updated BeautifulSoup scrapper-based Python library. The generator of scripts was custom-built using the various

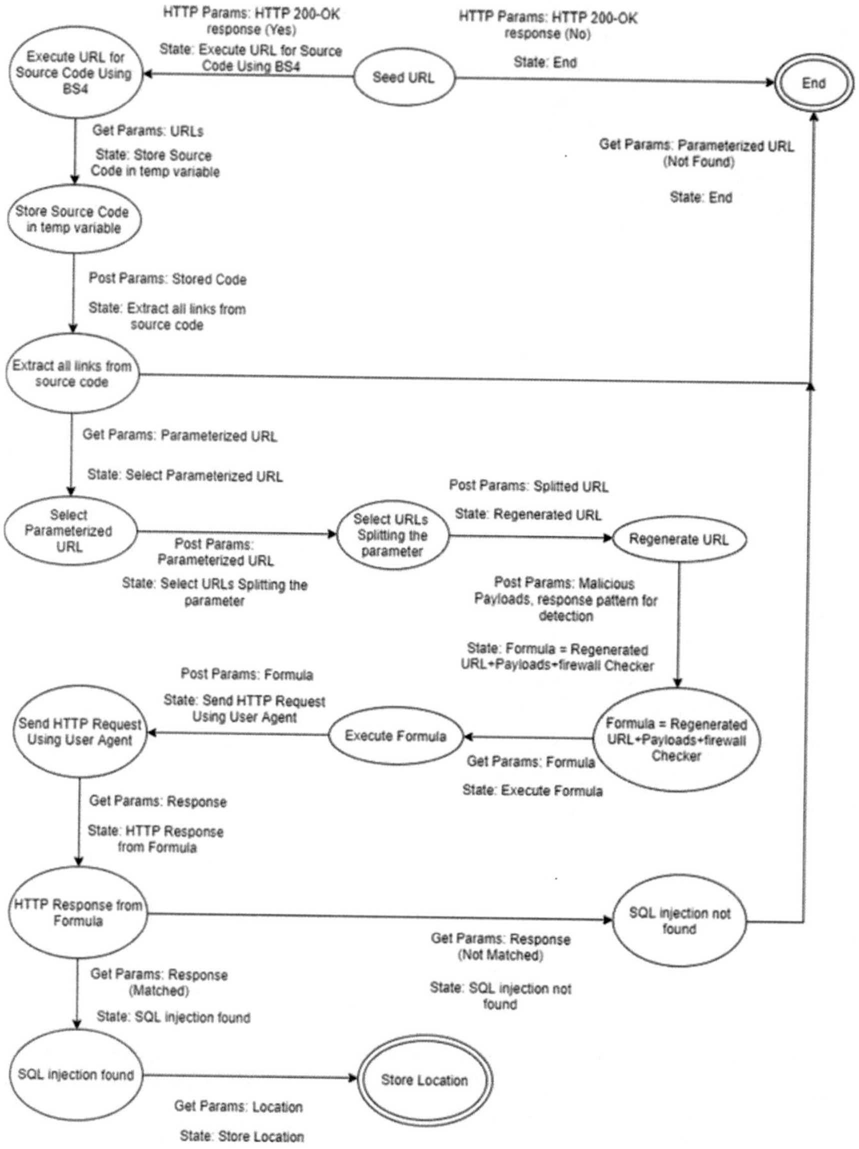

Fig. 2 ADT-SQLi model with FSM

Algorithm 1 SQL Injection Detection

```
Prerequisites for URL checking;
Take seed URL;
if seedURL = valid then
for loop:
    Execute URL for source code using BS4;
    tempVariable ← store source code
    Extract all links from source code
    if parameteizedURL = yes then
        Select and Split parameterized URLs
        formattedURL ← generate all possible vulnerable links
        maliciousscript ← generate from payloads
        predefinedresponse ← alert message
        formula ← formatted URL+malicious script+firewall checker
        Execute formula (malicious request) using useragent;
        response ← HTTP response from formula
        if predefinedresponse = response then return SQL Injection found and Store Location;
        else return SQL Injection not found;
goto loop.
close;
```

open-source libraries available in Python, which consisted of around 100 lines of code. The analyzer was also implemented in Python. Derived the scripts from the attack generator, the Python-based module transferred HTTP requests to the test server, received responses, and introduced a calculation algorithm for differential rankings.

2.3.1 Test Data Configuration

We chose four open-source applications for which we visited (http:/opensourcescripts.com) and discovered applications those have database-linked user input pages and are not using AJAX. Moreover, we also selected two live sites; we chose input fields that we used personally and that were possibly incomplete (e.g., one of the writers has an administrative account in a blog) to select living sites. Table 1 contains some background information for these applications.

3 Result and Discussion

Our experimental results are listed in Table 2. In column 1, application is presented, and columns 2 and 3 contain the number of attack requests and the number of successful attacks generated by ADT-SQLi. The last two columns indicate if vulnerability existed and detected for confirmation by this application. To assess any web appli-

Table 1 Test application details

Application	Type	Client side	Description	Confirm exploits
Magneto	Open source	HTML + JavaScript	Shopping	Yes
Book-fair	Authorized	HTML	Blog	Yes
Otelbor	Open source	HTML + JavaScript	Travel management	Yes
Auto muffler	Open source	HTML + JavaScript	Business management	Yes
Just Yatra	Open source	HTML	Travel management	No
Doctors	Authorized	HTML + JavaScript	Forum	Yes

Table 2 Number of attack requests and vulnerabilities recorded by ADT-SQLi

Application	Attack request		Successful attacks		Vulnerabilities (E)		Vulnerabilities (D)	
	Cred.	Dir.	Cred.	Dir.	Cred.	Dir.	Cred.	Dir.
Magneto	880	52	129	23	90	17	21	19
Book-fair	75	9	32	7	26	3	20	6
Otelbor	1250	48	217	18	162	17	54	11
Auto muffler	750	23	89	7	65	2	23	2
Just Yatra	250	16	55	11	3	1	0	1
Doctors	120	13	52	7	30	5	22	3

**E = Existing, D = Detected

cation, column 4 is the fundamental concern. The developer's sole obligation is to look at the server's hostile input and to determine whether the server is genuinely vulnerable or not manually. As for the testers (the security analyst team), carrying out more tests with agreed hostile information will validate exploitation.

Similarly, we have attempted to check that unwanted database access or directory access at least one exploit in each program. Efforts were made in five out of six applications to explore the exploits.

Table 3 exhibits the effectiveness of the ADT-SQLi tool. Our experiment conducted on six different platforms, including 1022 web URLs where false negative: 4 and false positive: 1 for Magneto; false negative and false positive for Book-Fair are accordingly: 2, 0; false negative and false positive for Otelbor are accordingly: 0, 2; false negative and false positive for Auto Muffler are accordingly: 2, 2; and false negative and false positive for Doctors are accordingly: 0, 1.

Performance. Attack code generator is the most costly computational part in ADT-SQLi. In less than a second, the HTML/JavaScript analyzer was executed to

Table 3 Effectiveness of ADT-SQLi tool

Application	Vul. page (E)	Vul. page (D)	False negative	False positive
Magneto	33	28	4	1
Book-fair	26	24	2	0
Otelbor	27	29	0	2
Auto muffler	23	21	2	2
Just Yatra	0	0	0	0
Doctors	24	25	0	1

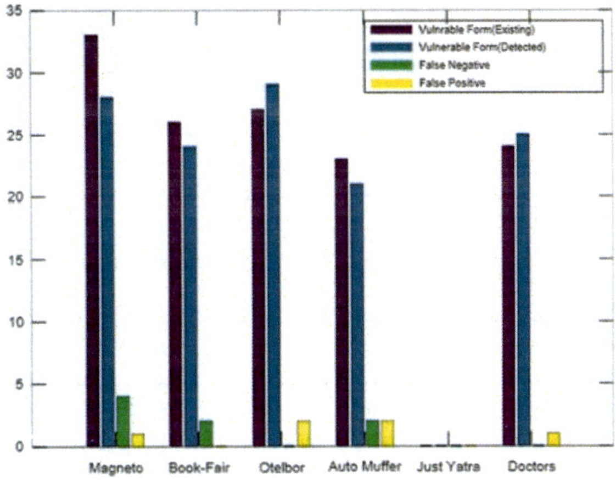

Fig. 3 Potency of ADT-SQLi

get the most accurate shape in our test suite. For each query, the analyzer was run the second time, ignoring the gaps between consecutive HTTP requests to prevent server overload. Figure 3 shows the potency of ADT-SQLi.

Comparative Analysis. The results of SQLMAP and manual black box testing are considered controlled environments, while the results of our tool ADT-SQLi are considered experimental environments. Throughout the research, we discovered SQLi vulnerability in six different platforms, including 1022 web URLs considering manual black box testing. SQLMAP, a well-known SQLi vulnerability scanner, identified 797 SQLi vulnerable URLs, whereas our automated vulnerability detection tool ADT-SQLi identified 914 web URLs with the same flaw. In terms of effectiveness, ADT-SQLi produced results with an accuracy of 89.47 percent which presents in Fig. 4.

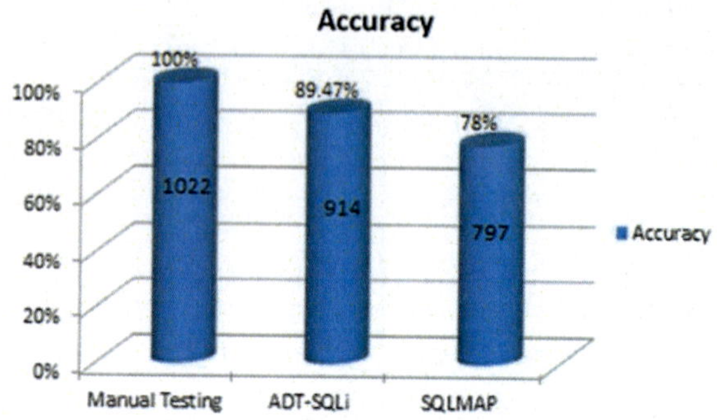

Fig. 4 Comparative analysis of ADT-SQLi

4 Conclusion

The insecure nature of the applications and careless coding practices were recognized as the biggest problem for data or information retrieval. Detection of vulnerabilities is required to reduce data or information extraction. This experiment introduced an automated detection system called ADT-SQLi that could be used to detect SQL injection vulnerabilities and bypass firewalls. The ADT-SQLi solution has been thoroughly tested in web applications and reported to work effectively with a limited number of false positives and false negatives. This tool can be quite useful in detecting SQL injection vulnerabilities with 89.47% effectiveness. Another contribution to this experiment is that the proposed model is a finite model simulated by finite state automata, as it has an initial and final state. However, in future, we will provide an algorithm to prove and validate the model ADT-SQLi with the automata theories and categorize the risk level of the asset when SQL injection occurred.

Acknowledgements The author of this work would like to acknowledge the Cyber Security Centre, DIU, for their assistance in carrying out the research and also appreciate the authorities of the organizations who have allowed us to investigate at their websites.

References

1. OWASP top ten web application security risks. Available at https://owasp.org/www-project-top-ten/. Accessed June 2020
2. SANS Institute (2017) Available at https://www.sans.org/top25-software-errors. Accessed June 2020
3. Liu X, Yu Q, Zhou X, Zhou Q (2018) OwlEye: an advanced detection system of web attacks based on HMM. In: 2018 IEEE 16th international conference on dependable, autonomic and secure computing, 16th International conference on pervasive intelligence and computing, 4th International conference on big data intelligence and computing and cyber science and technology congress (DASC/PiCom/DataCom/CyberSciTech)
4. Alam D, Kabir MA, Bhuiyan T, Farah T (2015) A case study of SQL injection vulnerabilities assessment of .bd domain web applications. In: 2015 Fourth international conference on cyber security, cyber warfare, and digital forensic (CyberSec)

5. Alazab A, Khresiat A (2016) New strategy for mitigating of SQL injection attack. Int J Comput Appl 154:1–10
6. Kyalo F, Otieno C, Njagi D (2018) Securing web applications against structured query language injection attacks using a hybrid approach: input filtering and web application firewall. Int J Comput Appl 182:20–27
7. Appelt D, Panichella A, Briand L (2017) Automatically repairing web application firewalls based on successful SQL injection attacks. In: 2017 IEEE 28th international symposium on software reliability engineering (ISSRE)
8. Hu H (2017) Research on the technology of detecting the SQL injection attack and non-intrusive prevention in WEB system
9. Dalai AK, Jena SK (2017) Neutralizing SQL injection attack using server side code modification in web applications. Secur Commun Netw 2017:1–12
10. Kar D, Sahoo AK, Agarwal K, Panigrahi S, Das M (2016) Learning to detect SQLIA using node centrality with feature selection. In: 2016 International conference on computing, analytics and security trends (CAST)
11. Maki M, Ahmed HM. Using Hash algorithm to detect SQL injection vulnerability. Int J Res Comput Appl Robot
12. Algaith A, Nunes P, Jose F, Gashi I, Vieira M (2018) Finding SQL injection and cross site scripting vulnerabilities with diverse static analysis tools. In: 2018 14th European dependable computing conference (EDCC)
13. Meo FD, Rocchetto M, Viganò L (2016) Formal analysis of vulnerabilities of web applications based on SQL injection. In: Security and trust management lecture notes in computer science, pp 179–195
14. Medeiros I, Neves N, Correia M (2016) DEKANT: a static analysis tool that learns to detect web application vulnerabilities. In: Proceedings of the 25th international symposium on software testing and analysis
15. Salih N, Samad A (2016) Protection web applications using real-time technique to detect structured query language injection attacks. Int J Comput Appl 149:26–32
16. Muhammad R, Muhammad R, Bashir R, Habib S (2017) Detection and prevention of SQL injection attack by dynamic analyzer and testing model. Int J Adv Comput Sci Appl 8
17. Steiner S, Leon DCD, Alves-Foss J (2017) A structured analysis of SQL injection runtime mitigation techniques. In: Proceedings of the 50th Hawaii international conference on system sciences
18. Katole RA, Sherekar SS, Thakare VM (2018) Detection of SQL injection attacks by removing the parameter values of SQL query. In: 2018 2nd International conference on inventive systems and control (ICISC)
19. Deepa G, Thilagam PS, Praseed A, Pais AR (2018) DetLogic: a black-box approach for detecting logic vulnerabilities in web applications. J Netw Comput Appl 109:89–109
20. Dong Y, Zhang Y, Ma H, Wu Q, Liu Q, Wang K, Wang W (2018) An adaptive system for detecting malicious queries in web attacks. Sci China Inf Sci 61
21. Bisht P, Hinrichs T. Skrupsky N, Venkatakrishnan VN (2011) Waptec. In: Proceedings of the 18th ACM conference on computer and communications security—CCS 11
22. Bisht P, Hinrichs T, Skrupsky N, Venkatakrishnan V (2014) Automated detection of parameter tampering opportunities and vulnerabilities in web applications. J Comput Secur 22:415–465
23. Mohosina A, Zulkernine M (2012) DESERVE: a framework for detecting program security vulnerability exploitations. In: 2012 IEEE sixth international conference on software security and reliability
24. Zainab SA, Manal FY (2017) Detection and prevention of SQL injection attack: a survey. In: 2017 Int J Comput Sci Mobile Comput
25. Gupta MK, Govil M, Singh G (2014) Static analysis approaches to detect SQL injection and cross site scripting vulnerabilities in web applications: a survey. In: International conference on recent advances and innovations in engineering (ICRAIE-2014)

Fuzzy Logic with Superpixel-Based Block Similarity Measures for Secured Data Hiding Scheme

Prabhash Kumar Singh, Biswapati Jana, Kakali Datta, Prasenjit Mura, Partha Chowdhuri, and Pabitra Pal

Abstract With time, humans have realized, the internet is the best alternative for exchange of messages through multimedia documents, especially images. Thus, its imperceptibility and security are big concerns. Recently, Ashraf et al. (Heliyon 6(5):e03771, [1]) evaluated similarity for image steganography using interval type-2 fuzzy logic, but the visual quality (PSNR) is not as good as current demand. To solve the problem, we have proposed a data hiding scheme by combining distinct and vague characteristics of an image through the techniques of superpixel and fuzzy logic respectively. Superpixel and TSK Fuzzy logic model with rule base is used to identify non-uniform regions of the image where secret bits are embedded in coefficients of quantized Discrete Cosine Transform (DCT). This technique can be used for secret data communications. With an average PSNR of 58 dB, the proposed approach provides excellent visual quality. Finally, in order to illustrate the efficacy of our technique, we compared the proposed system to existing methodologies.

Keywords Data Hiding · Fuzzy logic · Superpixel · Block Similarity · PSNR

1 Introduction

Internet is one of the wonderful things that mankind has come across. Its significance can be felt more during this time of ongoing pandemic where it has come as a relief to various areas such as education, communication, healthcare, e-commerce, and many more. Internet, the open media has become a lifeline for communication and exchange of data and information from one part of the world to the other. While most

P. K. Singh · B. Jana · P. Mura · P. Chowdhuri
Department of Computer Science, Vidyasagar University, Midnapore, West Bengal, India

P. K. Singh · K. Datta (✉)
Department of Computer and System Sciences, Visva-Bharati University, Santiniketan, West Bengal, India
e-mail: kakalidatta@hotmail.com

P. Pal
BSTTM, IIT Delhi, New Delhi, India

S. Basu et al. (eds.), *Proceedings of International Conference on Frontiers in Computing and Systems*, Lecture Notes in Networks and Systems 404,
https://doi.org/10.1007/978-981-19-0105-8_44

users have positive intent, some try to use open advanced technology to manipulate the exchanged documents. The manipulations can be done to show self-authority, copyright or to tamper the document with malicious intentions. Thus, it is required to protect these documents from unauthorized users through the techniques of data hiding so that at any desired time authenticity, copyright and tamper, if any can be proved.

Both the spatial and transform domains can be used to hide data. Techniques in the spatial domain take less time, are easier to apply, and provide higher imperceptibility. Bits are embedded by changing the pixel values directly [2]. The secret bits are buried in the transform domain by appropriate coefficients of discrete cosine transform (DCT), discrete Fourier transform (DFT), and discrete wavelet transform (DWT). It aids in the improvement of robustness and reversibility.

Traditionally, algorithms for data hiding in images have been designed to deal with specific information where it is tried to keep image distortion as low as possible. But nowadays, researchers have focused to design algorithms to deal with imprecise and ambiguous data. Genetic algorithm [3], Particle swarm optimization [4], and fuzzy logic [5] are employed to design data hiding algorithms. This helps to achieve more payload with minimum distortion.

While designing any data hiding techniques, researchers mainly concentrate more on payload capacity, imperceptibility and security. In this paper, a data hiding scheme has been designed with superpixel and fuzzy logic to measure the similar block in cover image. Then, embed secret data after partitioning and categorization of blocks. The major research contribution in this proposed system is the use of fuzzy logic and superpixel to facilitate a trade-off to deal with imprecise and precise information created due to the embedding of secret data and manipulations if any. The suggested scheme's goal is to strike a balance between the three opposing parameters of data hiding: capacity, imperceptibility, and resilience, because changing one impacts the others.

The significant contribution of this investigation are enlisted below:

1. A novel algorithm on Fuzzy logic with superpixel has been formulated that measures the similarity and dissimilarity of image block before data embedding.
2. Color difference (Δ_p) and Euclidean distance (δ_p) are measured in terms of linguistic variables low , $medium$ and $high$.
3. A fuzzy rule-based scheme takes linguistic variables as input and gives the result as the similarity of blocks.
4. The similarity of blocks has been measured considering Y-color component and data can be embedded in Cb and Cr color component of $YCbCr$ color model.
5. Data embedding has been performed using transform domain (DCT) to keep secret data more secure from adversary.
6. Comparative analysis demonstrated through visual quality and other statistical measure such as NCC and BER.

The following is how the paper is organized: Sect. 2 addresses various comparable works, followed by Sect. 3's preliminaries. The suggested scheme's data concealing mechanism is discussed in Sect. 4. Section 5 also includes an experimental analysis as well as a comparison. Finally, Sect. 6 brings the proposed strategy and evaluation to a close.

2 Related Work

In image data hiding, the embedding of information is done with respect to some modifications in the cover image. A pixel value adjustment scheme was proposed by Chao et al. [6] to embed data using diamond encoding. When compared to LSB embedding techniques, the proposed method is simple and capable of producing imperceptible stego images. The embedding scheme proposed by Hussaina et al. [7] employs parity-bit pixel value difference and enhanced rightmost digit replacement for efficient data concealment thus maintaining high visual imperceptibility, and so outperforms steganalysis assaults. LBP and weighted matrix has been used by Pal et al. [8–10] to hide data in image with high payload and robustness. LBP significantly depicts the local feature of an image while a weighted matrix improves the security and payload.

Rabie et al. [11] developed a DWT-LPAR (Discrete Wavelet Transform Laplacian Pyramid Adaptive Region) embedding method based on transform techniques that consider payload power as opposed to the imperceptibility trade-off problem. The Dual-Tree Complex Wavelet Transform was used by Kadhim et al. [12] to propose a better version of transform domain steganography (DT-CWT). Canny edge detection is used to classify the texture-rich and smooth portions of the cover image. After that, the cover image is separated into non-overlapping patches, with DT-CWT applied to each of them.

Miaou et al. [13] proposed a technique for medical pictures that combines HVS and fuzzy logic to keep the watermark information constant while keeping the image undetectable. SVD-based watermarking with fuzzy logic has also been shown by embedding the watermark information in DWT [14]. The author of [15] proposed a new fuzzy logic-based spatial masking technique for calculating picture characteristics including brightness, edge, and texture operations. Because evaluating these pixels for embedding the watermark improves the image's imperceptibility, fuzzy logic is utilized to build the algorithms that search for pixels with the best brightness, texture, and edge values.

Thus, analyzing the literature, it can be said that spatial domain provides mechanism for easy embedding with less robustness against stego attacks while transform domain succeeds in maintaining high robustness against stego attacks. As a result, the proposed study will combine superpixel and fuzzy logic in the transform domain using DCT to build an image steganography system with high resilience, payload, imperceptibility, security, error-free secret data extraction, and other features.

3 Preliminaries

In the proposed scheme, fuzzy logic and superpixel serve as two main utilities for data hiding. A brief idea about them is discussed in this sub sections:

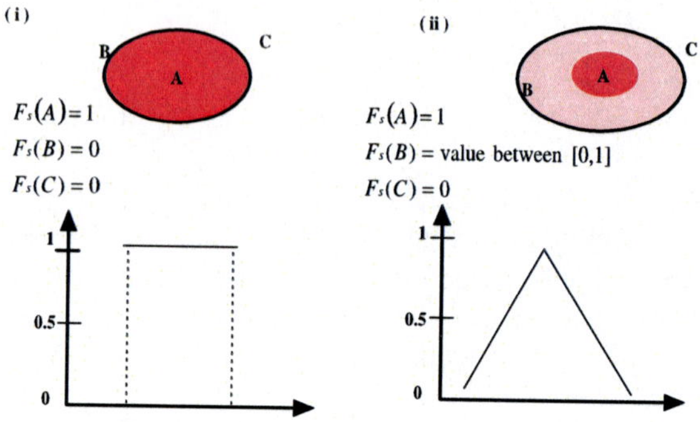

Fig. 1 Difference in representation of crisp and fuzzy set

3.1 *Fuzzy Logic*

The universal set can be described in such a way that all elements are categorized as members or nonmembers based on a predefined characteristic feature. If U denotes the universe set and u denotes the general elements, the characteristic function $F_S(u)$ maps all members of U into the set {0, 1}. The following is the general syntax of a characteristic function: $F_s(u) : U \rightarrow \{0, 1\}$. The classical sets can be represented mathematically as membership function by the following expression:

$$F_s(u) = \begin{cases} 1 & \text{if u belongs to } S \\ 0 & \text{if u does not belong to } S. \end{cases} \quad (1)$$

The fuzzy set theory is a mathematical approach to solving the vagueness and ambiguity details about the problem's knowledge, in contrast to the crisp set principle, which deals with concrete information and knowledge. Zadeh et al. [16] proposed the fuzzy set theory to describe a set of elements that is formulated using a fuzzy membership function. The membership function describes the relationship between an elementś value and its degree of set membership. The membership function of a fuzzy set S is denoted by F_S, where $F_S \colon U \rightarrow [0, 1]$. The difference between the crisp and fuzzy sets is seen in Fig. 1. The crisp set principle is illustrated in Fig. 1(i), where the characteristic function assigns a crisp value to each of the three elements A, B, and C, while the fuzzy set principle is illustrated in Fig. 1(ii), where element B is located on the crisp set's boundary and has a partial membership.

3.2 *Superpixel*

A superpixel is a set of pixels that are identical in color and proximity. Figure 2 shows how superpixels can be used for image recognition, semantic segmentation, and visual tracking. Many superpixel algorithms have been developed as the idea has

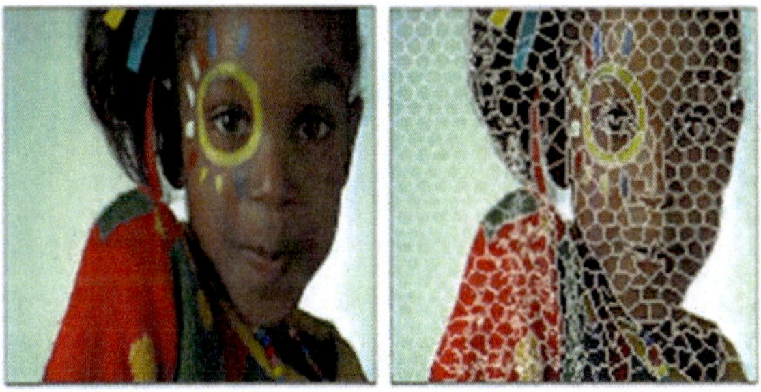

Fig. 2 Superpixel segmentation

grown in popularity; however, the Simple Linear Iterative Clustering (SLIC)-based superpixel algorithm has attracted the most attention from researchers all over the world. The algorithm begins with local pixel clustering by converting the original image of N pixels into a 5D CIELAB color space with color components L, a, b, and x, y pixel coordinates. For each cluster centers $C_k = [L_k, a_k, b_k, x_k, y_k]^T$, the Euclidean distance is obtained both with respect to pixel value differences and spatial pixel position distances. For details, refer [17].

4 Proposed Scheme

In the proposed scheme, the process of data hiding involves two fundamental phases, Embedding phase and Extraction phase. The embedding phase deals with methods of hiding secret data with minimum distortion in the cover image while the extraction phase recovers the secret message to verify the authenticity, copyright, and reversibility of the original image.

4.1 Embedding Phase

For embedding of data, a color image is at first converted to $YCbCr$ color model because this model depicts the native format and is better suitable to work with reversibility of the original pixel. A general outline of the embedding method is shown as a flowchart in Fig. 3. Further, the embedding phase undergoes four phases as mentioned:

4.1.1 Block Categorization with Superpixel

From the obtained $YCbCr$ color model, only Y component is considered and K number of superpixels are generated using SLIC algorithm. Each of the pixel

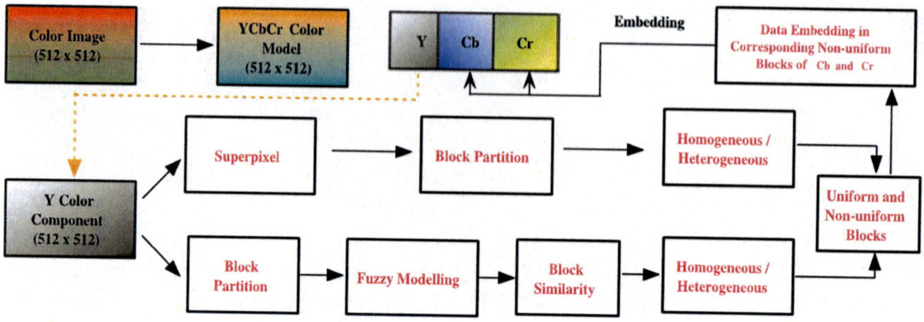

Fig. 3 Overall schematic diagram for the proposed scheme

$P(i, j)$ present in Y component is labeled with corresponding superpixel k, where $k = 1, 2, ..., K$ to form Y'. For example, if pixel $P(15, 28)$ is clustered under superpixel 10, such that $10 <= K$, then, it is labeled as 10. After labeling, Y' is partitioned into non-overlapping blocks (B_s) of size $b \times b$. Next, block $B_s(i, j)$ is categorized as homogeneous or heterogeneous as follows: if all the block elements present in $B_s(i, j)$ is labeled with the same superpixel, it is said to be homogeneous in nature, else termed as heterogeneous.

4.1.2 Block Categorization with Fuzzy Logic

In order to classify a block, a TSK fuzzy model is applied. First, Y component of the image is partitioned into non-overlapping blocks (B_f) of size $b \times b$. For each block, a normalized fuzzy similarity vector $O(i, j)$ is calculated by taking mean of pixel-pair similarity matrix of all P_i based on two fuzzy sets, color ($\tilde{C}$) and distance $\tilde{D}$. A fuzzy set ($\tilde{C}$) and ($\tilde{D}$) depicts the difference in color values (Δ_p) and Euclidean distance (δ_p) between two pixels respectively and is represented by three fuzzy linguistic variables low, $medium$, and $high$. This is computed for every pixel-pair P_i and P_j, $i \neq j, j = 1, 2, ..., b^2$, present in a block. So, for given two fuzzy inputs, the output in crisp through TSK fuzzy model is computed by a constant linear function $f(x)$ as shown in Eq. 2 and rule base mentioned in Table 1. Subsequently, if vector $O(i, j) >$ Threshold(Th), the corresponding block is identified as homogeneous else it is tagged as heterogeneous. Here, the value of Th is taken as 0.5.

$$f(x) = \alpha\Delta_p + (1 - \alpha)\delta_p;\ \text{where, } \alpha \in [0, 1] \text{ is a control parameter} \quad (2)$$

4.1.3 Block labeling

In this phase, each block $B(i, j)$ is relabeled as uniform or non-uniform blocks depending on the block category achieved in B_s and B_f. If a particular block has

Table 1 Fuzzy rules

Rule	Antecedent	Consequent	
R_1	If Δ_p is Low and δ_p is $High$	then	$f(x) = 0.8\Delta_p + 0.2\delta_p$
R_2	If Δ_p is Low and δ_p is $Medium$	then	$f(x) = 0.6\Delta_p + 0.4\delta_p$
R_3	If Δ_p is Low and δ_p is Low	then	$f(x) = 0.4\Delta_p + 0.6\delta_p$
R_4	If Δ_p is $Medium$ and δ_p is $High$	then	$f(x) = 0.6\Delta_p + 0.4\delta_p$
R_5	If Δ_p is $Medium$ and δ_p is $Medium$	then	$f(x) = 0.5\Delta_p + 0.5\delta_p$
R_6	If Δ_p is $Medium$ and δ_p is Low	then	$f(x) = 0.4\Delta_p + 0.6\delta_p$
R_7	If Δ_p is $High$ and δ_p is $High$	then	$f(x) = 0.6\Delta_p + 0.4\delta_p$
R_8	If Δ_p is $High$ and δ_p is $Medium$	then	$f(x) = 0.5\Delta_p + 0.5\delta_p$
R_9	If Δ_p is $High$ and δ_p is Low	then	$f(x) = 0.3\Delta_p + 0.7\delta_p$

been labeled the same, i.e., homogeneous or heterogeneous in both B_s and B_f, then the block is called uniform block otherwise, non-uniform. Data embedding is done in non-uniform blocks due to its implicit nature of inconsistency from the perspective of similarity in both fuzzy and non-fuzzy environments.

4.1.4 Secret Bit Embedding

The data embedding of the secret bits is done only in corresponding non-uniform blocks of Cb and Cr. A secret logo or message is first transformed into its equivalent binary form and then scrambled with key K_s. Next, bits are embedded into nonzero quantized DCT coefficients except the DC coefficient of the Cb and Cr. After embedding, inverse DCT is performed to obtain a modified block of Cb_{new} and Cr_{new}. Finally, the blocks of Y, Cb_{new}, and Cr_{new} are merged together and converted into a color stego image.

4.2 Extraction Phase

Extraction of data from the stego image is done with process similar to embedding phase and shared secret key K_s. The secret bits are collected from the quantized DCT nonzero coefficients of the non-uniform blocks. Thereafter, secret bits are positioned to its original position through key K_s.

5 Experimental Results and Comparisons

Experiments in MATLAB 2020 on the Windows platform were used to evaluate the proposed data hiding process. Intel Core i5 processor with 2.6GHz and 4GB RAM was chosen as the computing environment. The goal of the experiment was to compute the standard quality comparison metrics for the cover and stego image in terms of Q-Index, Structural Similarity Index (SSIM), Peak Signal-to-Noise Ratio (PSNR), Bit Error Rate (BER), and Normalized Cross Correlation (NCC) for the cover and stego image. The imperceptibility and robustness of the proposed methodology are then demonstrated by comparing values between the original cover image and the stego image. A set of six 512 × 512 standard color images "Aeroplane," "Baboon," "Pepper," "Barbara," "Boat," and "Tiffany" are acquired from standard USC-SIPI Image Database to draw an unbiased analogy against the advocated technique. The size of the secret data embedded varies depending on the available DCT coefficient values.

The experimental value of all the parameters considered between cover and stego image is shown in Table 2. It can be seen that for all the images, PSNR obtained is greater than 54 dB which indicates to better quality of the stego image. The maximum PSNR achieved is 61 dB. Higher the PSNR, better is the quality of the image. The quality can also be verified with parameter Q-Index which is mathematically characterized as a combination of three factors: loss of similarity, luminance distortion, and contrast distortion when compared to the reference picture. In all cases, the value is almost 1, which is a perfect score. The purpose of SSIM was to solve the major problem of imperceptibility. The SSIM analyzes the original and watermarked pictures to evaluate how similar they are. Its value varies between −1 and +1. Images with a +1 value are identical, but those with a −1 value are entirely different. This measure has a value greater than 0.98 for all of the pictures, indicating that the cover and stego image are substantially similar.

Table 2 shows the experimental value of all the parameters examined between the cover and stego image. It can be observed that the PSNR obtained for all of the images is larger than 54 dB, indicating that the stego image is of higher quality. 61 dB is the highest PSNR obtained. The higher the PSNR, the better the visual quality. When compared to the known standard, the quality may also be checked using Q-Index, which is mathematically defined with contrast and luminance distortion as well as loss of similarity. The value is almost 1 in all situations, which is a perfect score. The other metric, NCC, is used to determine the robustness of an image. NCC values obtained for all of the images are greater than 0.999, indicating that the stego image is robust. Both SSIM and NCC refer to the imperceptibility and robustness characteristics of the stego image. BER specify error in bits due to distortion in cover image after embedding. The average BER value calculated from the experiment is 0.02838, which is very much acceptable in comparison to deformation done for embedding in considered images.

Table 2 Results of experiment on standard USC-SIPI image database

Images	Non-uniform Blocks	Available Coefficient	Embedded Bits	PSNR	Q-Index	SSIM	NCC	BER
Aeroplane	1827	9810	9801	60.10	0.9999	0.9903	0.9999	0.0174
Baboon	2324	10326	10201	54.76	0.9999	0.9890	0.9999	0.0432
Peeper	2164	5674	5625	58.39	0.9999	0.9888	0.9999	0.0144
Barbara	2198	11413	11236	57.11	0.9999	0.9814	0.9999	0.0359
Boat	2215	10107	10000	56.46	0.9999	0.9801	0.9999	0.0457
Tiffany	2076	6473	6400	61.45	0.9999	0.9914	0.9999	0.0137

Table 3 Comparison of the proposed scheme with existing state-of-the-art schemes

Cover image	Tai et al. [18]		Chang et al. [19]		Parah et al. [20]		Tai and Liao [21]		Proposed	
	PSNR	Payload	PSNR	Payload	PSNR	Payload	PSNR	Payload	PSNR	Payload
Aeroplane	48.53	0.17	49.39	1.5	41.18	0.05	44.12	2	60.10	0.04
Baboon	48.21	0.04	49.39	1.5	39.60	0.05	44.14	2	54.76	0.04
Peepers	48.42	0.13	49.40	1.5	40.43	0.05	44.06	2	58.39	0.02
Boat	48.35	0.10	49.40	1.5	41.32	0.05	44.11	2	56.46	0.04
Tiffany	–	–	–	–	41.35	0.05	43.85	2	61.45	0.02
Average	48.38	0.11	49.39	1.5	40.78	0.05	44.06	2	58.23	0.03

A comparison of PSNR and payload value (bpp) is analyzed in Table 3 with four state-of-the-art schemes found in literature. The average PSNR of all the compared images is 58.23 dB which is relatively higher than all other existing schemes. This comparison tends to authenticate the potential of the proposed scheme in data hiding.

6 Conclusion

A new data hiding technique is designed using superpixel, fuzzy logic and DCT through which distinct and vague characteristics of an image are combined altogether to identify the proper region of embedding the secret bits. The secret bits are hidden at the LSB of nonzero coefficients of quantized DCT obtained at non-uniform blocks. This technique served to achieve high PSNR, better imperceptibility, and structural similarity. Also, the NCC and BER depicted through the experiment are reasonable. The scheme meets the objective of sharing the secret message, authentication and copyright protection however, it has some limitations to overcome. In the future, studies need to be conducted to increase the payload and robustness with the recoverability of the original cover image after attacks.

References

1. Ashraf Z, Roy ML, Muhuri PK, Lohani QD (2020) Interval type-2 fuzzy logic system based similarity evaluation for image steganography. Heliyon 6(5):e03771
2. Li B, Wang M, Li X, Tan S, Huang J (2015) A strategy of clustering modification directions in spatial image steganography. IEEE Trans Inf Forensics Secur 10(9):1905–1917
3. Roy R, Laha S (2015) Optimization of stego image retaining secret information using genetic algorithm with 8-connected PSNR. Procedia Comput Sci 60:468–477
4. Nipanikar SI, Deepthi VH, Kulkarni N (2018) A sparse representation based image steganography using particle swarm optimization and wavelet transform. Alexandria Eng J 57(4):2343–2356
5. Sajasi S, Moghadam AME (2013) A high quality image steganography scheme based on fuzzy inference system. In: 2013 13th Iranian conference on fuzzy systems (IFSC). IEEE, pp 1–6
6. Chao RM, Wu HC, Lee CC, Chu YP (2009) A novel image data hiding scheme with diamond encoding. EURASIP J Inf Secur 2009:1–9
7. Hussain M, Wahab AWA, Ho AT, Javed N, Jung KH (2017) A data hiding scheme using parity-bit pixel value differencing and improved rightmost digit replacement. Signal Process: Image Commun 50:44–57
8. Pal P, Chowdhuri P, Jana B (2018) Weighted matrix based reversible watermarking scheme using color image. Multimedia Tools Appl 77(18):23073–23098
9. Dey A, Pal P, Chowdhuri P, Singh PK, Jana B (2020) Center-Symmetric Local Binary Pattern-Based image authentication using local and global features vector. In: Computational intelligence in pattern recognition. Springer, Singapore, pp 489–501
10. Pal P, Jana B, Bhaumik J (2021) A secure reversible color image watermarking scheme based on LBP, Lagrange interpolation polynomial and weighted matrix. Multimedia Tools Appl 1-28
11. Rabie T, Baziyad M, Kamel I (2018) Enhanced high capacity image steganography using discrete wavelet transform and the Laplacian pyramid. Multimedia Tools Appl 77(18):23673–23698
12. Kadhim IJ, Premaratne P, Vial PJ (2018) Adaptive image steganography based on edge detection over dual-tree complex wavelet transform. In: International conference on intelligent computing. Springer, Cham, pp 544–550
13. Miaou SG, Hsu CM, Tsai YS, Chao HM (2000) A secure data hiding technique with heterogeneous data-combining capability for electronic patient records. In: Proceedings of the 22nd annual international conference of the IEEE engineering in medicine and biology society (Cat. No. 00CH37143), vol 1. IEEE, pp 280–283
14. Sridevi T, Fathima SS (2013) Digital image watermarking using fuzzy logic approach based on DWT and SVD. Int J Comput Appl 74(13)
15. Jamali M, Rafiei S, Soroushmehr SM, Karimi N, Shirani S, Najarian K, Samavi S (2017) Adaptive blind image watermarking using fuzzy inference system based on human visual perception. arXiv:1709.06536
16. Zadeh LA, Klir GJ, Yuan B (1996) Fuzzy sets, fuzzy logic, and fuzzy systems: selected papers, vol 6. World Scientific
17. Ren X, Malik J (2003) Learning a classification model for segmentation. In: Null. IEEE, p 10
18. Tai WL, Yeh CM, Chang CC (2009) Reversible data hiding based on histogram modification of pixel differences. IEEE Trans Circ Syst video Technol 19(6):906–910
19. Chang CC, Liu Y, Nguyen TS (2014) A novel turtle shell based scheme for data hiding. In: 2014 Tenth international conference on intelligent information hiding and multimedia signal processing. IEEE, pp 89–93
20. Parah SA, Sheikh JA, Loan NA, Bhat GM (2017) A robust and computationally efficient digital watermarking technique using inter block pixel differencing. In: Multimedia forensics and security. Springer, Cham, pp 223–252
21. Tai WL, Liao ZJ (2018) Image self-recovery with watermark self-embedding. Signal Process: Image Commun 65:11–25

Security Aspects of Social Media Applications

Ankan Mallick, Swarnali Mondal, Soumya Debnath, Sounak Majumder, Harsh, Amartya Pal, Aditi Verma, and Malay Kule

Abstract Social media applications are an integral part of human life nowadays. Starting from sharing personal information like text and pictures, we now share the latest news and its related photos, question papers, assignments, online surveys, and so many more things. With so much sharing of our data, hackers have found very easy ways to steal our personal information through various social sites. Although different platforms keep coming up with new versions for better security and experience of their users, this breach of personal information demands advances in security protocols to safeguard our data. This has become the basis of this research. In this paper, we compare the security aspects of different social media platforms to see how to fit our favorite social media applications for our users and how many of them keep their promise of providing us better security.

Keywords Social media · Hackers · Social networks · Security · Privacy · Encryption

1 Introduction

Social media [1] is the most obvious form of communication between people these days. They very much reflect the social image of a person and allow us to reach millions of people at the same time within seconds. The concepts of sharing our location and doing live streams make our presence felt more. Often we, as humans, forget that the more comfortable and attached we become with these social sites, the more careless and casual we become in sharing our personal details, and that is where the security aspect of various sites comes into the picture.

Most users are unaware of the privacy risks associated when they upload sensitive information on the Internet. They upload their personal information that commonly includes what they like, who they like to be with, some confidential information, and other personalized content. They very easily share such information with the whole

A. Mallick · S. Mondal · S. Debnath · S. Majumder · Harsh · A. Pal · A. Verma · M. Kule (✉)
Indian Institute of Engineering Science and Technology, Shibpur, India
e-mail: malay.kule@gmail.com

S. Basu et al. (eds.), *Proceedings of International Conference on Frontiers in Computing and Systems*, Lecture Notes in Networks and Systems 404,
https://doi.org/10.1007/978-981-19-0105-8_45

world through these social media applications, which are available with a single click of their devices [2–4]. The volume and ease of accessibility of private data available on networking sites make stealing data a child's play for people who seek to exploit such information.

Facts have revealed that most social media users post vulnerable information online and are unaware of the security threats and concerns. Social media sites are developed to act as a platform for worldwide users to interact in one place, but for attackers, it is much more than that and it is clearly not in the benefit of users [5].

This paper is organized into four different sections. Section 2 describes the background of social media sites followed by this introductory section. Section 3 elaborates the analysis of security issues of various social media sites followed by a conclusion in Sect. 4.

2 Background of Social Media Sites

The use of social media sites has increased over the past decade due to the introduction of various new technologies. To promote the use of social media, huge investments have been made, and it reflects all around us. It is now very strongly integrated into our daily lives, and we can no longer do without being on any such online social platform.

The Launch of Six Degrees in 1997 marked the first modern social media site. It allowed its users to create profiles and interact with each other. The website became quite popular and recorded around a million subscribers at its peak, but it is no longer functional. Friendster, which started in San Francisco in 2003, was the next site in the category [4]. It was created by Jonathan Adams. Some of the popular social media sites in 2021, as shown in Table 1, include LinkedIn, Facebook, Twitter, WhatsApp, Telegram, Signal, ShareChat, and Instagram. Figure 1 depicts how different social media are getting popular among people day by day.

Table 1 List of five biggest social media sites as of January 2021

Rank	Social media	Monthly visits (in millions)
1	Facebook	2740
2	YouTube	2291
3	WhatsApp	2000
4	Facebook Messenger	1300
5	Instagram	1221

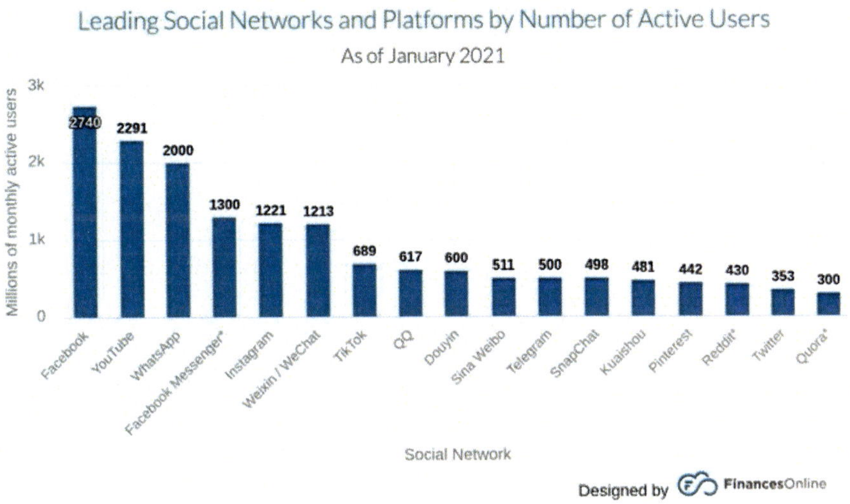

Fig. 1 Leading social media sites and platforms by the number of active users as of January 2021. *Source* statista.com

3 Analysis of Security of Various Social Media Sites

3.1 Facebook

Nowadays, Facebook is one of the most used social media sites by people with billions of users logged onto their servers, which further means a huge amount of data transfer, storage, and retrieval going on every second. The data center's threat model is a lot different from the one of the open Internet. Also, data encryption is not the only means of safeguarding services due to different constraints. Facebook controls almost everything, machines that execute clients and servers, the network connecting them, and the switches that transmit data across the machines, which they set up to add access control at different levels.

The communications services use Thrift as a call library for remote procedures. The orchestration system, referred to as Tupperware, is used to handle containers, and this is where the service generally runs [6]. They wanted to construct a solution that could be resilient even after the attackers had hypothetical access to the hosts. So, they started using the Kerberos System to encrypt the data, but after that, they switched over to the transport layer security (TLS) method. To ensure communications security on a TLS computing network and its now-deprecated predecessor, secure sockets layer (SSL), the cryptographic protocols were sketched. There is extensive use of many versions of the protocols in applications comparable to Internet browsing, instant messaging, email, and VoIP.

The main rationale of the TLS convention is to supply security and data integrity between two or more communicating computer applications. Associations secured by TLS between a client (any Web browser) and a server (wikipedia.org) ought to have at slightest one of them taking after properties:

1. The connection between client and server should be private.
2. The identity of the interacting sides should not be confirmed utilizing public-key cryptography.
3. The connection must be trustworthy because every single message transmitted incorporates a check for data integrity using message authentication code to avoid data corruption and modification.

A client and server must safely trade or concur upon an encryption key sometime recently starting to trade data that is secured by TLS [7]. The various methods used for key exchange are as follows: public and private keys formed using RSA, Diffie–Hellman, elliptic-curve Diffie–Hellman, pre-shared key, and secure remote password.

3.2 *WhatsApp*

With over 1.6 billion active users, today WhatsApp is the most used messaging app worldwide. Currently owned by Facebook Inc., WhatsApp allows its users to chat with each other, chat in groups, send images/videos/voice messages/files and make both voice and video calls all over the world. In recent updates, it also allows transferring monetary funds between its accounts.

End-to-end Encryption

WhatsApp chats are end-to-end encrypted which ensures that only the sender and the recipient can read the messages. No third party (including WhatsApp) can interpret them. This feature makes WhatsApp chats highly secure. Though payments through WhatsApp are not secured by end-to-end encryption. Because without the information related to the payments, financial institutions cannot process transactions. This end-to-end encryption is based on **The Signal Protocol** [8], designed by Open Whisper Systems [5].

The following steps describe the working of end-to-end encryption when two users communicate with each other [9].

1. When a user first installs and opens WhatsApp, a public key and a private key is generated. The private key remains with the user and the public key gets transmitted to the recipient through the centralized WhatsApp server. WhatsApp does not collect the private key.
2. When a user sends a message, the private key encrypts the message inside the sender's device. Only the encrypted message is transferred to the centralized server.

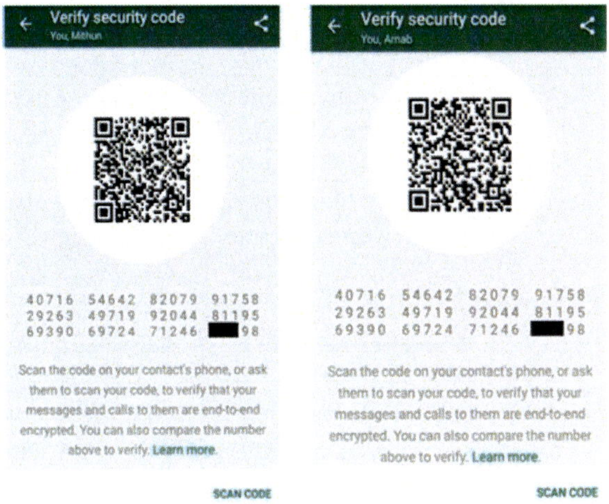

Fig. 2 WhatsApp provides a way to check whether the messages are end-to-end encrypted or not. The sender and the receiver can check their security codes. If they match, then the chat is end-to-end encrypted

3. The server relays the encrypted message to the receiver.
4. The receiver with the proper public key can only decrypt the message.

As both the process of encryption and decryption takes place inside the sender's and receiver's device, respectively, and not in the server, there is no way that a third party can have access to the original message.

The end-to-end encryption in WhatsApp is enabled by default, and there is no way to turn it off. As shown in Fig. 2, users have an option to verify the keys to ensure the integrity of their communication.

Two-Factor Authentication

WhatsApp also uses two-factor authentication that can be enabled manually. It adds an extra layer of security as the user has to enter a one-time password along with the username and password that further prevents any unauthorized access. It also allows you to change the WhatsApp pin if the account is compromised [6].

Disadvantages with WhatsApp

Though WhatsApp is considered among the most secure chatting and social media applications, it has some data security issues. Those are the following.

1. The personal information collected by WhatsApp is not hashed.
2. WhatsApp does not encrypt metadata.
3. Though the IOS version of WhatsApp encrypts messages while taking backup, the android version does not.
4. WhatsApp collects a lot of user's information including personal information, device information (e.g., Hardware model, OS version), performance information, contacts, location, etc. [5] So, there is a possibility of a massive data leak if its server somehow gets compromised.

3.3 ShareChat

ShareChat was started by Bhanu Pratap Singh, Ankush Sachdeva, and Farid Ahsan, and it is developed by Mohalla Tech Pvt. Ltd. ShareChat has approximately 160 million monthly active users in 15 Indic languages. It was incorporated on 8 January 2015. The application offers various features like private messaging, tagging, sharing videos, songs, and a personal messaging feature.

Privacy Policy

In order to provide new services to us, ShareChat collects specific information like our name, phone number, and gender. It may also request and store additional information [5], which is stored securely on Google Cloud Platform cloud servers and the Amazon Web Services cloud servers and is subject to the terms and conditions of the AWS and Google Cloud privacy policy.

ShareChat may share our information—including personal information—with third parties like business partners, subcontractors, and suppliers ("Affiliates") [5]. These affiliates use this information to provide and improve the services. It may share data with advertisers and advertising networks that use this data to improve their efficiency, by serving us relevant adverts. ShareChat does not reveal sensitive information about individuals, but they provide the advertisers with sufficient and relevant information about their users. It helps them to reach desired audiences and increase revenue. If it is completely required that personal data or information of users needs to be shared in order to comply with any legal obligation or any government request, or to prevent any damage to property, or safety of the company, the customers, or public, the company may share information with appropriate authorities.

3.4 Telegram

Unlike other secure messaging app options, Telegram does not support end-to-end encryption as a default feature. In order to achieve end-to-end encrypted transmission on Telegram, we have to enable the "secret chats" feature with each contact we wish to have the said feature. However, even then, not all messaging features are end-to-end encrypted as it is for other messaging apps. End-to-end encryption is not provided in group chats as well. All messages sent to a client by any person not in his/ her contact or for whom the secret chat is not enabled, it would not be end-to-end encrypted.

The **secret chats** feature leaves zero trace on its servers [5]. Messages also cannot be forwarded and can be sent with self-destruction of messages allowed. Secret chats can only be accessed by the sender's device and are separate from the cloud services.

All messages sent via telegram are securely encrypted via the protocol discussed. Messages in secret chats are client-to-client encrypted and non-tracked by cloud. The normal chats are client-to-server/server-to-client encrypted and stored encrypted in the cloud. Thus, in case of change of device, the messages can be retrieved in a fast and secure way. This is known as the Two Chat System.

Protocol

Telegram employs a symmetric encryption method MTProto [5] developed primarily by Nikolai Durov and the team. It is a combination of the 2048-bit RSA encryption method and Diffie Hellman key exchange algorithm and complies with the 256-bit symmetric AES [10] encryption standard. Pictorial representation of the method is shown in Fig. 3.

A session is set up for the transfer of messages between the client and the server. The session is attached to the sending/receiving client's device or the application to be precise. This is different from others where the session is attached to some http/https/tcp connection. Also, the client's user ID is used as a key to accomplish the authorization.

Before the transmission of a message over a network following some protocol, it is encrypted with a header applied. The header is obviously added to the top of the

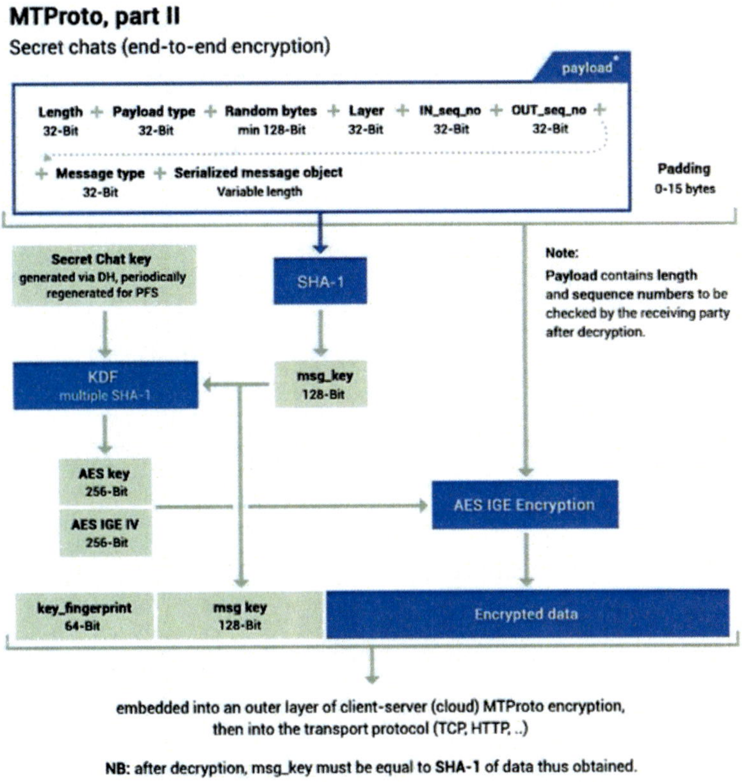

Fig. 3 Telegram's MTProto protocol

document transmitted. It comprises a 64-bit key **“auth_key_id”,** which is unique for the server and the client, and a 128-bit **“msg_key”** message key.

The message will then be encrypted using the AES-256 encryption method accomplished in Infinite Garble Extension (IGE) mode. This is done using a 256-bit key “**aes_key**” and a 256-bit “**aes_iv**” initialization vector. The authorization key is combined with the message key forming the aes_key. Variable data such as session ID, message ID, the server salt, and sequence number form the initial part of the message. These have a direct impact on the message key and hence on the encryption pattern. The latest MTProto defines the message key as the middle bits of the SHA-256 [8] of the message body, which includes padding, message ID, session ID, etc. It is then prepended by 32 bytes taken from the auth key.

Vulnerabilities

One of the major drawbacks that is faced in Telegram is the storage of data in the cloud. Any attack on the cloud can show all the personal data and messages of an individual. This storage of data in the cloud was done to ensure data retrieval on a change of device. But eventually can lead to severe data leaks.

3.5 Signal

“Signal Technology Foundation and Signal Messenger LLC” [5] is the company that developed the signal messaging app. It is used to send texts, files, folders, audio, images, and video, and even video and audio calls to a single person or to a group of people and receive from them as well.

Protocol

Encryption of messages sent via Signal is done using the Signal Protocol. This Signal protocol combines the utilization of prekeys, Extended Triple Diffie–Hellman (X3DH) handshake, and Double Ratchet Algorithm [5]. The protocol provides privacy, probity, corroboration, participant’s identity secrecy, destination validation, participation abrogation, unlinkability of messages, cancelation of messages, and asynchronicity. Message transfer and deposition of public key material require a server.

To secure and verify the actual user, Signal compares key fingerprints (or scans QR codes). “Trust-on-first-use mechanism” is used which notifies the user of any changes in the correspondent’s key. The decryption of incoming encrypted messages is done in the user’s local device and is stored in a local SQLite database after passing through the SQLCipher encryption method. The decryption key is also stored on the user’s local device.

The Protocol used by Signal also provides end-to-end message encryption. The group chat follows a protocol that is composed of multicast encryption and a pairwise double ratchet. It provides out-of-order resilience, speaker consistency, trust equality, computational equality, dropped message resilience, sub-group messaging, and various other features. The one-to-one protocol is also supported by the Signal Encryption Protocol.

Signal relies on centralized servers for their work. The servers route the Signal messages. The servers also check through the contact details of the user for users already registered on Signal and automatically exchange their public keys. The audio and video call features are normally peer-to-peer. If a call from an unknown user is received, the decision is routed via a server so as to cover the IP addresses of the users. In order to find Signal users from the contact list, the hashed values of the user's contact numbers are sent to the Signal server. Now, the server checks them for any match with the SHA256 hashed values of registered users in the database. The numbers are then erased from the server to ensure proper data security. They claim that the servers don't key log of the call details like sender, receiver, and time of the call.

In Table 2, we have compared the above-discussed social media platforms on different aspects of data security and user privacy from the information provided by them in their privacy policy and terms and conditions documents. It conveys a clear picture of the comparative advantages and disadvantages of using these social media applications.

4 Conclusion

In the light of all the above discussion and analysis, this study suggests the need to further improve the security methods and the cryptographic mechanisms that are employed by the various social media platforms, namely WhatsApp, Facebook, Telegram, Signal, and ShareChat. These platforms deal with the personal information of their users and need to guarantee the safety of their databases. Table 2 gives a detailed comparison of the various social media sites based on several factors like encryption method, data collection, etc. While WhatsApp and Signal use end-to-end encryption by default, it is not the case with Facebook and telegram, as we need to manually activate the feature, which may put the data at risk otherwise. Moreover, the Telegram cloud storage and Facebook data centers that store the user data, expose it to cyber-attacks. But overall, these applications are quite secure. Due to the recent buzz about data sharing with third parties in WhatsApp and Facebook messenger, a large audience shifted from these. Since Signal has end-to-end encryption and also does not backup messages on any cloud or third-party apps, it is considered one of the most secure personal messaging apps. There is absolutely no way the data can be compromised in Signal (unless the device is compromised). Additionally, it encrypts the metadata of the object so that no one can track the location and time of the messages being sent or delivered.

Table 2 Comparison among the social media platforms on different aspects of data security and user privacy

Feature	WhatsApp	Facebook messenger	ShareChat	Telegram	Signal
Chat privacy	WhatsApp uses end-to-end encryption that ensures that only the participants in the conversation can see the messages	Messenger does not use end-to-end encryption by default. The message is encrypted on the way to Facebook's server before being sent to the recipient	ShareChat does not support end-to-end encryption. It is written in their privacy policy that they collect chat data, device data (like media files, device specifications), etc	Telegram does not support end-to-end encryption for normal chats but allows for certain "secret chats"	Signal uses end-to-end encryption to provide a better security
Personal information	Personal information is not hashed	Personal information is not hashed	Personal information is not hashed	Personal information is not hashed	Personal information is not hashed
Two-factor authentication	Allows two-factor authentication	Does not allow two-factor authentication	Does not allow two-factor authentication	Allows two-factor authentication	Allows two-factor authentication
Backup encryption	IOS encrypts messages while taking backup but Android does not	Does not support backup to the cloud	NA	Does cloud backup with encryption	Supports backup encryption
Self-destructing messages	Sent messages can be deleted within a certain period of time	Sent messages can be deleted anytime	NA	Sent messages can be deleted anytime	Sent messages can be deleted
Encryption in transit	Does not have encryption in transit	Has encryption in transit	NA	Has encryption in transit	Has encryption in transit
Private key accessibility	The private key is not accessible by the provider	The private key is accessible by the provider	NA	NA	The private key is not accessible by the provider

(continued)

Table 2 (continued)

Feature	WhatsApp	Facebook messenger	ShareChat	Telegram	Signal
Screenshot detection	The screenshot detection feature is not supported	The screenshot detection feature is not supported	NA	The screenshot detection feature is supported for secret chats	The screenshot detection feature is supported for secret chats
Use of TLS (transport layer security)	Uses TLS to encrypt network traffic	Uses TLS to encrypt network traffic	NA	Uses TLS to encrypt network traffic	Uses TLS to encrypt network traffic
Encrypting metadata	Do not encrypt metadata	Do not encrypt metadata	NA	Do not encrypt metadata	Do not encrypt metadata
Enforcing perfect forward security	Enforces perfect forward security	Enforces perfect forward security	NA	Enforces perfect forward security	Enforces perfect forward security
Collecting users' data	Collects user's data	Collects user's data	Collects user's data	Collects user's data	Collects user's data
SIM card requirement	Requires sim card	Does not require a sim card	Does not require a sim card	Requires sim card	Requires sim card

References

1. Senthil Kumar N, Saravanakumar K, Deepa K (2015) On privacy and security in social media. In: ICISP 2015, Dec 2015
2. Kumar A, Gupta SK, Rai AK, Sinha S (2013) Social networking sites and their security issues. Int J Sci Res Publ 3(4). ISSN 2250-3153
3. Kumari S, Singh S (2015) A critical analysis of privacy and security on social media. In: 2015 Fifth international conference on communication systems and network technologies, Apr 2015
4. Prasansu H (2017) Social media: security and privacy. Int J Sci Eng Res 8(11). ISSN 2229-5518
5. Respective websites of social media. whatsapp.com/legal/updates/privacy-policy, telegram.org/privacy, facebook.com/policy.php, signal.org/legal, privacy.sharechat.com
6. A survey about the latest trends and research issues of cryptographic elements. Int J Comput Sci Issues (IJCSI) 8
7. https://www.cloudflare.com/learning/ssl/transport-layer-security-tls/
8. https://en.wikipedia.org/wiki/Signal_Protocol
9. Li C, Sanchez D, Hua S. WhatsApp security paper analysis
10. Kahate A. Cryptography and network security

Electronics, VLSI and Computing

Button Press Dynamics: Beyond Binary Information in Button Press Decisions

Peter Bellmann, Viktor Kessler, André Brechmann, and Friedhelm Schwenker

Abstract The analysis of button press dynamics (BPD) may reveal more detailed information about decisions in human–computer interaction. One huge advantage of a BPD-specific analysis lies in its tangible nature. More precisely, each button press (BP) constitutes a two-dimensional signal consisting of time and intensity coordinates, which can be easily illustrated. Moreover, one can also interpret BP signals by extracting several intuitive features, such as the duration and the maximum intensity. In this study, we analyse the characteristics of such intuitive BPD features. To this end, we conduct cluster analysis experiments with the following evaluation protocol. First, for each person-specific set of BPs, we will define ground truth (GT) clusterings by evaluating the BPs as time series and applying dynamic time warping (DTW) for the calculation of distances between two instances. Subsequently, we will extract a small set of intuitive features. Finally, we will compute the similarity between the DTW- and feature-based clusterings, based on two popular similarity measures. The outcomes of our experiments lead to the following observation. Extending binary BP information by one additional feature, i.e. the maximum press intensity, can already significantly improve the analysis of BP evaluations.

P. Bellmann (✉) · V. Kessler · F. Schwenker
Institute of Neural Information Processing, Ulm University, James-Franck-Ring, 89081 Ulm, Germany
e-mail: peter.bellmann@uni-ulm.de

V. Kessler
e-mail: viktor.kessler@uni-ulm.de

F. Schwenker
e-mail: friedhelm.schwenker@uni-ulm.de

A. Brechmann
Leibniz-Institute for Neurobiology, Combinatorial NeuroImaging, Brenneckestr. 6, 39118 Magdeburg, Germany
e-mail: andre.brechmann@lin-magdeburg.de

S. Basu et al. (eds.), *Proceedings of International Conference on Frontiers in Computing and Systems*, Lecture Notes in Networks and Systems 404,
https://doi.org/10.1007/978-981-19-0105-8_46

1 Introduction

Human–computer interaction (HCI) often requires the user to indicate a decision via button presses, which are often used as binary inputs, possibly enriched with information about response times. However, the recording of the dynamics of button presses offers to infer further information like the (un-)certainty of a decision [4]. This becomes important when HCI interactions involve a learning process that may be supported by a technical system [9]. In this study, we utilise behavioural data from a trial and error learning experiment and analyse the relationship between the dynamics of button presses and aggregated button press features by using hierarchical clustering.

In Sect. 2, we briefly recap the functionality of agglomerative hierarchical clustering, including two of the most popular similarity measures. The description of the analysed data set is presented in Sect. 3. In Sect. 4, we present and discuss the obtained outcomes, followed by the conclusion of the current work.

2 Cluster Analysis

In the current section, we will briefly recap basic agglomerative hierarchical clustering approaches, followed by the definition of two of the most popular similarity measures, i.e. the Rand Index as well as the Jaccard Index.

2.1 *Formalisation*

Let $X \subset \mathbb{R}^m$, $m \in \mathbb{N}$ be a set of data samples with n elements, i.e. $n = |X|$. A clustering of the set X is simply a family of k sets, $C = \{C_1, \ldots, C_k\}$, $k \in \mathbb{N}$, which constitutes a partition of X. More precisely, C consists of non-empty and disjoint subsets of X whose union is equal to X, i.e. for all $i, j \in \{1, \ldots, k\}$ it holds, $C_i \neq \emptyset$, $C_i \cap C_j = \emptyset$, and $\cup_{i=1}^{k} C_i = X$.

Each $C_i \in C$ is denoted as a cluster of clustering C. Moreover, by $P(X)$, we denote the set of all clusterings (partitions) of the set X. By definition, it holds $k \in \{1, \ldots, n\}$, for all $C \in P(X)$. The two trivial clusterings of each set $X \subset \mathbb{R}^m$ are equal to $C = X$ (for $k = 1$), and $C = \{\{x_1\}, \ldots, \{x_n\}\}$ (for $k = n$).

2.2 *Agglomerative Hierarchical Clustering*

There are two main categories of hierarchical clustering approaches, i.e. *divisive* (top-down), and the more common, *agglomerative* (bottom-up) methods. In this work, we focus on the latter type. In agglomerative clustering (AC) approaches, the initial situation is defined by assigning each of the data points to separate one-point clusters, i.e. $C^{\text{start}} = \{\{x_1\}, \ldots, \{x_n\}\}$. Then, in each step, two of the current clusters are merged together, based on a predefined distance criterion (clustering method). This process is repeated until all of the initial one-point clusters are merged to one single cluster, which is then equal to the set X.

In this work, we will focus on the approaches single-linkage clustering (SLC) (e.g. [6]), complete linkage clustering (CLC) (e.g. [2]), unweighted pair group method with arithmetic mean (UPGMA) (e.g. [7]) and the weighted pair group method with arithmetic mean (WPGMA) (e.g. [7]). Note that we will also, respectively, refer to the methods UPGMA and WPGMA simply as *unweighted/weighted average linkage clustering*, for the sake of readability. Let $d : X \times X \to \mathbb{R}_+$ be a distance function in the data space. With $D_{i,j}$ denoting the distance between the clusters $C_i, C_j \in C$, the aforementioned clustering approaches are defined as follows:

$$\begin{aligned} D_{i,j}^{\text{SLC}} &= \min_{x \in C_i, y \in C_j} d(x, y), \quad & D_{i,j}^{\text{UPGMA}} &= \frac{1}{|C_i| \cdot |C_j|} \sum_{x \in C_i} \sum_{y \in C_j} d(x, y), \\ D_{i,j}^{\text{CLC}} &= \max_{x \in C_i, y \in C_j} d(x, y), \quad & D_{i \cup j,l}^{\text{WPGMA}} &= \frac{1}{2} \left(D_{i,l}^{\text{WPGMA}} + D_{j,l}^{\text{WPGMA}} \right). \end{aligned} \tag{1}$$

Note that $D_{i \cup j,l}$ refers to the distance between the combined cluster $C_i \cup C_j \in C$ and cluster $C_l \in C$.

2.3 Similarity Measures

Let $C, C' \in P(X)$ be two clusterings of $X \subset \mathbb{R}^m$, with $N \in \mathbb{N}$ denoting the number of all pairs of n data points ($n = |X|$), i.e. $N = n(n-1)/2$. Moreover, let $N_{00}, N_{11} \in \{0, 1, \ldots, N\}$ be defined as follows:

$N_{00} \mathrel{\hat{=}}$ Number of pairs that are in different clusters for both clusterings, C and C',
$N_{11} \mathrel{\hat{=}}$ Number of pairs that are in the same cluster for both clusterings, C and C'.

The Rand Index (RI) and Jaccard Index (JI) are defined as follows (e.g. [8]):

$$\text{RI}(C, C') = \frac{N_{00} + N_{11}}{N}, \quad \text{JI}(C, C') = \frac{N_{11}}{N - N_{00}}. \tag{2}$$

Note that it holds $\text{RI}(C, C'), \text{JI}(C, C') \in [0, 1]$, for all $C, C' \in P(X)$. Moreover, in the case of $N_{00} = N$, one can simply set the value of $\text{JI}(C, C')$ to 0.

3 Data Set Description and Feature Extraction

In the current study, we analyse the data from an auditory category learning experiment in which nine participants had to learn a specific sound category by trial and error within 180 trials. Each sound had five different features with two values each, i.e. duration (short vs. long), pitch direction (rising vs. falling), loudness (low vs. high), frequency range (low vs. high) and speed of pitch change (slow vs. fast). The task-relevant features were the sound duration and pitch direction, resulting in four potential *target* categories. Note that the participants had no knowledge about the respective target category. On average, every ten seconds, the participants heard one sound and had to react within two seconds by pushing one of two buttons. The left button (index finger) corresponded to the target sounds and the right button (middle finger) to the non-target sounds. After each response, the participant received a verbal feedback stating *yes/correct* (if a target sound was correctly recognised, or a non-target sound was correctly rejected) and *no/incorrect* (in case of a miss for a target sound, or a false alarm for a non-target sound). The button press dynamics were recorded with 1000 Hz temporal resolution using the COVILEX ResponseBox 2.0 (COVILEX GmbH).

We defined the button press dynamic features as follows. Let $s^{(i)}$ be a sequence of length n_i, i.e. $s^{(i)} = (x_1^{(i)}, \ldots, x_{n_i}^{(i)}) \in \mathbb{R}_+^{n_i}$. Note that the signals that we are analysing constitute button press dynamics. Therefore, intuitive handcrafted features are the *duration* (Dur), *maximum press intensity* (Max), *averaged press intensity* (Mean) and the *accumulated press intensity* (Sum), i.e.

$$\begin{aligned} \text{Dur}(s^{(i)}) &= n_i, & \text{Max}(s^{(i)}) &= \max_{j=1,\ldots,n_i} \{x_j^{(i)}\}, \\ \text{Mean}(s^{(i)}) &= \frac{1}{n_i}\sum_{j=1}^{n_i} x_j^{(i)}, & \text{Sum}(s^{(i)}) &= \sum_{j=1}^{n_i} x_j^{(i)}. \end{aligned} \tag{3}$$

Moreover, we will denote the concatenated feature vector consisting of duration, maximum and averaged button press intensity by Comb, i.e. $\text{Comb}(s^{(i)}) = \big(\text{Dur}(s^{(i)}), \text{Max}(s^{(i)}), \text{Mean}(s^{(i)})\big)$. Note that we do not concatenate all of the four features, since the Mean feature is a combination of the features Sum and Dur, i.e. $\text{Mean}(s^{(i)}) = \text{Sum}(s^{(i)})/\text{Dur}(s^{(i)})$, for all button press signals.

Also note that, here, feature Dur represents the *relative* duration of a button press, which is simply the signal length. The actual duration of a button press is a combination of signal length and temporal resolution. Moreover, for the concatenated feature vector (Comb), we will analyse normalised feature values. More precisely, for each participant, we will divide each feature value by the Euclidean norm of all corresponding samples, e.g. $\text{Dur}(s^{(i)}) \mathrel{\hat{=}} \text{Dur}(s^{(i)})/\|(\text{Dur}(s^{(1)}), \text{Dur}(s^{(2)}), \ldots)\|_2$.

4 Results

In the current section, we will first define *ground truth* (GT) clusterings by applying dynamic time warping (DTW), e.g. [3, 5]. Subsequently, we will compute the similarities between the obtained GT and feature-based clusterings, followed by a discussion of the obtained outcomes. We used the MATLAB[1] software (R2019b) for the implementation, with the built-in functions for DTW distance computation and agglomerative clustering. Each of the methods was evaluated in combination with the Euclidean distance. Note that DTW is a method that is widely used to align two *time series* of different lengths, in general, for instance with applications in speech, music and motion. However, the general DTW approach can also be applied for any kind of *linear sequences*.

4.1 Dynamic Time Warping-Based Clustering

Applying the DTW approach, we computed the Euclidean distances between each button press pair, which we then used as input for the different AC methods (see Sect. 2.2). Table 1 depicts the results for $k = 2$ clusters, for each of the participants.

Note that each of the participants consists of 180 data points, i.e. button presses, except for one participant, who did not respond in one trial. Moreover, for the sake of readability, only the number of points in the smaller cluster is reported. For instance, applying the SLC approach in combination with the data corresponding to participant p1 leads to one cluster with two data points and one cluster with $180 - 2 = 178$ data points. Applying the CLC approach in combination with the same participant leads to one cluster with 17 data points and one cluster with $180 - 17 = 163$ data points, etc.

From Table 1, we can make the following observations. Firstly, the SLC approach seems to be useful for finding *outliers*. For seven out of the nine participants, the smaller of the two clusters contains less than three data points. This is an expected outcome, due to the construction of the SLC method, in which the distance between two clusters is defined as the minimum distance between all pairs from those clusters.

Secondly, the WPGMA and CLC approaches lead to the most balanced clusterings for four participants each. Moreover, the most balanced clusterings are obtained by the WPGMA approach for the participants p3 and p4, respectively, leading to 85 and 82 data points in the smaller cluster.

Thirdly, all of the four clustering approaches lead to a one-point cluster for the participants p6 and p9. Unfortunately, these cases are insignificant for our defined clustering similarity analysis. Therefore, we will remove the corresponding participants and the SLC approach completely from further evaluations. Moreover, we will include participant p7 only in the evaluation of the CLC approach.

[1] More information on www.mathworks.com

Table 1 Dynamic time warping distance-based clustering

Method	p1	p2	p3	p4	p5	p6	p7	p8	p9
Single	2	1	11	1	6	1	1	1	1
Complete	**17**	**44**	14	44	56	1	**15**	**14**	1
UPGMA	5	34	14	75	**65**	1	1	**14**	1
WPGMA	4	37	**85**	**82**	**65**	1	1	**14**	1

The number of clusters is set to 2. The total number of samples is 180 per participant (179, only for participant hr27). Depicted is the number of data points contained in the smaller cluster. The method leading to the most balanced clustering is denoted in bold. Column names represent participant IDs

4.2 Similarity Between DTW- and Feature-Based Clusterings

In the current section, we will present the similarity between DTW- and feature-based clusterings. As discussed above, the participant-specific DTW-based clusterings (with two clusters each) will be used as the ground truth. Tables 2, 3 and 4 depict the results for the CLC, UPGMA and WPGMA approaches, respectively. Each of these tables contains the RI and JI values (see Eq. (2)), which we used to measure the similarity between our defined ground truth (DTW-based clustering) and each of the intuitive features defined in Eq. (3). For instance, applying the CLC approach (Table 2) leads to a RI similarity value of 60 for participant p1 if we compare the clustering based solely on the feature Dur to the DTW-based clustering. Analogously, for the same approach and participant, we obtain a JI similarity value of 82 if we compare the clustering based solely on the feature Sum to the DTW-based clustering, etc. Note that the obtained values are multiplied by 100, for the sake of readability. The last rows of Tables 2, 3 and 4 state how often each of the features led to the highest similarity to the ground truth clustering.

From Table 2, we can make the following observations. Feature Max leads to the highest similarity to our defined ground truth, on four out of seven participants, followed by Sum and Mean with, respectively, three and two participants, based on both measures RI and JI. For participant p7, the Max- and DTW-based clusterings lead to exactly the same two clusters, with 15 data points in the smaller cluster and 165 data points in the bigger cluster (see Table 1). Moreover, features Dur and Comb never lead to the highest similarity to the DTW-based clusterings.

From Tables 3 and 4, we can observe similar outcomes. More precisely, feature Max leads to the highest similarity to our defined ground truth, on five out of six participants, based on both measures RI and JI in Table 3, and on four/five out of six participants based on the measures RI/JI in Table 4. In combination with the UPGMA approach and participant p1, the Max- and DTW-based clusterings lead to exactly the same two clusters, including five data points in the smaller cluster and 175 data points in the bigger cluster. Moreover, feature Dur never leads to the highest similarity to the DTW-based clusterings, whereas the concatenated feature vector Comb leads at most once to the highest similarity to the defined ground truth, for each of the clustering approaches.

Table 2 Complete linkage clustering (CLC)

CLC	Rand Index × 100					Jaccard Index × 100				
ID	Dur	Max	Mean	Sum	Comb	Dur	Max	Mean	Sum	Comb
p1	60	**89**	**89**	82	67	56	**89**	**89**	82	66
p2	54	54	50	**87**	53	47	51	43	**82**	50
p3	96	68	65	**98**	65	95	64	60	**97**	61
p4	66	51	65	**71**	52	62	41	53	**68**	42
p5	50	**98**	66	52	84	44	**96**	51	41	75
p7	50	**100**	90	55	86	46	**100**	90	50	85
p8	59	**91**	**91**	77	61	55	**91**	**91**	74	56
Σ	0	**4**	2	3	0	0	**4**	2	3	0

For each participant (row names), the similarity (RI and JI) is depicted between the DTW-based distance clustering and the corresponding feature-based (column names) distance clustering. The features leading to the highest similarity are denoted in bold. Dur: Duration. Comb: Combined feature vector consisting of (Dur, Max, Mean). Σ: Number of times that each of the features led to the highest similarity to the DTW-based distance clustering

Table 3 Unweighted Average Clustering

UPGMA	Rand Index × 100					Jaccard Index × 100				
ID	Dur	Max	Mean	Sum	Comb	Dur	Max	Mean	Sum	Comb
p1	91	**100**	92	94	98	91	**100**	92	94	98
p2	56	**98**	51	60	50	52	**97**	46	58	45
p3	90	97	96	**98**	**98**	89	96	95	**97**	97
p4	52	**99**	52	52	52	51	**98**	50	51	50
p5	51	**96**	51	52	53	38	**92**	48	52	53
p8	81	**87**	**87**	52	83	81	**86**	**86**	48	83
Σ	0	**5**	1	1	1	0	**5**	1	1	1

For each participant (row names), the similarity (RI and JI) is depicted between the DTW-based distance clustering and the corresponding feature-based (column names) distance clustering. The features leading to the highest similarity are denoted in bold. Dur: Duration. Comb: Combined feature vector consisting of (Dur, Max, Mean). Σ: Number of times that each of the features led to the highest similarity to the DTW-based distance clustering

4.3 Discussion and Conclusion

Often, button presses (BPs) are evaluated in a binary manner, i.e. by detecting whether a button was pressed or not, during a specific time period. However, analysing the dynamics of button presses opens up additional possibilities to capture different phenomena, such as uncertainty, during the BP-based interaction process (e.g. [4]).

In this study, we compared clusterings based on a set of intuitive features to clusterings based on sequence distances obtained by dynamic time warping (DTW). Our results indicate that, based on the evaluation task at hand, it might be sufficient to analyse the maximum force used to press the button. On the other hand, the obtained

Table 4 Weighted average clustering

WPGMA	Rand Index × 100					Jaccard Index × 100				
ID	Dur	Max	Mean	Sum	Comb	Dur	Max	Mean	Sum	Comb
p1	55	79	88	**97**	57	55	79	88	**97**	56
p2	50	56	56	**61**	54	41	**54**	53	50	51
p3	50	**74**	72	52	55	36	**60**	59	48	48
p4	51	**77**	50	74	74	41	**65**	48	59	61
p5	51	**63**	51	51	51	47	**48**	**48**	**48**	**48**
p8	68	**87**	**87**	63	61	68	**86**	**86**	62	59
Σ	0	**4**	1	2	0	0	**5**	2	2	1

For each participant (row names), the similarity (RI and JI) is depicted between the DTW-based distance clustering and the corresponding feature-based (column names) distance clustering. The features leading to the highest similarity are denoted in bold. Dur: Duration. Comb: Combined feature vector consisting of (Dur, Max, Mean). Σ: Number of times that each of the features led to the highest similarity to the DTW-based distance clustering

outcomes also indicate that there are significant differences between button press clusterings specific to intuitive features and specific to sequences. Therefore, based on the current results, as an extension of existing literature, we aim at including the DTW approach in the analysis of uncertainty detection in category learning experiments in order to infer optimal time points for feedback intervention to support learning, or to offer the opportunity to revise a decision. First basic results are presented in our corresponding concurrent study, in [1].

Acknowledgements This work is supported by the project *Multimodal recognition of affect over the course of a tutorial learning experiment* (SCHW623/7-1) funded by the German Research Foundation (DFG). We gratefully acknowledge the support of NVIDIA Corporation with the donation of the Tesla K40 GPU used for this research.

References

1. Bellmann P, Kessler V, Brechmann A, Schwenker F (2021) Dynamic time warping-based detection of multi-clicks in button press dynamics data. In: Proceedings of International conference on frontiers in computing and systems. In: Lecture notes in networks and systems. Springer
2. Defays D (1977) An efficient algorithm for a complete link method. Comput J 20(4):364–366
3. Gold O, Sharir M (2018) Dynamic time warping and geometric edit distance: Breaking the quadratic barrier. ACM Trans Algor 14(4):50:1–50:17
4. Kohrs C, Hrabal D, Angenstein N, Brechmann A (2014) Delayed system response times affect immediate physiology and the dynamics of subsequent button press behavior. Psychophysiology 51(11):1178–1184
5. Sakoe H, Chiba S (1978) Dynamic programming algorithm optimization for spoken word recognition. IEEE Trans Acoust Speech Signal Proces 26(1):43–49
6. Sibson R (1973) SLINK: an optimally efficient algorithm for the single-link cluster method. Comput J 16(1):30–34

7. Sokal R, Michener C (1958) A statistical method for evaluating systematic relationships. University of Kansas Science Bulletin, University of Kansas
8. Wagner S, Wagner D (2007) Comparing clusterings: an overview. Universität Karlsruhe, Fakultät für Informatik Karlsruhe
9. Wolff S, Brechmann A (2015) Carrot and stick 2.0: the benefits of natural and motivational prosody in computer-assisted learning. Comput Hum Behav 43:76–84 (2015)

Dynamic Time Warping-Based Detection of Multi-clicks in Button Press Dynamics Data

Peter Bellmann, Viktor Kessler, André Brechmann, and Friedhelm Schwenker

Abstract Dynamic Time Warping (DTW) is a method generally used to align pairs of time series with different lengths, which is for instance applied in speech recognition. In this study, we use a category learning experiment (CLE) as use-case, in which the participants have to learn a specific target from a pool of predefined categories within a certain amount of time. From a companion system-based point of view, it is important to detect certain anomalies related to affective states, such as surprise or frustration, elicited during the course of learning. In this work, we analyse the button press dynamics (BPD) data from an auditory CLE with the goal of detecting anomalies of the aforementioned type. To this end, we first select a small set of participants, for which we have definite ground truth labels. Subsequently, we apply DTW in combination with hierarchical clustering to separate the anomaly-specific data from the remaining samples. We compare the outcomes to clustering results based on the extraction of intuitive task-specific features. Our results indicate that applying the DTW approach in combination with Single Linkage Clustering in order to detect CLE-related anomalies is preferable to its feature extraction-based alternative, in person-independent scenarios.

Keywords Button press dynamics · Dynamic time warping · Cluster analysis

---

P. Bellmann (✉) · V. Kessler · F. Schwenker
Institute of Neural Information Processing, Ulm University, James-Franck-Ring, 89081 Ulm, Germany
e-mail: peter.bellmann@uni-ulm.de

V. Kessler
e-mail: viktor.kessler@uni-ulm.de

F. Schwenker
e-mail: friedhelm.schwenker@uni-ulm.de

A. Brechmann
Leibniz-Institute for Neurobiology, Combinatorial NeuroImaging, Brenneckestr. 6, 39118 Magdeburg, Germany
e-mail: andre.brechmann@lin-magdeburg.de

S. Basu et al. (eds.), *Proceedings of International Conference on Frontiers in Computing and Systems*, Lecture Notes in Networks and Systems 404,
https://doi.org/10.1007/978-981-19-0105-8_47

1 Introduction

Using the dynamics of button presses to infer more detailed information about a user's decision in human–computer interaction requires an algorithm that reliably extracts the binary information about the onset and offset of a button press, in order to isolate the separate button press events. A simple threshold may not be sufficient when an intermittent partial release of the button is to be classified as a multiple click or as a hesitant decision. The current paper presents a method for automatically identifying such anomalies in button press dynamics to avoid manual inspection, as applied in a previous study on the effects of delayed system response times on button press behaviour [3].

2 Single Linkage Clustering and Dynamic Time Warping

In this section, we will first recap the functionality of the single linkage clustering approach, followed by a brief summary of dynamic time warping.

The **Single Linkage Clustering** (SLC) [7] is an agglomerative hierarchical clustering (AHC) method. In general, the first step of AHC approaches consists of assigning each object (data point, time series, etc.) to a single-point cluster, which contains the object itself. Subsequently, the distance between each cluster pair is computed, based on a predefined distance measure. The cluster pair corresponding to the minimum distance is merged to a new cluster, followed by the recalculation of the cluster distances. These steps are repeated until all objects are contained in one single cluster. Each AHC method is defined by its individual inter-cluster distance update. In the SLC approach, the distance between two clusters C_i and C_j is defined as the *minimum* distance between the inter-cluster pairs, i.e. as $\min_{(x,y)\in C_i \times C_j} \text{dist}(x, y)$, whereby dist is a predefined distance function.

Dynamic Time Warping (DTW) [2] is a popular approach in the field of time series analysis, such as in speech recognition [6]. In general, DTW is applied to compute the distance between two sequences of different lengths. To this end, the sequences are aligned to a kind of *warped* sequences with the same length, in general by minimising a predefined distance function subject to a set of three conditions for both aligned sequences. (1) The first/last index of the sequence has to match the first/last index of the other sequence. (2) Each index of the sequence has to match at least one index of the other sequence without index-pair repetitions. (3) It is not allowed to rearrange the order of the sequence-specific elements. The distance between the equally long, aligned sequences, which corresponds to the optimised minimum value subject to the aforementioned conditions, is defined as the *DTW distance*. Note that the alignment of two original signals does not include any modification or removal of the sequence elements. Therefore, aligning two sequences implies that it is only allowed to repeat (copy) the elements from the original sequences. Moreover, the DTW approach is only useful in cases in which the original signals were recorded in equidistant manner.

The **choice of methods** is justified as follows. Since the data analysed in this work consists of time series of different lengths, we opted for a *distance-only*-based clustering method that does not require the computation of cluster centres. We apply DTW to compute the distances between pairs of sequences. Moreover, we chose the SLC approach due to its deterministic nature. In contrast to other popular clustering approaches, such as the k-means clustering [4], the SLC method leads to a unique solution if all object-pair distances are pairwise distinct. In addition, the SLC approach is able to separate single sequences in a set of button press dynamics [1].

3 Data Set Description and Annotation

In the current study, we analyse the data from an auditory category learning experiment in which the participants had to learn a specific sound category by trial and error within 180 trials (For details of the experimental design see [8]).

The participants had to indicate whether a sound belonged to the target category by left button press (index finger) or to the non-targets by right button press (middle finger). After each response, the participant received verbal feedback about the correctness of the decision. The button press dynamics were recorded with 1000 Hz temporal resolution using the COVILEX Response Box 2.0 (COVILEX GmbH).

Due to observed noise in the BPD signals, we defined a threshold θ of value 10 to automatically detect the participants' button-specific responses. Each button press constitutes a curve of parabolic shape in the intensity interval [0, 125]. In the case of oscillations, we replaced the affected signal parts by the corresponding interpolated values. We analysed the recordings specific to 15 participants, which we simply denote by P-01, ..., P-15. The data consists of 2666 button presses in total. Using the ATLAS tool [5], two of the authors independently plotted and annotated each button press, to ensure an unbiased manual detection of double and triple clicks in the data. Note that we will refer to such events as multi-responses. Figure 1 provides two examples of the 100 detected multi-responses in total. Moreover, to ensure a fair comparison, the starting and end points of each event were defined as the points in time at which the intensity values were less or equal to the predefined threshold θ for the last and first time, respectively.

4 Feature Extraction and Experimental Study

Each button press constitutes a sequence of intensity values, i.e. button press i is defined by $s^{(i)} = (s_1^{(i)}, \dots, s_{n_i}^{(i)}) \in \mathbb{R}^{n_i}_{\geq 0}$, with $n_i \in \mathbb{N}$. For each button press, we extracted the intuitive feature button press *duration* (Dur), as well as the *maximum* (Max) and *averaged* (Ave) button press intensity, defined as follows:

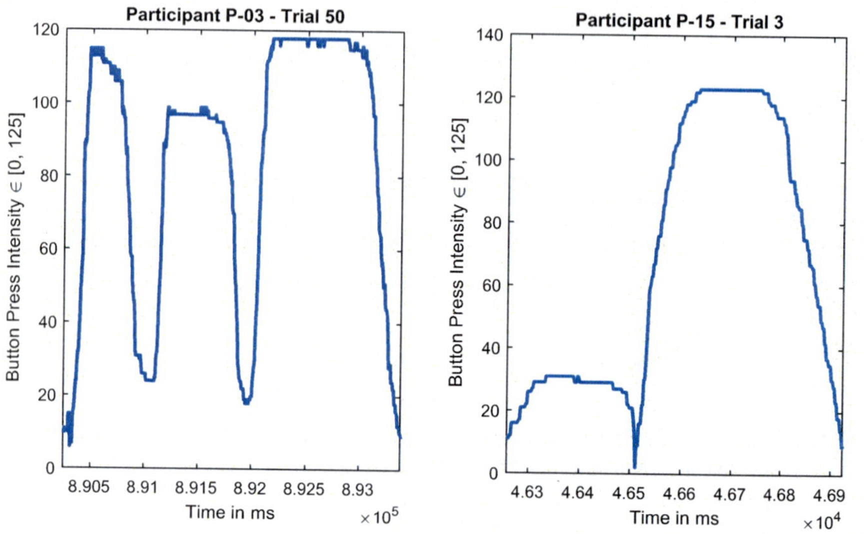

Fig. 1 Example illustration of two multi-responses specific to the participants P-03 and P-15

$$\text{Ave}(s^{(i)}) = \frac{1}{n_i}\sum_{j=1}^{n_i} s_j^{(i)}, \quad \text{Dur}(s^{(i)}) = n_i, \quad \text{Max}(s^{(i)}) = \max_{j=1,\ldots,n_i} \{s_j^{(i)}\}. \tag{1}$$

For each participant, we normalised the features as follows. Let N_i denote the number of button presses specific to participant i. We divided each feature value $\text{Ave}(s^{(i)})$, $\text{Dur}(s^{(i)})$, and $\text{Max}(s^{(i)})$ by $\left\| \left(\text{Fea}(s^{(1)}), \ldots, \text{Fea}(s^{(N_i)})\right) \right\|_2$, whereby Fea respectively denotes the features Ave, Dur and Max.

4.1 Experimental Settings

We used MATLAB[1], version R2019b, for the numerical evaluation of the experiments. For both, the computation of the dynamic time warping distances as well as the SLC method, we applied the built-in functions (*dtw* and *linkage*) in combination with the Euclidean distance.

We analysed two types of evaluation approaches, to which we will respectively refer to as *iterative* and *all-at-once* settings. In the iterative approach, we conducted M evaluation steps, with M denoting the number of multi-responses. In iteration j, the data consisted of all basic button presses and multi-response j. In the all-at-once approach, we conducted exactly one evaluation step including all button presses. In combination with both approaches, we set the number of clusters to the value 2. Note

[1] www.mathworks.com.

that the goal is to obtain two clusters, one cluster including all basic button presses and one cluster consisting of the multi-response(s) only.

4.2 Person-Dependent Multi-response Detection

Table 1 depicts the results for the person-dependent, iterative approach. The numbers in Table 1 indicate how many multi-responses were found with respect to the distances based on the listed features, feature combinations and DTW. Note that we classified a multi-response as *found* only if the SLC approach led to two clusters, including one one-point cluster consisting of the current multi-response.

From Table 1, we can make the following observations. First, features Max and Ave lead to the worst results, detecting only one and seven multi-responses, respectively. However, combining both features, as depicted in row (A, M), leads to a total amount of 56 detected multi-responses. Second, duration (Dur) is the best single feature for the detection of multi-responses. This is an expected outcome, since pressing the button(s) more than once, takes more time, in general, than pressing the button(s) one time. Third, the best feature-based results are obtained by (Ave, Dur) and the combination of all three features, i.e. (Ave, Dur, Max), simply denoted by All in Table 1. Fourth, the DTW distance-based approach leads to the best overall result of 91 detected multi-responses, significantly outperforming all of the feature distance-based approaches. Lastly, none of the approaches was able to detect the multi-response event of participant P-15, which is depicted in Figure 1, on the right-hand side.

Table 2 depicts the results for the person-dependent, all-at-once approach. The numbers in Table 2 show the constellation of the smaller clusters, for each participant approach combination. More precisely, the entries of the form *c-w* denote how many correctly detected (c) and wrongly detected (w) button presses are separated from the remaining data. We removed participant P-15 from the all-at-once setting. Note that both settings, iterative and all-at-once, are equal in cases where the participants' button press dynamics data contains only one multi-response.

From Table 2, we can make the following observations. First, features Max and Ave lead to the worst results, detecting only five and eight multi-responses, respectively. Moreover, feature Max leads to a total amount of 97 wrongly separated button presses, significantly underperforming all of the remaining approaches. Similar to the iterative setting, the combination of both features, Max and Ave, significantly improves the single feature-based detections from respectively 5-97 and 8-30 to 39-7. Second, feature Dur is the best single feature for the detection of multi-responses, leading to 60 correctly detected and only two wrongly detected button presses. Moreover, in contrast to the iterative setting, feature Dur is able to detect more multi-responses than all binary feature combinations, i.e. (Ave, Max), (Ave, Dur), and (Dur, Max). However, feature combination (Ave, Dur) provides only one wrongly detected button press, whereas feature Dur leads to a total number of two wrongly detected button presses. Third, the best feature-based result is obtained by the com-

Table 1 Iterative Multi-Response (MR) Detection Based on Single Linkage Clustering (SLC): ID: Participant ID. #MR: Number of Multi-Responses. Σ: Total number of (detected) MRs. Dur (D): Button press duration. Max (M)/Ave (A): Maximum/Averaged button press intensity. All: Combination of all 3 features, (A, D, M). DTW: Dynamic Time Warping. Bold figures indicate a non-tied maximum amount of detected MRs between extracted features and DTW

ID	P-01	P-02	P-03	P-04	P-05	P-06	P-07	P-08	P-09	P-10	P-11	P-12	P-13	P-14	P-15	Σ
#MR	1	1	23	4	6	1	2	1	10	2	5	18	17	8	1	100
Dur	0	1	1	4	2	1	2	1	7	1	2	**18**	17	2	0	59
Max	0	0	0	0	0	0	0	0	1	0	0	0	0	0	0	1
Ave	0	0	2	1	0	0	0	0	0	0	0	2	1	1	0	7
(A, M)	1	0	6	4	0	1	1	0	6	**2**	0	**18**	17	0	0	56
(A, D)	1	1	1	4	2	1	2	1	**10**	**2**	2	**18**	17	2	0	64
(D, M)	0	1	1	4	2	1	2	1	8	1	2	**18**	17	2	0	60
All	1	1	1	4	2	1	2	1	**10**	**2**	2	**18**	17	2	0	64
DTW	1	1	**23**	4	**4**	1	2	1	8	1	**3**	17	17	**8**	0	**91**

bination of all three features, i.e. (Ave, Dur, Max), leading to 61 correctly detected multi-responses in total, and only one wrongly detected button press. Fourth, in contrast to the iterative setting, the DTW distance-based approach leads to a low amount of only 35 detected multi-responses. However, this is the only approach leading to zero wrongly detected button presses.

Comparing the person-dependent outcomes in Tables 1 and 2 indicates that DTW-based multi-response detection is preferable in iterative settings, whereas feature-based multi-response detection in combination with all three features, i.e. (Ave, Dur, Max), is a good choice for all-at-once settings. Moreover, using the (Ave, Dur, Max) feature vector leads to similar outcomes in both settings, i.e. iterative and all-at-once, with respectively 64 and 61 multi-response detections.

4.3 *Person-Independent Multi-response Detection*

In the current section, we present the experimental evaluation of the person-independent scenario. More precisely, for the person-independent case, we combined the person-specific data subsets to one data set with a total amount of 2666 data points, i.e. 2566 basic button presses and 100 multi-responses. Note that the SLC method fails in the all-at-once setting for all feature-, feature combination-, and DTW-based distances. Feature Max led to one one-point cluster including a wrongly detected button press. Feature Ave led to one two-point cluster consisting of two multi-responses. All remaining approaches correctly separated only one single multi-response from the remaining data.

The results for the iterative person-independent approach are depicted in Table 3. Note that, based on the outcomes of the person-dependent approach, for the feature-based evaluation we focused on the combination of all three extracted features,

Table 2 All-at-Once Multi-Response (MR) Detection with Single Linkage Clustering (SLC): #MR: Number of MRs. Σ: Total number of (detected) MRs. Dur (D): Duration. Max (M)/Ave (A): Maximum/Averaged intensity. All: Combination of all 3 features, (A, D, M). DTW: Dynamic Time Warping. Bold figures indicate a non-tied maximum amount of detected MRs between extracted features and DTW. *c*-*w*: Number of correctly (c) and wrongly (w) detected MRs. The DTW method is the only approach leading to zero wrongly detected MRs, which is indicated by an asterisk (*)

ID	P-01	P-02	P-03	P-04	P-05	P-06	P-07	P-08	P-09	P-10	P-11	P-12	P-13	P-14	Σ
#MR	1	1	23	4	6	1	2	1	10	2	5	18	17	8	99
Dur	1-1	1-0	1-0	1-0	1-0	1-0	**2-0**	1-0	6-0	1-0	**2-0**	**17**-0	17-0	**8**-2	60-2
Max	0-3	0-1	1-4	0-2	0-9	0-19	1-1	0-2	1-0	0-47	1-2	0-1	0-5	1-1	5-97
Ave	0-1	0-1	2-0	0-1	0-2	0-19	0-1	0-1	1-1	0-1	1-2	1-0	1-0	1-0	8-30
(A, M)	1-0	0-1	1-0	**4-0**	0-2	1-0	1-0	0-1	1-0	1-0	1-2	1-0	17-0	0-1	39-7
(A, D)	1-0	1-0	1-0	1-0	1-0	1-0	**2-0**	1-0	6-0	1-0	**2-0**	**17**-0	17-0	5-1	57-1
(D, M)	0-1	1-0	1-0	1-0	1-0	1-0	**2-0**	1-0	6-0	1-0	**2-0**	**17**-0	17-0	5-1	56-2
All	1-0	1-0	1-0	**4-0**	1-0	1-0	**2-0**	1-0	**7-0**	1-0	**2-0**	**17**-0	17-0	5-1	**61**-1
DTW	1-0	1-0	**5-0**	1-0	1-0	1-0	1-0	1-0	1-0	1-0	1-0	2-0	17-0	1-0	35-0*

Table 3 Iterative Person-Independent Detection of Multi-Response (MR) Events with SLC: #GR: Total number of given responses. #MR. Number of Multi-Responses. (A, D, M): Combination of averaged intensity, button press duration and maximum intensity. DTW: Dynamic Time Warping

#GR	#MR	(A, D, M)	DTW
2666	100	14	97

i.e. (Ave, Dur, Max). From Table 3, we can make the following observations. First, the feature-based approach fails to detect the observed multi-responses. Only 14 out of 100 multi-responses were detected. Therefore, the detection of multi-responses decreased dramatically from 64 in the person-dependent case (see Table 1) to 14 in the person-independent scenario. Note that we only count a multi-response as detected if the SLC approach separates the current multi-response into a one-point cluster. Second, the DTW-based approach detects 97 out of 100 multi-responses in total. Thus, the detection of multi-responses even increased from 91 in the person-dependent case (see Table 1) by 6 in total.

Based on the outcomes presented in Table 3, it seems obvious that the sequence-based DTW distance computation is able to capture significant differences of button press characteristics. In particular, the DTW-specific approach seems to facilitate the detection of single multi-response events in sets that contain *many* basic button press dynamics, i.e. single clicks.

4.4 Discussion and Conclusion

In many use cases, button presses are analysed as binary signals, evaluating whether a specific button was pressed within a current time interval. In this work, we focused on the evaluation of button press dynamics. For the sake of simplicity, button press dynamics are generally reduced to a small set of intuitive features, such as the duration, maximum and averaged intensity of the button presses, for instance in the context of uncertainty analysis in button press-based human–computer interactions (HCIs) [3].

In the current work, we analysed the possibility of an automated multi-click/multi-response detection. Note that multi-clicks constitute interesting HCI cases, in which the participants likely experience some difficulties in decision making accompanied by affective reactions, such as frustration, uncertainty, or impatience. We formulated the multi-response detection as a clustering approach, applying the single linkage clustering (SLC) method. We evaluated the SLC approach with respect to the effectiveness of separating multi-responses from basic button presses, i.e. single clicks. Moreover, we analysed the detection performance specific to all of the aforementioned features, as well as all resulting feature combinations. Additionally, in this work, we proposed to analyse the provided button press dynamics as time sequences. To this end, we applied the SLC method in combination with Dynamic Time Warping (DTW).

As discussed above, the DTW-based multi-response detection seems to be highly preferable in person-independent scenarios. On the other hand, in person-dependent scenarios, the detection analysis choice depends on the type of the corresponding evaluation setting. More precisely, in the iterative setting, the DTW-based approach outperforms the feature-based multi-response detection. In contrast, in the all-at-once setting, the feature-based detection analysis is preferable to the DTW-based alternative.

As future work, we aim at combining both approaches, i.e. feature- and DTW-based multi-response detection, to benefit from the strengths of both methods. Moreover, as discussed in Sect. 4.3, we need to find an appropriate alternative for the proposed SLC-specific multi-response detection approach in person-independent all-at-once settings.

Acknowledgements This work is supported by the project *Multimodal recognition of affect over the course of a tutorial learning experiment* (SCHW623/7-1) funded by the German Research Foundation (DFG). We gratefully acknowledge the support of NVIDIA Corporation with the donation of the Tesla K40 GPU used for this research.

References

1. Bellmann P, Kessler V, Brechmann A, Schwenker F (2021) Button press dynamics: beyond binary information in button press decisions. In: Proceedings of international conference on frontiers in computing and systems, lecture notes in networks and systems. Springer (2021)
2. Gold O, Sharir M (2018) Dynamic time warping and geometric edit distance: Breaking the quadratic barrier. ACM Trans Algor 14(4):50:1–50:17
3. Kohrs C, Hrabal D, Angenstein N, Brechmann A (2014) Delayed system response times affect immediate physiology and the dynamics of subsequent button press behavior. Psychophysiology 51(11):1178–1184
4. Lloyd SP (1982) Least squares quantization in PCM. IEEE Trans Inf Theory 28(2):129–136
5. Meudt S, Schwenker F (2015) ATLAS—machine learning based annotation of multimodal data recorded in human-computer interaction scenarios. In: ISCT. Ulm University, pp 181–186
6. Sakoe H, Chiba S (1978) Dynamic programming algorithm optimization for spoken word recognition. IEEE Trans Acoust Speech Signal Processi 26(1):43–49
7. Sibson R (1973) SLINK: an optimally efficient algorithm for the single-link cluster method. Comput J 16(1):30–34
8. Wolff S, Brechmann A (2015) Carrot and stick 2.0: the benefits of natural and motivational prosody in computer-assisted learning. Comput Hum Behav 43:76–84

VLSI Implementation of Artificial Neural Network

Swarup Dandapat, Sheli Shina Chaudhuri, and Sayan Chatterjee

Abstract In this work, VLSI circuit-based artificial neural network has been implemented. The artificial neuron consists of three components such as multiplier, adder and neuron active function circuit, which perform arithmetic operations for realizing neural network. The focus on this work is linearity investigation in nonlinear artificial neural network as well as learning efficiency. A multiplier design has been proposed for reducing nonlinearity in an artificial neuron. Sigmoid circuit has been used for activation function. New weight update technique with both retrieving and on chip learning function has been proposed for better accuracy and improved learning ability. In this neural network, pulse width modulation technique has been used to compute the output signals using both multiplication and summation operations. This improves the linearity of the neural network. The learning operation of the neural network has been verified through simulation results by adopting digital function like "NAND". A high-speed and accurate error detection block has also been used for this purpose. Cadence Virtuoso has been used to perform circuit level simulation. Design has been done in TSMC 180 nm CMOS technology. In the proposed design, an error calculation time of 0–100 ps has been achieved, thereby making the overall operation fast and the learning efficiency 99%. Supply voltage is 1.8 V, and total power dissipation has been measured to be is 8 mW.

Keywords Artificial neural network (ANN) · Activation function (AF) · Pulse width modulation (PWM) · Very large-scale integration (VLSI)

1 Introduction

Artificial neural network (ANN) has been widely used in modern information processing [1]. Neural networks can solve complex function that are not feasible for conventional single-processors. Examples include speech and image processing,

S. Dandapat (✉) · S. S. Chaudhuri · S. Chatterjee
Electronics and Telecommunication Engineering, Jadavpur University, Kolkata, West Bengal 700032, India
e-mail: swarupdandapat92@gmail.com

S. Basu et al. (eds.), *Proceedings of International Conference on Frontiers in Computing and Systems*, Lecture Notes in Networks and Systems 404,
https://doi.org/10.1007/978-981-19-0105-8_48

medical opinion, and optimization problems. Most of the ANNs perceived are software simulations. Only hardware neural networks or HNNs [2] utilize the parallel computing/processing properties of ANN as well as they have energy and fault-tolerance properties. Also, HNN has been played an important role in high-level performance applications like recognition, mining, and synthesis (RMS).

Replicating the biological neuron, an artificial neuron has been modeled using back-propagation algorithm [3]; however, this algorithm cannot train different type of function in less time. Another approach is to simulate using hashing trick [4], which is space-efficient way for reducing product. Here, knowledge-based neural network has been approached; however, weight updating of the neural network is not mentioned. Moreover, there are lots of pins, thus making hardware implementation difficult. Fuzzy systems have even been used to design artificial neurons [5]. However, due to their limited number of inputs, the number of inputs of neural networks cannot be increased beyond a certain point. To overcome this input problem, analog circuit-domain approach for ANN has been given for fulfilling large number input requirement [6]. However, logical complexity has been increased exponentially. Also consume large area. On chip learning neural network design in analog VLSI [1, 7]. Here, multiplication has been done by using differential mixer circuit which is a nonlinear circuit. However, performing operation of the neural network has been degraded. Piecewise linear approximation has solved the nonlinear operation using lines [8]. However, it has been limited to number of segments. In actual application, the large number of segments is required for accuracy. Weight multiplication has been done by pulse width modulation technique [9]. Linearity and retrieving operations have been solved by this technique. However, only retrieving function has been implemented, to increase the accuracy of operation both retrieving, and on chip linear learning function should be present in neural network.

In this work, PWM technique has been used for multiplication and the hyperbolic tangent function for arithmetic operations of the main nonlinear functions in the artificial neuron node. Fast error calculating and weight updating techniques have been proposed to solve the linearity and accuracy issues. After compensation, the non-ideal property of the MOS transistors improves and the designed circuit has better linearity and a wide active rang. This increases the learning efficiency. In the proposed design, an error calculation time of 0–100 ps has been achieved, thereby making the overall operation fast and the learning efficiency 99%. It is to be noted that only digital operations have been performed, not analog operations.

2 Modeling of ANN from Biological Neuron

A biological neuron has three parts, namely dendrites, soma, and axon. Dendrites receive signal, soma collects input signal, and axon transmits output signal to surrounding neurons via their dendrites [10]. Dendrites are the connecting link between neurons, whereas soma is implemented using an adder and an activation function.

2.1 Forward Neural Network Design

Biological neurons are connected to each other through what is called synapse [11]. The synapse which is basically deciding the strength of the connection, that is, how strong or weak those connections are. A neuron could be connected to several other neurons or might be receiving input from several sources. Neurons are connected in this manner via the synapses. Artificial neurons have also been modeled where the strength of the connections is the synaptic weight. Synaptic weights will be defined for the connection between the input and neuron under consideration. Input may be received from other neurons also or directly from particular units.

Hence, neurons are modeled as having N number of inputs and connections or synaptic weights, $W_1 \ldots W_N$. Inputs defined as $X_1 \ldots X_N$ are linearly combined with the weights. So, on summing up all these inputs multiplied by their appropriate weights will give the sum total (Y) which is as follows:

$$Y = \sum_{i=1}^{N} W_i X_i \tag{1}$$

3 Implementation of Neural Architecture

Figure 1 shown is a neural network which consists one neuron, three synapses, one delta calculation block, and three weight update blocks. The neuron includes an activation function and a voltage to pulse converter.

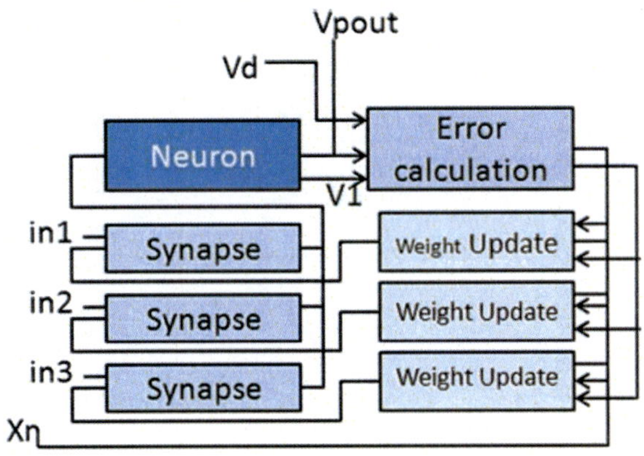

Fig. 1 Block diagram of ANN

3.1 Synapse Design and Circuit Descriptions

The synapse cell has been implemented using multiplier and summation operational blocks. The schematic of the synapse cell multiplier has been shown in Fig. 2. The output is linear because the circuit operates in saturation region. The corresponding output equation is as follows:

$$I_{\text{out}} = -\left[\mu_x C_{ox}\frac{W_9}{L}\left(V_{dd} - V_{thp}\right) + \mu_x C_{ox}\frac{W_{10}}{L}(V_{ss} - V_{thn})\right]$$
$$V_w + \left(\mu_x C_{ox}\frac{W_9}{L} - \mu_x C_{ox}\frac{W_{10}}{L}\right)V_w^2 \tag{2}$$

The synapse circuit uses a PWM signal to control the current. Here, the output charge is a product of the signal's pulse width and current. When the PWM signal is high, Devices $M11$ and $M12$ are switched on, but $M7$ and $M8$ are switched off. As a result, the output current becomes the difference of the currents of $M9$ and $M10$. When the PWM signal is low, Devices $M11$ and $M12$ are switched off, but $M7$ and $M8$ are switched on. So, the currents of $M9$ and $M10$ pass to the ground, and the output current turns out to be zero.

The rise and fall times of the PWM signal increase due to long distance transmission. So, a comparator is used to reduce the above parameters. Moreover, $M13$ and $M14$ have been used to reduce the non-ideal effects of $M11$ and $M12$. Device $M13$ balances the capacitor loads in the drain of Devices $M9$ and $M10$, whereas Device $M14$ reduces the feed through when the devices switch on and off.

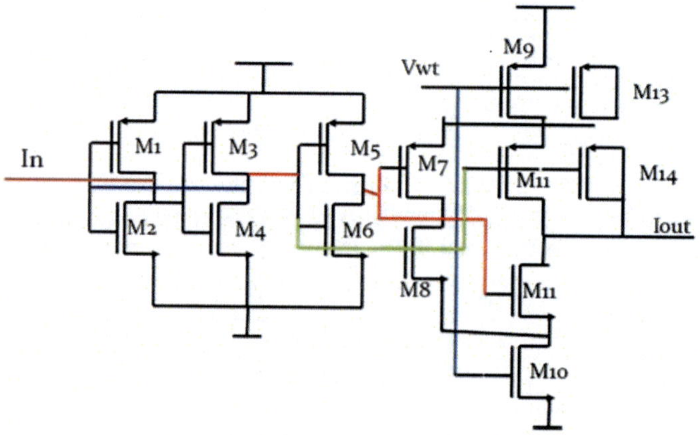

Fig. 2 Schematic of the synapse cell

3.2 Neuron Activation Function (NAF)

Nonlinear transfer function (integration) and summation operation have been done in NAF. In Fig. 3, integration cell has been shown. Synapse outputs in the form of current (I_{out}) have been integrated controlled by $Ø_2$ and converted into the output voltage (V_{out}).

Seeing as all synapse outputs have been summed in the neuron. NAF internal block is shown in Fig. 4.

The activation function (AF) block is modeled using two functions:

$$\text{tanh function} \quad \tanh(x) = \frac{e^x - e^{-x}}{e^x + e^{-x}} \tag{3}$$

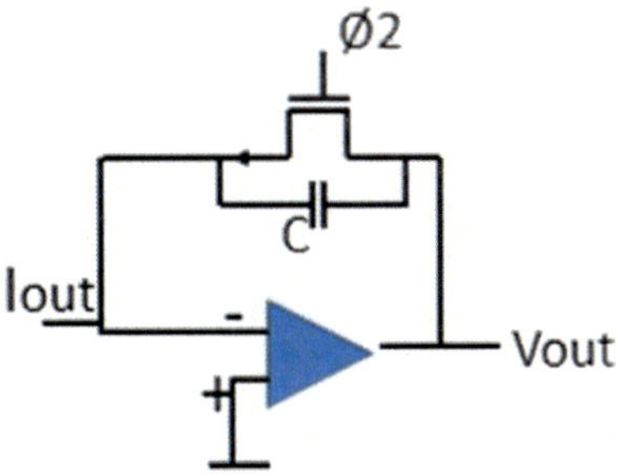

Fig. 3 Integration cell

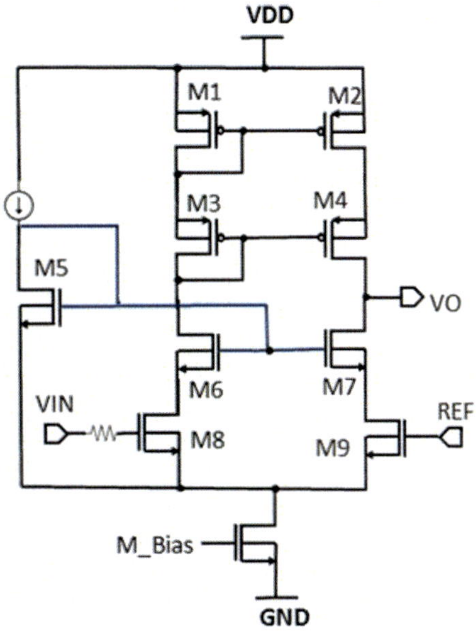

Fig. 4 Neuron activation function internal cell

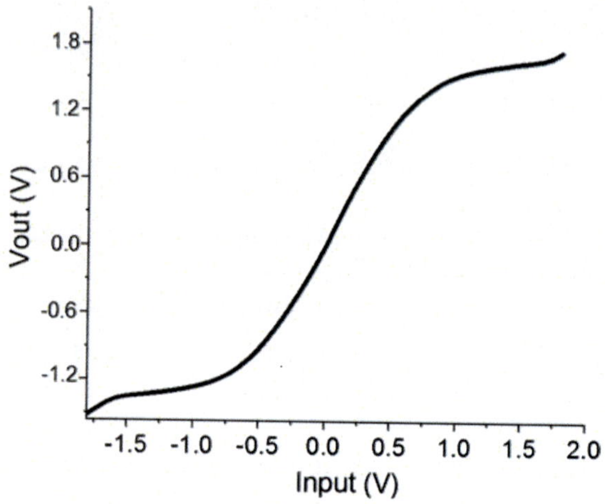

Fig. 5 tanh function obtained from sigmoid generator

$$\text{Sigmoid function} \quad s(x) = \frac{1}{1 + e^{-x}} \tag{4}$$

Both of the curves are similar in shape apart from their ranges. The tanh graph varies between −1.8 and 1.8 v, whereas the sigmoid curve varies between 0 and 1. Hence, tanh function is a modified version of sigmoid function. The simulating result shown in Fig. 5 is the tanh graph obtained from the schematic shown in Fig. 4.

The sigmoid function can be physically realized using a MOS in saturation region. We know that saturation current is

$$I_{\text{sat}} = I_0 e^{(\eta V_g - V_s)} \tag{5}$$

where I_0 is reverse saturation current, η is the emission coefficient, V_g is gate voltage, and V_s is source voltage.

The schematic of the sigmoid generator as shown in Fig. 4 is a differential amplifier. Transistors $M1$–$M7$ are always in saturation because they are diode connected. Transistor M_bias acts as a constant current source and so, and Transistors $M8$ and $M9$ remain in saturation. This ensures a constant current I_{out} through the circuit which is a function of the voltage difference between Transistors $M8$ and $M9$.

$$I_{\text{out}} = I_{\text{bias}} \tanh\left(\frac{k_B(V_{\text{in}} - V_{\text{ref}})}{2}\right) \tag{6}$$

where I_{bias} is the current through transistor M_bias, k_B is the Boltzmann constant, V_{in} is input voltage, and V_{ref} is reference voltage.

Hence, the difference between the two drain currents is

$$I_1 - I_2 = I_{\text{bias}}\left[\frac{e^{\frac{qV_1}{\eta k_B T}} - e^{\frac{qV_2}{\eta k_B T}}}{e^{\frac{qV_1}{\eta k_B T}} + e^{\frac{qV_2}{\eta k_B T}}}\right] \tag{7}$$

On Eq. (7), the functionality of differential amplifier as a tan sigmoid generator is proved.

3.3 Voltage to Pulse Converter (VPC)

In the PWM technique, neuron's state has been represented by pulse signal. Figure 6 shown below is a voltage to pulse converter.

In the circuit, capacitor charges in $Ø_1$ phase and discharges in $Ø_2$ phase when $M1$ is switched on. NAND gate logic is used to make sure Vp_{out} is active in $Ø_2$ phase. The simulation result has been shown in Fig. 7. Voltage Vp has been represented by idle integration of Vp_{out}, and linearity of the output is fine.

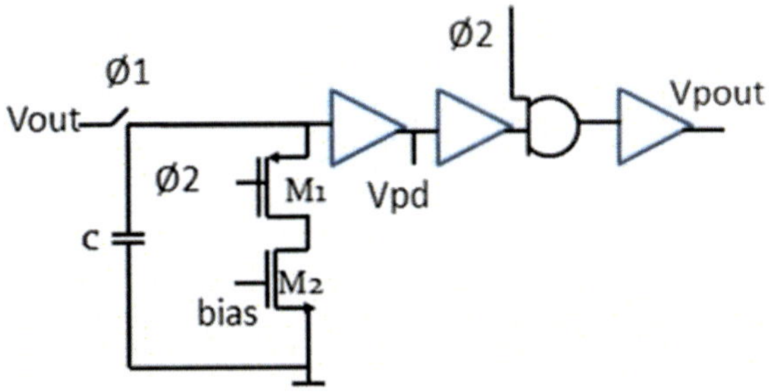

Fig. 6 Voltage to pulse converter circuit

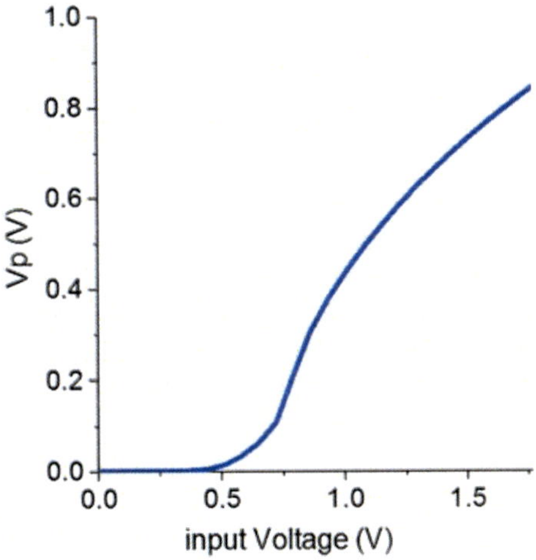

Fig. 7 VPC simulation result

4 Design of Learning Circuit

The delta learning rule has been used in this feedforward neural network. The learning expression is as follows:

$$w_i(t + \Delta t) = w_i(t) + \eta\delta x_i \quad (8)$$

where Δt is the learning cycle time, η is the gain term for adjusting the learning rate, and δ is the error term. The error term is same for all the synapses connected to the same neuron. It is given by

$$\delta = f'(z).(d - y) \quad (9)$$

where $f'(z) = \mathrm{d}f(z)/\mathrm{d}z$ is the differentiation of the neuron transfer function and d is the desired output. A small value of Δw_i will introduce nonlinearity in the operation of the learning circuit. Hence, a good linearity is required for proper functioning of the neural network. The learning circuit involves two function blocks—δ calculation block and weight update block.

4.1 δ (Error) Computation Block

Error detection block is the most important block which converts the error between desired output and real output that drives the weight updating block. The weight is updated to reach the desired output. This means accuracy of the neuron depends on error computing efficacy of this block. The error computation diagram has been shown in Fig. 8.

In the above circuit, when V_{out} is low and V_D is high for time t_1, for that time t_1, upper node becomes active and gives high output. As a result, $M8$ switches on, $M9$ switches off, and other components of the design behave as a NOT gate since output will be zero (low). When V_D is high, at that time, lower node goes from low to high. $M11, M12, M13$ have been on since the NOT gating input low and makes the output high. Hence, for time t_1, weight has been updated by decreasing the input weights voltage for getting the desired output and vice versa. There is no reset path which requires less time delay to compute the error. High-speed error detection has been done. Simulation output has been shown in Fig. 9. Delay of the error computation block is 0–100 ps, indicating that error computation is almost linear.

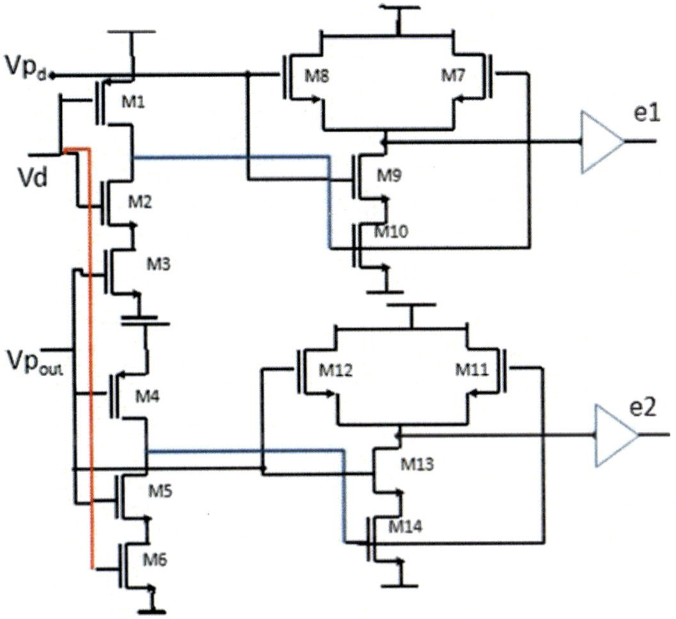

Fig. 8 Error computation circuit

Fig. 9 Error detection transfer curve

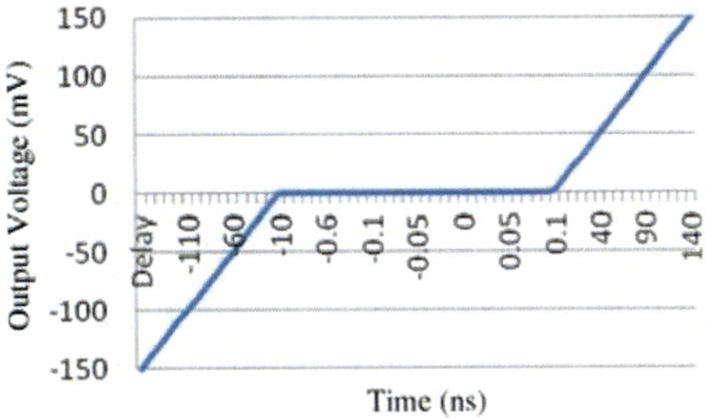

4.2 *Weight Update Block*

It converts the logic outputs of the error detection block to analog signal to drive the synapse and update the weights according to Eq. (8). A basic tristate buffer-based weight updating cell has been designed in TSMC 180 nm with 150 μA current. A basic weight updating configuration and the schematic have been shown in Fig. 10. The weight updating output for various error output combinations has been shown in Figs. 11 and 12.

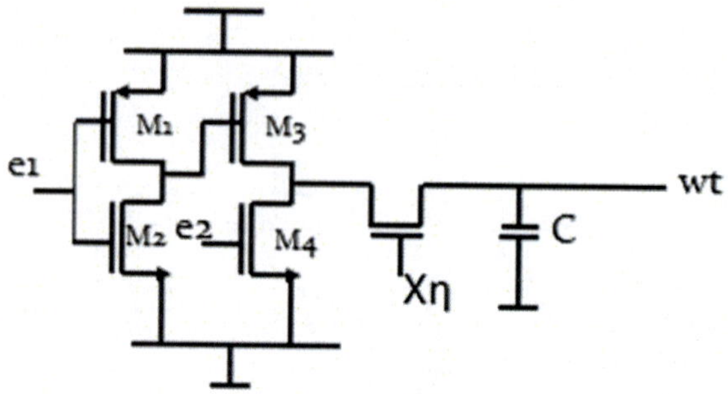

Fig. 10 Weight updates schematic

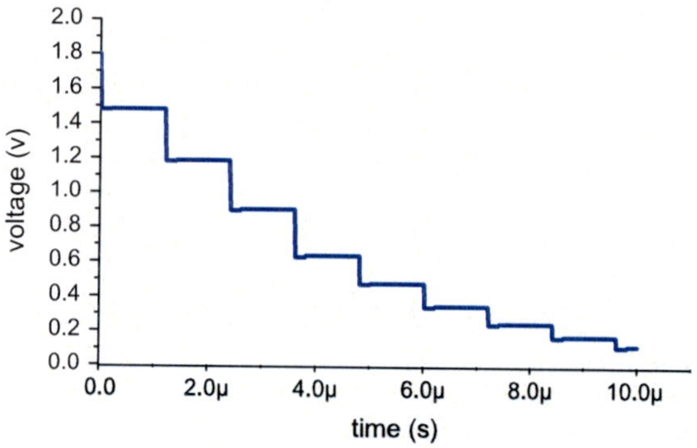

Fig. 11 Decreasing the voltage level

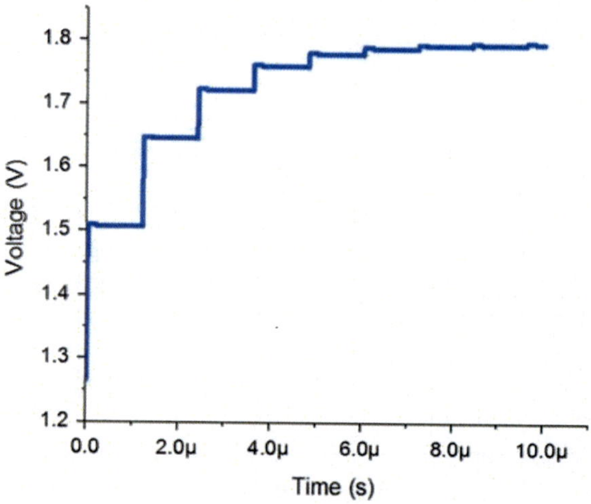

Fig. 12 Increasing the voltage level

5 Simulation Result and Dictation

Logical NAND operation learning procedure has been shown in Fig. 13. When $X\eta$ is

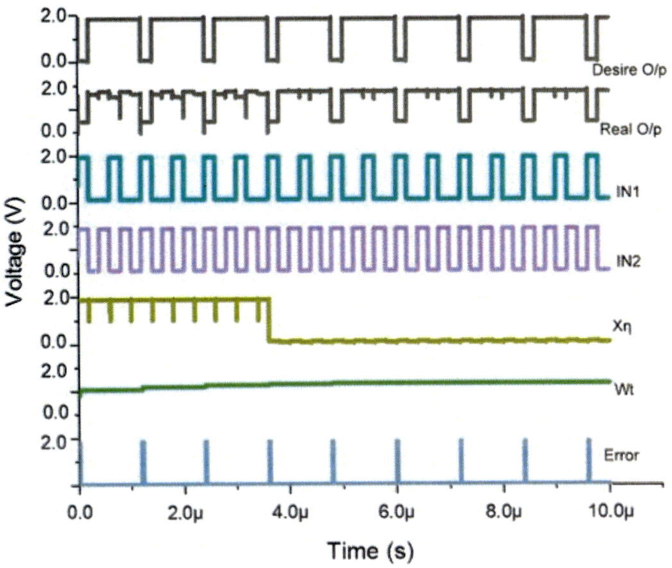

Fig. 13 Logical NAND operation learning procedure

active, it is in learning mode. Error calculation and weight updating have been done in this mode. Figure 13 shown in the learning mode becomes a learner after weight is updated. When $X\eta$ becomes inactive, it goes to output mode. The weight updating procedure stops and gives the output as fairly similar to the desired output. As seen from Fig. 13, the error calculation has been quick, resulting in proper weight update and high output accuracy.

6 Conclusion

Software simulation of scalable VLSI implementable architecture for artificial neural network (ANN) has been shown. An attempt has been made to replicate biological systems using VLSI circuit. PWM technique has been used for multiplication. The hyperbolic tangent function has been used for arithmetic operations of the main nonlinear functions in the artificial neuron node. Fast error calculating and weight updating techniques have been done to solve the linearity and accuracy issues. After compensating the non-ideal property of the MOS transistors, the design circuits have good linearity and wide active range. Delta learning rule has been perceived for learning ability. The same has been designed using Cadence Virtuoso in TSMC 180 nm technology. Efforts are being made to realize more operations using this ANN.

Acknowledgements The authors are extremely thankful to SMDP-III (C2SD) for providing them with the required laboratory facilities to carry on the research works, and they also extend their sincere gratitude toward IC CENTER, Jadavpur University.

References

1. Cyril Prasanna Raj P, Pinjare SL (2009) Design and analog VLSI implementation of neural network architecture for signal processing. European J Sci Res 27(2):199–216, © Euro Journals Publishing, Inc. ISSN 1450-216X
2. Chen T et al (2012) BenchNN: on the broad potential application scope of hardware neural network accelerators. In: IEEE international symposium on workload characterization (IISWC), pp 36–45. https://doi.org/10.1109/IISWC.2012.6402898
3. Rami MB, Bhatt HG , Shukla YB (2012) Design of neural architecture using WTA for image segmentation in 0.35 μm technology using analog VLSI. Int J Eng Res Technol (IJERT) 1(4)
4. Rajeswaran N, Madhu T (2016) An analog very large scale integrated circuit design of back propagation neural networks. World automation congress (WAC). IEEE, pp 1–4. https://doi.org/10.1109/WAC.2016.7583000
5. Suganya A, Sakubar Sadiq J (2016) An priority based weighted neuron net VLSI implementation. In: International conference on advanced communication control and computing technologies (ICACCCT)
6. Wilamowski BM, Binfet J, Kaynak MO (2000) VLSI implementation of neural networks. Int J Neural Syst 10(3):191–197
7. Shrinath K (2015) Analog VLSI implementation of neural network architecture. Int J Sci Res (IJSR) 4(2):2319–7064. ISSN (online)
8. Mada S, Mandalika S (2017) Analog implementation of artificial neural network using forward only computation. Asia Model Symp (AMS)
9. Xie X, Yang B, Chen Z (2009) A FPGA based artificial neural network prediction approach. Shang Dong Sci 22(1):7–12
10. Bor JC et al (1998) Realization of the CMOS pulse width modulation neural network with on-chip learning. IEEE Trans Circ Syst-11 45(1):96–107
11. Gavriloaia B, Militaru N, Novac M, Mara C (2018) Multi-band response antenna bio-inspired from biological neuron morphology In: 2018 10th international conference on electronics, computers and artificial intelligence (ECAI), Iasi, Romania, pp 1–4. https://doi.org/10.1109/ECAI.2018.8679034

A Novel Efficient for Carry Swipe Adder Design in Quantum Dot Cellular Automata (QCA) Technology

Suparba Tapna

Abstract The following research introduced a novel co-planer full adder circuit, executed in quantum dot cellular automata (QCA). The proposed novel full adder circuit was then afterward used with a new circuit for executing a novel 4-digit carry swipe adder (CSA) in QCA innovation. A designer device involving QCA form 2.0.1 is used to carry out the functioning of planned full adder QCA circuits. The execution results demonstrate that a planned QCA FA circuits show improved performance contrasted with other similar circuits.

Keywords Carry swipe adder · Coplanar circuit · Full adder · High-performance design · Quantum dot cellular automata

1 Introduction

PC number-crunching assumes a significant part in the data and correspondence applications, for example, ALU and cryptography [1–3]. A full adder has a significant part in PC number-crunching. In this way, the productivity of numerous PC number-crunching applications is basically dictated by the effectiveness of the full viper execution [1–3].

Then again, quantum dot cellular automata (QCA) innovation is another promising innovation, which can proceed with the Moore's law advancement. Such innovation utilizes charge arrangement to data change rather than current. Therefore, circuit plan in the QCA innovation has benefits in correlation with customary advancements, for example, CMOS innovation regarding little measurement, quick activity, and low force utilization [4, 5].

As of late, a few endeavors were also followed for improving the productivity of a full viper execution in QCA innovations [6–15]. Takala and Hanninen [6] came up with a full snake QCA which involved 0.1 μm^2 region and 102 QCA cells. Ramesh and Rani [7] also planned another QCA full snake system which was comprised of a

S. Tapna (✉)
Department of Electronics and Communication Engineering, Durgapur Institute of Advanced Technology and Management, Rajbandh, Durgapur 713212, India

S. Basu et al. (eds.), *Proceedings of International Conference on Frontiers in Computing and Systems*, Lecture Notes in Networks and Systems 404,
https://doi.org/10.1007/978-981-19-0105-8_49

0.038 μm^2 territory and 52 QCA cells. Abedi et al. [8] with 0.043 μm^2 territory and 59 QCA cells also designed a QCA system with full snake. Navi and Hashemi [9] in their research presented a model with 0.06 μm^2 region and 71 cells requirement. Mohammad et al. [10] introduced a QCA full viper which involved 0.02 μm^2 region and 38 cells of QCA. Thapliyal and Labrado [11] introduced another QCA full adder requiring 0.05 μm^2 region and 63 QCA cells. Ahmad [12] built a QCA full snake which is comprised of 0.04 μm^2 territory and 41 QCA cells. Balali et al. [13] planned a QCA full snake with 0.02 μm^2 territory and 29 QCA cell requirement. In any case, all of these full viper systems discussed above came with its own advantages, however, the intricacy as well as the required space for a full snake circuit in QCA innovation could be decreased with the novel strategy explained in this article.

A proficient circuit for the slightest bit QCA full adder was introduced. At that point, an effective circuit is intended for 4-bit QCA ripple carry adder (RCA). Usefulness of this planned circuits was confirmed utilizing QCA designer apparatus variant 2.0.1. The outcomes depict that the planned circuits come with its own benefits contrasted with ongoing adjusted the 4-bit QCA RCA and slightest bit QCA full adder circuits. The remainder of the research is coordinated as: Foundation of the created circuits was introduced in area 2. Planned circuits were introduced in Sect. 3. Section 4 assesses for the conclusion. At last, after Sect. 4 ends of this article.

2 Background

2.1 QCA Technology

QCA innovation is another arising innovation which can be used for improvement of computerized circuits dependent on the Moore's law. This new innovation uses a charge arrangement rather than the current for data progress. An essential component in such innovations is a 4-specks square, with two free electrons. Figure 1 demonstrates an arrangement of improved QCA cell [1–5].

The development of the free electron is used a zero state and one state rationale in the innovation. Utilizing these cells, rationale components, for example, majority

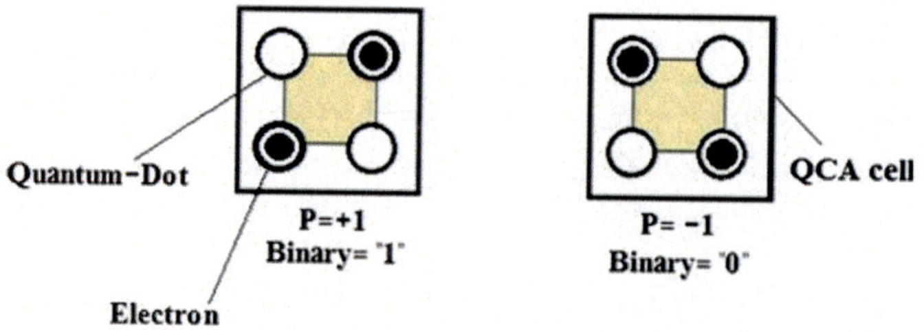

Fig. 1 Simplified QCA cell [1–5]

voter gate [4] can be created. It ought to be noticed that other rationale components, for example, OR entryway AND door can be created utilizing dominant part entryway [1, 4, 5]. Besides, complex advanced circuits, for example, full adder circuits [6–15] and multiplexer circuits [1–5] are created utilizing these rationale components.

2.2 QCA Full Adder Circuit

FA assumes an essential part in the complex computerized circuits. Accordingly, superior execution of this circuit is an appealing examination region. The legitimate capacity of the slightest bit full adder is appeared by the condition given below:

$$\text{Carry} = AB + AC_{\text{in}} + BC_{\text{in}} = \text{Maj}\,3\,(A, B, C_{\text{in}}) \quad (1)$$

$$\text{Sum} = A \text{ XOR } B \text{ XOR } C_{\text{in}} = \text{Maj3}(C_{\text{in}}, \text{Maj3}(A, B, C_{\text{in}}), \text{Carry}) \quad (2)$$

In the above conditions, A and B mean contributions of the slightest bit full viper. Carry is the convey yield and C_{in} means convey information, separately. Entirety signifies the yield of aggregate in the slightest bit full snake. Also, Maj3 signifies a 3-input dominant part work that is executed utilizing 3-input lion's share entryway in QCA innovation [1, 4, 9]. Figure 2 explains a circuit chart for slightest bit QCA full adder [6–9].

What is more, the 4-digit QCA CSA systems are accomplished by utilizing successively four the slightest bit FA [1, 6–9]. Figure 3 depicts a 4-bit QCA CSA circuit. In this circuit, "$A = (A3, A2, A1, A0)$ and $B = (B3, B2, B1, B0)$" are 2 inputs in four-bit

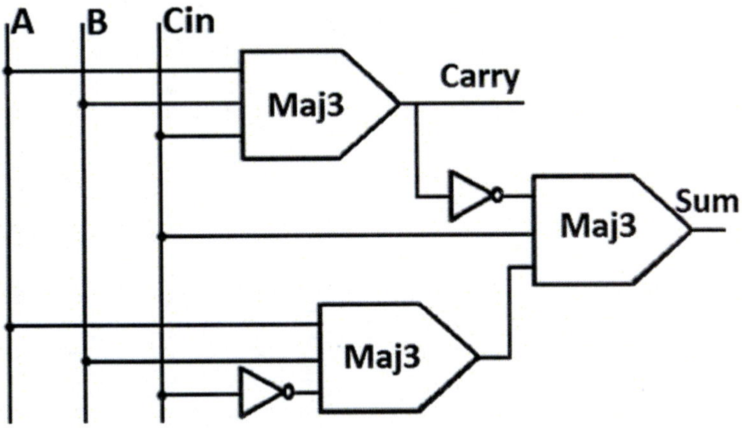

Fig. 2 Circuit of 1-bit QCA full adder [6]

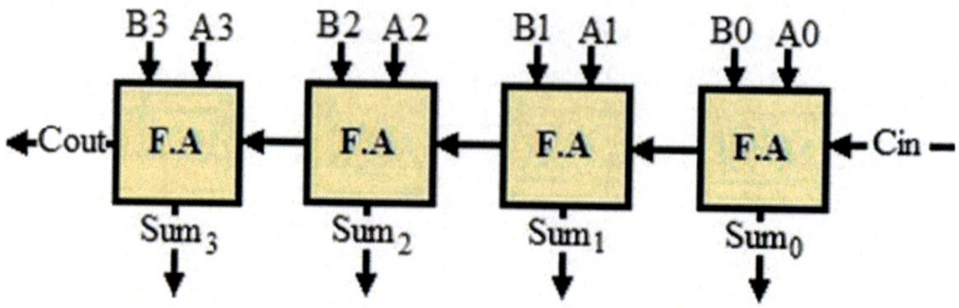

Fig. 3 Four-bit QCA CSA circuit [6]

arrangement. Here, C_{in} and C_{out} signify a slightest bit convey info and convey yield, and Sum = (Sum3, Sum2, Sum1, Sum0) is four-bit yield.

2.3 *Designed QCA Circuits*

This part traces the slightest bit QCA FA circuit. At that point, a new 4-bit QCA CSA circuit was planned dependent on the planned slightest bit QCA FA circuit.

2.4 *The Designed QCA Full Adder Circuit*

Figure 4 depicts a planned circuit for slightest bit QCA full adder system.

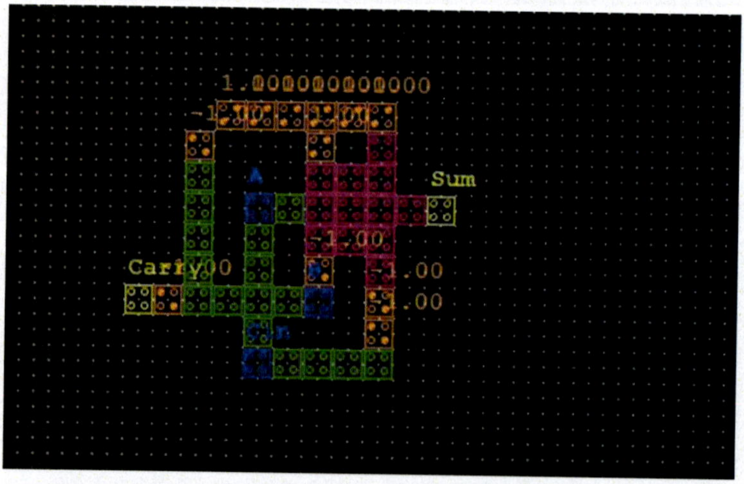

Fig. 4 1-bit QCA full adder circuit design

A and *B*, in the above circuit diagram, are the slightest bit information sources with *C* as the convey input. Convey means the yields of convey, and SUM means the aggregate. The circuit is comprised of 46 QCA cells. Another noteworthy point is that, 4 checking zones were used in the circuit:

- Green demonstrates clock zone 0,
- Violet demonstrates clock zone 1,
- Light blue shows clock zone 2,
- White demonstrates clock zone 3.

2.5 *The Designed QCA Carry Swipe Adder Circuit*

In this, *A* and *B* were the 2 four-bit sources of info and C_{in} is a slightest bit convey input. Convey and SUM mean yields of the slightest bit convey and 4-cycle total, separately. The circuit is comprised of 180 QCA cells. Carry swipe adder is additionally called an equal RCA snake. In CSA, the convey pieces are saved at each stage. Accordingly, the deferral of a CSA is steady on the grounds that there is no convey engendering is available. At the last stage, all the convey pieces ought to be added to get *N* bit aggregate. Consider an illustration of 4-bit, CSA having inputs: *A* = 1 0 1 and *B* = 1 1 0. The realistic outline of CSA as displayed in beneath Fig. 5.

3 Result and Simulation with Comparison

The planned circuits are carried out utilizing QCA designer device variant 2.0.1. The following segment presents the execution results.

A= 1 0 0 1
B= 1 1 1 0

IC= 1 0 0 0
IS= 0 1 1 1

SUM= 1 0 1 1 1

Fig. 5 Carry swiping operation of proposed adder

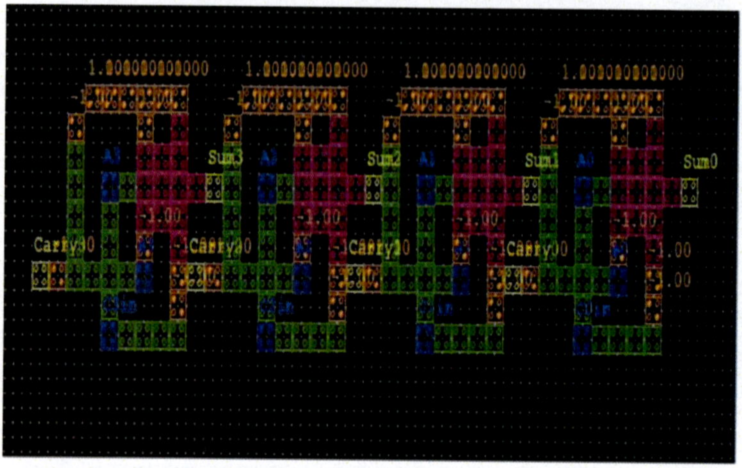

Fig. 6 Circuit designed for four-bit QCA carry swipe adder

3.1 The QCA 1-bit Carry Swipe Adder Circuit

Figure 6 demonstrates the execution after effects of a planned circuit for slightest bit QCA 1 bit carry swipe adder.

3.2 Simulation Results

The proposed carry swipe adder is implemented in QCA designer tool. Figure 7 mentioned is the simulation environment of the proposed adder circuit. At finally, we have to get the output waveform and it varies with respect to the clock. We have to consider the standard parameters during the simulation of the implemented adder circuit. The result discussed in detailed and also shown (Table 3) the strategy of this proposed work that makes the differences in existing literature.

The **simulation results** of a planned circuit for slightest bit QCA FA affirm the accuracy of the circuit. Table 1 sums up an execution after effect of a planned circuit for slightest bit QCA full adder contrasted with other the slightest bit QCA full viper systems designed in H¨anninen and Takala [6], Ramesh and Rani [7], Abedi et al. [8], Hashemi and Navia [9], Mohammadi et al. [10], Labrado and Thapliyal [11], Ahmad et al. [12], Balali et al. [13]. In view of our execution, the outcomes are depicted in Table 1 and Fig. 7, a planned circuit of slightest bit QCA full adder shows improvements as far as intricacy contrasted with other the slightest bit QCA FA circuits designed by H¨anninen and Takala [6], Ramesh and Rani [7], Abedi et al. [8], Hashemi and Navia [9], Ahmad et al. [12]. Albeit the cell includes in the slightest bit QCA FA circuits designed by Mohammadi et al. [10], Labrado and

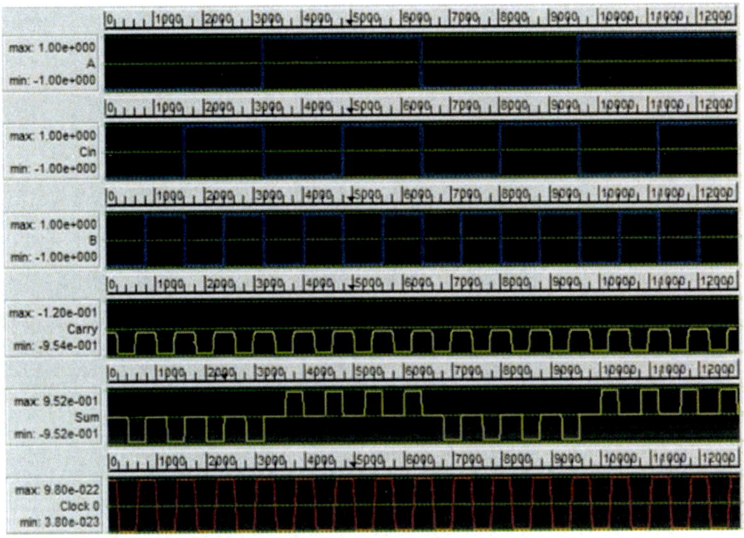

Fig. 7 Simulation of one-bit QCA CSA circuit design

Table 1 Simulation parameters

Parameters	Value
"Temperature	1.000000 K
Relaxation time	1.000000e−015 s
Time step	1.000000e−016 s
Total simulation time	7.000000e−011 s"
Clock high	9.800000e−022 J
Clock low	3.800000e−023 J
Clock shift	0.000000e+000
Clock amplitude factor	2.000000
"Radius of effect	80.000000 nm
Relative permittivity	12.900000
Layer separation"	11.500000 nm

Thapliyal [11], Balali et al. [13] were lower as compared to the planned the slightest bit QCA FA circuit, the planned four-digit QCA CSA circuit, that used the slightest bit QCA FA circuit as the fundamental square, shows benefits contrasted with 4-bit QCA CSA circuits designed by Mohammadi et al. [10], Balali et al. [13]. It was on the grounds that the info/yield ports in the created the slightest bit QCA full adder have appropriate spots. Thus, the spot and course bring about the created four-digit QCA CSA presents a superior outcome. In addition, the yield cells of the slightest bit QCA full adder circuit in Labrado and Thapliyal [11] were not reasonable put.

Along these lines, the execution of 4-cycle QCA CSA circuit utilizing a slightest bit QCA full adder circuit in Labrado and Thapliyal [11] is tough.

3.3 Designed QCA CSA Circuit

Figure 8 depicts execution after effects of a planned circuit for the four-cycle QCA carry swipe adder. The execution consequences of planned circuit for four-cycle QCA CSA affirm the rightness of the designed system. Table 2 sums up the execution consequences of this planned circuit of four-digit QCA CSA contrasted with rest of the conventionally found four-bit QCA CSA circuits in H¨anninen and Takala [6], Ramesh and Rani [7], Abedi et al. [8], Hashemi and Navia [9], Mohammadi et al. [10], Ahmad et al. [12], Balali et al. [13], Pudi and Sridharan [14]. In light of our execution, the results are showed in Table 2 and Fig. 8, for the planned circuit of a four-bit QCA CSA, showing improvement regarding intricacy, and region contrasted with other four-bit QCA CSA circuits in H¨anninen and Takala [6], Ramesh and Rani

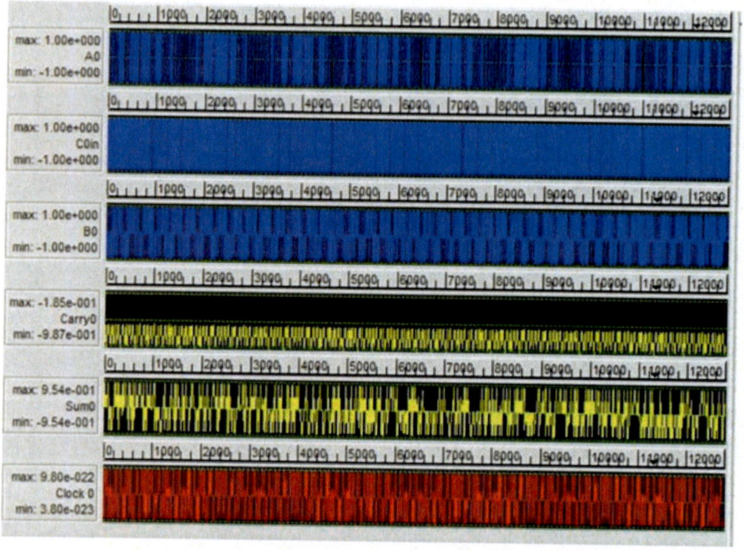

Fig. 8 Simulation results for the designed four-bit QCA CSA circuit

Table 2 Possible combination of input for 1 bit CSA circuit

A	B	C_{in}	Sum	C_{out}
0	1	0	1	0
1	0	1	0	1

Table 3 Comparison of 1-bit QCA carry swipe adder circuits

References	Design complexity (No. of cells)	Area (μm^2)	Delay (clock zone)
H¨anninen and Takala [6]	102	0.1	8
Hashemi and Navia [9]	71	0.06	5
Ramesh and Rani [7]	52	0.038	4
Abedi et al. [8]	59	0.043	4
Mohammadi et al. [10]	38	0.02	3
Labrado and Thapliyal [11]	41	0.04	2
Ahmad et al. [12]	63	0.05	3
Balali et al. [13]	29	0.02	2
Proposed work	45	0.05	1

[7], Abedi et al. [8], Hashemi and Navia [9], Mohammadi et al. [10], Ahmad et al. [12], Balali et al. [13], Pudi and Sridharan [14] (Table 3).

4 Conclusion

Full adders assume a significant part in PC number-crunching fields. In this way, proficient execution in full adders could expand the effectiveness of a PC math circuits. The paper introduced and assessed a proficient circuit of full adder system with QCA innovation. Also, we executed a 4-bit QCA CSA circuit dependent on the new slightest bit QCA full adder. Planned circuits are executed utilizing QCA designer device adaptation 2.0.1. The execution results affirmed that the planned circuits beat ongoing changed the four-bit QCA CSA circuits and slightest bit QCA full adder circuits [6–9, 12] regarding intricacy and required region (Tables 4 and 5).

Table 4 Possible combination of input for 4 bit CSA circuit

A	B	C_{in}	Sum	C_{out}
1	1	0	0	1
1	1	1	1	1

Table 5 Comparison of 4-bit QCA carry swipe adder circuits

References	Design complexity (No. of cells)	Area (μm^2)	Delay (clock zone)
H¨anninen and Takala [6]	558	0.85	20
Hashemi and Navia [9]	442	1	8
Ramesh and Rani [7]	260	0.28	10
Abedi et al. [8]	262	0.208	28
Mohammadi et al. [10]	237	0.24	6
Labrado and Thapliyal [11]	295	0.3	6
Ahmad et al. [12]	269	0.37	14
Balali et al. [13]	339	0.2542	7
Proposed work	180	0.16	4

References

1. Balasubramanian P (2015) A latency optimized biased implementation style weak-indication self-timed full adder. Facta Univ Ser Electron Energ 28:657–671
2. Rezai A, Keshavarzi P (2016) High-performance scalable architecture for modular multiplication using a new digit-serial computation. Micro J 55:169–178
3. Rezai A, Keshavarzi P (2015) High-throughput modular multiplication and exponentiation algorithm using multibit-scan-multibit-shift technique. IEEE Trans VLSI Syst 23:1710–1719
4. Balali M, Rezai A, Balali H, Rabiei F, Emadi S (2017) A novel design of 5-input majority gate in quantum-dot cellular automata technology
5. Rashidi H, Rezai A, Soltani S (2016) High-performance multiplexer circuit for quantum-dot cellular automata. Comput Electr 15:968–998
6. H¨anninen I, Takala J (2010) Binary adders on quantum-dot cellular automata. Sign Process Syst 58:87–103
7. Ramesh B, Rani MA (2015) Design of binary to BCD code converter using area optimized quantum-dot cellular automata full adder. Int J Eng 9:49–64
8. Abedi D, Jaberipur G, Sangsefidi M (2015) Coplanar full adder in quantum-dot cellular automata via clock-zone-based crossover. IEEE Trans Nanotech 14:497–504
9. Hashemi S, Navia K (2015) A novel robust QCA full-adder
10. Mohammadi M, Mohammadi M, Gorgin S (2016) An efficient design of full adder in quantum-dot cellular automata (QCA) technology. Microelectronic 50:35–43
11. Labrado C, Thapliyal H (2016) Design of adder and subtractor circuits in majority logic-based field-coupled QCA nano computing. Electron Lett 52:464–466
12. Ahmad F, Bhat GM, Khademolhosseini H, Azimi S, Angizi S, Navi K (2016) Towards single layer quantum-dot cellular automata adders based on explicit interaction of cells. Comput Sci 16:8–15
13. Balali M, Rezai A, Balali H, Rabiei F, Emadid S (2017) towards coplanar quantum-dot cellular automata adders based on efficient three-input XOR gate. Result Phys 7:1389–1395
14. Pudi V, Sridharan K (2012) Low complexity design of ripple carry and Brent-Kung adders in QCA. IEEE Trans Nanotech 11:105–119
15. Cho H, Swartzlander EE (2009) Adder and multiplier design in quantum-dot cellular automata. IEEE Trans Comput 58:721–727

A Novel Dual Metal Double Gate Grooved Trench MOS Transistor: Proposal and Investigation

Saheli Sarkhel, Riya Rani Dey, Soumyarshi Das, Sweta Sarkar, Toushik Santra, and Navjeet Bagga

Abstract Through this paper, we have propounded and investigated a novel structure of a grooved trench MOS transistor with double gate architecture using TCAD simulations. To date, only a single metal trench MOSFET has been reported which having weaker control of the gate bias over the channel charge, unlike improved in our propounded structure. The propounded dual metal double gate grooved trench (DMDGGT) structure incorporated the advantages of enhanced gate controllability and subdued the drain-induced barrier lowering (DIBL) owing to the presence of a bi-metal gate with dissimilar work functions. In addition, due to the inherent advantage of a grooved trench gate device of having a longer effective channel length by the trench gate geometry, the device results in a significant reduction in short-channel effects (SCEs). The acquired results from the SILVACO ATLAS simulation exhibit a significant improvement of the propounded DMDGGT MOSFET as compared to its single metal counterpart for surface potential, electric field, threshold voltage, and drain characteristics, thereby substantiating the efficacy of the propounded device structure.

Keywords Trench gate MOSFET · Double gate · Drain-induced barrier lowering · Short-channel effect · Threshold voltage roll-off

1 Introduction

The ever-increasing demand for improved speed and better as well as higher integration capability has enabled the scaling of the MOSFETs to satisfy the demands. This downscaling has brought down the device dimensions to sub-micron order, which has introduced us to some unavoidable problems in fabricating, realizing, and utilizing

S. Sarkhel (✉) · R. R. Dey · S. Das · S. Sarkar · T. Santra
Netaji Subhash Engineering College, Kolkata, West Bengal, India

N. Bagga
Pandit Dwarka Prasad Mishra Indian Institute of Information Technology Design and Manufacturing, Jabalpur, Madhya Pradesh, India

S. Basu et al. (eds.), *Proceedings of International Conference on Frontiers in Computing and Systems*, Lecture Notes in Networks and Systems 404,
https://doi.org/10.1007/978-981-19-0105-8_50

such a device. The first obvious problem irrespective of the downscaling is the existence of parasitic devices in the bulk substrate region, known as the latch-up effect. This problem gave rise to the silicon-on-insulator structure to perfectly insulate the substrate region from the channel and drain/source region combined.

Another problem is the case that the distance between source and drain is now close enough to be comparable to the depletion width of the respective regions, giving rise to anomalies like drain-induced barrier lowering (DIBL), threshold voltage roll-off, channel length modulation, increase in leakage current, etc., collectively known as short-channel effects (SCEs) [1].

For the last few decades, the entire VLSI research fraternity has channelized their focus toward innovating non-conventional device structures to address the above-stated issues associated with the scaling of device dimensions at a fanatic pace predicted by Moore. Some of the popular research avenues are gate material engineering, dielectric engineering, gate engineering, etc. In gate material engineering, two or more metals of varied work functions are placed one after another, deeming an exclusive gate electrode introducing a step-like feature into the channel surface potential profile, thereby screening the potential minima from the variations in drain bias which in turn reduces the DIBL effect [2–4].

The SCEs can be controlled to a larger extent by making the gate-controlled vertical electric field dominant over the horizontal drain electric field to suppress subthreshold conduction. This can be realized by replacing the simple single gate planar MOSFET with a double gate [5], triple gate [6], surrounding gate, and FinFET [7] structures where the increased effective gate surface area improves the command over the conduction channel. In addition to this, the concept of stacked gate oxide by using high-*k* dielectric materials on top of silicon dioxide reduces the effective oxide thickness, thereby further extending the scalability of gate oxide thickness while simultaneously preventing gate oxide damage [8]. Keeping pace with the research initiatives of proposing several innovative device structures, the concept of introducing a trench gate has become quite a popular research avenue in recent times [9, 10]. As the trench separates the source and drain, the trench gate structure effectively reduces the punch-through effect, improving the subthreshold slope while ensuring that the drain depletion layer penetration is masked by the trench and does not encroach into the channel. Motivated by the inherent advantage of this trench gate concept, in this work, we have taken an initiative to assimilate the benefits of gate material engineering, gate engineering, and dielectric engineering into the novel geometry of a trench gate device and propound a dual metal double gate grooved trench MOS transistor (DMDGGT). The propounded device structure has been studied in detail using the simulation platform of SILVACO ATLAS, and the results thus obtained are presented to establish the supremacy of this newly propounded structure in overcoming SCEs and thus opening a possibility for our propounded structure to replace the existing non-conventional device structures for future low-power applications.

2 Device Schematic and Simulation Framework

The illustration of the propounded dual metal double gate grooved trench (DMDGGT) MOSFET is shown in Fig. 1. In the propound structure, we have taken a double gate to improve the gate controllability which further improves by choosing two dissimilar metals shorted as the gate electrode. This results in the reduced lateral drain field and increase the vertical gate control over the channel charge, in turn, subdued the DIBL. In the propound structure, two metals with work function 5.1 and 4.8 eV have been taken, in the direction of source to drain, respectively, one after another. For investigation, the gate length is considered to be 34 nm with a stacked gate oxide (tox1 (SiO_2) = 1 nm and tox2 (HfO_2) = 2 nm) configuration. The negative junction depth (NJD) is 4 nm with a moderate channel doping of 10^{16} cm^{-3}.

In our simulation framework, we have incorporated a conventional drift/diffusion model to include the physics of charge flow. Further, mobility models, field dependent mobility (FLDMOB) for modeling any velocity saturation effect that may occur, concentration dependent mobility (CONMOB) for simple power-law temperature dependence model, recombination model like Shockley-Read-Hall (SRH) for fixed minority carrier lifetimes, and Auger for direct transmission model of three carriers that are important at high current densities, and carrier statistics model like the bandgap narrowing (BGN) for its importance at high-density region and quantum confinement models have been incorporated to govern the physics of the device [11]. The parameters and dimensions considered for this device simulation are mentioned in Table 1 and Fig. 1 unless stated otherwise.

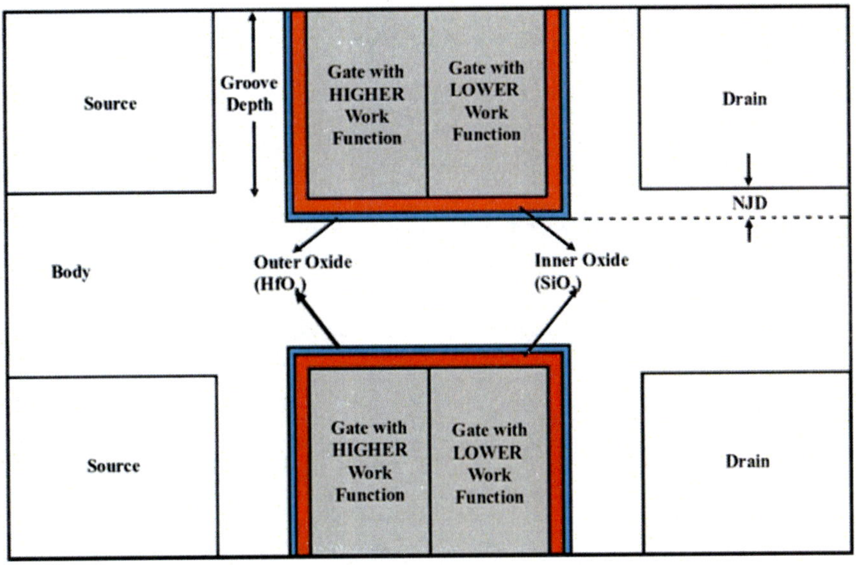

Fig. 1 Schematic of the propounded DMDGGT MOSFET structure

Table 1 Device parameters and the measurements

Device parameters	Value
Negative junction depth	4 nm
Groove depth	30 nm
Gate length	34 nm
Inner oxide thickness (tox1)	1 nm
Outer oxide thickness (tox2)	2 nm
Substrate doping (N_A)	1e16 cm^{-3}
Source/drain (N_D)	1e19 cm^{-3}
Body thickness	20 nm

3 Results and Discussion

Using TCAD simulations, we propounded a novel dual metal double gate grooved trench MOS transistor in this paper.

Figure 2 depicts that there exists a step in the dual metal double gate grooved trench MOS transistor near the source end in contrast to the single metal counterpart. This is owing to the presence of a bi-metal gate structure having varied work functions (metal having higher work function is present toward the source side). The step toward the source region screens the surface potential minima from drain bias variations which in turn significantly reduces the DIBL effect in the propounded DMDGGT device structure. Again, Fig. 3 shows that there is practically no visible change in the position of potential minima due to the drain bias variations. Moreover, in Fig. 2, the surface potential minima of the propounded DMDGGT device is shallower than that of its single gate counterpart, indicating lowering of the potential barrier, leading

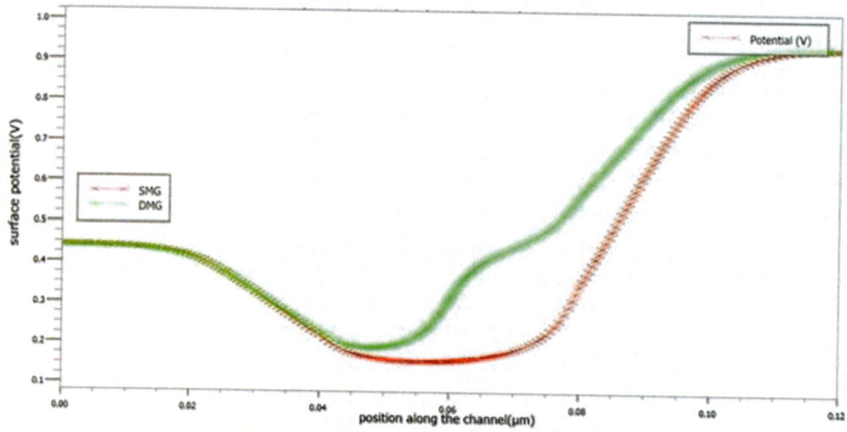

Fig. 2 Surface potential profile with respect to conduction channel length of the propounded DMDGGT transistor

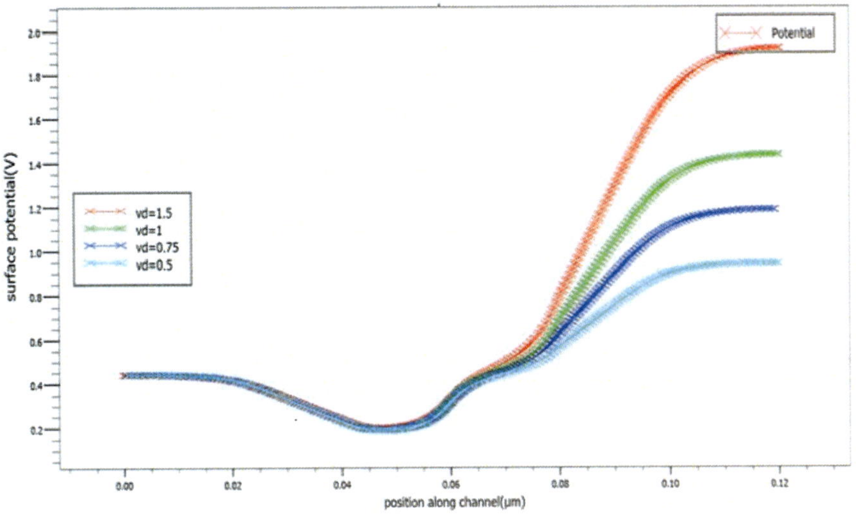

Fig. 3 Surface potential profile with respect to conduction channel length of DMDGGT structure for different drain bias

to a reduced value of threshold voltage of our propounded device, thereby having enhanced drain current and also reducing the power consumption.

Figure 4 shows the upward shift of potential with varying gate bias which indicates a reduction of the potential barrier at the source-channel junction leading to the formation of a conduction channel as gate bias is gradually increased. Moreover, a significant minimization in the peak electric field near the drain region is observed in the propounded DMDGGT structure compared to its single metal counterpart in Fig. 5. This shows the effectiveness of the propounded structure to mitigate impact ionization induced hot carrier effects. The results observed so far indicate the ability of the propounded DMDGGT device to mitigate various short-channel effects. It has already been discussed that the propounded DMDGGT structure has its potential minima position higher than that of its single gate counterpart, indicating lower threshold voltage and increased ON state drain current. The improvement in drain current is also shown in Fig. 6 which enables us to substantiate the supremacy of our propounded DMDGGT device in terms of current drivability too.

With the above-mentioned, structural and functional advantages come some practical manufacturing challenges. For a single metal gate, the propounded manufacturing challenges include the manufacturing of the trench in the very first place. The time duration between trench etching and then trench oxidation and filling must be intelligently considered, and the etched surface must not be exposed to any organic compounds such as derivatives of photoresists. Studies in this field prove that exposure to such conditions or extending the time between the trench etching and filling can cause IGSS gate failures or threshold unstableness in real-life testing [12]. Now for the dual metal gate, which is a concept propounded long back needs further steps

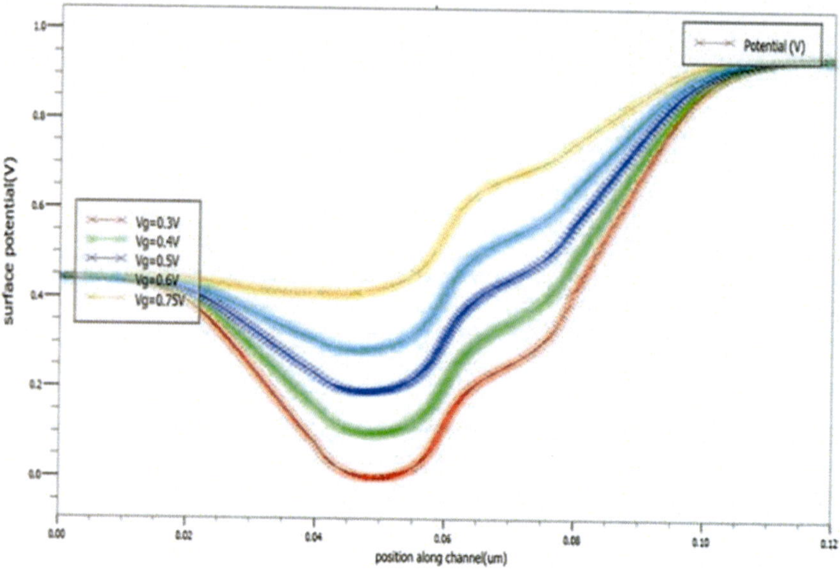

Fig. 4 Surface potential profile with respect to conduction channel length of DMDGGT structure for different gate bias

Fig. 5 Electric field profile with respect to conduction channel length of single metal (red) and dual metal (green) DGGT MOSFET

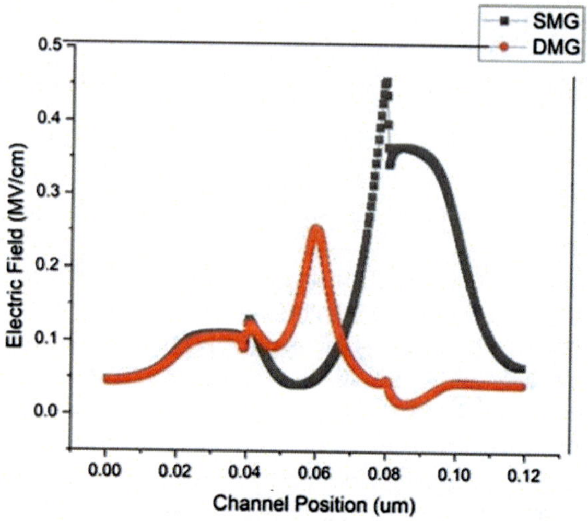

to laterally form two well-controlled contacting gate materials. Normal evaporation is used for second gate material deposition. E-beam writing and low drain doping (LDD) method can be implemented for refining the processing of the DMG structure in addition to the TAE method [13] (Table 2).

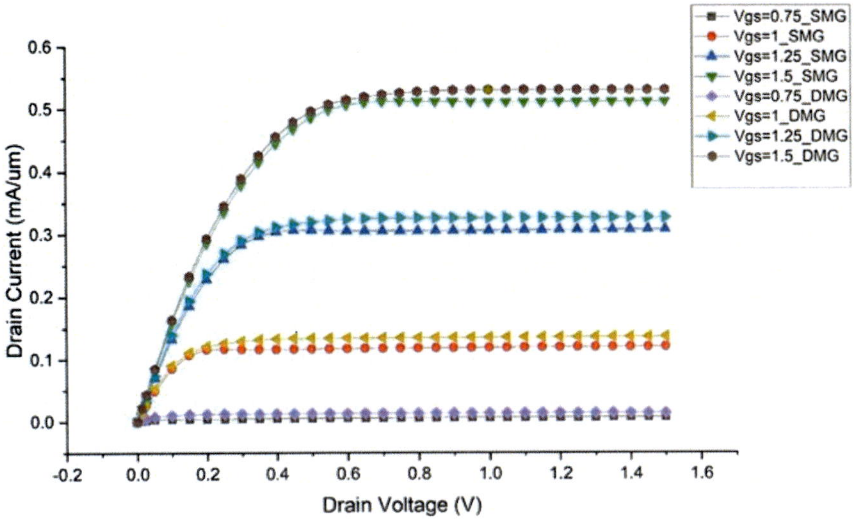

Fig. 6 I_d/V_d characteristics for different gate voltages in DMDGGT compared with similar characteristics of its single metal counterpart

Table 2 Comparison of results for different structures

Device name	Drain current (ION) (Amp)
Rectangular recessed channel SOI [14]	In the range of 10^{-3}
Up extended stepped drift SOI (UESD-SOI) [15]	In the range of 10^{-4}
The cross-sectional view of an asymmetric linearly graded work function trapezoidal gate (ASYLGTG) SOI MOSFET [16]	In the range of 10^{-4}
Dual material double gate nanoscale SOI MOSFET [17]	2×10^{-3} to 3×10^{-3}
Dual metal double gate grooved trench (DMDGGT) [propounded structure]	0.1×10^{-3} to 0.6×10^{-3}

4 Conclusion

This work presents an endeavor to conceptualize a novel dual metal double gate grooved trench transistor to wrap the dual benefits of enhanced gate control of a double gate device as well as subdued DIBL effect available from a dual metal gate electrode along with a notable increase in effective channel length under the grooved trench geometry and subsequent mitigation of SCEs. The results achieved from the simulation based on comparative analysis of the propounded DMDGGT structure with its single gate counterpart reveal significant performance improvements of the propounded DMDGGT structure according to the following parameters: subdued SCEs, excellent improvement in current characteristics by almost 13.445%, having a low threshold voltage of 0.5857 V and sufficiently improved subthreshold slope of

75 mV/dec., thereby establishing the propounded structure as a feasible alternative nano-dimensional device for applications in future.

References

1. Veeraraghavan S, Fossum JG (1989) Short-channel effects in SOI MOSFETs. IEEE Trans Electron Devices 36(3):522–528. https://doi.org/10.1109/16.19963
2. Chamberlain SG, Ramanan S (1986) Drain-induced barrier-lowering analysis in VLSI MOSFET devices using two-dimensional numerical simulations. IEEE Trans Electron Devices 33(11):1745–1753. https://doi.org/10.1109/T-ED.1986.22737
3. Lenka AS, Mishra S, Mishra SR, Banja U, Mishra GP (2017) An extensive investigation of work function modulated trapezoidal recessed channel MOSFET. Superlattices Microstruct 111. https://doi.org/10.1016/j.spmi.2017.07.043
4. Polishchuk I, Ranade P, King T-J, Hu C (2001) Dual work function metal gate CMOS technology using metal inter diffusion. IEEE Electron Device Lett 22(9):444–446. https://doi.org/10.1109/55.944334
5. Kang H, Han J, Choi Y (2008) Analytical threshold voltage model for double-gate MOSFETs with localized charges. IEEE Electron Device Lett 29(8):927–930. https://doi.org/10.1109/LED.2008.2000965
6. Park J-T, Colinge J (2002) Multiple-gate SOI MOSFETs: device design guidelines. IEEE Trans Electron Devices 49(12):2222–2229. https://doi.org/10.1109/TED.2002.805634
7. Jurczak M, Collaert N, Veloso A, Hoffmann T, Biesemans S (2009) Review of FINFET technology. IEEE Int SOI Conf 2009:1–4. https://doi.org/10.1109/SOI.2009.5318794
8. Razavi P, Orouji A (2008) Dual material gate oxide stack symmetric double-gate MOSFET: improving short channel effects of nanoscale double-gate MOSFET. In: 2008 11th international biennial baltic electronics conference, pp 83–86. https://doi.org/10.1109/BEC.2008.4657483
9. Mishra S et al (2020) Sub-threshold performance analysis of multi-layered trapezoidal trench gate silicon on nothing MOSFET for low power applications. In: 2020 IEEE VLSI device circuit and system (VLSI DCS). IEEE, pp 214–218
10. Mishra S, Bhanja U, Mishra GP (2019) Impact of structural parameters on DC performance of recessed channel SOI-MOSFET. Int J Nanoparticles 11(2):140–153
11. Atlas User's Manual: device simulation software, SILVACO Int. 2015
12. Williams RK, Darwish MN, Blanchard RA, Siemieniec R, Rutter P, Kawaguchi Y (2017) The trench power MOSFET: Part I—history, technology, and prospects. IEEE Trans Electron Devices 64(3)
13. Long W, Ou H, Kuo JM, Chin KK (1999) Dual-material gate (DMG) field effect transistor. IEEE Trans Electron Devices 46(5)
14. Singh M, Mishra S, Mohanty SS, Mishra GP (2016) Performance analysis of SOI MOSFET with the rectangular recessed channel. Adv Nat Sci Nanosci Nanotechnol 7(1):015010
15. Saremi M, Arena M, Niazi H, Saremi M, Goharrizi AY (2017) SOI LDMOSFET with up and down extended stepped drift region. J Electron Mater 46(10):5570–5576
16. Mishra S, Mishra GP (2019) Influence of structural parameters on the behaviour of an asymmetric linearly graded work-function trapezoidal gate SOI MOSFET. J Electron Mater 48(10):6607–6616
17. Venkateshwar Reddy G, Jagadesh Kumar M (2005) A new dual-material double-gate (DMDG) nanoscale SOI MOSFET—two-dimensional analytical modeling and simulation. IEEE Trans Nano-Technol 4(2)

Application of Undoped ZnS Nanoparticles for Rapid Detection of *E. coli* by Fabricating a Mem-Mode Device Sensor After Conjugating Antibody

Himadri Duwarah, Neelotpal Sharma, Kandarpa Kumar Saikia, and Pranayee Datta

Abstract Zinc sulphide (ZnS) nanoparticles are known for their biological sensors. Conjugation of *E. coli* bacteria to these ZnS nanosamples are done as well as these bacteria are sensed on dried samples based on memristor-based property. We have varied the concentration of bacteria, and as a result, a variation of current occurs like a hysteresis (Figure eight pattern) loop. ZnS is prepared successfully by the chemical precipitation method. Characterization was done by various techniques to confirm its specifics. In this study, the conjugation of *E. coli* with the antibody-conjugated nanoparticle is confirmed by taking absorbance before and after adding of *E. coli* in UV–Vis spectroscopy that results in a significant shift in the wavelength of absorption. Electrical characterization leads to a change in voltage gap among different molar concentration of bacterial conjugated ZnS samples. Results show that process basically leads to a mem-mode observation (either memresistive, memcapacitive or meminductive) in nature. Almost orderly declining pattern of voltage gap with bacterial *E. coli* concentration is obtained for the as-fabricated devices using ZnS quantum dot in bacterial estimation after conjugating antibody *E. coli*.

Keywords Voltage gap · Mem-mode · Conjugation · Antibody · Hysteresis · Nanodevice

H. Duwarah (✉) · P. Datta
Department of Electronics and Communication Technology, Gauhati University, Guwahati, Assam, India
e-mail: himadriduwarah@yahoo.co.in

N. Sharma · K. K. Saikia
Department of BioEngineering and Technology, Gauhati University, Guwahati, Assam, India

S. Basu et al. (eds.), *Proceedings of International Conference on Frontiers in Computing and Systems*, Lecture Notes in Networks and Systems 404,
https://doi.org/10.1007/978-981-19-0105-8_51

1 Introduction

It has been found that zinc sulphide (ZnS) is the one of the important II-IV compound semiconductors having a wide band gap (3.54 eV) and a large exciton binding energy [1]. Moreover, ZnS semiconductor nanoparticles are among one of the most studied classes of nanoparticles because of its wide range of application in optical sensor, solid-state solar window layers, photoconductors, phosphors and catalysts. ZnS is prepared successfully by the chemical precipitation method [2]. *E. coli* was grown in nutrient broth, and the nanoparticle with different molar concentration (0.1–0.3M) is added on the bacterial broth culture. In this study, the conjugation of *E. coli* with the antibody-conjugated nanoparticles is confirmed by taking absorbance before and after adding of *E. coli* in UV–Vis spectroscopy that results in a significant shift in the wavelength of absorption. Also SEM images were taken before and after the addition of *E. coli* in the antibody-conjugated nanoparticle confirms the nanostructures as well as presence of ***E. coli***. Earlier study tells that ***E. coli*** to be the best bacterial indicator of faecal pollution in drinking water [3]. Study found that there are different transduction methods like fluorescence, surface plasmon resonance, surface-enhanced Raman spectroscopy used through which detection limit of such bacterial sample ranges from minimum 1 cell to maximum as 10^4 cfu/ml, but still there arise some problems like non-specific binding, particle size variation, nanoparticle aggregation and nanoparticle stability in sensing [4–6]. These lead to develop some other signal transduction methods which may be expected to be less sensitive to the above variations. Therefore, in this regard, a novel approach is adopted based on the evergreen experiment by Strukov et al. at HP Labs in 2008 [5, 6] which led to realize "Memristor (MR)" proclaimed by Leon Chua in 1971. Carra et al., in 2012, presented their in-depth investigations on memristive devices based on silicon nanowires and functionalized with rabbit antibodies in order to sense antigens [7, 8]. They observed that memristive (MR) behaviour [8] changes with antibody–antigen interaction. Here, an attempt is made to estimate or detect ***E. coli*** bacteria using a mem-mode (memristive, memcapacitive or inductive) of nanodevice based on zinc sulphide nanoparticles using antibody of *E. coli*.

2 Experimental procedure

2.1 Chemicals and Reagents

The chemicals used were of pure and best suited for analytical experiments. Zinc acetate, sodium sulphide, PVA and nutrient agar were purchased from Sigma-Aldrich. Department of Bio Engineering, IST, Gauhati University, provides microbial strain *Escherichia coli* as well as *E. coli* antibody is bought from Abcam which is conjugated with the nanoparticles as to make specific binding with the *E. coli* bacteria and to make a detection device for *E. coli* in any sample of concern.

2.2 *Fixation of* E. coli *Cells*

A pure colony of Escherichia coli (*E. coli*) cells from Petri plate is inoculated and then cultivated in nutrient broth. Then incubation of the cells in broth is done at 37 °C, for 24 h. The cells are harvested and are rinsed with 0.01M phosphate buffered saline (PBS), and thereafter, fixation is done with ethanol/acetic acid (3:1) solution. The fixed cells then tried to attain an absorbance of 0.9 at 420 nm [9]. These cells are then stored for further experiments. Four different dilution concentrations of (***E. coli***) i.e. 10^{-1} to10^{-4} are made by the same procedure by Himadri et al. [10].

2.3 *Bioconjugation of QDs with Antibody* E. coli

EDC/NHS chemistry is used for the preparation of ZnS QD polyclonal *E. coli* antibody conjugation. Before conjugation, 40 μl N-Hydroxysuccinimide and 40 μl 1-ethyl-3-(3-dimethylaminopropyl) carbodiimide (EDC) are added to 1 ml of ZnS nanosample [11]. The prepared solution was then shaken for two hours duration to initiate the coupling reaction at room temperature. The solution was then centrifuged at 4500 rpm for 15 min, discarded the supernatant and collected the pellet.

2.4 *Detection of* E. coli *with the Antibody-Conjugated Nanoparticles*

The fixed *E. coli* cells were added in the antibody-conjugated nanoparticles. The solution was then incubated at room temperature for about 2 *h*. After incubation, the conjugation of *E. coli* cells with the antibody-conjugated nanoparticles was analysed via taking SEM images and by taking UV spectroscopy to observe any shift of peak prior and after treating the antibody-conjugated nanoparticles with *E. coli* cells. Rabbit anti-***Escherichia coli*** **antibody** recognizes **Escherichia coli** and is basically reactive with all somatic and capsular (O and K) antigenic serotypes.

3 Scanning Electron Microscopy (SEM)

SEM has been done to examine the topography and the composition of the nanoparticles prepared via the chemical precipitation method. It is also used to analyse the binding of *E. coli* bacteria with the antibody-conjugated nanoparticles. EDAX has also done for the detection of elemental composition of the nanoparticles. SEM image of ZnS nanorods before bacterial conjugation is shown in Fig. 1. EDAX of ZnS nanosamples with *E. coli* in phosphate buffer saline (PBS) solution is also shown

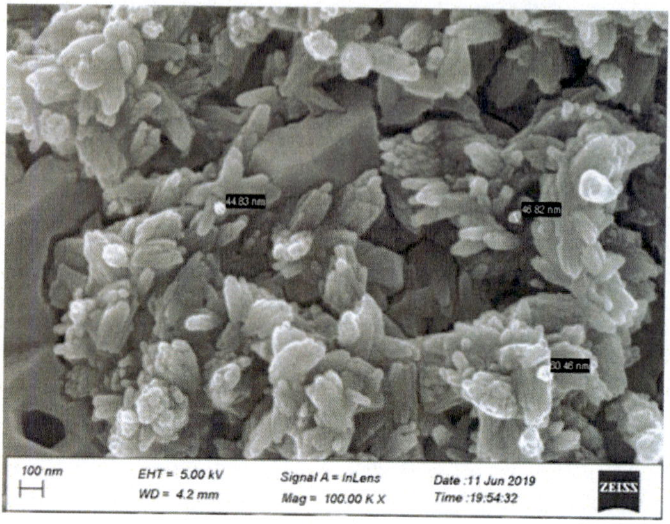

Fig. 1 SEM image of ZnS nanorods fore bacterial conjugation

in Fig. 2 as well as ***E. coli***-conjugated ZnS nanoparticles are shown in Fig. 3. SEM images show that prepared ZnS nanorods were regular in shape, and distribution is even in nature. The diameter of nanorods was smaller than 50 nm. The capping agent covalently is bound to the surface of the nanorods via disulphide bond and reduces the surface tension [12]. SEM images of ZnS tagged anti-rabbit *E. coli* polyclonal antibody conjugation are shown in Fig. 4. These are known as bionanoconjugates or revealing antibody [12].

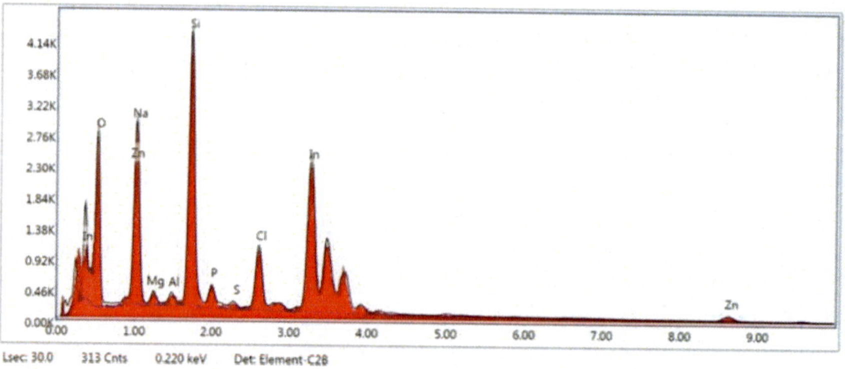

Fig. 2 EDAX of ZnS nanosamples with *E. coli* in PBS solution

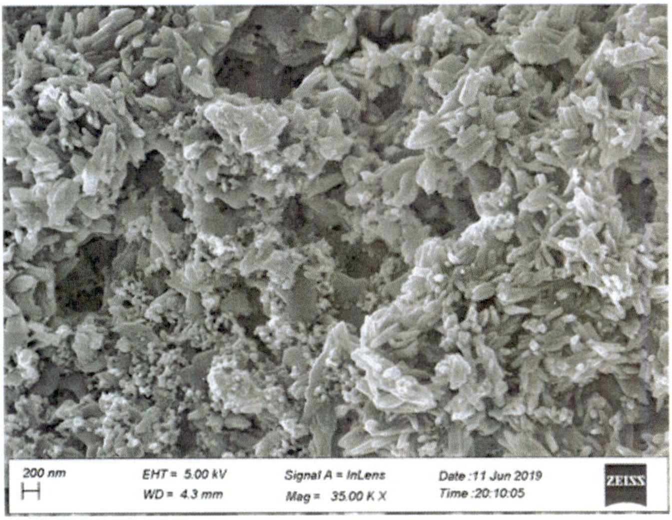

Fig. 3 SEM of ***E. coli***-conjugated ZnS nanoparticles

Fig. 4 SEM of antibody ***E. coli***-conjugated ZnS nanoparticles

4 E. *c*oli with the Antibody-Conjugated Nanoparticles

Here, the antibody-conjugated ZnS nanoparticles (Ab-ZnS) are used as to capture the *E. coli* cells which can help to confirm the correct sensing of *E. coli* in any sample, be it in water sample or human urine sample. The SEM image in Fig. 4 shows the presence of *E. coli* bacteria successfully bound with the antibody-conjugated nanoparticles as well as change in UV-visible spectra in Fig 5.

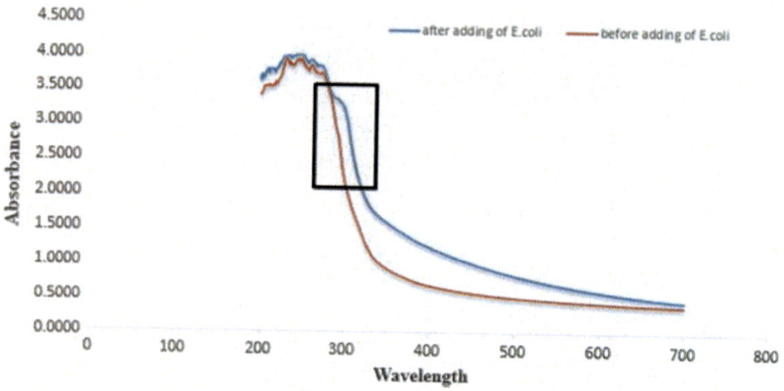

Fig. 5 UV–Vis spectra for before and after conjugation of *E. coli* to ZnS nanosamples

5 Experimental Observations

The proposed ZnS nanosamples of molarity 0.1M ($S^{0.1}$) are mixed with varying dilution concentrations (10^{-1}, 10^{-2}, 10^{-3}, 10^{-4}) of ***E. coli*** for 2 h at room temperature. Thereafter, an amount of 30 μl of bacteria mixed sample is taken, a sandwiched like device Cu/ZnS nanosample-bacteria-anibody/ITO is fabricated, and I–V characteristics are measured by using Keithley 2450 Sourcemeter up to ±5 V within 10 μs. Experiments are conducted based on the molarity of the sample with concentration at fixed temperature and rotation speed. In this way, four devices (D_1, D_2, D_3, D_4) of molarity 0.1M are fabricated to sense antigen.

6 Results and Discussion

Thus, fabricated devices are tested for application as biosensor using circuit as shown in Fig. 6. The electrical characterization is done at room temperature and controlled humid environment. The change in current is observed for changing voltages in forward as well as in backward directions. Current flows through the non-conductive polymer matrix due to tunnelling [13]. Figure 7 shows pinched hysteresis loop in bare ZnS nanosample after electrical characterization. The effect of biofunctionalization, i.e. antibody-conjugated ZnS nanodevice, is found distinct in logarithmic graphs as voltage gaps appearing in Fig. 11 for 0.1 M ZnS nanosample. Memristive effect of the fabricated nanodevice (D_0) as shown in Fig. 8 is confirmed by the pinched hysterics loop ("Figure Eight" pattern) in bare quantum dot. This characteristic is altered after bio-functionalization of the device (giving memcapacitive or meminductive behaviour i.e. not pinching at zero). Bacterial pathogen ***E. coli*** provides a group of charged particles and that leads to develop a gate-like structure by producing an electric field around the source–drain channel of the biosensor [9, 14]. This leads to

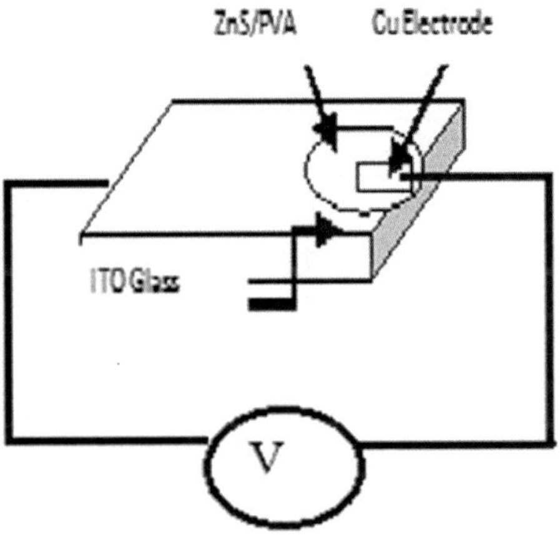

Fig. 6 Schematic circuitdiagram of proposed device for current–voltage characteristics

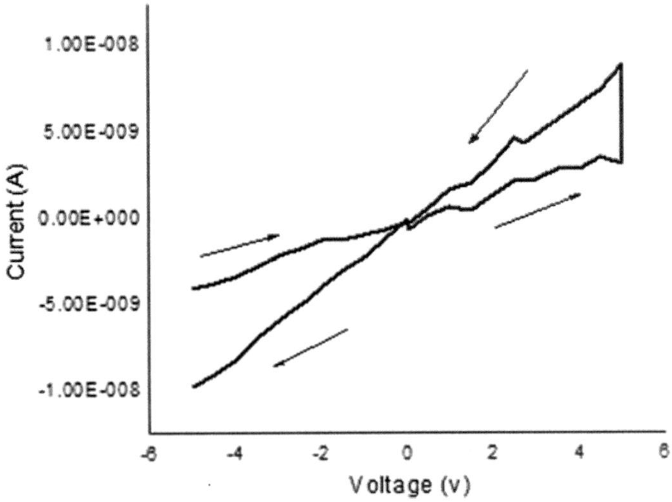

Fig. 7 Bare Zns nanosample (Device D_0)

a difference in voltages in between the minimas for forward and backward current (voltage gap) as shown in Fig. 11

7 Conclusion

An experimental relationship is established between voltage gap and concentrations of bacteria in the nanocomposite as we conjugate *E. coli* antibody. Specifically,

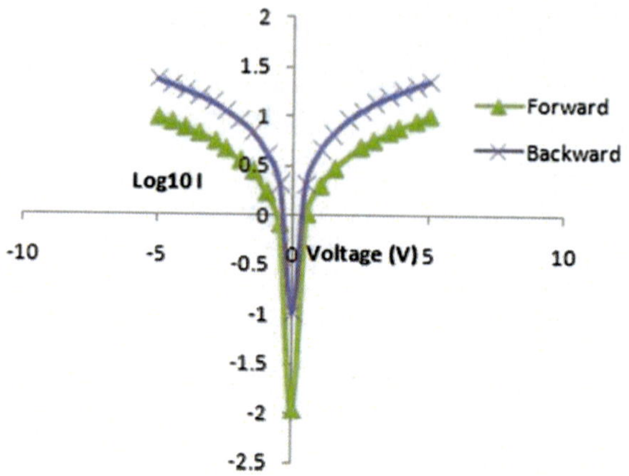

Fig. 8 Current (*I*) versus voltage (*V*) characteristics of bare Zns nanosample with antibody *E. coli*

detection of bacteria can be observed after conjugating antibody. It has also been found that voltage gap of neutral solution (PBS) with antibody shows zero value as it is seen in ZnS nanosample with *E. coli* at zero concentration of bacteria (Fig. 9). Earlier research shows that voltage gap increases as increase in *E. coli* bacterial concentration increases [6]. Here the study shows the opposite trend, as antibody *E. coli* binds specifically *E. coli* antigen, hence detects *E. coli* bacteria, and voltage gap decreases as *E. coli* concentration increases, i.e. antibody tries to neutralize/act

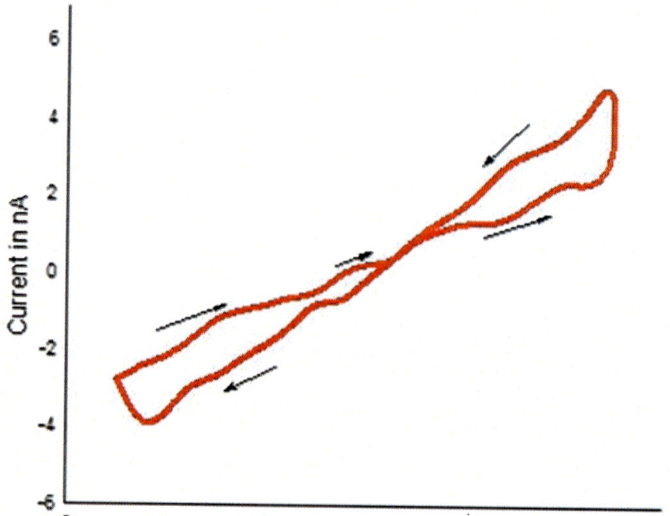

Fig. 9 Log *I* versus *V* curve of antibody-conjugated Zns nanosample with PBS solution

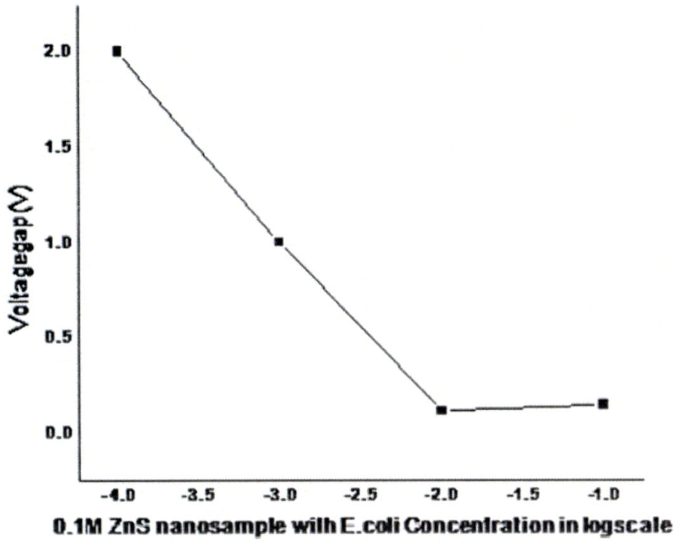

Fig. 10 Voltage gap versus Ab-*E. coli* concentration of Zns nanosample (in log scale) for device 1

on antigen [8, 15, 16]. These observations can be found from Fig. 10, where change in voltage gap with respect to concentration of *E. coli*-conjugated Ab-ZnS samples is given. Thus, the simple fabricated undoped ZnS nanodevice can detect *E. coli* bacteria in the samples by mem-mode observation. From Fig. 10, we can get the minimum sensitivity as $S = 2$ V/0.0001 ml. The device is also stable over more than 200 μs (maximum delay for 50 μs per cycle) duration.

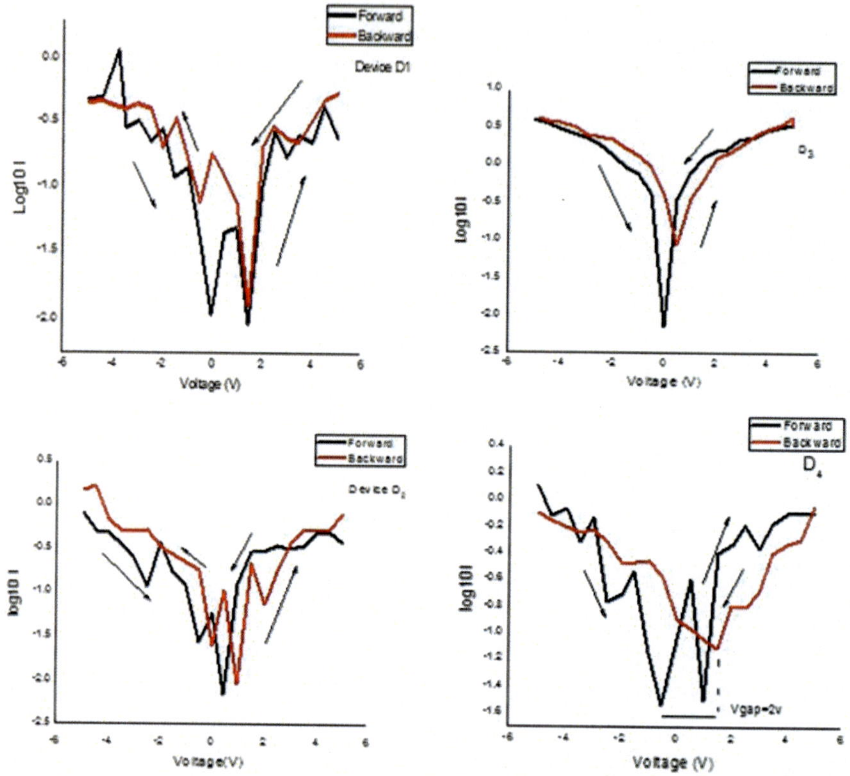

Fig. 11 Log I versus V curve of antibody-conjugated 0.1M ZnS nanosample with $10^{-1}/10^{-2}/10^{-3}/10^{-4}$ *E. coli* concentration (Device $D_1/D_2/D_3/D_4$)

References

1. Chen B, Zhong HZ, Li R, Zhou Y, Ding Y, Li YF et al (2012) Sci Adv Mater 4:342–345
2. Kaur N, Kaur S, Singh S, Rawat M (2016) J Bioelectron Nanotechno 1(1)
3. Odonkor ST, Ampofo J (2013) Microbiol Res 4:e2
4. Vikesland PJ, Wigginton KB (2010) J Am Chem Soc Environ Sci Technol 44(10):3656–3669
5. Chua LO, Kang SM (1971) IEEE Trans Circ theory 18:507
6. Duwarah H, Sharma N, Devi J, Saikia KK, Datta P (2021) Mater Today Proc
7. Sacchetto D, Doucey MA, Micheli GD, Leblebici Y, Carrara S (2011) Bio NanoSci 1:1–3
8. Carrara S, Sacchetto D, Doucey MA, Baj-Rossi C, Micheli GD, Leblebici Y (2012) J Sens Actuators B 171–172:449–457
9. Barua S, Ortinero C, Shipin O, Dutta J (2012) J Fluoresce 22:403–408
10. Duwarah H, Devi J, Sharma N, Saikia KK, Datta P (2019) 2nd international conference on innovations in electronics, signal processing and communication (IESC)
11. Vo NT, Ngo HD, Vu DL, Duong AP, Lam QV (2015) J Nanomaterials
12. Sarma A, Rao VK, Kamboj DV, Gaur R, Upadhyay S, Shaik M (2015) Biotechnol Rep, 129–136
13. Bhadra R, Singh VN, Mehta B, Datta P (2009) Chalcogenide Lett 6(5):189–196
14. Puppo F, Doucey MA, Ventra M, Micheli GD, Carrara S (2014) IEEE Trans 2257–2260
15. Antigen and antibody interaction—immunology. Study material, lecturing notes, Assignment, reference, Wiki description explanation, brief data
16. Izumi K, Tsumoto K, Tsumoto https://doi.org/10.1038/npg.els.0001117

Swift Sort: A New Divide and Conquer Approach-Based Sorting Algorithm

Tanmoy Chaku, Abhirup Ray, Malay Kule, and Dipak Kumar Kole

Abstract Sorting implies the task of presenting a specific type of data in a specific order. These tasks are getting accomplished by different algorithms proposed by researchers. Researchers are trying to achieve this task in minimal space and time complexity with improved stability, correctness, finiteness and effectiveness. Sorting is used in wide range of fields namely, in Operating systems, Data Base Management systems, in searching and in various other data science related areas. In this paper, a divide and conquer approach-based algorithm is proposed to sort the data in a specific order using min–max searching. The time complexity of the proposed Swift Sort algorithm is $O(n\log n)$ and $O(n^2)$ in the average and worst cases, respectively. Moreover, time complexity of the proposed algorithm is comparable to Quick Sort, Merge Sort, Heap Sort and TimSort but at the same time Randomised Quick Sort, Merge Sort and Heap Sort produces a better Time Complexity in their worst cases than Swift Sort. The experimental results prove the correctness of the proposed algorithm.

Keywords Sorting · Swift sort · Time complexity · Space complexity · Divide and conquer · Data science

1 Introduction

An algorithm that is used for rearranging a set of unordered items into a finite sequence or lexicographical ordering, lowest to highest or vice versa is called as a sorting algorithm [1]. Given a set of items as input, the algorithm returns the user an ordered set of items. These sorting algorithms are used in different types of

T. Chaku · D. K. Kole
Jalpaiguri Government Engineering College, Jalpaiguri, West Bengal, India

A. Ray (✉)
RCC Institute of Information Technology, Kolkata, West Bengal, India
e-mail: abhirupray14@gmail.com

M. Kule
Indian Institute of Engineering Science and Technology, Shibpur, Howrah, West Bengal, India

S. Basu et al. (eds.), *Proceedings of International Conference on Frontiers in Computing and Systems*, Lecture Notes in Networks and Systems 404,
https://doi.org/10.1007/978-981-19-0105-8_52

applications like in online shopping websites that are used to sort a specific type of items from price range of high to low. Even if in case of searching on any search engine, it provides results on an order of most visits for a certain period of time. Moreover, in a huge database, efficient searching requires data to be kept in sorted order. There are two types of sorting, namely, Internal Sorting [2] and External Sorting [3], based on extra space required to sort a set of data. Internal sorting is the type of sorting technique that store the sorted dataset into the same input array [2] and external sorting [4] is defined as the type of sorting in which the sorted set of data are store into a newly initialised array [3].

In this paper, we consider several other comparable sorting techniques [5] like Merge Sort [6], Quick Sort [7], Insertion Sort, Bubble Sort, Heap Sort [8] and Tim Sort [9] to compare and analyse our experimental results for different input size of the array.

In this work, we propose a new sorting approach, where swapping is not done as it consumes a certain amount of memory. Moreover, this algorithm sorts the data in much faster way than other comparable algorithms mentioned above as size of array increases. Again, our proposed algorithm works fastest with the growth of number of duplicate elements within the input array.

This paper is organised in four sections. Section 2 elaborates the proposed sorting algorithm followed by this introductory section, Sect. 3 Describes experimental results and analysis, Sect. 4 concludes the proposed work.

2 Swift Sort: The Proposed Sorting Method

2.1 *Description of the Working Procedure of the Proposed Algorithm*

At first, as shown in Fig. 1, an array is taken as input. Here the variable *max* is defined as the maximum valued data present in the array and the variable *min* is defined as the minimum valued data present in the array. This type of sorting algorithm can be called as max-min sorting algorithm [10, 11] where at each and every iteration we are performing a search for finding the maximum and minimum element from the array. Here, the integer variable *div* is computed by taking average of the *max* and *min* $\left(div = \frac{(max+min)}{2}\right)$. An array named *Result* is the vacant array whose length will be equal to that of the *input* array.

Steps

Step 1: At first, all the values equal to max are added at the first vacant cells of the result array.

Step 2: All the values equal to min are added at the last vacant cells of the result array.

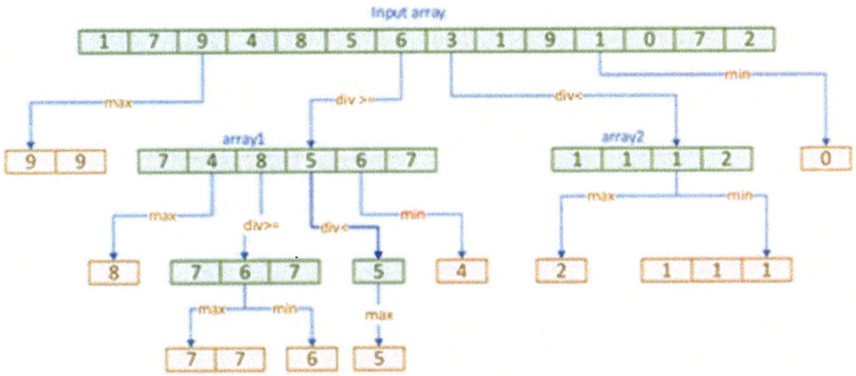

Fig. 1 Working procedure of swift sort (proposed)

Step 3: A new array is created and named array1 which will contain all the elements greater than or equal to the div variable.
Step 4: A new array is created and named array2 which will contain all the elements less than that of div variable.
Step 5: The above four steps are repeated for both of array1 and array2.

2.2 *Pseudocode of Swift Sort*

Firstly, the algorithm consists of three arguments namely, *array*, *max* and *min* where.

- *Array*—Contains all the elements of the input array.
- *Max*—Contains maximum value of the input array.
- *Min*—Contains minimum value of the input array.

Variables used within the algorithm are described below.

- *div*—It stores the computed average value of max and min.
- *Index*—Index ranges from zero to length of the array and it will be incremented by one within the loop.
- *array1[]*—It stores the elements that have values greater than or equal to *div*. "*array1*[](append)" means that the elements are getting added at the end of the array.
- *array2[]*—It stores the elements that have values lesser than *div*. "*array2*[](append)" means that the elements are getting added at the end of the array.
- *temp[]*—It stores all the minimum valued elements present in the array. "*temp[]*(append)" means that the elements are getting added at the end of the array.
- *result[]*—It is the final sorted array. "*result[]*(append)" means that the elements are getting added at the end of the array.

Some basic functions used in the proposed algorithm are described below.

- *length()*—It computes the length of the array.
- *max()*—It computes the maximum valued element from the array.
- *min()*—It computes the minimum valued element from the array.

Pseudo-Code of Swift Sort

Input: array, max, min

Output: result[]

```
SwiftSort(array, max, min):
div← (max+min)/2
for index in range(0 to length(array))
        if array[index] = max then
                result[](append) ← array[index]
        else if array[index] >= div then
                array1[] (append) ← array[index]
        else if array[index] <div then
                array2[] (append) ← array[index]
        else
                temp[] (append) ← array[index]
        end if
end for
if length(array1) > 0 then
        SwiftSort(array1, max(array1), min(array1))
end if
if length(array2) > 0 then
        SwiftSort(array2, max(array2), min(array2))
result[](append) ← temp[]
```

3 Results and Analysis

3.1 Experimental Results

This experiment has been done in a computer with core i5 9th generation CPU and 8 GB 4th generation 26,000 MHz RAM using python 3.8 programming language. During the execution of the experiment, 20% CPU has been used.

In this experiment arrays are created using numpy array of Python 3.8. The default max and min function of python has been used here.

Some of the other sorting algorithms are considered to compare the results of Swift Sort. Here, the experimental results are obtained by comparing Swift Sort with Quick Sort, Merge Sort, TimSort, Bubble Sort and Insertion Sort. The average time complexities of Quick Sort, Merge Sort, TimSort and the proposed algorithm are same and each of them are equal to $\Theta(n\log n)$. The best case of TimSort and proposed algorithm is same and equal to $\Omega(n)$. The worst case of Quick sort and proposed algorithm is same and equal to $O(n^2)$.

Firstly, an array of size n (where n = 100, 500, 1000, 5000, 10,000, 50,000, 100,000, 500,000, 1,000,000) is created randomly by taking values ranging from 0 to 10,000,000,000. This array is used in each and every individual algorithm to compute sorting time in seconds. This process is repeated 1000 times for each and every input of size n and thereby calculating the average time required by every algorithm is given in Table 1. Figure 2 graphically represents the array size versus time comparison of different algorithm.

When input size is 100, then all the algorithms except bubble and insertion sort are taking comparable time to sort the array of given size. Swift Sort is taking less time compared to all other algorithms with increase of array size.

Table 1 Execution time (in second) based on input size

Input size	Swift sort (proposed)	Quick sort [7]	Merge sort [6]	Timsort [9]	Heap sort [8]
100	0.000326	0.000298	0.000361	0.000556	0.000633
500	0.001667	0.001958	0.002122	0.003505	0.004277
1000	0.003317	0.004282	0.004460	0.007192	0.009221
5000	0.019162	0.027049	0.027297	0.032697	0.059934
10,000	0.042502	0.061135	0.061376	0.072612	0.136261
50,000	0.231324	0.349136	0.340734	0.432077	0.795340
100,000	0.487364	0.749016	0.722906	0.911002	1.710103
500,000	2.816433	4.303240	4.045736	5.441350	9.810348
1,000,000	9.810348	9.016068	8.592704	11.262387	20.884943
5,000,000	39.672084	80.091238	47.717037	53.845601	118.087293
10,000,000	72.269198	105.388523	98.330298	111.013686	246.655054

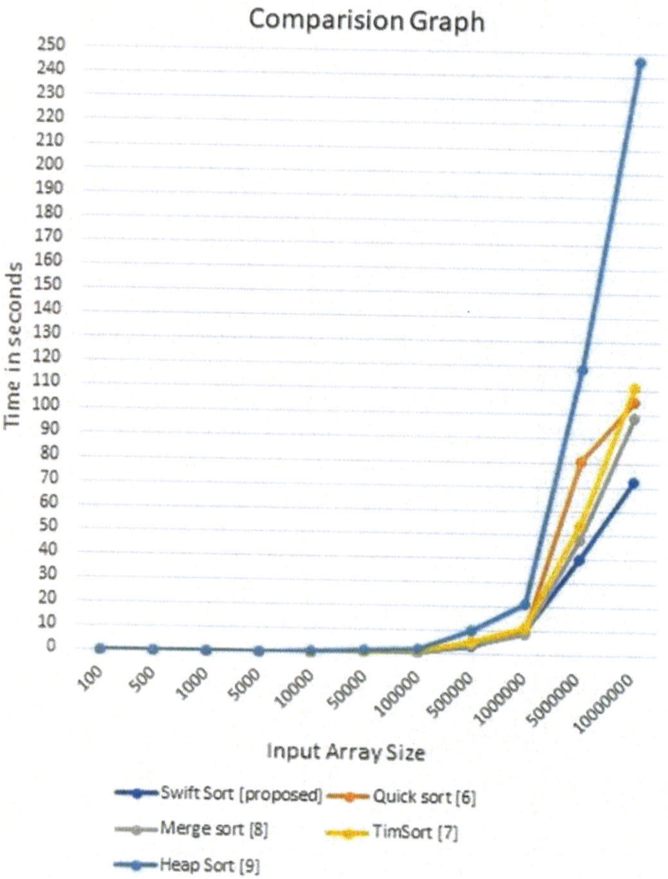

Fig. 2 Array size versus time comparison of different algorithms

3.2 *Time Complexity Analysis*

Best Case: This algorithm works best and takes least time when all the elements in an n-sized array is same. For this case all the elements will get stored in max and the array will get sorted in $\Omega(n)$ time.

So, $T(n) = \Omega(n)$.

Average Case: This case will occur when the two arrays after every iteration will contain equal number of elements.

$T(n) = 2 * T((n - 2)/2) + n.$

This $(n - 2)$ is considered as n because when n is very large then, $(n - 2) \approx n$.

So,

$$\begin{aligned} T(n) &= 2 * T(n/2) + n \\ &= 2 * [2 * T(n/4) + n/2] + n \\ &= 2^2 * T\left(n/2^2\right) + 2n \\ &= 2^3 * T\left(n/3^3\right) + 3n \\ &= 2^k * T\left(n/2^k\right) + kn \end{aligned}$$

Let, $T(1) = 1$.

So $n/2^k = 1$, when $k = \log_2 n$.

Therefore,

$$\begin{aligned} T(n) &= n + n + n + \cdots \left(\log_2 n \text{ times}\right) \\ &= n * \log_2 n \end{aligned}$$

$$T(n) = O(n\log_2 n)$$

Worst Case: This algorithm will take maximum time when array containing elements lesser than div will contain no elements and the array containing elements greater than or equal to div will contain all the remaining element except maximum and minimum for every iteration or when array containing elements greater than or equal to div will contain no elements and the array containing elements lesser than div will contain all the remaining element except maximum and minimum for every iteration.

So therefore,

Let, $T(0) = 0$

$$\begin{aligned} T(n) &= T(0) + T(n-2) + n \\ &= 0 + T(n-4) + (n-2) + n \\ &= 0 + 2 + 4 + 6 + \cdots + (n-2) + n \\ &= 2 * [1 + 2 + 3 + \cdots + (n/2)] \\ &= 2 * (n/2) \times ((n/2 + 1)/2) \\ &= \left(n^2 + 2 * n/4\right) \end{aligned}$$

$$T(n) = O\left(n^2\right)$$

Table 2 Performance-based analysis on time and space complexity, stability and inplace sorting

Sorting algorithm	Time complexity			Space complexity (worst)	Stable	Inplace
	Best	Average	Worst			
Swift sort (proposed)	$\Omega(n)$	$\Theta(n\log_2 n)$	$O(n^2)$	$O(n)$	Yes	No
Quick sort [7]	$\Omega(n)$	$\Theta(n\log_2 n)$	$O(n^2)$	$O(\log_2 n)$	No	Yes
Merge sort [6]	$(n\log_2 n)$	$\Theta(n\log_2 n)$	$O(n\log_2 n)$	$O(n)$	Yes	No
Timsort [9]	$\Omega(n)$	$\Theta(n\log_2 n)$	$O(n\log_2 n)$	$O(n)$	Yes	Yes
Inserting sort [12]	$\Omega(n)$	$\Theta(n^2)$	$O(n^2)$	$O(1)$	Yes	Yes
Bubble sort [13]	$\Omega(n)$	$\Theta(n^2)$	$O(n^2)$	$O(1)$	Yes	Yes

3.3 *Space Complexity Analysis*

We know that Total space = Extra Space + Stack Space.

Extra Space: Extra Space is defined as the extra space required to store the newly formed sorted array which is initialised within the algorithm.

Stack Space: Stack Space is the space required for the recursive calls implemented inside the algorithm.

Therefore,

$$S(n) = n + \log_2 n$$
$$S(n) = O(n)$$

Table 2 shows the comparitive studies of time and space complexities of different sorting algorithms, where it is observed that the proposed algorithm is comparable to other popular algorithms.

4 Conclusion

A sorting technique can be defined as good, based on certain parameters like definiteness, finiteness, correctness, low time complexity, space complexity and simplicity. This proposed Swift Sort algorithm produces a time complexity of $O(n\log_2 n)$ in average cases which is comparable to the average case time complexity of renowned algorithms like quick sort and merge sort. Moreover, this work can further be extended to find out more optimised and simple approach that can be applied in real life aplications.

References

1. Python Program for Heap Sort. https://www.geeksforgeeks.org/python-program-for-heap-sort/. Accessed 18 Nov 2020
2. Sorting Algorithms. https://www.geeksforgeeks.org/sorting-algorithms/. Accessed 12 Oct 2020
3. Internal Sort. https://en.wikipedia.org/wiki/Internal_sort. Accessed 11 Nov 2020
4. External Sorting. https://en.wikipedia.org/wiki/External_sorting. Accessed 25 Oct 2020
5. Data Structure—Sorting Techniques. https://www.tutorialspoint.com/data-structures-algorithms/sortingalgorithms.htm. Accessed 22 Nov 2020
6. https://www.geeksforgeeks.org/timsort/
7. Python Program for QuickSort. https://www.geeksforgeeks.org/python-program-for-quicksort/. Accessed 25 Nov 2020
8. Python Program for Merge Sort. https://www.geeksforgeeks.org/python-program-for-merge-sort/. Accessed 15 Nov 2020
9. TimSort. Accessed 01 Dec 2020
10. Paira S, Chandra S (2014) Max min sorting algorithm: a new sorting approach. Int J Technol Explor Learn
11. Chharchhodawala M, Mendapara B (2013) Min-max selection sort algorithm-improved version of selection sort. Int J Adv Res Comput Sci Softw Eng 2(5). IJERTV2IS50210
12. Cormen TH, Leiserson CE (2009) Introduction to algorithms, 3rd edn. The MIT Press, Cambridge
13. CormenTH, Leiserson CE, Rivest RL, Stein C (2003) Introduction to algorithms, 2nd edn. Paperback

Natural Language Processing

Speech Recognition System of Spoken Isolated Digit in Standard Khasi Dialect

Fairriky Rynjah, Bronson Syiem, and L. J. Singh

Abstract This paper aims to analyze the performance of a speaker independent spoken isolated digit recognition system in standard (Sohra) dialect of Khasi Language. Initially, we have prepared an ideal set of word-based digit pronunciations of 15 digits for the said dialect. The standard approach was used to collect speech data in an open room using Zoom H4N handy portable digital recorder from native language speakers of diverse age groups and gender. Each spoken isolated data has been processed and segmented using a WaveSurfer software tool. Sampling frequency was set at 16 kHz and bit resolution at 16 bits per sample. We extracted the relevant speech feature vectors using Mel-frequency cepstral coefficient (MFCC) and perceptual linear prediction (PLP) that are compatible with the hybrid acoustic models of conventional (Gaussian mixture model-hidden Markov model)-based models (monophone and triphone) and hidden Markov model-deep neural network (HMM-DNN). Using MFCC and PLP features, we found that the HMM-DNN model performs better than GMM-HMM-based models with a word error rate (WER) of 6.67% and 7.40%, respectively.

Keywords Isolated digit recognition · Khasi dialect · MFCC · PLP · GMM-HMM · HMM-DNN

1 Introduction

Recognition of isolated speech activated digits is an important system for a wide and significant number of applications mainly in voice-activated telephone dialing, automated banking system, employee attendance using codes and comparison of quoted rates. The exceptional results of automatic speech recognition on smartphones, which currently require no physical connection, have inspired speech processing researchers

F. Rynjah (✉) · B. Syiem · L. J. Singh
North Eastern Hill University, Mawkynroh, Shillong 793022, India
e-mail: fairriky.rynjah@gmail.com

L. J. Singh
e-mail: jplairen@nehu.ac.in

S. Basu et al. (eds.), *Proceedings of International Conference on Frontiers in Computing and Systems*, Lecture Notes in Networks and Systems 404,
https://doi.org/10.1007/978-981-19-0105-8_53

[1]. This has motivated us to build a speech recognition system that is specifically for isolated speech of spoken digits. The speech recognition systems produced better performance under no-noise conditions, but regardless of significant improvements in technology, the performance of every system is still not yet achieved due to the uncontrolled noise, which cannot be neglected.

Kaldi is an open-source software developed by Daniel Povey in 2011 [2]. It has a collection of several tools required for building speech recognition system. The toolkit is programmed in C++, but the overall toolkit includes some scripted languages also like Bash, Perl and Python. It has a matrix library that includes standard linear algebra package(LAPACK) and basic linear algebra subprograms(BLAS) routines. In order to obtain better system performance, it uses open finite state transducer(FST) as a library [3].

The main objective of the Kaldi toolkit is to provide cutting-edge, versatile code that is simple to comprehend, update and develop. The tools will compile on most Unix-like systems as well as Microsoft Windows.

2 Related Works

This section of the paper highlights literature survey on the research work that has been done in isolated digit recognition in different languages as well as for isolated word recognition. A Kaldi toolkit-based approach automatic speech recognition system in Odia language was implemented by Karan et al. [4]. GMM-HMM was used as acoustic modeling, and a considerable reduction in WER was observed in triphone model as compared to monophone model. Jeeva Priya K et al. discuss the Kannada speech recognition system with 200-word dictionary for tourism applications. When utilized for recognition in online and offline modes, the system, which was created using HMM acoustic modeling and the HTK toolkit, had an accuracy of roughly 90.6 percent and 83.2% [5]. Georgescu et al. discuss how increasing the size of the corpus and replacing the classic GMM-HMM approach with DNN-based acoustic models improved the quality of a Romanian ASR system [6]. Piero Cosi et al. discuss an ASR system in Italian that uses the Kaldi toolbox [7] and records the WERs of children's speech samples using the DNN model.

3 Methodology

Feature Extraction Techniques: To represent the system compactly and efficiently, the speech waveforms are processed using extraction techniques to generate speech features without losing important information, and then only it can be used in the acoustic modeling. We are using two commonly used features for our work to observe the best performance.

3.1 Mel-Frequency Cepstrum Coefficients (MFCC)

The purpose behind MFCC being most ordinarily utilized for extricating highlights is that it almost takes after the real human auditory speech perception. It contrasts from the genuine cepstral that a nonlinear frequency scale is utilized, which approximates the conduct of the auditory system behavior [8].

3.2 Perceptual Linear Prediction(PLP)

PLP provides a representation consistent with a smooth short-term spectrum equalized and compressed similar to the human auditory system, bringing it closer to the Mel-cepstrum-based features [9]. PLP provides minimized resolution at elevated frequencies that means auditory filter bank-based strategy, yet provides orthogonal outputs that are comparable to cepstral assessment. It utilizes linear projections for spectral smoothing; hence, the name is linear perceptual prediction [10].

4 Experimental Approach

4.1 Database Collection

In this paper, we have used a speech database related to isolated digits in the Khasi standard (Sohra) dialect. In the Khasi script, the digits are symbolized in English, but the pronunciation of each digit is native even with the same district. Initially, an ideal set of word-based digit pronunciation of 15 digits (0 to 11,100,1000 and 100,000) in both the dialects of Khasi language was prepared. As per the standard procedure, a recorder-based continuous speech database was collected, and each spoken data was processed and segmented using WaveSurfer with sampling frequency 16 kHz and 16 bits per sample with mono channel. Waveform representation of each digit was captured using WaveSurfer which are shown in Figs. 1 and 2. For the training phase, we have collected speech files of spoken isolated digits in standard format around 72 native speakers of Khasi standard dialect. However, for the testing phase, separate data of 18 native speakers of Khasi standard dialect have been collected in the same format. We have recorded spoken data from speakers of different age groups and gender in an open room. To build a spoken isolated digit recognition system, training and testing files were taken. The details of the digit base words equivalent in English and database description are shown in Tables 3 and 4, respectively (Tables 1 and 2).

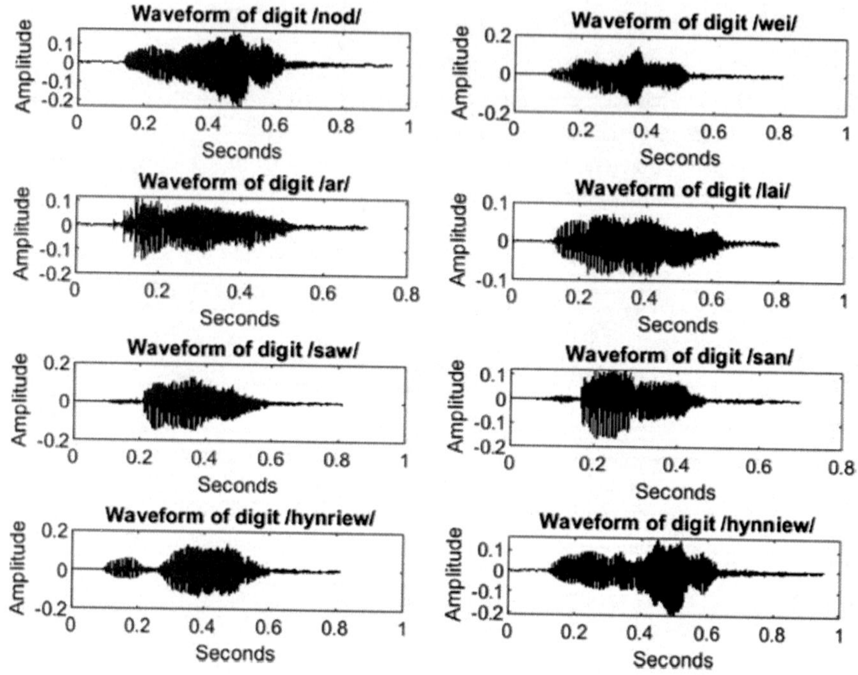

Fig. 1 Waveform of digits nod (zero) to hynniew (seven)

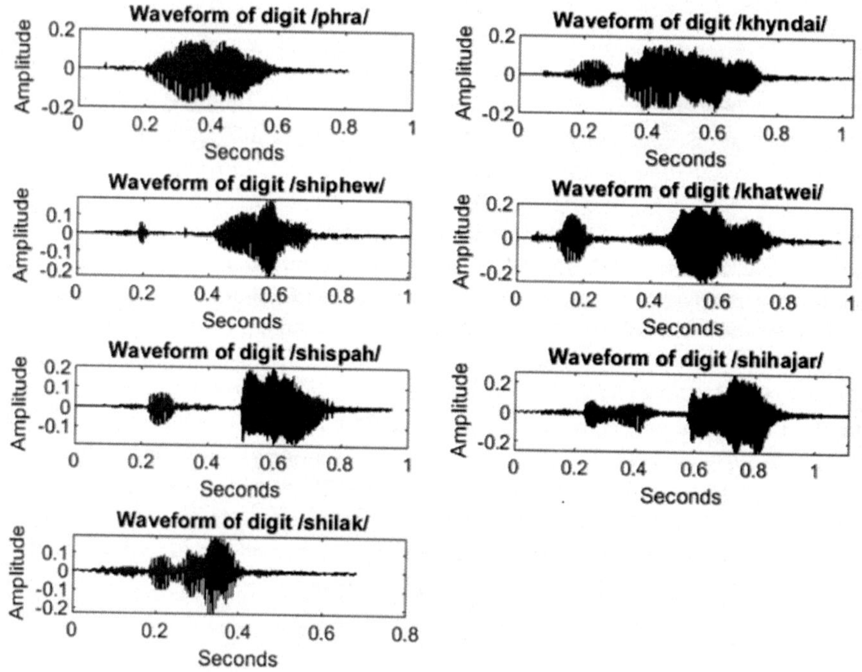

Fig. 2 Waveform of digits phra (eght) to shilak (one lakh)

Table 1 Khasi to English pronunciation

Digits	Text in Khasi	Text in English
0	nod	Zero
1	wei	One
2	ar	Two
3	lai	Three
4	saw	Four
5	san	Five
6	hynriew	Six
7	hynniew	Seven
8	phra	Eight
9	khyndai	Nine
10	shiphew	Ten
11	khatwei	Eleven
100	shispah	One hundred
1000	shihajar	One thousand
100000	shilak	One lakh

Table 2 Database description of isolated digits

Database description	
Tool used for recording	Zoom H4N handy portable Digital recorder
Sampling frequency	16 KHz
Distance between microphone and speaker's mouth	12 in.
Channel	Mono
Duration of wave file	1–2 s
Age of speakers	18–55 years
Language	Khasi
Dialect	Sohra (Khasi standard dialect)
No. of speakers	72
No. of male speakers	42
No. of female speakers	30
Total no. of wave files per speaker	15

Table 3 Evaluation of Performance of different models

Model	Feature type	WER(%)	Recognition accuracy(%)
Monophone	MFCC	17.78	82.22
	PLP	18.89	81.49
Triphone	MFCC	14.81	85.19
	PLP	13.33	86.67
DNN	MFCC	6.67	93.33
	PLP	7.40	92.6

4.2 *Experimental Set-up*

Our experiment was conducted on the Kaldi toolkit which runs on Ubuntu 18.04 LTS (64-bit operating system) having the unique advantage of the flexibility to support any existing acoustic model. Using Kaldi toolkit, we have selected MFCC and PLP which are the most commonly used features extraction method to provide the best possible representation of information for the given acoustic model to produce the corresponding result. The speech features for both the methods are extracted by applying Hamming window of 25ms. Approximately, around 400 samples are filtered to 13 cepstral coefficients. We have used acoustic parameters of 39 dimensions for both the methods by appending the 13 static coefficients with 13 delta (first-order derivative) and 13 acceleration (second-order derivatives) coefficients, respectively.

Acoustic modeling uses the speech data and their corresponding transcriptions to train the system by generating statistical representations of each isolated speech file. After generating the acoustic features, the system uses the standard GMM-HMM design to perform the task of speech recognition, i.e., the system was then trained and tested using the monophone model, and then the accuracy of recognition of the system is assessed using the triphone model which was later followed by DNN model using the hybrid classifier HMM-DNN. The N-gram (i.e., N = 2) language model was used for monophone, triphone and DNN acoustic modeling. Using MFCC and PLP parameterization techniques on the Kaldi toolkit, the system performance was analyzed which was evaluated using word error rate(WER).

4.3 *Results and Discussion*

We have observed the system accuracy of MFCC and PLP feature vectors using monophone, triphone and DNN model. The word error rate (WER) was evaluated using equation (1), and recognition accuracy was evaluated using equation (2). It was reported that better result was obtained with hidden layer of 5 [11]. In the instance of DNN, we experimented with different hidden layers sizes ranging from 2,3,4 and 5

Table 4 Confusion matrix for digit recognition (in %) using DNN

Digits	nod	wei	ar	lai	saw	san	hynriew	hynniew	phra	khyndai	shiphew	khatwei	shispah	shihajar	shilak
nod	66.7	0	0	0	0	0	0	0	0	0	0	0	33.3	0	0
wei	0	100	0	0	0	0	0	0	0	0	0	0	0	0	0
ar	0	0	66.7	0	0	0	0	0	33.3	0	0	0	0	0	0
lai	0	0	0	66.7	0	0	0	0	33.3	0	0	0	0	0	0
saw	0	0	0	0	100	0	0	0	0	0	0	0	0	0	0
san	0	0	0	0	0	100	0	0	0	0	0	0	0	0	0
hynriew	0	0	0	0	0	0	100	0	0	0	0	0	0	0	0
hynniew	0	0	0	0	0	0	0	100	0	0	0	0	0	0	0
phra	0	0	0	0	0	0	0	0	100	0	0	0	0	0	0
khyndai	0	0	0	0	0	0	0	0	0	100	0	0	0	0	0
shiphew	0	0	0	0	0	0	0	0	0	0	100	0	0	0	0
khatwei	0	0	0	0	0	0	0	0	0	0	0	100	0	0	0
shispah	0	0	0	0	0	0	0	0	0	0	0	0	100	0	0
shihajar	0	0	0	0	0	0	0	0	0	0	0	0	0	100	0
shilak	0	0	0	0	0	0	0	0	0	0	0	0	0	0	100

to evaluate the recognition performance. Best result was achieved with hidden layer size of 3. And the reason might be because of the amount of data used. Further, it was observed that DNN model produced better results than monophone and triphone. The system performance using MFCC and PLP feature vectors with quality of speech is nearly equivalent in monophone, triphone as well as in DNN. Table 3 shows the performance of WER for MFCC and PLP features using monophone, triphone and DNN.

$$\text{WER} = \frac{(S + D + I)}{N} \times 100\% \tag{1}$$

$$\text{Recognition Accuracy} = \frac{N - D - S - I}{N} \times 100\% \tag{2}$$

where N is the total number of words, D is the number of deletions, S is the number of substitutions, and I is the number of insertions [12].

We elaborate our observations with confusion matrix for all the digits in (%) using the HMM-DNN modeling, which is shown in Table 4. The matrix element Cij denotes how often the digit in row i (reference digits) is classified as the digit in column j(recognized digits). From the analysis, it is found that most of the digits are recognized except for some barely confusing digits.

5 Conclusion

From the experiment, firstly, we observed that the system performance using MFCC and PLP feature vectors with quality of speech is nearly equivalent in monophone, triphone as well as DNN. Secondly, we observed that the system accuracy of HMM-DNN is much better than the GMM-HMM-based acoustic modeling. Future work in this field is by adding more speech data and additional parameters in order to develop a robust system which can give better performance even under noisy condition.

References

1. Mahadevaswamy, Ravi DJ (2019) Performance of isolated and continuous digit recognition system using Kaldi toolkit. Int J Recent Technol Engi 8(2S2). https://doi.org/10.35940/ijrte.B1047.0782S219
2. Povey D et al (2011) The Kaldi speech recognition toolkit. IEEE, workshop on automatic speech recognition and understanding
3. http://kaldi-asr.org . Kaldi ASR Org. Available https://kaldi-asr.org/doc
4. Karan B, Sahoo J, Sahu PK (2015) Automatic speech recognition based Odia system. In: International conference on microwave, optical and communcation engineering. https://doi.org/10.1109/ICMOCE.2015.7489765
5. Priya J, Sree K, Navya SS, Gupta D (2018) Implementation of phonetic level speech recognition in Kannada using HTK. In: International conference communication signal processing (ICCSP)

6. C H Georgescu CBA (2017) SpeeD's DNN approach to Romanian speech recognition. In: International conference speech technol. Human Computer Dialogue (SpeD)
7. Cosi P (2015) A KALDI-DNN-based ASR system for Italian. In: International Jt. conference neural networks (IJCNN), Kill
8. Gupta K, Gupta D (2016) An analysis on LPC. RASTA and MFCC techniques in Automatic Speech recognition system. https://doi.org/10.1109/CONFLUENCE.2016.7508170
9. Hermansky H (1989) Perceptual linear predictive (PLP) analysis of speech. J Acoust Soc Am 87(4):1738–1752. https://doi.org/10.1121/1.399423
10. Markel JD, Gray AH (1976). Linear prediction of speech Springer. https://doi.org/10.1007/978-3-642-66286-7
11. Syiem B, Joyprakash Singh L (2019) Deep neural network-based phoneme classification of standard Khasi dialect in continuous speech. Int J Appl Pattern Recogn 6(1). https://doi.org/10.1504/IJAPR.2019.104288
12. Renjith S, Joseph A, Babu KK (2013) Isolated digit recognition Malayalam An application perspective. In: International conference on control commuincation and computing. https://doi.org/10.1109/ICCC.2013.6731648

Sentiment Analysis on COVID-19 News Videos Using Machine Learning Techniques

S. Lekshmi and V. S. Anoop

Abstract Coronavirus disease (COVID-19) has affected all walks of human life most adversely, from entertainment to education. The whole world is confronting this deadly virus, and no country in this world remains untouched during this pandemic. From the early days of reporting this virus from many parts of the world, many news videos on the same got uploaded in various online platforms such as YouTube, Dailymotion, and Vimeo. Even though the content of many of those videos was unauthentic, people watched them and expressed their views and opinions as comments. Analysing these comments can unearth the patterns hidden in them to study people's responses to videos on COVID-19. This paper proposes a sentiment analysis approach on people's response towards such videos, using text mining and machine learning. This work implements different machine learning algorithms to classify people's sentiments and also uses text mining principles for finding out several latent themes, from the comments collected from YouTube.

Keywords COVID-19 · Sentiment analysis · Computational social science · Machine learning · Text mining

1 Introduction

The whole world is fighting against a deadly virus called coronavirus that impacted all walks of human life across the globe. As of 13 June 2021, a total of 17,53,06,598 confirmed cases and 37,92,777 deaths were reported globally as per the statistics of the World Health Organization (WHO) [1]. From the early reporting of COVID-19, many news in the form of videos and text were being added to the public domains, and also, the people were expressing their views, concerns, and opinions on the same. Many videos that discuss the symptoms, causes, and treatments were published on platforms such as YouTube [2] and Vimeo [3], along with the news on how different countries were combating COVID-19. Even though many videos shared fake news on

S. Lekshmi · V. S. Anoop (✉)
Rajagiri College of Social Sciences (Autonomous), Kochi, Kerala, India
e-mail: anoop@rajagiri.edu

S. Basu et al. (eds.), *Proceedings of International Conference on Frontiers in Computing and Systems*, Lecture Notes in Networks and Systems 404,
https://doi.org/10.1007/978-981-19-0105-8_54

the same, there were official YouTube and Vimeo channels of reputed news agencies such as BBC and New York Times that discussed accurate and factual information on COVID-19 [4]. These videos are being viewed by many people not only to get updates on COVID-19 but also to share their comments and views on the same [5, 6].

Sentiment analysis is an active research area of natural language processing (NLP) that attempts to classify sentiments into positive, negative, and neutral polarities. Sentiment analysis helps to quantify the attitude, opinions, and emotions of people. In the recent past, this area of NLP has grown exponentially, and many innovative tools and techniques were reported in the literature. From the initial rule-based approaches, sentiment analysis has evolved through different phases, with deep learning approach showing significant performance over other machine learning algorithms. Computational social sciences make heavy use of sentiment analysis to find out the opinions and emotions of people towards many activities related to government policies, decision-making, and other administrative reforms [7]. The COVID-19 pandemic was not different from this as people shared their views and emotions towards many resources related to the pandemic, through various platforms such as social media and other content-sharing websites [8, 9].

This work proposes an approach using machine learning techniques to analyse the sentiment of the general public towards the online news videos published in YouTube, one of the very popular online video sharing and social media platforms. The comments posted by the people for various videos were collected and classified using different classification algorithms to understand how people were reacting to the videos. The remainder of this paper is organized as follows: Sect. 2 discusses some of the very prominent and recent works reported in the literature for sentiment analysis, and Sect. 3 introduces and details the proposed approach. In Sect. 4, the experiment details are given, and Sect. 5 presents the results and discusses it. In Sect. 6, the authors present the conclusions and the future work.

2 Related Work

Sentiment analysis uses natural language processing techniques to extract the polarity of a given data. There are multiple techniques and approaches reported in the recent literature on sentiment analysis, with varying degrees of success. These methods include lexicon-based, traditional machine learning algorithms, deep learning-based algorithms, and also hybrid methods [10]. This approach compares the techniques in the tourism domain. Sentiment analysis is widely used on social media data to gauge the opinion of people on a particular topic [11]. This paper analyses important events in real-time to extract people's opinions on the issue under consideration by building a custom system called "Polaris". Since a humongous amount of real-time data is generated, AsterixDB is used to process the data, and labelling of the data was done using the emoticons. In addition to analysing the sentiment, the proposed system also predicts the sentiment trajectory. Deep learning techniques like CNN and LSTM were employed to achieve this.

The YouTube comment section is a major source to extract user opinion on particular video content. Numerous researchers have worked on this domain to derive meaningful insights into the user's opinion and behaviour [12]. This paper explores the gender gap that exists in hosting STEM-related videos in YouTube. It also exhibits the user comment behaviour and subscription and views pattern for a female-hosted STEM videos. Apart from sentiment analysis, methods such as topic modelling can be utilized [13]. Fuzzy lattice reasoning is utilized to detect and identify polarity and useful hidden topics based on user comments on a movie video. The method suggested in this paper can be utilized in recommendation systems to suggest movies to similar groups of users. Aspect-based sentiment analysis can be used to improve the relevancy and quality of recommended videos [14]. In addition to the likes and dislikes count, aspects extracted for the user's comments can be utilized to identify videos with similar content. In addition to the English language, recent years have seen sentiment analysis exploration in other languages too [15]. The objective of this paper is to create a corpus for code-mixed English–Tamil comments extracted from YouTube. Sentiment analysis on YouTube comments based on Dravidian languages Tamil and Malayalam was explored in [16]. Comments extracted were classified into positive, negative, or neutral polarity and also checked the ability of the system to recognize if a particular comment was not from the intended language.

Deep learning techniques can also be employed in sentiment analysis to improve the accuracy [17]. The above paper explores the performance improvement when we utilize different deep learning methods and also their applications on various tasks of sentiment analysis. These methods were implemented on real-world data sets extracted from YouTube, Imdb, Amazon etc., and the results were analysed. There were also studies that analysed the scope of deep learning NLP techniques in Arabic Subjective Sentiment Analysis (ASSA) [18]. Large number of research studies were analysed, and it was observed that CNN and RNN models were the commonly used models for ASSA. Hybrid deep learning models were also proposed to improve the overall accuracy [19]. This paper implements a hybrid model (RNN-LSTM-BiLSTM-CNN) to calculate sentiment polarity. The model was tested on existing data sets such as SST-1, SST-2, and MR and yielded better performance when compared to other traditional approaches. This paper proposes an approach that attempts to collect publicly available comments that the users have shared for news videos related to COVID-19 and train machine learning classifiers that can classify the YouTube comments.

3 Proposed Approach

This section details the proposed approach for analysing the sentiment of people on the COVID-19 news videos published online on YouTube. The overall workflow of the proposed approach is depicted in Fig. 1. There are mainly five steps involved in the proposed approach that starts with collecting the comments from the online videos published on YouTube. Since each video contains a large number of public

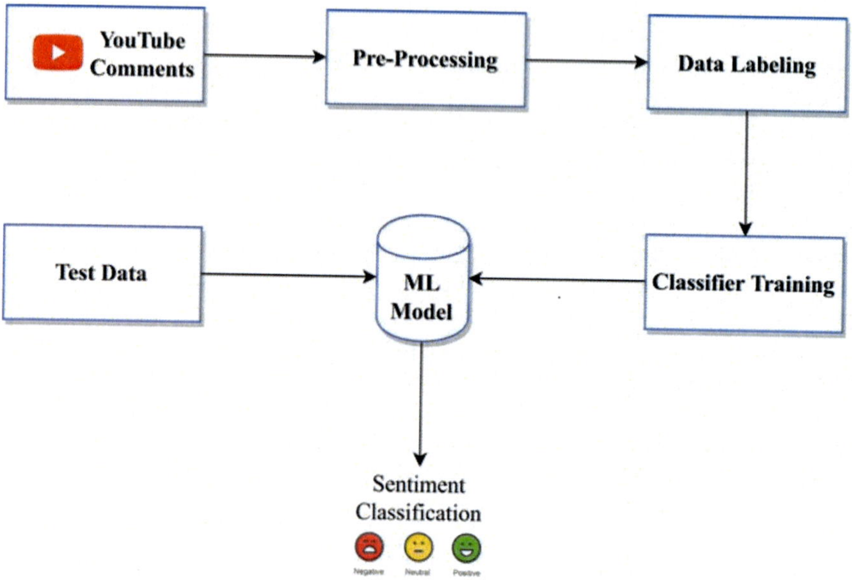

Fig. 1 Overall workflow of the proposed approach

comments, it is nearly impossible to collect them using the manual process. The proposed approach used Web scraping to compile the data set for the purpose. The comments collected from YouTube contain many noises and unwanted signals and eliminating them from the experiment-ready copy of the data set is very important. The comments mainly will have emojis and smileys along with website URLs and special symbols. The second step of the proposed approach deals with removing such unwanted data to prepare the experiment-ready version of the data set. In machine learning, classification is an algorithm that comes under the category of supervised learning, which requires large samples of labelled data for the training. As the comments scrapped from YouTube are unstructured in nature that do not contain any labels, labelling the comments into positive, negative, and neutral is an important step in training the classifier. This requires great human effort to examine the comment and label them into pre-defined labels. Once the labelled data is ready, it can be used for training the classifier. In this proposed approach, we employ logistic regression (LR), linear support vector machine, random forest (RF), K-nearest neighbour (KNN), and multinomial naïve Bayes (MNB) classifiers. A short description of each of these classifiers is discussed here.

- **Logistic Regression (LR)**: Logistic regression can be seen as a statistical learning algorithm categorized in the supervised learning methods used for classification tasks. LR uses a more complex cost function that can be called a sigmoid function or logistic function rather than a linear function as its base. LR function hypothesizes that it will limit the cost function between the values 0 and 1.

- **Support Vector Machine (SVM)**: SVM comes under the category of supervised learning which is heavily used for classification tasks. SVM shows high accuracy improvements with less computational power when compared with other classification algorithms. SVM finds a hyperplane in an n-dimensional space that distinctly classifies the data points. To separate any two classes of data points, there exists many hyperplanes, but SVM finds the hyperplane that has a maximum margin.
- **Random Forest (RF)**: RF algorithm comes under the category of ensemble model, and as the name suggests, it consists of a large number of decision tree classifiers as its core. Random forest algorithm uses various sub-samples of the data set and uses averaging to improve the predictive accuracy and control overfitting. RF algorithm assumes that a large number of models working together (ensemble) will always outperform the individual models.
- **K-Nearest Neighbour (KNN)**: KNN is a classification algorithm that comes under the category of non-parametric classification method. KNN computes the similarity between the new data points and the available data points and puts the new data point into a class that is most similar to the available class. It is a lazy learning algorithm and does not make any assumption on the underlying data points.
- **Multinomial Naïve Bayes (MNB)**: This is one of the popular machine learning algorithms for the classification of categorical text data. MNB has its theoretical foundation from the Bayes theorem and assumes that each feature being classified is not related to any other feature. One great advantage of MNB is that it is highly scalable and can easily handle large data sets.

The proposed approach implements the machine learning algorithms mentioned for classifying the sentiment into any of the classes such as positive and negative. The details on the implementation are discussed in Sect. 4.

4 Experiments

The details of the experiments conducted using the proposed approach are discussed here. The details on the experimental testbed, data set used, and the implementation details of the different machine learning algorithms are described in detail. This work uses comments from YouTube https://www.youtube.com posted by the public for different video news on COVID-19. We have scrapped a total of 12,335 public comments from YouTube. After collecting the data set containing user comments in the comma-separated value, we have pre-processed the data set. People tend to include emojis or smileys while writing a comment, so the first step was to remove them from the comments. We have used the *demoji* library available at https://pypi.org/project/demoji/ for removing emojis from the text. The cleaned set of comments are extracted using the demoji library applied to all the comments extracted. This work only focuses on analysing the sentiment of English comments. So out of all the

12,335 comments scrapped, we have only considered the comments that are written in English, and this was done with the help of *langdetect* library available at https://pypi.org/project/langdetect/. After this step, we have got 10,512 comments in total which are written in English.

The next set of pre-processing was done to convert the comment text into lowercase, to remove unwanted white spaces, and also to remove numbers. We have also expanded the contractions in the pre-processed comments to enhance the same. Contractions are words or combinations of words that are shortened by dropping letters and replacing them with an apostrophe. For example, "might've" should be replaced with "might have", "we've" should be replaced with "we have" and so on. This has been done by collecting common contractions replacement details available for the English language. The punctuation in the text is also removed in the pre-processing step. After this step, the polarity value for each comment is calculated using the TextBlob library, and the polarity category value is computed in such a way that if the polarity value is greater than zero, then the polarity category value is assigned as 1; otherwise, −1. After this step, there were 4843 comments with polarity category value as −1 and 3157 comments with a polarity value of 1. In the pre-processing stage, we have also removed all the stopwords, and also the lemmatization was performed on the comment text. For these purposes, Natural Language Toolkit (NLTK) library in Python is used. Finally, we are left with 103,956 comments in our experiment-ready data set. The most frequently occurring bi-grams in positive polarity comments and most frequently occurring bi-grams in negative polarity comments are shown in Fig. 2a, b, respectively.

5 Results and Discussions

The results obtained from the experiment conducted using the proposed framework are detailed here. All the classification algorithms outlined in the proposed method section have been implemented in four trials. In the first trial, 4000 comments were considered, and the logistic regression, SVM, random forest, KNN, and naïve Bayes classifiers were trained on the 4000 samples.

Later, the sample size has been increased to 6000, 8000, and 10,512 (all samples), and the classifiers are trained on them. The training and test accuracy obtained from each of these trials are given in Tables 1, 2, 3, and 4 (Fig. 3).

6 Conclusions and Future Work

This work attempted to analyse the sentiment of people towards the online video news on COVID-19, published on YouTube. The publicly available comments were collected and classified based on their polarity values using machine learning techniques. As the results are promising, the authors would like to conduct further

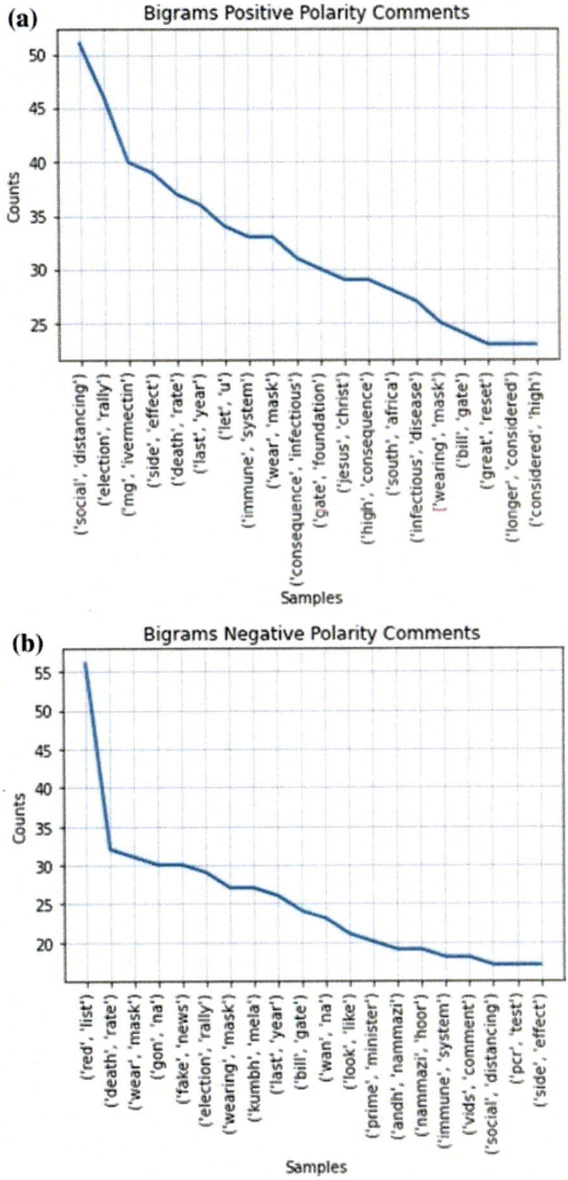

Fig. 2 **a** 20 most frequently occurring bi-grams in positive polarity comments and **b** 20 most frequently occurring bi-grams in negative polarity

research on applying deep learning techniques for better sentiment classification of YouTube comments.

Table 1 Training and test accuracy for **4000** samples on five classification algorithms

Classifier	Train accuracy	Test accuracy
Logistic regression	80.1	80
Support vector machine	80.9	81
Random forest	80.2	80
K-nearest neighbour	64	63
Naïve Bayes	73.5	73

Table 2 Training and test accuracy for **6000** samples on five classification algorithms

Classifier	Train accuracy	Test accuracy
Logistic regression	81.4	81
Support vector machine	81.7	83
Random forest	82.3	82
K-nearest neighbour	63.7	64
Naïve Bayes	73.2	74

Table 3 Training and test accuracy for **8000** samples on five classification algorithms

Classifier	Train accuracy	Test accuracy
Logistic regression	82.8	83
Support vector machine	82.3	84
Random forest	82.7	83
K-nearest neighbour	64.4	65
Naïve Bayes	74.9	75

Table 4 Train and test accuracy for **10,512** samples on five classification algorithms

Classifier	Train accuracy	Test accuracy
Logistic regression	83.7	84
Support vector machine	84.1	84
Random forest	83.1	83
K-nearest neighbour	65.1	65
Naïve Bayes	75.1	75

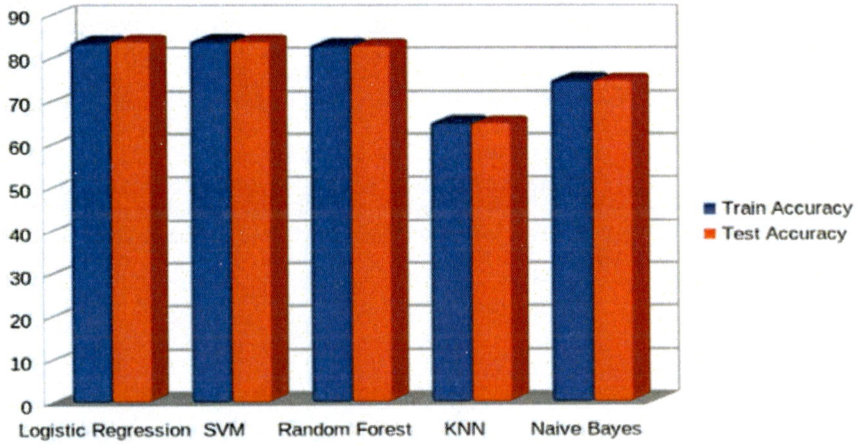

Fig. 3 The train and test accuracy of the proposed approach for **10,512** data samples

References

1. https://covid19.who.int/. Last accessed on 25 Aug 2021
2. https://www.youtube.com/. Last accessed on 25 Aug 2021
3. https://vimeo.com/. Last accessed on 25 Aug 2021
4. Serrano JCM, Papakyriakopoulos O, Hegelich S (2020). NLP-based feature extraction for the detection of COVID-19 misinformation videos on Youtube. In: Proceedings of the 1st workshop on NLP for COVID-19 at ACL 2020
5. Cinelli M, Quattrociocchi W, Galeazzi A, Valensise CM, Brugnoli E, Schmidt AL, Zola P, Zollo F, Scala A (2020) The covid-19 social media infodemic. Sci Rep 10(1):1–10
6. Szmuda T, Syed MT, Singh A, Ali S, Özdemir C, Słoniewski P (2020) YouTube as a source of patient information for Coronavirus Disease (COVID-19): a content-quality and audience engagement analysis. Rev Med Virol 30(5):e2132
7. Barkur G, Vibha GBK (2020) Sentiment analysis of nationwide lockdown due to COVID 19 outbreak: evidence from India. Asian J Psychiatry 51:102089
8. Manguri KH, Ramadhan RN, Amin PRM (2020) Twitter sentiment analysis on worldwide COVID-19 outbreaks. Kurdistan J Appl Res 54–65
9. Rustam F, Khalid M, Aslam W, Rupapara V, Mehmood A, Choi GS (2021) A performance comparison of supervised machine learning models for Covid-19 tweets sentiment analysis. Plos one 16(2):e0245909
10. Shi Y, Zhu L, Li W, Guo K, Zheng Y (2019) Survey on classic and latest textual sentiment analysis articles and techniques. Int J Inf Technol Decis Mak 18(04):1243–1287
11. Dang NC, Moreno-García MN, De la Prieta F (2020) Sentiment analysis based on deep learning: a comparative study. Electronics 9(3):483
12. Amarasekara I, Grant WJ (2019) Exploring the YouTube science communication gender gap: a sentiment analysis. Public Underst Sci 28(1):68–84
13. Jelodar H, Wang Y, Rabbani M, Ahmadi SBB, Boukela L, Zhao R, Larik RSA (2021) A NLP framework based on meaningful latent-topic detection and sentiment analysis via fuzzy lattice reasoning on youtube comments. Multimedia Tools Appl 80(3):4155–4181
14. Chauhan GS, Meena YK (2019) YouTube video ranking by aspect-based sentiment analysis on user feedback. In: Soft computing and signal processing. Springer, Singapore, pp 63–71
15. Chakravarthi BR, Muralidaran V, Priyadharshini R, McCrae JP (2020) Corpus creation for sentiment analysis in code-mixed Tamil-English text. arXiv:2006.00206

16. Chakravarthi BR, Priyadharshini R, Muralidaran V, Suryawanshi S, Jose N, Sherly E, McCrae JP (2020) Overview of the track on sentiment analysis for dravidian languages in code-mixed text. In Forum for information retrieval evaluation, pp 21–24
17. Habimana O, Li Y, Li R, Gu X, Yu G (2020) Sentiment analysis using deep learning approaches: an overview. Sci China Inf Sci 63(1):1–36
18. Nassif AB, Elnagar A, Shahin I, Henno S (2020) Deep learning for Arabic subjective sentiment analysis: challenges and research opportunities. Appl Soft Comput 106836
19. Aslam A, Qamar U, Saqib P, Ayesha R, Qadeer A (2020) A novel framework for sentiment analysis using deep learning. In: 2020 22nd International conference on advanced communication technology (ICACT). IEEE, pp 525–529

Bengali POS Tagging Using Bi-LSTM with Word Embedding and Character-Level Embedding

Kaushik Bose and Kamal Sarkar

Abstract Part-of-speech tagging (POS) is an important and very fundamental process in natural language processing (NLP). POS tagging is required as a preprocessing task in many types of linguistic research such as named entity recognition (NER), word sense disambiguation, information extraction, natural language translation, and sentiment analysis. In this paper, we propose a practical Bengali POS tagger, which takes as input a text written in Bengali and gives a POS tagged output. In recent times, Bi-LSTM networks have been proven effective in sequential data processing but not very much tested on resource-poor and inflectional languages like Bengali. This paper addresses the issues of the POS tagging task for the Bengali language using Bi-LSTM with transfer learning by applying pre-trained word embedding information. The POS tagged output from our proposed model can be used directly for other applications of Bengali language processing as our proposed tagger can also handle out-of-vocabulary (OOV) words. Our experiment reveals that Bi-LSTM with transfer learning is effective for tagging Bengali documents.

Keywords POS tagging · Word embedding · Character embedding · Bi-LSTM

1 Introduction

Part-of-speech tagging is usually used at the preprocessing step in many natural language processing applications. It is the process of automatic annotation of lexical categories (verb, adjective, noun, etc.) to words. A word's tag in a sentence depends on the word's syntactic property in the context it occurs in. Therefore, tagging a word with the right POS tag is an important activity. Vital information about a particular word and its surrounding words can be identified by POS tagging because the grammatical properties of a word are also determined by the words to which it is related. Therefore, in various NLP applications like shallow parsing, NER,

K. Bose (✉)
Govt. General Degree College, Narayangarh, India

K. Sarkar
Computer Science and Engineering Department, Jadavpur University, Kolkata, India

S. Basu et al. (eds.), *Proceedings of International Conference on Frontiers in Computing and Systems*, Lecture Notes in Networks and Systems 404,
https://doi.org/10.1007/978-981-19-0105-8_55

word sense disambiguation, information extraction, machine translation, question answering, chunking, etc., parts-of-speech tagging has a very important role. Hence, POS tagger is considered a very effective tool for the linguistic study of any language [1].

Unknown words or out-of-vocabulary words are one of the main problems of POS tagging tasks for resource-poor language like Bengali because it is highly inflectional. Here we tried to overcome the OOV problems by applying character sequence-based word embedding. In recent times, deep learning [2] has become very effective in various NLP tasks. Recurrent neural networks (RNN) and its improved versions such as long short-term memory (LSTM) [3], Bi-LSTM [4] are being widely used for solving various NLP problems because of their sequential information processing capability. Although some work [5, 6, 7, 8] has already been proposed on POS tagging based on Bi-LSTM for different languages but not implemented on Bengali. This fact motivates us to develop a Bi-LSTM-based Bengali POS tagger using publicly available datasets.

2 Related Work

In a broader sense, automatic POS tagging can be of two types- rule-based and stochastic. The rule-based approach works by designing the rules by hand. This technique is laborious and requires detailed linguistic knowledge. But the stochastic approach is based on statistical methods where lexical and contextual probabilities are learned by training on a tagged corpus. Stochastic models do not require deep linguistic knowledge [9]. Different POS tagging models for natural languages were proposed in the past based on rules or stochastic approaches but stochastic approaches gained more popularity because linguistic knowledge is not required. Among stochastic models, different variations of the hidden Markov model (HMM) [10, 11], support vector machine (SVM) were very popular at the early stage of POS tagger design. Later deep learning approaches gained huge popularity. All of these approaches require a large POS tagged corpus for training that is suitable for European languages like English.

As Bengali is very morphologically rich as well as resource-poor language, researchers face difficulties in developing a stochastic POS tagger for Bengali. That is the reason most of the earlier works for Bengali POS tagging [12, 13, 14, 15] experimented on researchers' dataset which is not available publicly. An HMM-based model and maximum entropy (ME)-based model for Bengali POS tagging were proposed in [12]. They used morphological checker to boost up the capabilities of the tagger specifically in cases when a large tagged dataset is not available. SVM- and CRF-based model is proposed in [13, 14], respectively. From their research work, it is evident that the SVM model exceeds the available models on HMM, ME, and CRF at that time. Research work which uses the public dataset for testing POS tagger [16, 17, 18] for Bengali is limited. A second-order HMM and Trigram HMM-Based POS tagger is proposed in [16] and [17], respectively. A CRF-based model for Bengali is proposed in [18]. An artificial neural network (ANN)-based model for Bengali POS tagging [19] was also proposed where they used manual feature identification approach for their model. But most of the above-mentioned models used handcrafted feature selection approaches for training which require a substantial size of annotated data. Because of the unavailability of large annotated datasets, models struggle with the unfamiliar dataset or dataset with more OOV words.

3 Proposed Methodology

We propose a sequential deep learning model for Bengali POS tagging. We have used bidirectional long short-term memory [3, 4] for this task. The architecture for the proposed Bi-LSTM-based Bengali tagger is shown in Fig. 1. As shown in this figure when raw Bengali text is entered into the system, preprocessing of the data by separating sentences is needed to be carried out first. After sentence separation, we need to perform word separation. Then word-level embedding has been carried out by using default Keras [20] word embedding or pre-trained word embeddings [21]. The use of pre-trained word embedding is the concept of transfer learning [22]. It is the process of using the knowledge of an already trained machine learning model on large-scale data to solve a different or related problem. But embedding vectors for all words are not available in the list of pre-trained word vectors [21]. The main reason behind this is that the pre-trained word embedding model was trained on

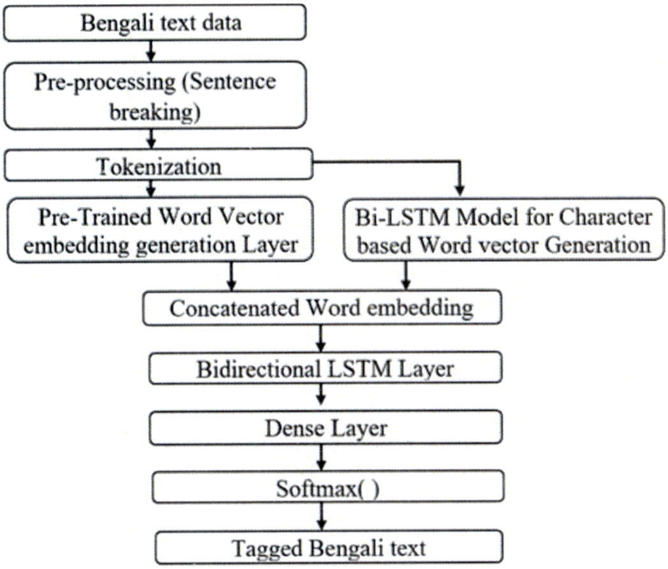

Fig. 1 Proposed system architecture

different corpora. This problem is known as the out-of-vocabulary (OOV) problem. To overcome this problem, character-level word embedding is useful. To obtain character-level word embedding, we have used a separate Bi-LSTM sub-network. Then we represent each word by concatenating the character-level word embedding with the word embedding taken from a pre-trained word embedding model or Keras embedding. Finally, the concatenated word embedding sequences are given to the main Bi-LSTM whose outputs are connected to the dense layer. Finally, the dense layer's output is passed through the softmax() layer that produces the tagged output of the corresponding input data.

3.1 Detailed Model Description

In this section, the proposed POS tagger is described in detail. The model is implemented using Bi-LSTM [4] cells represented in Fig. 2. It is a specific type of RNN cell which is implemented using two LSTM [3] cells that operate in the opposite directions- forward direction and backward direction to process past and future dependencies of a sequence.

We can consider LSTM as a memory cell block [3] which takes previous hidden state (h_{t-1}), previous cell state (C_{t-1}), and current input (x_t). as input and produce

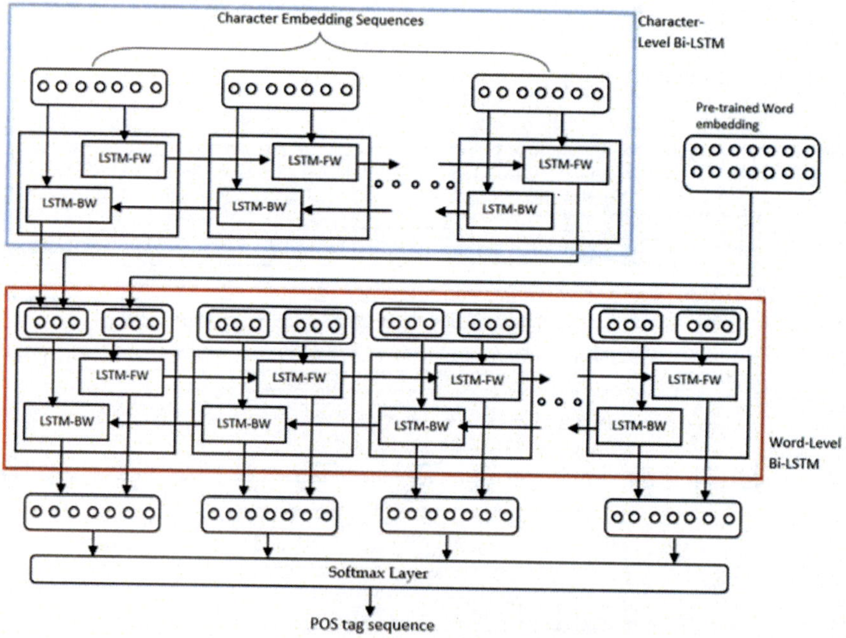

Fig. 2 Bi-LSTM-based POS tagging model with character-level and pre-trained word embeddings

current cell state (C_t) and current hidden state (h_t) as output. Equation-1 represents the hidden state of a Bi-LSTM [4] at time step t, that is a concatenation of hidden states from forward and backward LSTM cells, respectively:

$$h_t = \left[\vec{h}_t + \overleftarrow{h}_t \right] \tag{1}$$

In our proposed model, Bi-LSTM is used in two contexts: the character-level Bi-LSTM is used to generate character-sequence-based word embedding which assists our model in predicting the tags of OOV words. The character sequence-based word embedding h_k^i conforming to the input word x_k^i at any time step t can be defined as.

$$h_k^i = \text{BiLSTM}(c_{k1}, c_{k2}, \ldots, c_{kl}) \tag{2}$$

Here c_{kj} denotes the keras embedding vector of j th character of k th word in ith sentence. The second Bi-LSTM layer produces the output h_t^i corresponding to k th word x_k^i in the i th input sentence at time step t is defined as.

$$h_t^i = \text{BiLSTM}\left(\left[h_k^i, V_{\text{Pre-trained}}\left(x_k^i\right)\right]\right) \tag{3}$$

Here $V_{\text{Pre-trained}}\left(x_k^i\right)$ gives the contextual 300-dimensional pre-trained word vector and h_k^i is a 150-dimensional character sequence-based word vector for the of the kth word in the ith input sentence at time step t. In our model, we use pre-trained word vectors[1] [21] for word embedding, which we retrieved in 2020. The word vectors provided in [21] were trained on Wikipedia and Common Crawl. It uses CBOW [23, 24] with positional weight to produce 300-dimensional vectors with character n-grams of length 5, a window size of 5 and 10 negatives.

4 Experiment

The proposed Bi-LSTM-based Bengali POS tagger is implemented using Keras [20] a neural networks API, which runs on TensorFlow with GPU acceleration. In our proposed model, we have used one bidirectional LSTM layer with 128 hidden units along with the dropout of 0.5 for main POS tagging network and another bidirectional LSTM with 150 units for character-based word embedding task. The model is trained with 'Adam' (a variation of stochastic gradient descent) optimizer with 'categorical cross-entropy' loss function. While training, we have set the batch size to 32, learning rate to 0.001, and the number of epochs to 60 as hyperparameter, which gave us the best result.

[1] https://fasttext.cc/docs/en/crawl-vectors.html.

4.1 Dataset

We have developed our model using ICON 2013 [25] dataset which is basically POS tagged named entity recognition (NER) dataset. We removed all named entity (NE) tags and kept words and POS tags to generate a new dataset for Bengali POS tagging. Finally, our dataset contains 3611 lines with 43 tags, which includes one extra padding tag. The extra padding tag is required for the training of Bi-LSTM model.

5 Evaluation and Results

5.1 Evaluation

Our proposed POS tagger model is evaluated using accuracy, weighted $F1$-score, weighted precision, and weighted recall:

$$\text{Accuracy} = \frac{\text{Number of matched tags}}{\text{Total number of tags in the testing corpus}} \tag{4}$$

$$F1\,\text{Score} = \frac{2{*}(\text{Recall} * \text{Precision})}{(\text{Recall} + \text{Precision})} \tag{5}$$

$$\text{Precision} = \frac{\text{TP}}{\text{TP} + \text{FP}} \tag{6}$$

$$\text{Recall} = \frac{\text{TP}}{\text{TP} + \text{FN}} \tag{7}$$

Our model is cross-validated by applying tenfold cross-validation method. The final results are acquired by averaging results overall ten folds.

5.2 Results

We have evaluated the performance of the proposed Bi-LSTM-based POS tagger using the POS tagged dataset extracted from the ICON 2013 dataset described in Section 4.1.

The final result reported in this paper is obtained by taking the average of the results obtained by our system over ten folds. We have shown in Table 1 the POS tagging performance of the proposed Bi-LSTM Bengali POS tagger in terms of accuracy, precision, recall, and F1-score.

Table 1 Experimental result

Model	Features	Accuracy	Precision	Recall	F1-Score
Bi-LSTM (Model—1)	Keras Embedding	85.3533	85.1895	85.3533	85.0354
Bi-LSTM (Model—2)	Pre-Trained Word Embedding	86.4647	86.2173	86.4647	86.1765
Bi-LSTM (Model—3)	Keras Embedding + Character Embedding	86.8391	86.5463	86.8391	86.5813
Bi-LSTM (Model—4)	Pre-Trained Word Embedding + Character Embedding	87.0143	86.7358	87.0143	86.7244

As we can see from Table 1, our proposed Bi-LSTM model with concatenation of the character sequence-based word embedding and the pre-trained word-level embedding performs better than other models that use only word-level embeddings. It is also evident from the table that the pre-trained embedding is more effective than Keras embedding for the POS tagging task for the Bengali language.

5.3 Sample Output

In this section, we have shown a sample output of our proposed Bengali POS tagger. Figure 3 shows the sample input documents. Figure 4 shows the actual tags given by the human annotator for the words of the sample input documents and Fig. 5 shows tags for the words of the sample input documents predicted by our proposed model.

কিছুদিন আগেও পৃথিবীর মাত্র শতকরা কুড়ি ভাগ অঞ্চলে আবহাওয়া সম্পর্কে মোটামুটি নির্ভরযোগ্য পূর্বাভাষ জোগান সম্ভবপর ছিল।

Fig. 3 Sample text

কিছুদিন NN আগেও NST পৃথিবীর NN মাত্র QF শতকরা QF কুড়ি QC ভাগ NN অঞ্চলে NN আবহাওয়া NN সম্পর্কে PSP মোটামুটি INTF নির্ভরযোগ্য JJ পূর্বাভাষ NN জোগান VM সম্ভবপর JJ ছিল VM । SYM

Fig. 4 Actual tags

কিছুদিন NN আগেও NST পৃথিবীর NN মাত্র RB শতকরা QF কুড়ি QC ভাগ NN অঞ্চলে NN আবহাওয়া NN সম্পর্কে PSP মোটামুটি INTF নির্ভরযোগ্য JJ পূর্বাভাষ NN জোগান VM সম্ভবপর JJ ছিল VM । SYM

Fig. 5 Predicted tags

6 Error Analysis

The dataset used in our model is not balanced, as there are some words for which the correct tags are most likely confusing that is why the proper tags were not given by the annotator and those words has been tagged with CC:?, INJ:?, JJ:? These kinds of tags. This type of tags occurs very few in numbers, and we got very low score for these tags. As we did not change these tags during our experiment, number of tags also increased. So, we think these kinds of confusing tags hamper our model score to some extent.

7 Conclusion

A Bi-LSTM-based parts-of-speech tagger is proposed in this paper for the Bengali language. The model is evaluated using word-level encoding of the dataset along with the character sequence-based word embedding. The proposed POS tagging model that represents each word as a concatenated vector of pre-trained word embedding and character sequence-based word embedding produces better results than other models. The dataset used in our model is not balanced for all tags and may also contain some outliers, which might affect some amount of accuracy. Although the proposed Bi-LSTM-based Bengali POS tagger is an effective approach, there are still limitations compared to human taggers. The output from the proposed POS tagger can be useful as inputs to different Bengali language processing tasks like machine translation, and information retrieval, and this model can be applied to other Indian languages with minor modification. In the future, we will compare the proposed model with other established methods for POS tagging even for other languages and also explore attention-based deep neural networks, transformer, and different language models like BERT for learning the best POS sequence for the Bengali language.

References

1. Jurafsky D, Martin JH (2009) Speech and language processing, Pearson
2. Young T, Hazarika D, Poria S, Cambria E (2018) Recent trends in deep learning based natural language processing. IEEE Comput Intell Mag 13(3):55–75

3. Hochreiter S, Schmidhuber J (1997) Long short-term memory. Neural Comput 9(8):1735–1780
4. Schuster M, Paliwal KK (1997) Bidirectional recurrent neural networks. IEEE Trans Signal Process 45(11):2673–2681
5. Horsmann T, Zesch T (2017) "Do LSTMs really work so well for PoS tagging?—a replication study." In Empirical methods in natural language processing. Copenhagen, Denmark
6. Plank B, Søgaard A, Goldberg Y (2016) "Multilingual part-of-speech tagging with bidirectional long short-term memory models and auxiliary loss." In 54th annual meeting of the association for computational linguistics, Berlin, Germany
7. Wai TT (2019) Myanmar language part-of-speech tagging using deep learning models. Int J Sci Eng Res 10(3):1020–1024
8. Kumar S, Kumar MA, Soman K (2019) Deep learning based part-of-speech tagging for Malayalam twitter data. J Intell Syst 28(3):423–435
9. Anbananthen KSM, Krishnan JK, Sayeed MS, Muniapan P (2017) Comparison of stochastic and rule-based pos tagging on Malay online text. Am J Appl Sci 14(9):843–851
10. Huang Z, Eidelman V, Harper M (2009) "Improving a simple bigram HMM part-of-speech tagger by latent annotation and self-training." In Proceedings of human language technologies: the 2009 annual conference of the North American Chapter of the association for computational linguistics, companion volume, Short papers, Boulder, Colorado
11. Lee SZ, Tsujii JI, Rim HC (2000) "Part-of-speech tagging based on hidden Markov model assuming joint independence." In Proceedings of the 38th annual meeting on association for computational linguistics, Hong Kong
12. Dandapat S, Sarkar S, Basu A (2007) "Automatic part-of-speech tagging for Bengali: an approach for morphologically rich languages in a poor resource scenario." In Proceedings of the 45th annual meeting of the acl on interactive poster and demonstration sessions, Prague, Czech Republic
13. Ekbal A, Bandyopadhyay S (2008) "Part of speech tagging in Bengali using support vector machine." In International conference on information technology, Bhubaneswar, India
14. Ekbal A, Haque R, Bandyopadhyay S (2007) "Bengali part of speech tagging using conditional random field." In 7th international symposium of natural language processing (SNLP), Pattaya, Thailand
15. Ekbal A, Hasanuzzaman M, Bandyopadhyay S (2009) "Voted approach for part of speech tagging in Bengali." In 23rd Pacific Asia conference on language, information and computation, Hong Kong.
16. Sarkar K, Gayen V (2012) "A practical part-of-speech tagger for Bengali." In 2012 third international conference on emerging applications of information technology. Kolkata, India
17. Sarkar K, Gayen V (2013) "A trigram HMM-based POS tagger for Indian languages." In International conference on frontiers of intelligent computing: theory and applications (FICTA), Berlin, Heidelberg
18. Sarkar K (2016) A CRF based POS tagger for code-mixed Indian social media text. arXiv
19. Kabir MF, Abdullah-Al-Mamun K, Huda MN (2016) "Deep learning based parts of speech tagger for Bengali." In 5th international conference on informatics, electronics and vision (ICIEV), Dhaka, Bangladesh
20. Chollet F et al "Keras," keras.io, 2015. [Online]. Available: https://keras.io. Accessed 04 Jun 2020
21. Grave E, Bojanowski P, Gupta P, Joulin A, Mikolov T (2018) Learning word vectors for 157 languages. arXiv:1802.06893
22. Pan SJ, Yang Q (2010) A survey on transfer learning. IEEE Trans Knowl Data Eng 22(10):1345–1359
23. Mikolov T, Sutskever I, Chen K, Corrado G, Dean J (2013) "Distributed representations of words and phrases and their compositionality." In Proceedings of the 26th international conference on neural information processing systems, vol 2, Lake Tahoe, Nevada

24. Mikolov T, Chen K, Corrado G, Dean J (2013) Efficient estimation of word representations in vector space, arXiv
25. ICON (2013) "NLP tools contest on named entity recognition in Indian languages, 2013." Icon 2013, 2013. [Online]. Available: https://ltrc.iiit.ac.in/icon/2013/nlptools/

A Comparative Study on Effect of Temporal Phase for Speaker Verification

Doreen Nongrum and Fidalizia Pyrtuh

Abstract In this paper, the temporal phase influence on speech signal is demonstrated through different experimental models, notably for speaker verification. Feature extraction is a fundamental block in a speaker recognition system responsible for obtaining speaker characteristics from speech signal. The commonly used short-term spectral features accentuate the magnitude spectrum while totally removing the phase spectrum. In this paper, the phase spectrum knowledge is extensively extracted and studied along with the magnitude information for speaker verification. The Linear Prediction Cepstral Coefficients (LPCC) are extracted from speech signal temporal phase and its scores are fused with Mel-Frequency Cepstral Coefficients (MFCC) scores. The trained data are modeled using the state-of-art speaker specific Gaussian mixture model (GMM) and GMM-Universal Background Model (GMM-UBM) for both LPCC and MFCC features. The scores are matched using dynamic time warping (DTW). The proposed method is tested on a fixed-pass phrase with a duration of <5 s in a speech signal. The score level fusion technique helps in the reduction of equal error rate (EER) and improves recognition rate.

1 Introduction

Speaker recognition is a technique to recognize a person from one's own voice which is accomplished by extracting information from a person's speech samples [1–3]. Speaker recognition is classified into speaker verification and speaker identification. In speaker identification, an unknown speaker does not claim an identity whereas in speaker verification, a claim is associated with the unknown speaker's input utterance. During testing phase, the unknown speaker's input utterance is compared to the

D. Nongrum (✉) · F. Pyrtuh
Department of Electronics and Communication Engineering, North-Eastern Hill University, Shillong, Meghalaya, India

F. Pyrtuh
e-mail: fpyrtuh@nehu.ac.in

S. Basu et al. (eds.), *Proceedings of International Conference on Frontiers in Computing and Systems*, Lecture Notes in Networks and Systems 404,
https://doi.org/10.1007/978-981-19-0105-8_56

model of the registered speaker whose identity is claimed [4]. The first stage in recognizing a person from speech samples is to extract speaker unique information/features through Mel Frequency Cepstral Coefficients (MFCC) [5] and Linear Prediction Cepstral Coefficients (LPCC) [6] which are explored extensively in this paper. In the following stage, the extracted features are further processed using speaker models such as Gaussian mixture model (GMM) [7], adapted Gaussian mixture models known as Gaussian mixture model-Universal Background Model (GMM-UBM) [8], i-vector modeling [9], deep neural network (DNN) approach [10] and convolutional neural network (CNN) model [11]. In this study, the speaker verification system is implemented using GMM and GMM-UBM speaker modeling techniques. The back end processing of the speaker verification system utilizes dynamic time warping (DTW) algorithm as a score matching technique, by aligning test data and trained data through mapping of the time axes [12]. The resultant scores are fused with weighted sum combination approach to effectuate a single score which can be used as a deciding factor for accepting or rejecting the claim [13].

The current ASV systems are gaining recognition in different security sectors like surveillance, banking, finance, etc., where short utterances signals are used as biometric [14–17]. A current trend for short utterances used practical scenario is 15 s [18] but fixed-pass phrase of <5 s is considered in this paper to replicate the existing pin number or fingerprint validation systems.

The phase extraction techniques such as All Pass Cepstral Coefficients (APCC) [19], Instantaneous Frequency Cosine Coefficients (IFCC) [20] and Linear Prediction Coefficients (LPC) [21] are extensively studied in literature, whereas in this paper, Linear Prediction Coefficients (LPCC) technique is adopted as the time domain least squares error criterion translates into frequency domain spectral matching criterion. Also, frequency-domain analysis gives the signal's energy distribution over a range of frequencies, which in turn interpret into speaker's information/content. On the contrary, comprehensive review on various applications of automatic speech/speaker recognition for Indian languages [22], human emotion and music mood recognition study [23], forensic study [24] has also been explored.

The human auditory system is observed to be sensitive to magnitude spectrum while phase spectrum is generally discarded. In this paper, the phase spectrum is studied extensively, and it is proven to contain some speaker specific information which can augment the recognition rate of a speaker.

The feature selection approaches include statistical voice activity detection, periodicity based voice activity detection, self-adaptive VAD and vowel and non-vowel-like region selection [25]. In this paper, vowel-like region (VLR) selection is performed as it is more effective in capturing high energy regions of speech signal. The vowel-like regions include vowels, semi-vowels and diphthongs. The vowels contain high energy and are spectrally well defined which can be recognized easily. Vowel-like regions [26] are detected by using vowel-like region onset point (VoP) [27] which denotes the point in the speech signal when a vowel appears. An energy-based threshold (0.06*average energy) is used to identify these regions. Only regions above the threshold are examined for feature extraction, and any remaining features below the threshold are eliminated. In this work, the phase spectrum is being considered

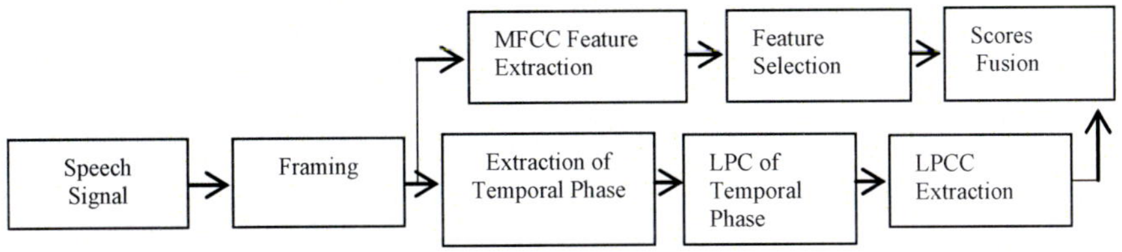

Fig. 1 Block diagram of the proposed method

for extracting speaker specific information from the speech signal [28–30]. LPCC are extracted from the temporal phase of the speech signal and scores obtained by using temporal phase [31–33], as a feature extraction block in the speaker verification system, are fused with MFCC features.

The block diagram representation is as shown in Fig. 1. To obtain the phase spectrum, the Discrete Fourier Transform (using FFT) is calculated for each frame (25 ms). The time domain phase is computed and LPC are extracted at the frame level. LPCC are then computed from the temporal phase feature vector. At the score level, the MFCC features obtained from the VLR are fused with temporal phase using the weighted sum approach. Because MFCC is given more weight than the other feature set, the weighting factor, λ, is set to 0.8.

$$S_{\text{combined}} = \lambda S_{\text{MFCC}} + (1 - \lambda) S_{T-\text{Phase}}$$

where S_{combined} are the fused scores of S_{MFCC} and $S_{T-\text{Phase}}$, obtained from MFCC and temporal phase model, respectively. To evaluate the performance of the speaker verification system, the fused scores are utilized to calculate the equal error rate and depict the detection error tradeoff curve.

2 Temporal Phase

The Fourier transform $S(e^{j\omega})$ of the speech signal $s[n]$ is given as,

$$S\left(e^{j\omega}\right) = \sum_{n=-\infty}^{\infty} s[n] e^{-j\omega n}$$

where $S\left(e^{j\omega}\right) = \left|S\left(e^{j\omega}\right)\right| e^{j\measuredangle S\left(e^{j\omega}\right)}$,

$$e^{j\measuredangle S\left(e^{j\omega}\right)} = \frac{S\left(e^{j\omega}\right)}{\left|S\left(e^{j\omega}\right)\right|}$$

Table 1 Proposed algorithm

Step 1: Frame the speech signal into 25-ms-long segments that do not overlap (number of frames = M)
Step 2: for i = 1 to M do
Compute Fourier transform of ith frame $s_i[n]$, i.e., $S_i(e^{jw})$
Obtain phase spectrum of $S_i(e^{jw})$
Compute frequency domain phase $e^{j\measuredangle S(ej\omega)}$
Compute the inverse Fourier transform of $e^{j\measuredangle S(ej\omega)}$ to obtain time domain phase
Extract LPC from time domain phase
end for
Step 3: Compute LPCC from T-phase LPC feature vector
Step 4: Extract MFCC with vowel-like regions
Step 5: Fuse the scores obtained from Step 3 and 4

The quantities $\left|S\left(e^{j\omega}\right)\right|$ and $\measuredangle S\left|e^{j\omega}\right|$ are the magnitude spectrum and phase spectrum of $s[n]$, respectively.

The phase signal $y[n]$ is generated using

$$y[n] = F^{-1}\left\{\frac{S\left(e^{j\omega}\right)}{\left|S\left(e^{j\omega}\right)\right|}\right\}$$

which can also be expressed as

$$y[n] = F^{-1}\left(e^{j\measuredangle S\left(e^{j\omega}\right)}\right)$$

where F^{-1} is the Inverse Discrete Fourier Transform.

The LPCC are extracted from the temporal phase $y[n]$ to generate a feature vector with 13 coefficients for each frame. The extracted features are used to model the GMM and GMM-UBM modeling techniques.

The proposed algorithm is as shown in Table 1 and realized using Matlab 2017 version with Voicebox Toolbox [34] and MSR Toolkit [35] (Table 2).

3 Experimental Setup

The dataset used for the speaker verification system consists of 200 speakers (100 female and 100 male) with 30 utterances each, all of which are used for training the background model. For speaker specific model (GMM) training and testing, there are 60 genuine speakers (30 female and 30 male), each with 9 speech recordings, of

Table 2 Experimental results

Feature set	Speaker model	EER(%)	
		Female	Male
MFCC	GMM	11.3333	16
MFCC + VLR		7.3333	10.6667
T-Phase		38.6667	34
MFCC + T-Phase		13.333	14
MFCC + VLR + T-Phase		6.6667	10
MFCC	GMM-UBM	4.0741	6.6667
MFCC + VLR		3.3333	4.8148
T-Phase		30	27.7778
MFCC + T-Phase		5.1852	7.7778
MFCC + VLR + T-Phase		2.963	4.8148

which 4 utterances are for training and 5 utterances are utilized for testing. There are 60 imposter speakers (30 female and 30 male), with 50 speech recordings each, all of which are used for testing. The training and test segments contain fixed-pass phrase of <5 s duration (Fig. 2).

4 Conclusion

This work studies the temporal phase effect on speech signal for speaker verification. The results show that when using the Gaussian mixture modeling (GMM) approach, the score level fusion of temporal phase and MFCC (with VLR) decreases the EER by 0.67% compared to MFCC (with VLR) for both male and female speakers. Using Gaussian mixture modeling-Universal Background Modeling (GMM-UBM), score level fusion of temporal phase and MFCC (with VLR) EER improves by 0.37% for female speakers. It is observed that the phase spectrum of the speech signal can provide additional speaker information when it is combined with spectral features such as MFCC.

Currently, the exploration of neural networks to enhance speaker recognition has been widely explored in ANN, CNN and DNN. The future scope this paper, will focus on developing a system using these state-of-art neural networks [10, 11].

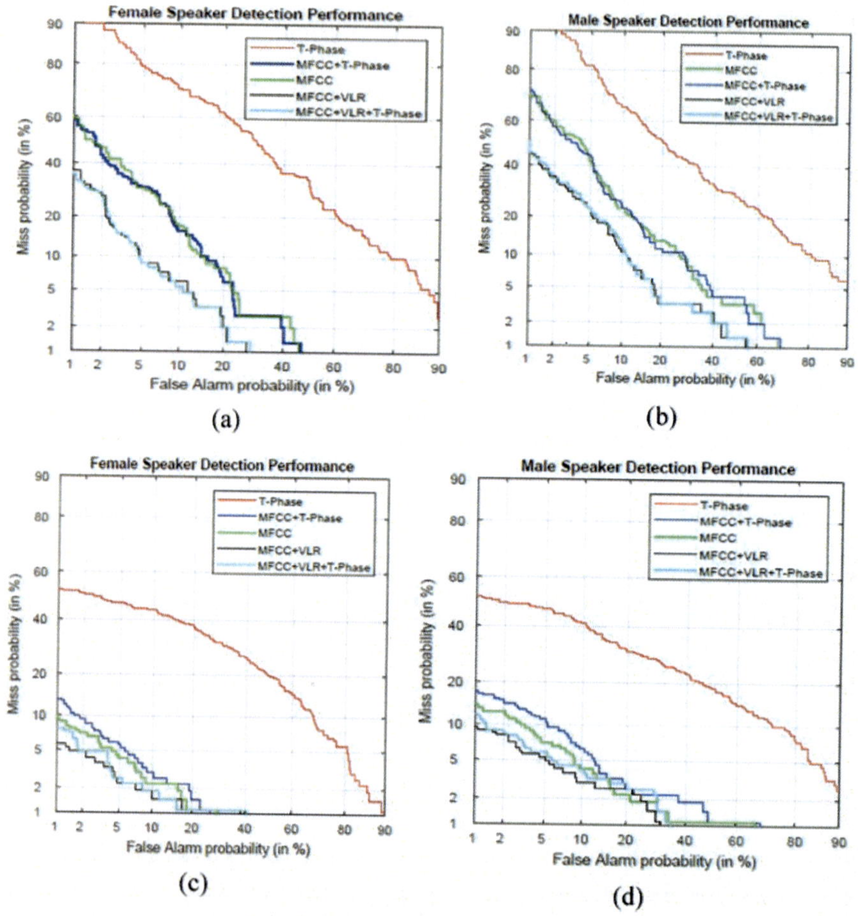

Fig. 2 Depicts the DET curve for **a** GMM model for female **b** GMM model for male **c** GMM-UBM model for female **d** GMM-UBM model for male

References

1. Atal BS (1976) Automatic recognition of speakers from their voices. Proc IEEE 64(4):460–475
2. Rabiner L, Juang B (1993) Fundamentals of speech recognition. Prentice-Hall, Inc, Upper Saddle River, NJ, USA
3. Doddington G (1985) Speaker recognition identifying people by their voices. Proc IEEE 73(11):1651–1664
4. Naik J (1990) Speaker verification: a tutorial. IEEE Commun Mag 28(1):42–48
5. Furui S (1981) Cepstral analysis technique for automatic speaker verification. IEEE Trans Acoustics Speech Signal Process 29(2):254–272
6. Makhoul J (1975) Linear prediction: a tutorial review. Proc IEEE 63(4):561–580
7. Reynolds DA, Rose RC (1995) Robust text- independent speaker identification using Gaussian mixture speaker models. IEEE Trans Speech Audio Process 3(1):72–83
8. Reynolds DA, Quatieri TF, Dunn RB (2000) Speaker verification using adapted Gaussian mix-ture models. Digital Signal Proces 10(1):19–41

9. Dehak N, Kenny P, Dehak R, Dumouchel P, Ouellet P (2011) Front-end factor analysis for speaker verification. IEEE Trans Audio Speech Lang Process 19(4):788–798
10. Richardson F, Reynolds D, Dehak N (2015) Deep neural network approaches to speaker and language recognition. IEEE Signal Process Lett 22(10):1671–1675
11. Jagiasi R, Ghosalkar S, Kulal P, Bharambe A (2019) "CNN based speaker recognition in language and text-independent small scale system." In 2019 third international conference on i-smac (iot in social, mobile, analytics and cloud) (I-SMAC), pp 176–179
12. Sakoe H, Chiba S (1978) Dynamic programming algorithm optimization for spoken word recognition. IEEE Trans Acoust Speech Signal Process ASSP-26(1):43–49
13. Răstoceanu F, Lazăr M (2011) "Score fusion methods for text-independent speaker verification applications." In 2011 6th conference on speech technology and human-computer dialogue (SpeD), pp 1–6. https://doi.org/10.1109/SPED.2011.5940740
14. Shetty M 'ICICI bank to roll out voice authentication'. Available at: https://timesofindia.indiatimes.com/business/india-business/icici-bank-to-roll-out-voice-authentication/articleshow/46818823.cms. April 6, 2015 [Online]
15. Loshin P 'Barclays replaces passwords with voice authentication'. Available at: https://searchsecurity.techtarget.com/news/450301866/Barclays-replacespasswords-with-voice-authentication. 3 Aug 2016 [Online]
16. Larcher A, Lee KA, Ma B, Li H (2012) 'RSR2015: database for text-dependentspeaker verification using multiple pass-phrases.' Proc Interspeech, pp 1580–1583
17. Kanagasundaram A, Vogt R, Dean D, Sridharan S, Mason M (2011) "i-vector based speaker recognition on short utterances." In Interspeech 2011
18. Griffiths J (2017) "Citi tops 1 million mark for voice biometrics authentication for Asia Pacific consumer banking clients". www.citigroup.com/citi/news/2017/170321b.htm
19. Vijayan K, Kumar V, Murty KSR (2014) "Allpassmodelling of Fourier phase for speaker verification". In Proceedings of ODYSSEY 2014: the speaker and lan-guage recognition workshop, Joensuu, Finland, pp 112–117
20. Soong FK, Rosenberg AE (1998) On the use of instantaneous and transitional spectral information in speaker recognition. IEEE Trans Acoustics Speech Signal Process 36(6):871–879
21. Gandhi A, Patil HA (2018) "Feature extraction from temporal phase for speaker recognition." In SPCOM 2018–12th international conference signal processing communications, pp 382–386
22. Singh A, Kadyan V, Kumar M, Bassan N (2020) ASRoIL: a comprehensive survey for automatic speech recognition of Indian languages .53(5), Springer Netherlands
23. Chaturvedi V, Kaur AB, Varshney V, Garg A, Chhabra GS, Kumar M (2021) "Music mood and human emotion recognition based on physiological signals: a systematic review." Multimed Syst 0123456789
24. Jessen M, Meir G, Solewicz YA (2019) Evaluation of nuance forensics 9.2 and 11.1 under conditions reflecting those of a real forensic voice comparison case (forensic_eval_01). Speech Commun 110:101–107
25. Graf S, Herbig T, Buck M et al (2015) Features for voice activity detection: a comparative analysis. EURASIP J Adv Signal Process 2015:91
26. Jelil S, Das RK, Sinha R, Prasanna SM (2015) "Speaker verification using gaussian posteriorgrams on fixed phrase short utterances". In 2015 ISCA, Dresden, Germany, pp 1042–1046
27. Prasanna SRM, Zachariah JM (2002) "Detection of vowel onset point in speech." In 2002 IEEE international conference on acoustics, speech, and signal processing, pp IV-4159-IV-4159. https://doi.org/10.1109/ICASSP.2002.5745575
28. Paliwal K, Alsteris L (2003) "Usefulness of phase spectrum in human speech perception". In Proceedings eighth European conference on speech communication and technology (EUROSPEECH2003), Geneva, Switzerland, pp 2117–2120
29. Oppenheim AV, Lim JS (1981) The importance of phase in signals. Proc IEEE 69(5):529–550
30. Schluter R, Ney H (2001) "Using phase spectrum information for improved speech recognition performance". In Proceedings of IEEE international conference acoustics speech and signal process (ICASSP), Salt Lake City, UT, USA, vol 1, pp. 133–136

31. Shi G, Shanechi MM, Aarabi P (2006) On the importance of phase in human speech recognition. IEEE Trans Audio Speech Lang Process 14:1867–1874
32. Vijayan K, Reddy P, Murty KSR (2016) Significance of analytic phase of speech signals in speaker verification. Speech Comm 81(C):54–71
33. Murty KR, Yegnanarayana B (2006) Combining evidence from residual phase and MFCC features for speaker recognition. IEEE Signal Process Lett 13:52–56
34. Brookes M et al (2018) "VOICEBOX: speech processing toolbox for MATLAB," Software available [April 2018]
35. MSR identity toolkit, Microsoft research, http://research.microsoft.com/. Available Online 2013. Last Accessed 28 Feb 2015

An Acoustic/Prosodic Feature-Based Audio Dataset for Assamese Speech Summarization

Priyanjana Chowdhury, Swarnav Das Barman, Catherina Basumatary, Sandeep Deva Misra, Sanghamitra Nath, and Utpal Sharma

Abstract The spoken form is the most predominant form of communication laden with rich information than its textual counterpart. Development of summarization techniques for spoken data in a resource-poor language is a challenging task, especially due to the unavailability of appropriate datasets. In this paper, we focus on creating a dataset for speech summarization in Assamese, a low-resource language of Northeast India, the genre of speech being broadcast news. The contribution of this paper is twofolds: Firstly, a dataset of the audio samples of the broadcast news comprising of acoustic/prosodic features is developed, where each tuple is the manually segmented utterance representing a sentence. Each utterance is labeled as either summary-worthy or not summary-worthy, in accordance to the handcrafted summaries. Secondly, the relation of the prosodic features in indicating the informativeness of an utterance is evaluated. Binary classification is performed with the help of various classifiers, and a statistical analysis of the performance of the classifiers is presented. This approach for speech summarization is highly relevant for resource-poor language which does not have adequate resources to generate the textual representation of the audio. To the best of our knowledge, this is the first effort to develop a dataset in Assamese language for speech summarization.

Keywords Assamese speech dataset · Speech summarization · Acoustic/Prosodic features · Low-resource language · Broadcast news dataset · Binary classification

---

P. Chowdhury (✉) · S. Das Barman · C. Basumatary · S. D. Misra · S. Nath · U. Sharma
Tezpur University, Tezpur, Assam, India
e-mail: priyan@tezu.ernet.in

S. Nath
e-mail: s.nath@tezu.ernet.in

U. Sharma
e-mail: utpal@tezu.ernet.in

S. Basu et al. (eds.), *Proceedings of International Conference on Frontiers in Computing and Systems*, Lecture Notes in Networks and Systems 404,
https://doi.org/10.1007/978-981-19-0105-8_57

1 Introduction

The goal of automatic summarization system is to discern the main theme and the core information that represents the original text or spoken document. The need for speech summarization is potent since most of the data available are not restricted to only text. The spoken form is the most predominant form of communication which is laden with more information than its textual counterpart [9]. Some of the most common spoken communications are news broadcast, lectures, conversations, meetings etc.

Extractive summarization is one of the ways of summarizing where the important utterances are identified and aggregated into a summary representing the main theme of the spoken document [2]. Development of text summarization methods has been flourishing for quite a few years; text-based approaches for English, Japanese, Chinese, and other resource-rich languages exist, where transcripts are generated using an automatic speech recognizer (ASR) upon which text-based methods are employed for speech summarization [3–5, 10–12, 15]. This approach is fairly popular since it only requires an ASR and existing text methods, but also suffers from ASR errors like noise, disfluencies, recognition errors that degrade summarization results. It also misses out on the underlying prosodic information of an utterance. Moreover, this approach is not feasible for languages lacking resources such as acoustic models, speech recognizers, and other text computational tools. Audio-based summarization which is dependent on purely acoustic/prosodic data of the spoken document is a solution to this circumstance [7, 14].

For the above reasons, we have developed a dataset of utterances represented by its acoustic/prosodic data, which is associated to the corpus of broadcast news in Assamese to address the task of extractive speech summarization. Assamese, belonging to the Indo-Aryan language family, is one of India's national languages and spoken widely in Northeast India. It is a low-resource language, and there are several ongoing efforts to develop corpus for computational linguistics [16, 17] but none dedicated to summarization. To the best of our knowledge, this work is the first attempt of developing a dataset for speech summarization in Assamese language. This paper is organized in the following way: Sect. 2 presents the corpus details, the development of a dataset of prosodic features is described in Sect. 3, experimental results, and discussions in Sect. 4, conclusions and future work in Sects. 5 and 6, respectively.

2 Corpus

The corpus comprises of radio news speech read by a professional newsreader broadcasted in All India Radio Akashbani Guwahati. The read news is a continuous speech having little or no pauses between words and sentences. The corpus is roughly three hours in duration, and the news stories collected here are chosen such that greater range of intonational variation is expressed. Each news story has a recurrent structure in its narration. The speaker first introduces himself, reads out all the headlines of the topics going to be covered, followed by a brief discussion of those topics. Some other topics not mentioned in the headlines are also discussed, followed by an ending note.

2.1 *Preprocessing of Speech Data*

Background and static noise of the radio news broadcasts were removed by using a built-in tool in Audacity software. The audio samples are then converted to mono channel and stored in WAV format. All the news stories are manually segmented into sentences. There are approximately 1.5k audio samples after the manual segmentation, the average duration of a segment being 10–15 s.

3 The Dataset

A typical summarization method is to select the important utterances and aggregating them into a summary; this method is called extractive summarization. In extractive speech summarization, the task of selecting important utterance can be formulated as a binary classification problem, where an utterance is labeled as either to be included in a summary or not to be included [1, 18]. In order to develop a summarizer in the form of a classifier, there must be training instances with corresponding handcrafted summaries or labeled data which denotes the summary. Thus, we have built a dataset congruous with the corpus, where each instance is the segmented utterance, the attributes are the acoustic/prosodic features of the segments, and the class-label is 1 and 0 for in-summary and not-summary, respectively.

3.1 *Features*

Humans use intonational variation—which is indicated by variations in pitch range, prominence or stressing an utterance—to characterize the importance of an utterance while speaking; these elements of speech are called prosody [8]. Prosody provides important information which are not encapsulated in a sentence or other lexical units; for example, when an utterance is stressed, it is intuitively considered to have a important context irrespective of lexical meaning. The 'prosodic stress' element is used in this dataset, and the acoustic features which are associated to it are pitch variations, duration, and intensity. In some cases, timbre or voice quality is also considered to represent stress in an utterance, which are associated acoustically with spectral features [6]. Thus, the features that are extracted from the utterances and

Prosodic elements	*Acoustic feature*	*Description*
Pitch variations	Pitch	F_0 (fundamental frequency)
	MinPitch	Minimum value of F_0
	MaxPitch	Maximum value of F_0
	PitchRange	Difference between minimum and maximum F_0
	PitchSlope	Slope of the curve of F_0
Loudness	Intensity	Perceived energy of the audio
	MaxIntensity	Maximum value of intensity
	MinIntensity	Maximum value of intensity
	MeanEnergy	Average value of intensity
	RMSEnergy	Root-mean-square of intensity
Duration	SentenceDuration	Time length of each spoken utterance
Voice quality	MFCC features	Timbre of the voice

Fig. 1 Acoustic features associated with stressed utterance used in the dataset

included in the dataset are depicted in Fig. 1. There are 11 features for prosodic elements of pitch, intensity, and duration; and 13 MFCC features for timbre. In total, there are 24 features in this dataset and 1 class-label.

3.2 Labeling

The class-labels of the dataset are 1 denoting summary-worthy sentence, and 0 for not summary-worthy sentence. The dataset is labeled manually. Several volunteers comprising of both native speakers and non-native speakers of Assamese took part in handcrafting the summaries and selecting the important sentences. The native speakers made their choice by considering both the meaning of the sentence and the intonational variation of the speaker in that sentence, while the non-native speakers relied only on the intonational variation as the content of the continuous speech was difficult for them to understand. The final label of that utterance was decided by taking the majority of the votes into account based on the decisions of both the native and

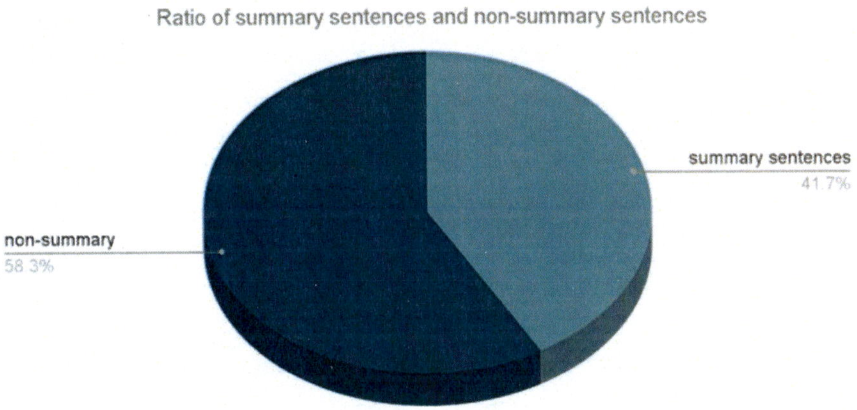

Fig. 2 Class distribution of summary-worthy (41.7%) and non summary-worthy (58.3%) utterances in the dataset

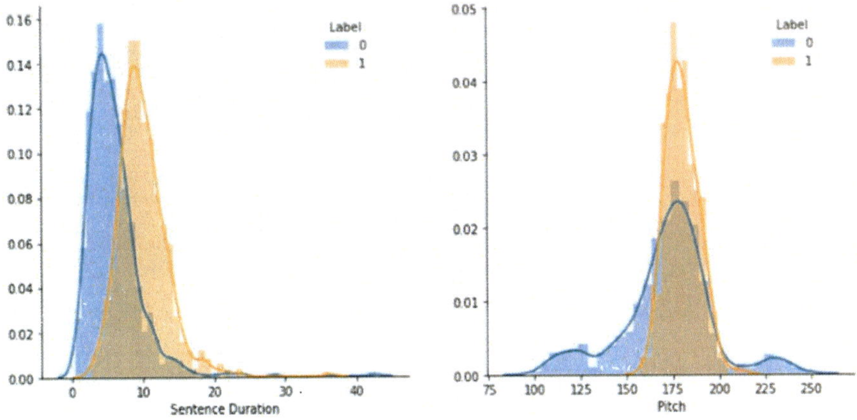

(a) Data distribution per Sentence duration feature

(b) Data distribution per Pitch feature

Fig. 3 Data distribution of acoustic/prosodic features

non-native speakers of the language. Figure 2 shows the ratio of summary-worthy sentences to non summary-worthy sentences to illustrate the class distribution.

4 Experimental Results and Discussion

The acoustic data of the utterances collected in the dataset can be visualized in the heatmap in Fig. 4 where the relationship between the features are characterized by the gradient of the color. At the initial stage, the acoustic features of pitch, intensity,

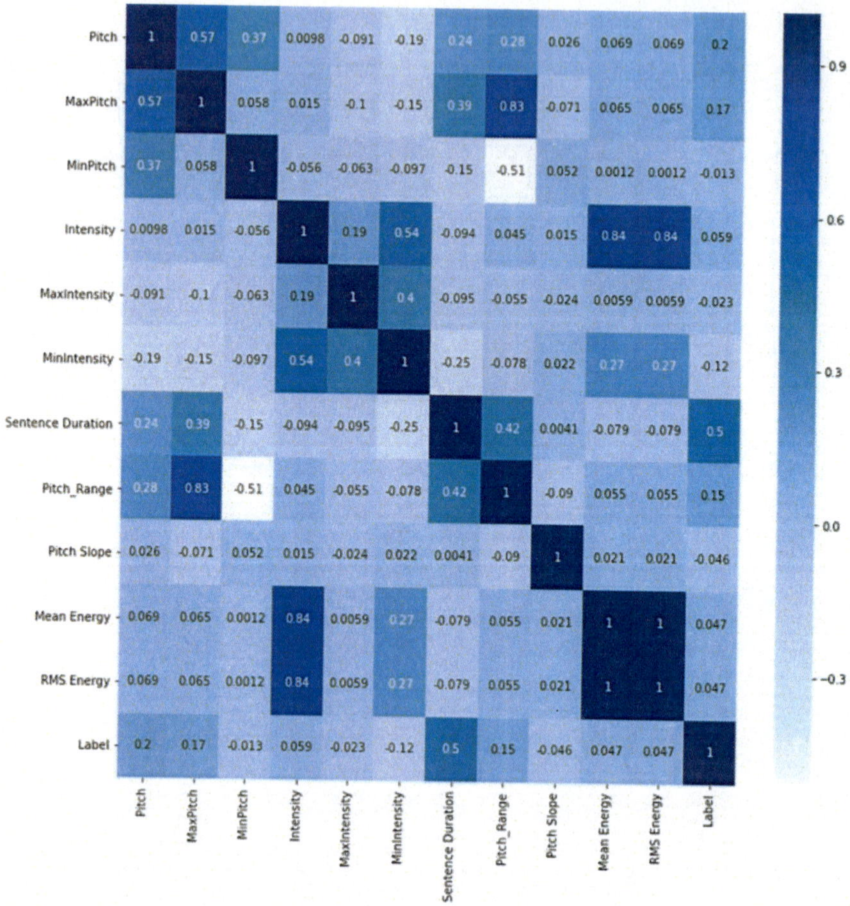

Fig. 4 Correlation of features in the dataset (features from the top: pitch, MaxPitch, MinPitch, Intensity, MaxIntensity, MinIntensity, SentenceDuration, PitchRange, PitchSlope, MeanEnergy, RMSEnergy, Label)

and duration are utilized in the dataset. From this figure, it is evident that the feature *sentence duration* is the most predictive feature since it has a high correlation with the class-feature *label* that indicates the inclusion of that utterance in the summary. The feature *pitch* is also fairly predictive in the same context. The class distribution of these two features are denoted in Fig. 3a, b. The sentence duration of our audio segments are concentrated in a range of a few seconds to 20 s, and it is interesting to note that class 1 denoting summary-sentences audio segments have longer duration compared to class 0 denoting non-summary sentences, which indicates that sentence duration plays an important role in deciding the utterance importance.

Classifiers like random forest, decision tree, gradient boosting, naive Bayes, k-nearest neighbor, logistic regression, and support vector machine (SVM) are used

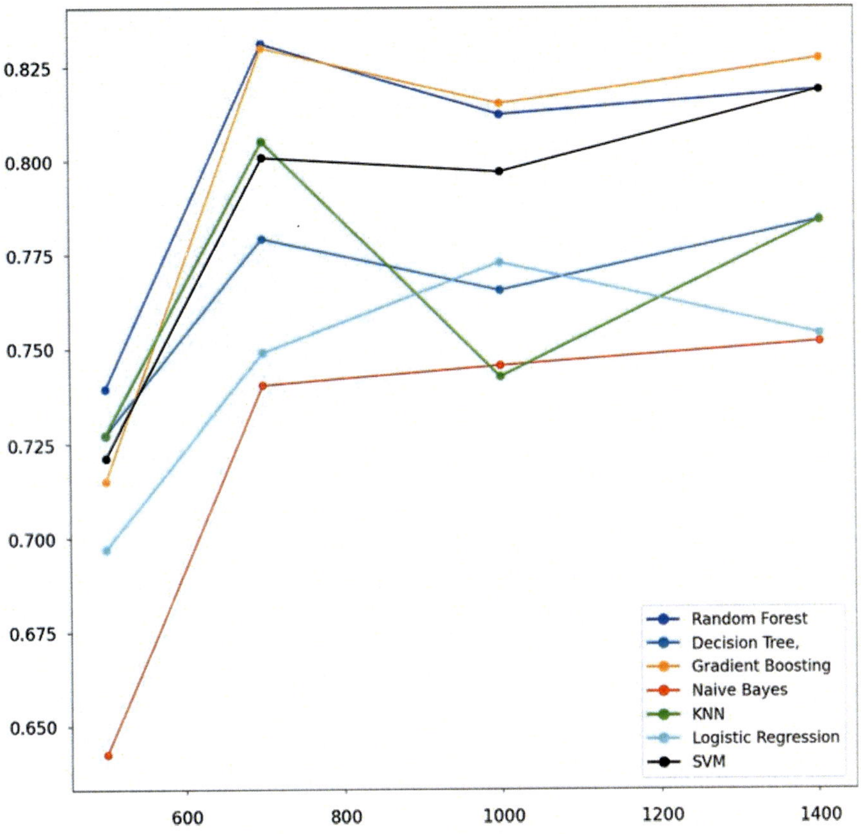

Fig. 5 Comparison of performances of various classifiers on different size of dataset (legend from top to bottom: random forest, decision tree, gradient boosting, naive Bayes, KNN, logistic regression, SVM)

Table 1 Comparing the three best performing models at various stage of dataset

Stage of dataset	Accuracy using Gradient boosting	Accuracy using Random forest	Accuracy using SVM
Original dataset	82.72	81.86	81.86
Feature selection on original dataset	85.96	86.61	86.72
Updated dataset (after including MFCC features)	86.83	86.18	88.55
Feature selection on revised dataset	87.69	87.47	90.28

to perform binary classification, and their performances are compared in Fig. 5. It is observed from this graph that the performance of the models improves with increasing size of the data.

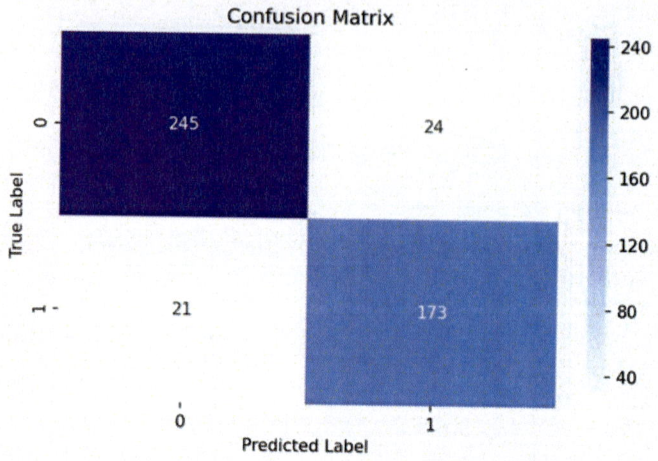

Fig. 6 Confusion matrix of the performance of SVM on the dataset, with precision = 0.88, recall = 0.89, and f score = 0.88

After adding the acoustic features of timbre represented by MFCC features, the performance of the models improved even more. Table 1 presents a comparison of the accuracy achieved by the three best-performing models. Feature selection is done using correlation statistics, and the best performance is achieved by SVM for a set of 21 features, with precision= 0.88, recall= 0.89, and f score = 0.88, as depicted in Fig. 6.

5 Conclusion

In this paper, we built a Assamese radio news speech corpus and a corresponding dataset to address the task of speech summarization as a binary classification problem. This approach can be used for other low-resource languages for which ASRs are not easily available to generate transcripts for summarization. This corpus of read continuous speech and its corresponding dataset would be valuable for various speech processing tasks due to the inclusion of a wide array of prosodic/acoustic features, applicable in domains like emotion recognition, continuous speech recognition, etc. Performance of benchmark classifiers on our dataset, including the best performing ones and the feature set for which the classifiers give the highest accuracy, is discussed in detail for comparison and analysis. It is found out that sentence duration and pitch are among the most predictive acoustic feature for indicating a summary-worthy utterance, and SVM performed the best with an accuracy of 0.90 and f-score of 0.88 in a scale of 0 to 1.

6 Future Work

The corpus will be expanded by adding transcripts of the each broadcast session. Speech-to-text tool for continuous speech in Assamese is scarcely available, so the transcripts needs to be generated manually. The transcripts would then provide further lexical information that will be beneficial for selecting the important utterances. The lexical information can be leveraged in the dataset through features like number of words in the current and previous sentence, respectively, number of named entities in that sentence, ratio of named entities of that sentence to total unique named entities, etc., [13] to analyze the degree of dependency of a summarizer to different modes of information. In addition, the acoustic features associated with other prosodic elements like intonation can also be extracted from the utterances to include in the dataset which would allow us to compare which prosodic features are more predictive of informative utterances.

References

1. Bokaei MH, Sameti H, Liu Y (2016) Extractive summarization of multi-party meetings through discourse segmentation. Nat Lang Eng 22(1):41
2. Chen B, Lin SH, Chang YM, Liu JW (2013) Extractive speech summarization using evaluation metric-related training criteria. Inf Process Manage 49(1):1–12
3. Chen KY, Liu SH, Chen B, Wang HM, Chen HH (2016) Exploring the use of unsupervised query modeling techniques for speech recognition and summarization. Speech Commun 80:49–59
4. Chen KY, Liu SH, Chen B, Wang HM, Chen HH (2016) Novel word embedding and translation-based language modeling for extractive speech summarization. In: Proceedings of the 24th ACM international conference on multimedia, pp 377–381
5. Chen KY, Liu SH, Chen B, Wang HM, Jan EE, Hsu WL, Chen HH (2015) Extractive broadcast news summarization leveraging recurrent neural network language modeling techniques. IEEE/ACM Trans Audio Speech Lang Process 23(8):1322–1334
6. Collins B, Mees IM (2013) Practical phonetics and phonology: a resource book for students. Routledge
7. González-Gallardo CE, Deveaud R, Sanjuan E, Torres JM (2020) Audio summarization with audio features and probability distribution divergence. arXiv:2001.07098
8. Hirschberg J (2002) Communication and prosody: functional aspects of prosody. Speech Commun 36(1–2):31–43
9. Hori C, Furui S (2001) Advances in automatic speech summarization. In: Seventh European conference on speech communication and technology
10. Liu SH, Chen KY, Chen B, Wang HM, Yen HC, Hsu WL (2015) Combining relevance language modeling and clarity measure for extractive speech summarization. IEEE/ACM Trans Audio Speech Lang Process 23(6):957–969
11. Liu SH, Lee HS, Hung HT, Chen KY, Chen B, Wang HM, Yen HC, Hsu WL (2015) Incorporating proximity information in relevance language modeling for extractive speech summarization. In: 2015 Asia-Pacific signal and information processing association annual summit and conference (APSIPA). IEEE, pp 401–407
12. Liu TE, Liu SH, Chen B (2019) A hierarchical neural summarization framework for spoken documents. In: ICASSP 2019—2019 IEEE international conference on acoustics, speech and signal processing (ICASSP). IEEE, pp 7185–7189

13. Maskey S, Hirschberg J (2005) Comparing lexical, acoustic/prosodic, structural and discourse features for speech summarization. In: Ninth European conference on speech communication and technology
14. Maskey S, Hirschberg J (2006) Summarizing speech without text using hidden Markov models. In: Proceedings of the human language technology conference of the NAACL, companion volume: short papers, pp 89–92
15. Ogawa A, Hirao T, Nakatani T, Nagata M (2019) ILP-based compressive speech summarization with content word coverage maximization and its oracle performance analysis. In: ICASSP 2019—2019 IEEE international conference on acoustics, speech and signal processing (ICASSP). IEEE, pp 7190–7194
16. Sarma H, Saharia N, Sharma U, Sinha SK, Malakar MJ (2013) Development and transcription of Assamese speech corpus. arXiv:1309.7312
17. Sarma SK, Bharali H, Gogoi A, Deka R, Barman A (2012) A structured approach for building assamese corpus: insights, applications and challenges. In: Proceedings of the 10th workshop on Asian language resources, pp 21–28
18. Zhang J, Fung P (2009) Active learning of extractive reference summaries for lecture speech summarization. In: Proceedings of the 2nd workshop on building and using comparable corpora: from parallel to non-parallel corpora (BUCC), pp 23–26

Influential Node Detection in Online Social Network for Influence Minimization of Rumor

Maitreyee Ganguly, Paramita Dey, Swarnesha Chatterjee, and Sarbani Roy

Abstract In any online social media platform, it is necessary to reduce the effect of rumor data from original information as it may cause harm to society. Influential users can be detected through different centrality measures. When the rumor is generated through some influential users, they have more impact on society. Here, we have proposed a prognostic method to distinguish those influential users of online social media, based on network analysis. The susceptible-infectious-recovered (SIR) model has been used for simulation of the propagation of information. For the particular seed nodes which have been chosen by practicing different centrality measures, detailed relative learning in terms of infected nodes is also exhibited. Selection of seed nodes through centrality measure is computationally exhaustive; therefore, we form a composite model, where the original social network is decomposed using the k-core, and centrality nodes are obtained from that decomposed network. Centrality measurements thus derived from the generated network are used as the seed of information propagation. Another important result derived from the empirical study is that not only the influential nodes, but neighbors of the influential nodes also have a greater impact on maximizing the effect of rumors.

Keywords Centrality analysis · Online social network · Information propagation · SIR model

1 Introduction

As online social networking platforms are becoming significant in today's life by enacting an indispensable role in gauging the opinion of people, it is also essential to notice what kind of content is present there. A rumor is an unverified piece of content

M. Ganguly (✉) · P. Dey · S. Chatterjee
Government College of Engineering & Ceramic Technology, Kolkata, India
e-mail: maitreyee12aug@gmail.com

S. Roy
Jadavpur University, Kolkata, India
e-mail: sarbani.roy@jadavpuruniversity.in

S. Basu et al. (eds.), *Proceedings of International Conference on Frontiers in Computing and Systems*, Lecture Notes in Networks and Systems 404,
https://doi.org/10.1007/978-981-19-0105-8_58

because there is no proper evidence that can support it, and hence, the truthfulness behind it remains unresolved, while it still keeps circulating in the network [11]. It has been observed that information propagated from the seed nodes, or having seed nodes in their path, maximizes the flow of information; in this case, rumor. Therefore, it is really important to find out all the individual influential nodes in a network because even if they are rare, but still this limited number of nodes can rapidly propagate any content in a network. Propelled by this fact, we have proposed an algorithm (Fig. 1) to find out those seed nodes in a network and determine the correct path of information propagation based on the top five centrality analysis processes [4] together with the SIR model. Centrality-based heuristic models [5] run faster on these kinds of huge-scale networks and provide us a satisfactory result in less time.

On the other hand, the mode of information propagation in a network is very uncertain in nature as it relies on the relationship of the nodes with each other in the present network. The SIR model is used to detect the propagation of information. For each node that belongs to the network, this method is practiced in the k-core of the huge network which replicates the same as the nodes in the original network. From the outcomes, it can be clearly seen that the eigenvector centrality and page rank outperform other centrality measures for the diffusion of information within a network.

The foundation of the paper is prepared as follows: In Sect. 2, the related studies are briefly discussed. In Sect. 3, preliminaries of network properties are exhibited. The data collection and processing are presented in Sect. 4. In Sect. 5, the results of the study followed with discussion are shown broadly. Finally, the paper ends in Sect. 6 with the conclusion.

2 Related Studies

Over the last few years, different approaches have been suggested to recognize the influential user, and among those, most common is the degree centrality-based approach, which indicates the significance of a node in a network [1]. The influence of bridge nodes, which connect several communities can be measured using betweenness centrality [3]. If the impact of a node relies upon the number of its neighbors as well as on the significance of each neighbor node, eigenvector centrality is used [2]. Pairs of distinct influential node ranking algorithms are introduced by Zareie et. al [9], which use the heterogeneity of the neighbors of each node to get its ranking value. The SIR model is used to detect the propagation of rumor in complex networks [7]. To illustrate the dynamics of the SIR model on homogeneous networks, Ya et al. [8] offers a trust mechanism, which minimizes the final rumor dimension and the pace of rumor circulation. The SIR rumor spreading model with its likelihood of a spreader convert into a stifler by contacting a stifler, or another spreader is adapted in this study [10].

Here, in this present study, the identification of influential users using centrality analysis processes has been discussed. The SIR propagation model is used to detect the path information flow to minimize the effect of rumor on a network.

3 Network Preliminaries

In this paper, the SIR model together with centrality measurements from social network analysis (SNA) has been used to derive influential nodes and simulation of propagation. Data collected from Twitter is used to generate network structures in terms of nodes where every node is an individual actor or human within the network, and the relationship between them is represented with the help of edges, or links that connect them, only if they have interacted by some means. With the help of network analysis and graph theory, the importance of a node or edge is derived in terms of connectivity or information flow in the network. In this section, the centrality measurements and the SIR model are being presented briefly.

3.1 Degree Centrality

Degree centrality is applied to detect the popular or the most connected node in a network. This is the simplest way to find or measure the connectivity of any node in a network.

Degree of a vertex, η, in a graph $\Delta = (\nu, \epsilon)$, is defined by, $d_\eta(\Delta)$, where $|\nu| =$ vertices, and $|\epsilon| =$ edges, and

$$d_\eta(\Delta) = d_{\text{in}}(\eta) + d_{\text{out}}(\eta)$$

Then, the degree centrality of the node η is determined by, $C_{\text{dc}}(\eta; \Delta)$, and

$$C_{\text{dc}}(\eta; \Delta) = \frac{d_\eta(\Delta)}{(|\nu| - 1)}$$

3.2 Betweenness Centrality

Betweenness centrality counts the number of times a node is appeared in the shortest path with other nodes and behaves like the bridge node present in a network connecting another set of nodes. If a node or in this case an individual has a higher betweenness centrality value that means it has more influence on the connection or interaction between different clusters in a network.

Betweenees centrality of a vertex, η, can be expressed as,

$$C_{bc}(\eta; \Delta) = \sum \frac{\Theta_{\text{sp}}(\eta)}{\Theta_{\text{sp}}}$$

where $\Theta_{\text{sp}}(\eta)$ defines the shortest path passing through η, and Θ_{sp} defines the total count of shortest paths presented in that network.

3.3 Closeness Centrality

Closeness centrality detects the important node in a network elicited from how a node is adjacent to other nodes, in terms of the shortest path connecting each individual in a network. Nodes having higher closeness centrality within a particular network are excellent broadcasters of any news. So, it is essential to discover the influential nodes presented in the network within a single cluster. Closeness centrality of a vertex, η, is decepited as,

$$C_{\text{cc}}(\eta; \Delta) = \sum \frac{1}{\Phi_{xy}}$$

Here, y is the focal node, x is another node in the network, and is Φ_{xy} the shortest distance between these two nodes.

3.4 Eigenvector Centrality

To estimate the influence of a selected node within a particular network, eigenvector centrality is applied. All the nodes in a particular network are assigned with some relative score, based on the concept that nodes with high scores contribute more to a network in comparison to a low-scoring node. The node with high eigenvector scores exhibits that it is connected to such nodes who themselves are defined with high scores. For this fact here, in this study, eigenvector centrality is directly connected with the SIR Model.

Table 1 Comparative analysis of different centrality measures

Degree centrality	Betweenness centrality	Closeness centrality	E.V Centrality
56 (77)	277 (14607.8250)	277 (0.088)	56 (1.0)
344 (75)	175 (13502.8113)	290 (0.0868)	344 (0.9964)
271 (72)	338 (9826.7015)	322 (0.0866)	271 (0.9774)
172 (71)	188 (7155.2398)	344 (0.0864)	276 (0.9407)
283 (68)	144 (4690.3484)	315 (0.0863)	213 (0.8941)

The eigenvector centrality of a vertex, η, is represented by the weighted aggregation of the centralities of all nodes that are connected to it by an edge, Υ_{xy}, for graph $\Delta = (\nu, \epsilon)$.

$$e_x = \omega^{-1} \sum \Upsilon_{xy} e_j.$$

where Υ_{ij} defines the correlation between two nodes x and y and e is the eigenvector associated to the eigenvalue ω of Υ.

A comparative analysis of top five nodes with their centrality values of the nodes from the network is presented on Table 1.

3.5 SIR Model

In this work, we have adopted the SIR model to capture the information spreading rate by a node in a network. By analyzing the value S, I, and R of each node, the information spreading rate is represented as a form of a graph. In a network, rumor can be interpreted as an infectious epidemic disease, whose propagation is directly proportional to infection rate β. Here, we have considered two facts; one, those nodes having high information spreading rate, and two, the nodes whose immediate neighbors are very influential, because by chance if these non-influential nodes share any fake news, these can also spread rapidly.

$$S_{i+1} = S_i - (\beta S_i I_i)\delta t$$

$$I_{i+1} = I_i + (\beta S_i I_i - \gamma I_i)\delta t$$

$$R_{i+1} = R_i + \gamma I_i \delta t$$

If there is a rumor in a network, the node (person) who has heard it and wishes to spread it further is the infectious, defined as I, the one who has not heard that rumor is the susceptible, S, and the one who has heard it but is not interested in spreading it further is in recovered status, defined as R. Initially, as the starting value, I_0 is taken as 0.001, a very low number that worked as initial rumor spreaders. The value is very less than 0.001 just to show that how fatal it can be even with such a low initial value. S as $(1.0 - I_0)$, S is defined as the initial suspect, basically, those people who can spread more. The total number of days is considered as 20 because a rumor spreads in a short period, and after that, people came to know that whether a piece of information is a rumor or not. δt which represents the time step in days is taken as 0.125. β is the infection rate, and here, it worked as the rumor spreading rate which has been taken from the eigenvalue directly. γ is the recovery rate, taken as 0.2. In this process, the nodes having low eigenvalue but have an influential node as an immediate neighbor are also being considered. We have considered that the rumor spreading rate will be high in those cases also. In these cases, β is taken as an average of their eigenvalues.

4 Data Collection and Processing

We have used a dataset available at Stanford University (ID, sentiment of the text, text), where 0,1 represents positive and negative sentiment, respectively [6]. This dataset consists of node features in terms of user profiles, connecting edges presenting relationship between users, circles, and ego networks from Twitter. The total number of nodes and edges presented in the dataset are 81306 and 1768149, respectively. Along with this, tweets based on hashtags and tweets from a particular person's Twitter handle has been extracted using Tweepy API to check whether they are rumor or not. To form the network with the help of graph theory and then finding the important nodes in that network using centrality analysis and SIR model, this dataset is also needed.

5 Result and Discussion

In this paper, a precise and sustainable model has been discussed for finding out the influential nodes with the help of the SIR model. After derivation of eigenvector centrality of the nodes, the SIR propagation model has been used for the assessment of the rumor spreading rate. A flow diagram for influential user detection is shown in Fig. 1.

The information spreading rate of any node in the network is being found after propagation. We have analyzed three types of cases here, such as information spreading rate of low influential user, second is the information spreading rate of high influential user, and lastly, information spreading rate of user who has high influential neighbor.

In every graph, the blue line is showing the susceptible (S), which means people who can get affected, the red line is showing the infectious (I), represents the nodes who are spreading the information, and the green line is showing the recovered (R), which represents who have resisted the spreading of that information. The X-axis represents the days over which the tweet has been propagated in the network; we have taken at most 20 days, and the Y-axis represents the information spreading rate.

Here, in this first graph of Fig. 2 the information spreading rate of the node 122 from the network, which has a comparatively low eigenvalue 0.657 is considered, where β is 0.65, total days is 20, γ is 0.2 as mentioned earlier. Here, the neighbors of node 122 have not been considered, though it has a neighbor 56 with a high eigenvalue. So, here, the information spreading rate little low, which is $I_{\max} = 0.333$, and the recovered rate is $R_{\max} = 0.505$.

In the second graph, it is shown that the information spreading capacity of node 56 which has a high eigenvalue of 1.00, where β is 1.00. In this graph, it is visible that the I-curve has a high peak, as the spreading capacity is very high, $I_{\max} = 0.485$, and the recovery rate is $R_{\max} = 0.861$.

In the third graph, node 122 has been considered again as a node, but this time, considering the neighbors with high eigenvalue which is node 56, among its other neighbor nodes, such as (3, 30, 50, 56, 71, 97, 138, 267, 282). So, here, β is taken as the average of the two eigenvalues of node 56 and node 122 (1.00 and 0.65), which is 0.82. Here, it can be seen that the peak of the S-curve is less than the graph of node 56 (second graph) but greater than node 122 (first graph), infection rate $I_{\max}$ is 0.417, and the recovery rate $R_{\max}$ is 0.755.

A comparative result is shown in Table 2.

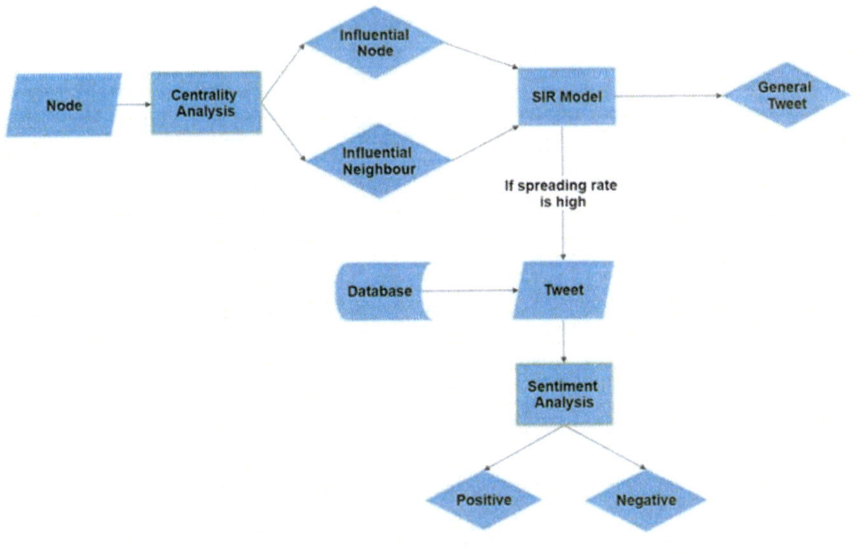

Fig. 1 Finding influential node using centrality measure

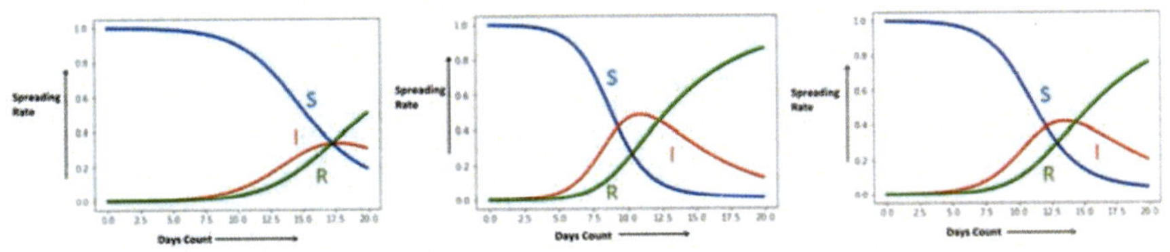

Fig. 2 Information spreading rate of a low influential node, high influential node, and considering the neighbours, respectively

Table 2 Information spreading rate of different types of nodes

Node type	γ	δt	β	I_{max}	R_{max}
Low-influential node (122)	0.2	20	0.65	0.333	0.505
High-influential node (56)	0.2	20	1.0	0.485	0.861
Low-influential node with high-influential neighbor (122 with 56)	0.2	20	0.82	0.417	0.755

6 Conclusion

Because of the enormous attention gained by social media and its contents concept, finding influential nodes with the ability to spread information is a crucial criterion, and this paper proposes an algorithm in this direction. According to the results from the Table 2 and graphs shown in Fig. 2, it can be empirically concluded that nodes with the highest eigenvalue generate the highest spreading rate ($I_{max} = 0.485$), compared to the nodes having low eigenvalue ($I_{max} = 0.333$). Even if the nodes having low eigenvalue, but have neighbors with high eigenvalue as shown in the third graph becomes more infectious ($I_{max} = 0.417$) compared to the first one with a low eigenvalue. In future variance of the SIR models, classified Twitter data can be used in the aforementioned way for faster detection of the source of rumor.

References

1. Albert R, Jeong H, Barabási A (2000) Error and attack tolerance of complex networks. Nature 406(6794):378–382
2. Amani AM, Jalili M, Yu X, Stone L (2017) Finding the most influential nodes in pinning controllability of complex networks. IEEE Trans. Circ. Syst. II: Express Briefs 64(6):685–689
3. Brandes U (2001) A faster algorithm for betweenness centrality. J Math Soc 25(2):163–177
4. Dey P, Chaterjee A, Roy S (2019) Influence maximization in online social network using different centrality measures as seed node of information propagation. Sādhanā 44(9):205
5. Kundu S, Murthy CA, Pal SK (2011) A new centrality measure for influence maximization in social networks. In: International conference on pattern recognition and machine intelligence. Springer, pp 242–247
6. Leskovec J, Krevl A (2014) SNAP datasets: stanford large network dataset collection. http://snap.stanford.edu/data
7. Moreno Y, Nekovee M, Pacheco AF (2004) Dynamics of rumor spreading in complex networks. Phys. Rev. E 69(6):066130
8. Ya-Qi W, Xiao-Yuan Y, Yi-Liang H, Xu-An W (2013) Rumor spreading model with trust mechanism in complex social networks. Communi Theor Phys 59(4):510
9. Zareie A, Sheikhahmadi A, Jalili M (2019) Influential node ranking in social networks based on neighborhood diversity. Future Gener Comput Syst 94:120–129
10. Zhao Laijun, Cui Hongxin, Qiu Xiaoyan, Wang Xiaoli, Wang Jiajia (2013) Sir rumor spreading model in the new media age. Physica A: Stat Mech Appl 392(4):995–1003
11. Zubiaga A, Aker A, Bontcheva K, Liakata M, Procter R (2018) Detection and resolution of rumours in social media: a survey. ACM Comput Surv (CSUR) 51(2):1–36

Author Index

S. Basu et al. (eds.), *Proceedings of International Conference on Frontiers in Computing and Systems*, Lecture Notes in Networks and Systems 404,
https://doi.org/10.1007/978-981-19-0105-8